Hubris:

The Troubling Science,
Economics, and Politics
of Climate Change

Hubris:

The Troubling Science, Economics, and Politics of Climate Change

Michael Hart

Compleat Desktops Publishing

2015

ISBN 978-0-9949038-0-8 Pb
          978-0-9949038-1-5 digital
          978-0-9949038-2-2 pdf

Compleat Desktops Publishing, a division of
Compleat Desktops R&D Inc.
27 Saddle Crescent
Ottawa K1G5L4
Canada

Cover design: Edward Hart, based on "The fall of Icarus," 1891,
Ceiling mural by Merry Joseph Blondel (Louvre Museum, Paris),
photo by Marie-Lan Nguyen, Wikimedia Commons

Book design: Edward Hart
Set in Book Antiqua and Calibri
Figures: derived from sources as indicated for each figure and
adapted in gray scale by Edward Hart

For more information about this book, the author, and other titles:
Omniadisputanda.com or Compleatdesktops.com

Special discounts are available on quantity purchases by corporations, associations, educators, and others. For details, contact the publisher at the above listed address.

Joseph Ratzinger, Pope Benedict XVI Address to the Congress held on the 10th Anniversary of Pope John Paul II's Encyclical *Fides at Ratio*, Rome, October 16, 2008.

The hubris of reason ... can acquire characteristics that are dangerous to humanity itself. Science, moreover, is unable to work out ethical principles; it can only accept them and recognize them as necessary to eradicate its potential pathologies. ... This does not mean restricting scientific research or preventing technology from producing the means for development; rather, it consists in maintaining vigilance about the sense of responsibility that reason possesses in regard to science, so that it stays on track in its service to the human being.

G.K Chesterton *The Flying Inn* (1914)

There is no great harm in the theorist who makes up a new theory to fit a new event. But the theorist who starts with a false theory and then sees everything as making it come true is the most dangerous enemy of human reason.

Contents

Boxes

Tables

Figures

Contents

❧❧❧❧❧

Preface

Imagine a movement so bent on achieving its political objectives that it is willing to corrupt science to meet them. Imagine governments around the globe, first adopting and then promoting this *official* science for more than two generations. Imagine that they are willing to use their regulatory power to implement a massive program of social engineering in order to "save" the planet. Imagine the United Nations leading this movement and insisting that a global effort is required. Imagine the movement's leaders believing that people around the globe must *change* their eating, heating, cooling, lighting, toilet, transportation, manufacturing, entertainment, even housing habits and reject values that are critical to their prosperity, happiness, and welfare, confident that humans can adapt and revert to simpler, more primitive, more local lifestyles, have fewer children, and embrace lives presumed to be more in harmony with nature.

Imagine thousands of scientists engaged at public expense in developing a convincing rationale for this unprecedented project. Imagine that these scientists are willing to compromise their integrity in pursuit of the role of a single factor that they insist controls the most complex and chaotic earth system, a molecule – carbon dioxide – that is literally the building block of all of life. Imagine that they believe that by reducing its miniscule – .04 percent – presence in the atmosphere, the planet will cool and climate will stabilize at an optimum level, a level seen only in micro-seconds of geological time. Imagine scientists who dismiss the work of hundreds of their colleagues and believe that their work must be suppressed. Imagine a scientific movement dominated by greedy grant farmers and cheered on by the media, insisting that there is no further need to study the science and that governments need to start implementing its preferred policy of worldwide social engineering.

Imagine that many leaders of this movement believe that the world's population needs to be thinned down to a billion people within a generation or two. Imagine that some of the movement's most revered leaders, even as they advocate that ordinary people must curb their consumption and live simpler lives, pursue lifestyles that consume more energy and other commodities in a year than an ordinary family of four would need over its lifetime. Imagine a movement whose leaders habitually dissemble and mislead and justify this on the claimed greater good they are pursuing. Imagine politicians, civil servants, scientists, activists, and the media flying from one exotic location to another as they plan what must be done to coerce changes in our lifestyles, even to the point of sacrificing human freedom and democracy.

Most thoughtful people would conclude that only Hollywood could come up with such a bizarre plot. A little more thinking, however, and they might connect the dots. There *is* such a movement, and it has demanded our attention for more than thirty years. It has devoured billions of dollars in public money and has inserted its menacing tentacles into every aspect of modern life. The UN and all its organs are the leading force behind it, but most governments of the world support it in one way or another. Elites, the media, and even religious leaders, have embraced it, even though they seem poorly informed and ignore its demands while urging others to adopt sharply reduced lifestyles.

The public face of this science, climate science, is part of a worrying new trend: the emergence of "official" or consensus science. In this perversion of real science, policy becomes the goal of scientific enquiry rather than its result. Over the last thirty years and more, public policy has focused increasingly on dealing with risks to health, safety, and the environment. Much of that policy ostensibly relies on scientific findings. In their decision-making, governments increasingly look to scientists and have resorted to funding science that meets their political need for certainty. Consensus on controversial issues is critical to governments. Ever since Rachel Carson published *Silent Spring* in 1962, activists have stood ready to convince governments of all manner of risks to humanity and nature, and scientists have obliged by reporting findings that satisfy activist political needs. Once governments acquiesce, it is critical that scientists not undermine their decisions with awkward new findings. Public policy is not easily reversed. The result is a potential monster spewing out more and more regulations, presumably making us safer and healthier and safeguarding the environment, but also substituting social for personal responsibility, reducing freedom and choice, and creating an ever larger, more costly, and intrusive public footprint.

For many years it seemed that the public agreed that there was a need to take action to control the globe's climate, but that support has steadily eroded as people have begun to realize the enormity of what is being demanded, the flimsy ground on which this demand is based, and the impact of what would need to be imposed. Public support has declined further as sceptical scientists have pointed out more and more problems with the underlying scientific hypothesis, as engineers have indicated the extent to which purported energy substitutes are not up to the job, and as economists have calculated the enormous costs and minimal benefits. Only general scientific illiteracy has kept the project afloat.

The movement advocates fundamental changes in lifestyles and succeeds by spreading *alarm* based on the alleged adverse consequences resulting from human-caused (anthropogenic) changes in the climate system. The movement points to the prospect of catastrophic results if those changes are not imposed by public authorities. *Alarmism* best describes this phenomenon. Critics of the movement are characterized as *sceptics* because their criticism is grounded in scepticism of various aspects of the science, the economics, and/or the politics of climate change alarmism. Some alarmists refer to sceptics as deniers, a term used in the pejorative sense. Sceptics do not *deny* climate change or the role of greenhouse gases (GHGs). As explained further in the scientific chapters, many sceptics consider the role of anthropogenic GHGs to be relatively minor and the prospect of catastrophic changes to the climate to be minuscule, thus obviating the need for anything other than appropriate adaptive measures.

This book dissects the global warming/climate change movement in all its ramifications. It analyzes the evolving science of climate change and places its pursuit and findings within the broader context of modern scientific praxis, identifying its strengths and weaknesses and areas of agreement and disagreement. Unlike the popular meme that the science of climate change is settled, the book demonstrates that in proper scientific practice, no issue is ever settled; scepticism is at its heart. Climate science is no exception. The book further argues that, as with other ambitious UN agendas, from the New International Economic Order (NIEO) of the 1960s to sustainable development in the 1990s, embrace of these movements by governments has a predictable life cycle, starting slowly, building momentum, and then gradually fading as a more realistic appreciation of the issues intrudes. While the primary movement is withering on the vine, its effects linger for generations. Governments may never meet the primary objectives of the global warming movement, but they have succeeded in embedding many of its tentacles into public regulatory policies and programs. Multiple interests have become dependent on these policies and will fight to maintain them, including thousands of officials whose careers are wedded to them. As so often happens in public policy, the unintended and harmful consequences become accepted practice, despite their costs and annoyance.

The world will be a better place when governments agree to tame this monster and refocus their energies on issues within their competence; when religious leaders and other elites accept that they have fallen prey to a movement whose motives are much darker and more damaging than they realize; and when the media adopt a more balanced approach and provide the public with the critical assessment that is often missing from their reporting. It is time for all three to accept that the UN is pursuing a path that can only result in a less prosperous and more divided world.

♦♦♦♦♦

This book is the product of nearly a decade of reading, thinking, and discussion. It originated in a request in the fall of 2007 from some of my graduate students at the Norman Paterson School of International Affairs at Carleton University to discuss the implications of climate change policy for Canadian trade. This led to a draft paper outlining some of the issues and a special class at the end of the term to discuss those issues. The discussion was lively and indicated that much more reading and thinking might be in order.

During my final years as an official in Canada's Department of External Affairs and International Trade, climate change came up occasionally in the morning meetings of the executive committee as an issue being followed by officials in the Legal Bureau who were involved in negotiating the 1992 UN Framework Agreement on Climate Change. At the time, I thought the idea of an international accord governing climate change rather bizarre; I had learned that UN-sponsored regimes on such topics were also often of little importance. Over the following years, however, climate change gained momentum with the media and the public, including my students, hence their interest in a special class.

By 2010, I had read much more and discussed the issue more widely. Seventeen graduate students joined me in a summer seminar, reading a wide range of sources, making presentations, and discussing the science, economics, and politics of climate change. I enjoyed the seminar, as did they, and I repeated it in the summers of 2011, 2012, and 2013, and in the winter of 2015, involving 48 more students. For the last seminar, I circulated a draft version of this book as the basis for class discussion, much of which had been prepared during my 2013-14 sabbatical. In all five seminars, class discussion was lively, and the book benefited immensely from that discussion and from the presentations and reports prepared by the 65 students. I also benefited from the presentations made to each seminar by my colleague Tim Patterson of Carleton's Earth Science Department. He introduced the students to the basic science of climate change and described the contours of some of his research on climate change in the high Arctic and on Canada's west coast. Tim is one of those scientists who gets his hands dirty and his feet frozen as he researches climate change over both geological and historical time. Like many geologists, he is deeply sceptical of the alarmist claims made by some climate scientists.

By the end of 2014, as the text took final shape, I had added over 300 books to my research library and had downloaded more than 3,000 articles on every aspect of the issue. Many of these are mentioned in the notes in this book, but many more added to my education on the multiple dimensions of this complex issue. I read material from a wide range of perspectives, from the highly sceptical to the deeply committed. Never before has an issue of science and public policy commanded such detailed attention from so many different angles.

Once the text took shape, I benefited from comments from Derek and Joan Burney, Keith Cassidy, Tony Halliday, Fen Hampson, Peter and Helen Hart, Chris Maule, Tim and Liz Patterson, and Cornelis (Kees) van Kooten, all of whom read the text in whole or in part and provided invaluable comments. Kees van Kooten was kind enough to provide me with detailed, constructive comments, in addition to the many details and insights I gained by reading his own thoroughly researched book on the economics of climate change policy. Three anonymous readers provided comments for UBC Press, which decided not to proceed to its publication, despite their earlier successful experience with three of my previous books. As discussed in chapter three, peer review no longer serves the constructive role it may once have played.

My wife, Mary Virginia, went through the text with a fine tooth comb not once, not twice, not three, but four times. Together with our friend Gus Heidemann, a retired colleague from the English Department at Carleton, they ensured that the text is as clear, readable, and unambiguous as possible. No author could ask for a better pair of editors. Any remaining errors or omissions are my fault, the product of my stubborn Dutch temperament.

With the decision self-publish, it became critical to ensure that not only the text but also the overall appearance of the book would be clear and attractive. My son Edward helped in designing the book, enhancing the quality of the graphics, and ensuring a faithful rendering in the format used by Lulu. In today's world of political correctness and squeamish publishers, services such as Lulu have become indispensable, and more and more authors are turning to them.

Ottawa
September 2015

1 | The Problem Stated

*To the improver of natural knowledge, scepticism is the
highest duty; blind faith the one unpardonable sin.*

Thomas Henry Huxley, 1860

*We have found it of paramount importance that in order
to progress, we must recognize our ignorance and leave
room for doubt. Scientific knowledge is a body of statements of varying degrees of certainty – some most un-
sure, some nearly sure, but none absolutely certain.*

Richard Feynman, *The Value of Science*, 1955

In 1977, a month after his 70th birthday, my father succumbed to his
fifth and final heart attack, a victim of cardiovascular disease. After
his first heart attack in 1962 his doctor had prescribed a strict diet
aimed at reducing his intake of animal fats, considered at the time to
be the prime contributor to high levels of serum cholesterol, which in
turn was considered to be one of the leading causes of cardiovascular
disease. For the final 15 years of his life my mother watched his diet
like a hawk but continued to feed him cookies and cake with his
morning coffee and afternoon tea; carbohydrates were not proscribed
He quit smoking cigarettes but continued smoking a pipe and an occasional cigar. He also maintained his life-long aversion to exercise.
My mother survived him by 18 years and died of complications from
a stroke at the age of 87. For the last 20 years of her life, she also maintained a diet low in animal fats but not in carbohydrates. Since both
of my parents died of cardiovascular-related causes, my genes suggest a predisposition to cardiovascular disease.

Having learned from my father's experience, I have been careful all my adult life to limit my intake of animal-based fats. Unlike my father, I maintained a moderate exercise regime and quit smoking in 1967. Nevertheless, measurements of my blood chemistry in the early 1970s indicated higher than desirable levels of cholesterol and other lipids – perhaps a genetic factor at work, since five of my siblings exhibited the same traits. After a diagnosis of Crohn's disease a few years later, I was motivated to adhere to a balanced, low-fat, alcohol-free diet. My GP monitors my blood chemistry annually and, in 2006, prescribed Lipitor – a powerful statin – to reduce my "bad" cholesterol and triglycerides. He was satisfied with the results. In 2009, however, I suffered a mild stroke. A CT scan indicated two earlier mini-strokes. This came as a surprise, as did the neurologist's conclusion that I suffered from an advanced case of arteriosclerosis. I listened to her admonitory lecture, which included many of the same things my father had been told nearly five decades earlier. She also quadrupled my dosage of Lipitor and added blood pressure and blood thinning drugs to my regimen.

Further investigation using some of modern medicine's advanced diagnostic tools revealed that neither my carotid nor cardiac arteries were blocked or narrowed to any appreciable degree and that the state of my cardiovascular system was normal for someone my age, i.e., perhaps some hardening of the arteries but no signs of narrowing or blockage. Reassuring as this information was, I determined that I needed to know more about cardiovascular disease and the role of diet in causing and controlling it.

Health and environmental risks, public policy, and science

Among the books that I read was Gary Taubes, *Good Calories, Bad Calories*.[1] Taubes had spent a decade investigating the extent to which science understood the relationship between diet, obesity, diabetes, and cardiovascular disease. He concluded that scientists had made significant progress in understanding these modern killer diseases[2]

---

1.   Gary Taubes, *Good Calories, Bad Calories: Fats, Carbs, and the Controversial Science of Diet and Health* (New York: Alfred A. Knopf, 2007). See also Harvey Levenstein, *Fear of Food: A History of Why We Worry About What We Eat* (Chicago: University of Chicago Press, 2012) and Nina Teicholz, *The Big Fat Surprise: Why Butter, Meat, and Cheese Belong in a Healthy Diet* (New York: Simon & Schuster, 2014).

2.   They are characterized as "modern" killer diseases because they are considered to be the products of modern civilization's lethal combination of sedentary life styles and easy access to rich, refined food. They have replaced the lethal infections that killed our ancestors but that have been conquered with antibiotics.

and had isolated some of the principal culprits in their growing modern incidence: smoking, lack of regular exercise, the increase in our diets of refined carbohydrates, and the human body's penchant for storing the metabolites of carbohydrates as fat. Their investigation had clarified that animal fats, rather than leading to high levels of serum cholesterol and other lipids, are critical to providing our bodies with energy and necessary nutrients for various bodily functions. Serum cholesterol does not increase because we eat fatty foods but because the body manufactures it for one reason or another. Nevertheless, public health authorities and the medical profession have remained firmly attached to ideas about diet and disease that owe little to science and much more to aggressive marketing of ideas first advanced in the 1950s and never subjected to rigorous scientific investigation. (See Box 1-1) Taubes concludes:

> The urge to simplify a complex scientific situation so that physicians can apply it and their patients and the public embrace it has taken precedence over the scientific obligation of presenting the evidence with relentless honesty. The result is an enormous enterprise dedicated in theory to determining the relationship between diet, obesity, and disease, while dedicated in practice to convincing everyone involved, and the lay public most of all, that the answers are already known and always have been – an enterprise, in other words, that purports to be a science and yet functions like a religion.[3]

Conventional expert wisdom, even when wrong, can be remarkably "sticky." Not surprisingly, the lay public remains thoroughly confused about the role of diet in the incidence of modern killer diseases. The interaction between science and diet has become a commonplace of journalism, frequently in ways that are not helpful. The need for funding has disposed scientists to announce every new discovery and insight based on their laboratory investigations, while the media's need for sensation and alarm has led to newspaper stories transforming minor insights into major stories. Industry groups are quick to seize on those they find helpful and to discredit those that are not. Missing from the media's enthusiasm for reporting research that may not be ready for prime time is the surge in retractions of articles published in scientific and medical journals, not because subsequent research has brought earlier conclusions into question but because of fabrication, plagiarism, error, and irreproducible results.

---

3. Taubes, *Good Calories, Bad Calories*, 451-2.

Box 1-1: Lipophobia and Cardiovascular Disease

In the years immediately after World War II, Ancel Keys, a University of Minnesota physiologist, was struck by the number of heart attacks among middle-aged, professional men. After testing a cohort of 286 men, he suspected that the presence of high levels of serum cholesterol was the culprit causing arteriosclerosis – hardening of the arteries leading to fatty deposits and narrowing of arterial walls. In his mind, the hypothesis was confirmed when he examined data from the World Health Organization (WHO) showing that countries with diets rich in saturated (animal) fat had higher levels of cardiovascular disease.

He discussed his theory with Paul Dudley White, a prominent Harvard cardiologist, who agreed that this was exactly the common sense explanation he was seeking. In the years to come, White and Keys proved a convincing team. Not everyone shared their enthusiasm, particularly after some researchers looked at the WHO data and learned that Keys had used only 7 (8 in a later study) of the countries for which the WHO had gathered relevant data. Had he used data from all 22 countries available in the WHO files, the correlation he observed between diet and cardiovascular disease would have disappeared. He ignored, for example, evidence that the Inuit in the Arctic and the Masai in Africa, both of whose diets consisted largely of animal fats, exhibited very low levels of cholesterol and rarely suffered from heart disease.

But these reservations came too late: the word had spread and became the gospel for medical practitioners. Sixty years later it remains deeply embedded in popular discourse, reinforced by public health authorities and heart and stroke societies in virtually every country.

Lipophobia becomes official science

Lipophobia became further embedded in medical practice as the basis for dietary advice promoted by public health authorities. In the 1970s, the US federal government began to prescribe recommended diets and, with strong media support, launched what became a steady drumbeat for the next sixty plus years: saturated fat in your diet will lead to obesity and heart disease and kill you!

By following this dietary advice, people were getting more of their energy from processed carbohydrates rather than from fats. In order to make low-fat food more palatable, the food industry used more salt as well as highly processed vegetable fats and carbohydrates (sugars and starches). This diet leaves people less satisfied and craving for snacks, which the food industry also supplied, most based on highly processed carbohydrates: sugary drinks, fatty chips, and similar products. Portions in restaurants became larger at the same time that changing lifestyles disposed people to get more and more of their calories from restaurant meals, whether sit-down or take-out, or in fully prepared form from food retailers such as Whole Foods.

The results were predictable, but the denial of the public health authorities continued: the recommended diet was leading to higher intakes of processed carbohydrates that were being metabolized and stored by the body as fat. Even more damning, studies repeatedly showed that diets higher in animal fats led to weight loss while those low in animal fats led to weight gain. Advances in biochemistry that allowed scientists to study the metabolic process at the molecular level confirmed that much of the conventional wisdom that went into public health policies related to nutrition and chronic diseases such as type II diabetes, obesity and cardiovascular disease was misguided.

Pharmaceutical solutions

The diet/cholesterol/heart disease theory became even more deeply embedded in medical practice when the pharmaceutical industry introduced drugs that could lower serum cholesterol. The discovery of a group of drugs called statins proved efficacious in lowering serum cholesterol and, by implication, in reducing cardiovascular disease.

Based on intensive marketing, drug firms convinced doctors to test cholesterol levels in all their older patients and to put them on statins if those levels were high. The medical profession cooperated, and soon millions of men, and then women, were saddled with a life-long commitment to statin therapy. The level of serum cholesterol and other lipids was indeed successfully lowered, and the number of patients mushroomed. Within a decade, statins had become the pharmaceutical industry's most prescribed and profitable drugs.

Confounding research

Bio-medical research, however, kept rolling along and identified a number of problems with the Keys theory and with the widespread use of statins. With increasingly sophisticated measuring techniques, researchers identified the much more complicated structure of serum lipids, fractioning them into "good" and "bad" cholesterol. Research also showed that the body manufactures lipids in the liver, brain, and elsewhere because it needs them for a variety of functions. Interfering with the chemical process that leads to the production of lipids also interferes with the production of other chemicals that the body needs. Increasing reports of undesirable side effects – including muscle deterioration, nerve damage, and memory loss – suggested that interfering with that chemical process was not necessarily beneficial.

Bio-chemical research into human metabolism was demonstrating in much more detail the relationship between diet and levels of serum cholesterol. The simple, "common sense" explanation favoured by Keys and White had no basis in science. The many types of fat in the human diet all fulfill a need, some more critical than others. In addition, the body's ability to manufacture lipids does not appear to be closely related to dietary intake of fat. At the same time, growing use of highly processed vegetable fats, particularly transfats, were creating their own problems.

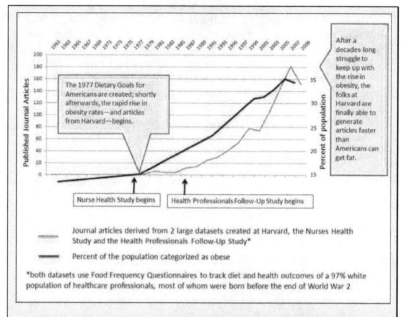

The 1977 Dietary Goals for Americans are created; shortly afterwards, the rapid rise in obesity rates—and articles from Harvard—begins.

After a decades-long struggle to keep up with the rise in obesity, the folks at Harvard are finally able to generate articles faster than Americans can get fat.

Nurse Health Study begins

Health Professionals Follow-Up Study begins

Journal articles derived from 2 large datasets created at Harvard, the Nurses Health Study and the Health Professionals Follow-Up Study*

Percent of the population categorized as obese

*both datasets use Food Frequency Questionnaires to track diet and health outcomes of a 97% white population of healthcare professionals, most of whom were born before the end of World War 2.

Obesity and the Output of Articles from the Harvard Nurses Health Study. Source: Adele Hite, "As the calories churn" at eathropology.com, July 2013.

Research also stubbornly failed to confirm the relationship between diet, serum cholesterol, and cardiovascular disease. A 2010 meta analysis of 21 earlier studies covering 347,747 subjects failed to find any association between saturated fat consumption and cardiovascular disease. A 2014 meta analysis in the March issue of the *Annals of Internal Medicine* covering 72 studies involving over 600,000 individuals in 18 countries similarly found no evidence to support current public health guidelines recommending that people restrict their intake of saturated fats.

Research into the relationship between serum cholesterol and cardiovascular disease was also steadily undermining the Keys-White hypothesis. "Good" and "bad" cholesterol were indeed present in the fatty deposits in arterial walls thought to be critical to heart attacks and strokes, but in a beneficial rather than a malign way. Cholesterol serves to heal lesions that arise in the artery wall from other mechanisms.

Finally, the results of epidemiological studies looking at the association between statins and morbidity due to strokes and heart attacks were also ambiguous, with different studies reaching different conclusions. Studies sponsored by drug companies seemed to be more supportive than independent studies. The drug companies were now admitting that the cholesterol-lowering effect of statins might be less prophylactic than claimed. Fearing loss of the golden goose, there was now a new claim that statins also fight inflammation of the arteries, a more important effect. Time will tell whether or not this new claim survives continuing scrutiny.

No doubt there are other factors contributing to cardiovascular disease, such as our modern sedentary life style. Although changes in diet based on epidemiological studies may be key to understanding the rise in obesity and type II diabetes, they have failed to provide useful information on cardiovascular disease. In a turn-around-is fair-play moment, nutritionist Adele Hite provides a wonderful graphic showing the correlation between the output of the Harvard Nurses Health Study – instrumental in convincing public health authorities of the benefits of a low saturated fat diet – and the increase in obesity (See Figure below). Did the Harvard study "cause" the increase in obesity? Perhaps not, but the association is certainly more robust and suggestive than many epidemiological studies that have gained considerable credence.

From four directions, research has steadily undermined the Keys-White hypothesis. As an object lesson in the stickiness of established wisdom, however, none of this research has in any way dulled the enthusiasm of public health authorities, heart and stroke societies, and many medical practitioners for prescribing statins and telling cardiovascular patients to pursue an even stricter version of the low-fat diet. Is the continued reliance on statins and low-fat diets good science or bad science? If nothing else, the issue is highly suggestive of the difficulties encountered in trying to reverse "official" science.

For more detail, see Gary Taubes, *Good Calories, Bad Calories,* Harvey Levenstein, *Fear of Food,* Nina Teicholz, *The Big Fat Surprise* (New York: Simon and Schuster, 2015), Malcolm Kendrick, *The Great Cholesterol Con* (London: John Blake, 2007), Mary Enig, *Know Your Fats* (Bethesda, MD, Bethesda Press, 2000), and Adele Hite, "As the Calories Churn," Episodes 1-3. each with copious notes pointing to the medical and scientific literature.

Weary of the oversimplification and selective reporting that characterize many of the stories about diet and disease, the general public has learned to discount much of the hype. Who can blame them? Warnings one year about the need to restrict our intake of eggs is followed a year later by a new study indicating that moderate consumption of eggs poses no significant issue for cardiovascular health. Similar examples abound, from oat bran to corn oil. The general population has learned to discount such dire warnings, and not just those about health and diet. On the whole, consumers would be well served to ignore most media stories about scientific issues. The media have long played a leading role in perpetuating medical and other supposedly science-based myths, even frauds. More particularly, however, as Stanford's John Ioannidis has demonstrated, "for many current scientific fields, claimed research findings may often be simply accurate measures of the prevailing bias."[4] Unfortunately, the media have also become attracted by what can only be described as junk science, and the dividing line between it and serious evidence-based science has become increasingly difficult to find as research scientists politicize their findings and make common cause with activist nongovernmental organizations, industrial interests, and others with an agenda.

The media's sensationalist reporting would be of little moment except for the extent to which it feeds into activist agendas and the perceived need for public policy responses. Over the course of the last five or so decades, the focus of government regulatory activity has shifted from seeking to determine economic outcomes to addressing risk, perceived or real, arising from health, environmental, and safety concerns. Over that period, risk-related regulation has become a dominant concern of national, state, and provincial governments. Much of that regulatory activity is, we are told, grounded in science, but much of that science is rarely as clear as claimed by activists or as clear-cut as governments assert. Governments have increasingly become captive of what economist Ross McKitrick characterizes as "official" science: science that is in many ways a caricature of real science.[5] Rather than a dynamic quest grounded in a perpetual attitude of scepticism, it asserts that the claims of science can bring certainty and point confidently to solutions. Public health policy provides a troubling example as does much of environmental policy.

Social tolerance for risk has declined markedly in recent years, further skewing the equation. In response, governments now frequently rely on the so-called precautionary principle as the basis for making difficult decisions, responding to a perception that the public would rather be safe than sorry. The implications of this approach for economic well-being and material progress are profound. In these circumstances, the role of science has become critical. Not surprisingly, interest groups have learned to manipulate the work of scientists in order to press their political and economic agendas,

---

4.   John P. Ioannidis, "Why Most Published Research Findings Are False," *Public Library of Science (PLoS): Medicine* 2:8 (August 30, 2005), 124. His findings are discussed further in chapter 2.

5.   Ross McKitrick, "Bringing Balance, Disclosure and Due Diligence into Science-based Policymaking," in Jene M. Porter and Peter W.B. Phillips, eds., *Public Science in Liberal Democracy* (Toronto: University of Toronto Press, 2007).

and scientists have learned to manipulate public discussion in order to enhance funding for their research. Management of risks to public health and the environment has, of course, always had to grapple with making decisions under uncertain conditions. Science deals in probabilities; much of science-based public policy seeks to address fears and uncertainties by finding a socially acceptable balance between risks and benefits, a judgment that requires governments to make assessments about risks, costs, and benefits, informed by science, politics, and economics.[6]

Environmentalism and public policy

My encounter with the neurologist coincided with reading that I was doing to prepare for a new course I planned to introduce in the summer of 2010 about scientific uncertainty and the public policy implications of the debate about anthropogenic global climate change. Even more than public health, concerns about the environment have become a favourite focus of sensationalist journalism. For more than two generations, the public has been fed a steady diet of stories about the role of humans in the alleged deteriorating state of the global commons, statements that are more political than scientific in nature. The natural world is in a constant state of flux, and human interaction with it is but one of many influences. Whether those influences are benign, malign, or indifferent is a matter of values, not science.[7] The concept of an ideal state of nature is, again, political.

Among these stories, concerns about global warming have succeeded in grabbing the greatest attention but not for lack of effort by activists concerned with other environmental issues. Today's generation of students has been force-fed from kindergarten the dire consequences of human interaction with nature. Universities have gotten on the bandwagon with new programs, new departments, or newly revamped and refocused disciplines. Geography departments, for example, have morphed into environmental studies programs. "Sustainability" has become the mantra of university admin-

---

6. See, generally, G. Bruce Doern and Ted Reed, eds., *Risky Business: Canada's Changing Science-Based Policy and Regulatory Regime* (Toronto: University of Toronto Press, 2000); H.W. Lewis, *Technological Risk* (New York: W.W. Norton, 1990); and Howard Margolis, *Dealing with Risk: Why the Public and the Experts Disagree on Environmental Issues* (Chicago: University of Chicago Press, 1996).

7. Jeffrey Foss provides an excellent overview of the unscientific basis of much of modern environmentalism in *Beyond Environmentalism: A Philosophy of Nature* (Hoboken, NJ: John Wiley & Sons, 2009).

istrators and has also permeated the core curriculum.[8] Not unlike public health, much of the resulting activism, public policy, and even some of the university-based teaching rest on the slimmest of evidence-based research. Much of it is based on the false allure of scientific certainty, a reality that rarely exists in the world of serious scientific investigation.

In the early 1990s political scientist Aaron Wildavsky, together with his graduate students at Harvard University, compiled a fascinating study on the problem of scientific certainty and public policy. They surveyed a series of public policy concerns that had originated in alarming scientific conclusions about environmental health and safety issues and that had succeeded in capturing significant media attention and activist concern. For each problem, they asked: "But is it true?"[9] Closer scrutiny revealed that many of the scientific claims being made, each of which had been critical to the case for public action, were open to serious doubt based on subsequent, evidence-based scientific investigation. Wildavsky's quest rested on the premise that a well-educated citizenry should be able to make an informed judgment about most health- and environment-related scientific issues. In his words, his quest involved "understanding the scientific bases for rival claims, engaging in informed discussion, and making reasoned judgments ... [not] as apprentice scientists ... [but] as reasoned deliberators capable of taking informed action in fields not necessarily determined by but infused with conflicting scientific and technological assertions."[10]

His case studies, which ranged from DDT and PCBs to the role of rodent studies in predicting cancer in humans, indicated that many of the claims were based on shoddy or incomplete work, much of which had been sensationalized by the media and activists. Political responses were often premature, lacked informed risk analysis, and were predicated on ensuring political survival rather than averting serious harm. In the twenty years since Wildavsky's book appeared, society's aversion to risk and the media's attraction to alarm and sensation have increased the field for further case

---

8.   See Rachelle Peterson and Peter W. Wood, *Sustainability: Higher Education's New Fundamentalism* (New York: National Association of Scholars, 2015) for a thorough discussion of the extent to which the totalitarian demands of the sustainability movement have permeated the modern university.

9.   Aaron Wildavsky, *But Is It True? A Citizen's Guide to Environmental Health and Safety Issues* (Cambridge: Harvard University Press, 1995).

10. Wildavsky, *But Is It True?* 1.

studies. The activist preference for the precautionary principle has made political authorities even more prone to act.

One of the issues that Wildavsky studied was anthropogenic climate change or, as it was framed at that time, global warming. His examination, and that of the graduate students who worked on the issue, indicated that the case for public action was very weak. In the intervening years the case has seemingly become more robust, or has it? Those calling for action have clearly succeeded in creating a very broad base of public support and, in response, governments have created an international, UN-based public policy process that has taken on a life of its own, increasingly divorced from scientific principles and focused on progressive goals. Distinguished MIT atmospheric physicist Richard Lindzen, one of the leading critics of global warming alarm, concludes that "global warming is about politics and power rather than science. In science, there is an attempt to clarify; in global warming, language is misused in order to confuse and mislead the public."[11] In the quest for power, the UN is one of the key players, using alarm about climate change as a prime vehicle for advancing its progressive agenda.

The United Nations and the progressive agenda

The founding of the UN in 1945 was a response to the outrage of two world wars, and the UN's charter provides it with a mandate to broker peace and global order and to limit war and disorder. That mandate includes an economic and social dimension predicated on the idea that a more prosperous and socially cohesive world will be a more peaceful one. On the political side of its mandate, the UN has had moderate success. On the economic and social side, however, it has focused on an increasing array of fads, cajoling member governments into adopting a wide range of agendas and resolutions favoured by the so-called international community, a term adopted by the media to describe activists with agendas.[12] The UN adopted the climate change agenda in the 1980s and has since developed it into an all-consuming, ambitious framework through which to tackle a number of earlier progressive causes, from gender inequality to sustainable development.

---

11. Richard Lindzen, "Global warming, models, and language," in Alan Moran, ed., *Climate Change: The Facts* (Melbourne, Australia: Institute of Public Affairs, 2015), kindle edition.

12. I explore these themes in more detail in Michael Hart, *From Pride to Influence: Towards a New Canadian Foreign Policy* (Vancouver: UBC Press, 2008).

The UN's efforts to establish a New International Economic Order (NIEO) in the 1960s and 1970s and its current focus on "saving" the planet give one a strong sense of déjà vu. Then too, there was much earnest talk by governments at a succession of major international conferences, and much written by "experts", but in the end the NIEO suffered the fate of most simplistic ideals based on weak intellectual foundations: the dustbin of history. The focus for much of this effort was the UN Conference on Trade and Development (UNCTAD), established in 1964. Its high point was the fourth conference in Nairobi in 1976, at which governments agreed to establish the Integrated Program for Commodities (IPC). By the time of the Manila (1979) and Belgrade (1983) conferences, governments were losing their enthusiasm for the NIEO. Over the course of the 1980s, more and more developing-country governments realized that their development goals were more likely to be met working through the trade regime embedded in the General Agreement on Tariffs and Trade (GATT) and its successor, the World Trade Organization (WTO), than through UNCTAD. The WTO now has twice as many members as GATT had during UNCTAD's heyday, and many developing countries have embarked on serious efforts to open their markets to global competition. Most have turned their backs on the *dirigiste* policy preferences of the NIEO and are beginning to see positive results. UNCTAD still exists but is now a mere shadow of its earlier self, its many meeting rooms standing idle for much of the year.

In the 1970s, the intellectual foundations of economics were being bent to meet the requirements of UNCTAD's agenda. Governments rode with it but kept a wary eye on the real world. Over the past few decades, the focus on sustainability and the climate change agenda raise suspicions that a similar phenomenon is at play. This time it is science that is being harnessed to meet the political objectives of those committed to warding off the alleged crisis of global warming and the equally alleged deteriorating state of the global commons, with the goal of ushering in a world more to their liking. Climate change professor Mike Hulme from the UK's University of East Anglia baldly asserts: "We need to ask not what we can do for climate change, but to ask what climate change can do for us. ... Rather than trying to 'solve' climate change, ... we need to approach climate change as an imaginative idea, an idea that we develop and employ to fulfill a variety of tasks for us. Because the idea of climate change is so plastic, it can be deployed across many of our projects and can serve many of our psychological, ethical and spiritual needs."[13] For Hulme, the climate change "crisis" provides a convenient basis upon which to tackle such UN perennials as population control, income redistribution, and sustainable development. Christiana Figueres, the Executive Secretary of the UN Framework Convention on Climate Change (UNFCCC), has enthusiastically taken up this theme. At a Brussels press conference, she explained that climate negotiations are "probably the most difficult task we have ever given ourselves. ... This is the first time in the history of mankind that we are setting ourselves the task of intentionally, within a defined period of time to change the economic development model that has been reigning for at least 150 years, since the industrial revolution."[14] The NIEO appears to have morphed into the quest for sustainable development, with all its questionable economic assumptions intact but now covered with a new coat of environmentalist paint.

The UN and climate alarmism

Over the last 25 years a whole industry has developed at universities and multiple specialized research institutions devoted to the study of climate change in all its ramifications. The literature has become immense and is well beyond the capacity of any single individual to grasp. It is also relentlessly one-sided in its orientation. Research funds flow to those prepared to examine the assumed malign aspects of climate change – imagined or real. Funding for research that takes a more critical approach or that examines the possibility that there may be benefits to climate change is much more difficult to obtain. Studies that reach findings at odds with mainstream dogma routinely add a sentence or two in the abstract or conclusions to reassure readers that the findings do not invalidate the politically correct orthodoxy, even when they do. Rent-seeking behaviour has become one of the distinguishing features of the modern academy. Much of the best critical literature comes from older, even retired, academics who are no longer as involved in the chase for research funding. Its dissemination and discussion take place at a few conferences and, more importantly, in the blog-

---

13. Mike Hulme, *Why We Disagree About Climate Change* (Cambridge: Cambridge University Press, 2009), 326 and 329.

14. "Figueres: First time world economy is transformed intentionally," UN Regional Information Centre for Europe, February 3, 2015.

osphere. The weight of numbers, however, should not be taken to be dispositive. A discerning eye can differentiate quality from rote.

Climate alarmism is based on four interrelated assertions: global temperatures are climbing to unprecedented levels; human activity is largely responsible for this increase; climate change of this order is exceptional and will have catastrophic impacts on the earth's biosphere; and policy-induced changes in human behaviour can stabilize the climate and ward off calamity. From this perspective, "climate change represents a 'tragedy of the commons' on a global scale. The nations of the world, and individuals within them, over-exploit the planet's atmosphere because they gain all the material advantages from the activities that contribute to global warming but suffer only a fraction of the environmental costs. In turn, nations and individuals typically are unwilling to reduce their greenhouse gas emissions unilaterally, because in doing so they would pay the full price of abatement but gain only a fraction of the benefits."[15]

Contrary to popularly held views, the scientific underpinnings of many of these assertions are far from settled. Additionally, many researchers are not convinced that the technology needed to replace current sources of energy exists or that the cost of implementing even a modest version of the preferred policy prescriptions is justified by any benefits that could reasonably be attained. Indeed, many are convinced that most of the solutions offered would have either catastrophic effects of their own or remain technologically impossible. Nevertheless, through the work of the UN's Intergovernmental Panel on Climate Change (IPCC) and the 1992 UNFCCC, many of the world's governments have committed to pursuing costly national programs and international mitigation strategies with the goal of essentially altering global climate patterns. Ironically, the

---

15. Kathryn Harrison and Lisa McIntosh Sundstrom, "The Comparative Politics of Climate Change," *Global Environmental Politics* 7:4 (November 2007), 1. Similar paragraphs can be found introducing books or articles by thousands of ecologists, geographers, political scientists, economists, sociologists, and other social scientists who have all accepted the mantra of global warming and hitched their research programs to exploring its malign effects. In addition to the millions poured annually into climate research programs, these ancillary research programs swallow up millions more. The claimed thousands of scientists who contribute to the work of the IPCC include many more social scientists than climate and other natural scientists. See Donna Laframboise, *The Delinquent Teenager Who Was Mistaken for the World's Top Climate Expert* (Toronto: CreateSpace, 2011) and *Into the Dustbin: Rajendra Pachauri, the Climate Report & the Nobel Peace Prize* (Toronto: CreateSpace, 2013). Cheeky as her titles and her website are, Laframboise is a seasoned researcher whose findings are well-documented.

same environmentalists who rail against human interference with natural systems have enthusiastically embraced an effort to alter fundamentally one of nature's most complex and critical systems, a fact indicating that much more than the environment is involved.

The stakes in both national and international discussions are very high because the policies advocated by many in the alarmist community would require substantial changes in lifestyles and standards of living and necessitate fundamental changes in the nature of modern economies and in the prosperity they provide. As suggested by the UN's Figueres, only a social engineering program of unprecedented dimensions can "save the planet." Efforts by the alarmist community to reduce doubt have met with considerable success at official levels but at the expense of scientific integrity. Ethicist Thomas Sieger Derr observes: "One would never know there are dissenters of distinguished credentials in the scientific community. Even where their existence is admitted, they are thoroughly marginalized, accused of being in the pay of the oil companies (Gore slyly and meanly implies this in his movie, *An Inconvenient Truth*), or dismissed as over-the-hill retirees out of touch and perhaps a bit senile. Their articles are denied publication in *Science* and *Nature*, those two so-called flagship science journals of high reputation despite some embarrassing lapses."[16]

The increasing intensity of the campaign, however, has sown doubt in the general public. Commitment to action remains most pronounced in Europe, consistent with broader European preferences for risk aversion and statist solutions. Governments in developing countries, while perhaps sceptical about the issue, are prepared to milk it as a new source of financial aid. Governments in North America initially limited their commitments to politically calculated lip service rather than to action, a situation that changed with the Obama administration's increasing preference for a more activist approach. To date, neither the Canadian nor the US government has implemented a policy approach responding to the demands of the alarmist community. In its second term, however, the Obama administration has begun aggressively pursuing steps within its executive powers, stepping around the continued reluctance of Congress to mandate the preferred policies. In Canada, only provincial governments, particularly in Ontario, have taken significant

16. Thomas Sieger Derr, "The Politics of Global Warming," *First Things*, August/ September, 2007.

steps to pursue a climate mitigation strategy with results that pose a sober lesson for future governments.

A major milestone in international discussions took place over the course of 2009, culminating in Copenhagen, Denmark, in December when parties to the 15th of the UNFCCC's annual Conference of the Parties (COP) tried to see if a broader policy consensus on action could be developed than had been achieved at the previous two such meetings in Bali, Indonesia (2007) and Poznan, Poland (2008). The change of administrations in the United States injected a high level of expectation into the conference – and its preparatory meetings – and added to pressures on the governments of Canada and others to respond more enthusiastically than they had to date. At the same time, lack of any perceptible warming for over a decade made the case for immediate action increasingly difficult to sustain.

In any event, more than 45,000, counting media, officials, and activists came, but little was accomplished. The prospect of an overreaching, new climate change treaty faded quickly as government leaders wiggled out of the mess of expectations they had created. Instead, they issued an anodyne political statement and a promise of cash for developing countries. In effect, governments provided themselves with the political breathing room to stand back and take a fresh look at the issues rather than be stampeded by the activist community at a time of economic turmoil. They also resorted to another familiar UN technique: agree to meet again, this time in Cancun, Mexico in 2010. In the UN world, *process is progress.*

Cancun has come and gone, as have subsequent conferences. The faithful still come, but the media have decided that these annual festivals of alarmist hype no longer provide sufficient news to warrant full coverage. As in their dealings with UNCTAD, ministers and heads of state are now also much more circumspect, keeping expectations low and attendance to the bare minimum that is politically acceptable. The meetings themselves have done little more than confirm what had been agreed at Copenhagen: more talk and promises of more cash for developing countries. Based on past UN form, the Copenhagen results may have marked the high point of international activism; follow-up meetings suggest a slow decline into the same obscurity that has befallen UNCTAD, notwithstanding a full court press by the UN to insist that the December 2015 meeting in Paris will witness a major breakthrough.

Science and public policy

Public policy is a matter of identifying problems and opportunities that would benefit from government attention and action, of developing appropriate policies and programs, and of weighing their costs and benefits. As Lindzen points out, the fact that something has been identified as an issue does not necessarily lead to a need for public action.[17] Nevertheless, the public in the prosperous, industrialized countries have become so accustomed to activist government that few stop to think whether or not climate change is an issue that governments can or should address. In order to warrant action, therefore, governments need to consider such questions as:

- To what extent is climate change natural? Are current patterns of change outside the boundaries of previous human experience? Do we know enough about climate change to warrant decisive action?

- What are the real, long-term effects of climate change – natural and/or anthropogenic – and to what extent will adaptation and voluntary changes in behaviour reduce negative effects? Are there any offsetting benefits associated with these problems?

- What tools and instruments are available to control climate change, mitigate its negative effects, or facilitate adaptation? How effective are they likely to be?

- What are the costs and benefits of deploying such tools and instruments? How do the costs of mitigation and adaptation compare?

- How do these costs and benefits stack up against the costs and benefits of addressing other global and national problems?

By framing the issue in apocalyptic terms, alarmists have sought to avoid consideration of these questions and to rush governments into considering radical approaches to what increasingly appears to be a non-problem or one easily addressed through gradual adaptation and supportive policy measures. Suggesting that the IPCC has answered all questions borders on the risible. The release of thousands of emails in late 2009 demonstrated the extent to which the IPCC and the climate scientists associated with it were deeply committed to a single perspective and were working assiduously to freeze out all who questioned that perspective. The *Sum-*

---

17. Richard Lindzen, "Issues in the Current State of Climate Science," SPPI, March 2006, 27.

maries for Policy Makers of its five assessment reports – 1990, 1995, 2001, 2007, and 2013-14 – have become steadily shriller, while their scientific foundations have become less and less convincing.

Climate change is inevitable; it is an integral part of the chaotic natural world. Understanding the extent of that change, its time frame, and its causes may be critical, however, to determining whether there is a need for a public policy response. Until the 1980s climate change was not part of public consciousness; to the extent that governments considered climate change at all, the focus was on resilience, i.e., ensuring that infrastructure was suited to current and possible future circumstances. Since the 1980s, however, alarmists have succeeded in raising public anxiety, and governments have become much more focused on the issue. Over the past 25 years, alarmists have managed to seize the commanding heights of media and public policy discussion and have convinced many govern-ments that human activity *is* a major driver of climate change and that there is, therefore, an urgent need to impose solutions. But the solutions being considered are fraught with difficulties and show all the hallmarks of haste, alarm, and religious zeal rather than a care-ful weighing of costs, benefits, and alternatives.

The scientific basis for alarmist claims is grounded in the green-house gas (GHG) theory of climate change. To many IPCC-affiliated scientists, GHG-induced climate change is a problem to which they claim to have an appropriate, science-based solution. Much of the science, however, has been politicized in order to strengthen the case for action because the ability to implement responses is em-bedded in the political economy of nation states. The IPCC hypoth-esis may be based on science, but it lacks observational confirma-tion; rather, it is based on computer models, and the numbers gen-erated by models do not reflect anything more than the built-in as-sumptions and data fed into computers and manipulated with sta-tistical programs. The resulting numbers may be useful to scientists in their research, but they do not provide a basis for making policy. Under these circumstances, it is critical that policy makers reach in-formed assessments of the certainty with which some climate scien-tists are willing to attribute the planet's ever-changing climate to human factors, i.e., to factors that are amenable to policy measures.

IPCC-affiliated scientists, in simplifying the climate system for analytical purposes, focus much of their attention on quantifying fluxes in the Earth's radiation budget – the difference between in-coming short-wave radiation from the sun and outgoing long-wave radiation from the Earth's surface. They tell policy makers that the global climate is controlled by the balance between these fluxes, referring to changes in that balance as the result of the impact of a forcing agent (positive or negative). The IPCC process was set up to investigate the role of humans as the principal forcing agent of change. The models upon which IPCC scientists rely are designed and parameterized on the basis of minimal natural variation, whether resulting from changes in total insolation, changes in coupled atmospheric-oceanic circulation systems, or changes in cloud cover, all three of which play a critical role in the way the climate system distributes heat. Surprisingly, careful reading of the underlying scientific reports issued by the IPCC over the past 25 years indicates that the evidence to justify the alarm set out in the IPCC's press releases and *Summaries for Policymakers,* in media reporting, in political speeches, and at ENGO websites, is simply not there. In these advocacy documents, the basic science is being twisted and exaggerated to support a political agenda. Butos and McQuade, after a review of the funding and scientific output of climate science, conclude that:

> A confluence of scientific uncertainty, political opportunism, and ideological predisposition in an area of scientific study of phenomena of great practical interest has fomented an artificial boom in that scientific discipline. The boom is driven and sustained by the actions of Big Players, the IPCC and various government entities, in funding the boom and singularly promoting one among a number of plausible hypotheses describing the relevant phenomena. Given the scientific uncertainties inherent in the system under study and the incentives for continued political involvement (even in the face of widespread failures in government-supported businesses whose activities were premised on the reliability of the AGW hypothesis), it is possible, even likely, that the boom will persist for a considerable time, not unlike previous booms in eugenics and nutrition science.[18]

Critics of the IPCC hypothesis, on the other hand, see earth systems – including climate – as dynamic and chaotic. They regard climate science as relatively immature with understanding of the various chaotic processes involved still at an early stage. Computer models that show the effects of increases in greenhouse gases remain crude, simplified versions of the global climate system and are

---

18. William N. Butos and Thomas J. McQuade, "Causes and Consequences of the Climate Science Boom," forthcoming in *The Independent Review.* Prepublication text at *The Libertarian Alliance Blog,* March 6, 2015.

not capable of providing the certainty that the IPCC community generates in its projections or story lines. In each instance, too many assumptions are required to reach these kinds of projections, as well as "tuning" with fudge factors to get the right answers. This problem is compounded by the poor quality of data and the questionable nature of some of the statistical methods employed. The kind of data needed to feed models is only now being generated, starting with the satellite era and gaining sophistication and detail over the years. Thirty-five-plus years of data are not enough to understand a system as complex and chaotic as the climate system, nor are the data good enough to provide the precision claimed by IPCC scientists. The usefulness of models as research tools is clear, but their usefulness in providing governments with policy advice is still highly overrated, breeds misunderstanding, and suggests unrealistic levels of certainty.

Critics of the IPCC perspective concede that human activity may well contribute to climate system dynamism and chaos but point out that it is difficult to separate the human signal from the many natural forcings and feedbacks. Their challenge to IPCC scientists, therefore, is to demonstrate that late 20th century warming – which on geological timescales is exceedingly small – is outside natural boundaries and not part of natural ever-changing climate patterns. The scientific controversy boils down to the sensitivity of the climate system to small changes in the composition of the atmosphere, particularly the fraction made up of the greenhouse gas carbon dioxide (CO_2), which has risen from about 0.03 to 0.04 percent since the beginning of the Industrial Revolution. Many scientists insist that increased understanding will only come when observations confirm model results based on the normal path of science as it advances hypotheses and tests them against real-world observations.

A major and very inconvenient fact for the alarmists is that following the giant El Niño of 1997-98, there has essentially been no global warming. While the global temperature anomaly – the metric of choice for alarmist scientists – has fluctuated from year to year and month to month, it has done so within a 1.3°C boundary around a trend-line of zero, a boundary that is much narrower than experience over the Holocene, the geologic period since the end of the last ice age 12,000 years ago. This has made a mockery of the models on which alarmists rely and has forced them to scramble for explanations, including accepting that perhaps their assumptions and conclusions may need to be refined. They have not, however, conceded that there may be a fundamental flaw in their theory and continue to press for remedial action to control climate change. [19]

The alarmist movement often refers to its critics as "deniers." In truth, there are few scientists who insist that there has been little or no global warming over the past century and a half or who claim that emissions of CO_2 are immaterial. Their scepticism rather focuses on the extent of prospective warming, on the fact that insufficient attention is paid to the interplay of natural factors in the climate system, on the extent of the threats that may result from any future warming, and on the capacity of government action to change climate patterns. In any case, scepticism should be at the very heart of the scientific process. Among scientists, characterizing someone as sceptical is normally considered an accolade.

Similar to that of the eugenics movement at the beginning of the 20th century, future generations will study global warming and the role of the IPCC as one of the prime examples of the corrupting influence of motivated reasoning and official science. Among progressive thinkers at that time, eugenics was widely held to be the sane thing to do in the face of a rapidly growing population, particularly the numbers of poor people and other "undesirables." Israeli historian Jacob Talmon observed more than half a century ago:

> [this kind of movement] is based upon the assumption of a sole and exclusive truth in politics. ... It recognizes ultimately only one plane of existence, the political. It widens the scope of politics to embrace the whole of human existence. It treats all human thought and action as having social significance, and therefore as falling within the orbit of political action. ... Politics is defined as the art of applying this philosophy to the organization of society, and the final purpose of politics is only achieved when this philosophy reigns supreme over all fields of life. ... This is the curse of salvationist creeds: to be born out of the noblest impulses of man, and to degenerate into weapons of tyranny. An exclusive creed cannot admit opposition."[20]

Climate change, costs, and benefits

Alarmist scientists often take the view that climate change is a scientific/technical issue for which they have a diagnosis, a prognosis,

---

19. Climatologist Roy Spencer provides a monthly update of the global temperature as calculated from data derived from satellite observations. See "Latest Global Temp Anomaly," at droyspencer.com.

20. Jacob L. Talmon, *The Origins of Totalitarian Democracy* (London: Secker and Warburg, 1952), 1-2 and 253.

and a solution. Chapters four to seven discuss why their diagnosis and prognosis are undermined by inconsistencies and controversy. Alarmist science is largely grounded in computer models rather than in observational evidence. The discussion in chapters eight and nine indicates that their solutions are equally fraught with problems. Alarmists see the problem as being similar to that of a patient with severe cardiac failure facing imminent death and being advised to prepare for a heart transplant; there is little time for debate, for further study, or for cost-benefit analysis. Climate change, however, is not such a straightforward issue: even a patient suffering from severe cardiac failure may wish to seek a second or third opinion and discuss with family members the risks, the costs, and the benefits in terms, for example, of remaining years of good health. The patient may even discover on further examination that the problem is not heart failure but a much more benign and curable condition.

Climate change policy is a complicated issue with many conflicting opinions and a wide range of possible options. Additionally, climate policy needs to be situated within the context of broader societal values and priorities. Alarmists have tried an end run around this inconvenience by insisting that climate change is the most threatening challenge humanity has ever faced. Others may see things differently and believe that there are challenges more deserving of public attention and the expenditure of scarce resources. In these circumstances, political leaders should seek advice from a broader spectrum of expert opinion.

Governments will need more than scientific advice if they are to avoid costly mistakes. They will need advice on the costs and benefits of various alternative responses, including benign neglect. It may well be that the most appropriate response will be to live with natural adaptation and let open markets work as necessary to mitigate any undesirable effects. Making sense of the climate change file is not just a matter of understanding the *science* of global warming, but also of assessing its *political economy*. Whether anthropogenic or natural, benign or malign, the climate will change, and governments will need to decide, individually and collectively, to what extent they are prepared to devote scarce resources at a cost to other societal needs in order to alter the path of climate change (assuming that this is within the realm of scientific, technological, and economic possibility), or to adapt to changes as they arise. In analyzing this issue, officials will learn that those who accept the scientific case for catastrophic anthropogenic climate change (CAGW) are ready to believe that urgent public action is needed and possible while those who are sceptical about the scientific case find the need for public action either unnecessary or less urgent and prefer a more balanced and analytical approach to the study of climate change. Not surprisingly, given the pervasive scaremongering by IPCC scientists, the UN, and ENGOs, the literature is heavily tilted in favour of proactive public policy on climate.

To date, the public policy response to the issue has suffered from lack of systematic analysis of the available options and from the stifling of open debate. Policy that is hastily conceived and inadequately discussed is unlikely to succeed in meeting its objectives. No government should entertain policy choices with such momentous negative consequences without a much firmer basis in both science and economics. Few national governments have made the effort to provide a comprehensive and credible basis for their decisions. Instead, they have relied on an international process that from the beginning has been dominated by an overwhelming conviction that the "crisis" is too great and urgent to allow time for analysis and debate. Instead, as discussed in chapter five, governments have relied on an IPCC process marred from the outset by commitment to the alarmist side of the debate.

In their assessments of the options governments can accept that there is broad agreement in the scientific community that the global climate has warmed over the past century and a half and that human activity is a contributing factor, but the extent of both and their impacts on the biosphere are hotly debated as is the capacity of humans to control climate change. Given that we live in an age in which the voice of "experts" is very powerful, the argument from authority has proven one of the most effective instruments available to the alarmist community. Having gained control of the commanding public heights of the issue, from government environment and meteorology departments to some of the leading science journals and the two key UN agencies – the UN Environmental Program (UNEP) and the World Meteorological Organization (WMO) – the "experts" have resorted to demonizing their critics as cranks and shills in the pay of questionable business interests, no matter how false the charges. Sceptics have had to fight an up-hill battle. And yet, their numbers have grown, and the claims of the alarmist community have become ever more shrill.

It is not difficult to conclude that the rise of climate alarmism to the top of the global anxiety agenda has been a matter of design.

The means by which a broad section of the public has been con-vinced that dangerous global warming is occurring are not subtle and include: reports from the United Nations, principally through the IPCC; incessant lobbying by environmental NGOs and allied scientists, political groups, and businesses; pleas for funding from climate scientists who have found that work tied to the anxiety agenda is more likely to be funded; and the obliging promulgation of selectively alarmist climate information by the media. The media are particularly prone to broadcasting correlation studies, omitting the many caveats that accompany them as well as the warning that correlation points to issues ripe for further investigation, not to cause and effect.

The pages that follow place discussion of the public policy of climate change into a broader and more balanced perspective. First, it builds on the idea developed by Wildavsky that sufficiently moti-vated citizens should be able to understand the basic contours of the underlying science and come to grips with both the theories sup-porting anthropogenic climate change and the criticisms that have been lodged against it. Second, it insists that many of the investiga-tions of the effects of climate change need to be reconsidered for the simple reason that they have been premised on the most extreme scenarios and have reached conclusions that are not supported by the evidence of past and current experience. Third, the book devel-ops the idea that the economics and politics of public policy are as important as the science. Even if the science case is stronger than I believe it is, global climate change responses still have to satisfy the requirements of sound cost-benefit analysis and technological feasi-bility. Wishful thinking has no place in public policy. Fourth, cli-mate change has become the driving force of the UN sustainability agenda pursuing the dream of a world governed by technocrats based on progressive, socialist principles and dedicated to sustaina-ble development. Finally, it is important that the wider public gain a better appreciation of the extent to which the climate change movement has become a cult bent on implementing a utopian agenda.

*****

2 | Science and Public Policy

Yet, in holding scientific research and discovery in high respect, as we should, we must be alert to the equal and opposite danger that public policy could itself become the captive of a scientific technological elite.

US President Dwight Eisenhower, *Farewell Address*, 1961

Science, risks, costs, benefits, and public policy

In the half century since President Eisenhower expressed his wise concern about the potential tyranny of a scientific-technological elite, his words have gained increasing currency as the regulatory state has grown by leaps and bounds. The authority of science is invoked routinely by both proponents and opponents of an increasing range of public policy preferences. Governments regularly look to scientists to help them address those issues. More and more experts provide governments with advice as governments attempt to keep up with the numerous perceived problems that have become the preoccupation of the modern state. Governments have become the prime funders of much basic science, and government priorities and preferences permeate scientific research and outcomes. Never before have science and technology played larger roles in the making of public policy, and never before has government played such a large role in shaping the pursuit and outcome of scientific enquiry. Butos and McQuade document the increasingly negative effects of the symbiotic relationship between government and science, suggesting that it affects "the structure and function of the scientific arrangements themselves: government funding affects science in a way analogous to the ways price controls, subsidies, credit expansion, and central planning affect markets, ... [leading] to an institutional structure in science that is made more unstable and maladapted to its environment as a consequence."[1]

In an earlier era, governments focused on society's welfare by seeking to ensure politically desirable outcomes in many areas of economic life. That broad social concern gradually faded and led to the deregulation movement of the 1970s as economists made a convincing case that society would be better off with fewer economic regulations and restrictions. Results-oriented economics gave way to the economics of opportunity. The liberalization of international trade through intergovernmental trade agreements was an important product of that movement as were the deregulation and privatization of many services that were formerly delivered by heavily regulated public or private monopolies. In their place, however, advanced societies have seen a rapid rise in quality-of-life regulations, i.e., efforts by governments to address risk-based concerns. Every year national, provincial, state, and local governments in Canada, the United States, Europe, and elsewhere introduce or amend, at tremendous cost, thousands of regulatory requirements affecting their citizens. Many of these regulations serve little purpose other than to satisfy the bureaucratic hunger for a more orderly and controlled world. Columnist George Will suggests that "climate alarmism [and other alarms] validates the progressive impulse to micromanage others' lives – their light bulbs, showerheads, toilets, appliances, automobiles, etc."[2]

Many of these regulatory requirements originate in alarming claims about environmental, health, and safety risks and are ostensibly grounded in science. In many cases, the claims have succeeded in capturing significant media and activist attention. As Aaron Wildavsky has suggested, however, one question needs to be addressed much more often: "But is it true?" Finding truth may be a stretch, but scientists can be more certain about some things than others; the extent of scientists' confidence in their findings has critical implications. In science, confidence is more than an informed opinion but rather a matter of evidence, methodology, probability, and integrity. Wildavsky and his students discovered that most of the alarming scientific claims they examined were open to serious doubt based

---

1.   William N. Butos and Thomas J. McQuade, "Government and Science: A Dangerous Liaison," *The Independent Review* 9:2 (Fall 2006), 205.

2.   "Satiating the feel-good crowd," Washington *Post,* February 27, 2014.

upon subsequent, evidence-based scientific investigation, a conclusion confirmed in each instance by well-credentialed and respected scientists. In a more recent assessment, legal scholar Jason Scott Johnston concludes that "modern regulatory regimes continue to insist that they are based on science even as regulatory science – the science that underlies actual regulatory decisions – has lost credibility. The lack of confidence in regulatory science has, with considerable justification, translated into a lack of confidence in modern environmental, health, and safety regulations."[3] This new reality goes to the heart of modern public policy.

Most new regulations are related to matters of health, safety, and the environment; rather than being based on science, many are grounded in irrational fears. All involve an implicit trade-off between a perceived risk and the social and economic costs of the regulatory response.[4] As Ottawa *Citizen* reporter Dan Gardner asks in his best seller, *Risk*: "Why are the safest and healthiest people in history living in a culture of fear?"[5] Much of that fear, and the responses to it, are presumably based on the work of scientists and their ability to assess risk as a matter of scientific probability, or are they? Risk-based analysis has made significant progress over the past thirty years as both public and private analysts have gained experience with the many facets and challenges of risk. Scientists have learned to provide governments with an assessment of the probability of physical risk that may result from human, animal, or environmental exposure to a particular product or process. They have learned that

3. Jason Scott Johnston, "Introduction," in Johnston, ed., *Institutions and Incentives in Regulatory Science* (Lanham, MD: Rowman and Littlefield, 2012), 1.

4. The extent of these costs is generally underappreciated by activists, officials, and the public. Clyde Crews, in his annual survey of US federal regulatory costs, points out that in 2012 "63 federal departments, agencies and commissions had just completed or were at work on 4,062 rules and regulations at various stages of planning and implementation. Of those, 224 were classified as 'economically significant,' meaning they have $100 million or more in annual economic impact. All in all, 3,708 final rules hit the books in 2012 – more than 10 daily." "The real cost of federal regulations," *Washington Times*, October 2, 2013. Canada and other OECD countries are equally committed to expanding the regulatory state.

5. Dan Gardner, *Risk: Why We Fear the Things We Shouldn't – and Put Ourselves in Greater Danger* (Toronto: McClelland and Stewart: 2008), frontispiece. For an extended discussion of risk and its management in modern society, see H. W. Lewis, *Technological Risk* (New York: W.W. Norton, 1990) and, for its evolution as a concept, Peter L. Bernstein, *Against the Gods: The Remarkable Story of Risk* (New York: John Wiley & Sons, 1996).

risk is the product of both probability and consequence; assessing risk is thus a matter of finding credible ways of measuring both. Some risks are relatively straightforward to assess while others pose serious challenges. The probability of harm from direct exposure to a noxious substance, for example, is typically a matter of the dose and can be addressed on the basis of controlled use. The probability of harm due to long-term exposure from a very low dose of the same substance, on the other hand, is more difficult to verify with observation or experiment due, for example, to the confounding influences of other factors or the moral implications of some forms of testing. Such long-time exposure also needs to be balanced by a scientific assessment of the benefits that use of the substance in question may provide.

Once scientists have provided governments with an evidence-based risk assessment, economists can provide a cost-benefit analysis of the risk, possible remedies, and benefits. Only then can policy makers and their advisers determine from a range of options how best to manage the risk at a reasonable cost to society.[6] Society is best served when policy makers are provided with sound, evidence-based assessments of risk by scientists and of costs and benefits by economists and other experts. Policy makers must then be capable of making a judgment on the quality of both, a difficult challenge when the risk being assessed is characterized by high levels of uncertainty in the underlying scientific knowledge. In such instances economic assessments will be largely meaningless because the range of possible risks is very wide. Johnston points out that "current laws and institutions for regulatory science have made science so significant in regulatory policy that they have created both an in-

---

6.   This is not to suggest that policy makers always conscientiously assess the trade-offs between costs and benefits. Once an activist campaign succeeds, costs become a secondary factor. The US Environmental Protection Agency, for example, is notorious for its cavalier attitude toward the cost of the many regulations it imposes on society. See, for example, James E. McCarthy and Claudia Copeland, *EPA Regulations: Too Much, Too Little, or on Track?* (Washington: Congressional Research Service), R41561, July 16, 2013. More generally, as Lewis points out, many mandates governing regulatory agencies in the United States are written or have been interpreted by the courts to forbid the agency from balancing the social and economic costs of risk reduction. *Technological Risk*, 93. Strikingly, the US Supreme Court's decision in *Michigan v the EPA* (June 29, 2015), reverses this trend, invalidating the EPA's regulatory fiat regarding mercury emissions from coal-fired power plants because it failed to take compliance costs into account. It remains to be seen whether this decision will change the EPA's – and other agencies' – regulatory culture.

centive for regulators to cloak policy decisions in the mantle of science, and have also – by failing to understand the uncertainty inherent in most scientific findings – put scientists in the position where their own policy preferences guide important choices of scientific methodology and interpretation."[7] Science thus plays a critical role in informing much of modern public policy. As scientists become more and more dependent on governments to fund their research, the findings that support public regulatory activity may need greater independent scrutiny than ever before.

The reliability of science never was, and certainly is not now, as unimpeachable as popularly imagined. Scientists are as prone to the frailties of ambition, careerism, misconduct, greed, and exaggeration as the rest of humanity. The academic pressure to publish is so intense that much of what passes for new research and insights needs to be considered with great care. Often, conscientious policy makers, not well versed in the arcane details of complex scientific issues, need to make special efforts to get behind the headline-grabbing sensationalism provided by the media on breakthroughs in environmental and health science and to consider the broader literature more diligently in order to gain a balanced appreciation of what scientists know, what they suspect, and what they do not know. Even at their best, however, scientists are not well suited to the role of authorizing policy. By relying on such mantras as "research shows" or "science says" to justify policy measures, decision makers are evading responsibility for a proper assessment of values and interests that go beyond science.

Scientific findings or claims on their own may be useful, but their value derives from assessing them within the broader context of societal needs, priorities, and values. Expertise on risk may be a critical input, but its assessment is ultimately a matter of political judgment subject to both democratic and judicial oversight. This was a point already clear to organizational theorist Luther Gulick in 1937: "The expert knows his stuff. Society needs him, and must have him more and more as man's technical knowledge becomes more and more extensive. But history shows us that the common man is a better judge of his own needs in the long run than any cult of experts."[8] That was true then, and even more true today, when

---

7.   Johnston, "Introduction," *Institutions and Incentives in Regulatory Science*, 1.

8.   Luther Gulick, "Notes on the Theory of Organization," in Gulick and Lyndall Urwick, eds., *Papers on the Science of Administration* (New York: Institute of Pub-

experts have saddled society with more and more rules and requirements, insufficiently sensitive to the fact that the economy has limits, that knowledge is imperfect, and that some risks need to be tolerated if we are not to stifle all innovation and enterprise.[9]

The validity and objectivity of science were much debated in the 1980s between natural scientists and social scientists in what became known as the science wars. As Daniel Sarewitz notes, "in the heat of battle, all nuance was lost in the quest for victory, and a single, black-and-white question came to dominate the contest: Does science achieve an objective view of nature, or are all scientific facts constructed by social interactions? The latter, 'constructivist' view considers the 'truth' or 'falsity' of scientific claims ... as deriving from the interpretations, actions, and practices of scientists rather than as residing in nature." Sarewitz finds that the two opposing sides were both right and wrong. He argues that "facts are both objective (that is, representations of something real) and constructed (that is, products of social context). ... As for the public debate, the natural scientists, not surprisingly, soon had the constructivist social scientists on the run."[10] Sarewitz concludes that in the context of the science-policy interface, we "are not suffering from a lack of objectivity, but from an excess of it. Science is sufficiently rich, diverse, and Balkanized to provide comfort and support for a range of subjective, political positions on complex issues such as climate change, nuclear waste disposal, acid rain, or endangered species. ... The problem is not one of good science versus bad, or 'sound' science versus 'junk' science. The problem is that nature can be viewed through many analytical lenses, and the resulting perspectives do not add up to a single, uniform image, but a spectrum that can illuminate a range of subjective positions."[11] Once scientists insist on a single, consensus perspective on any issue, it is likely that that perspective is more a matter of politics than of science.

---

lic Administration, 1937), as quoted in Sheila Jasanoff, *The Fifth Branch: Science Advisers as Policymakers* (Cambridge, MA: Harvard University Press, 1990), 10.

9.   See Jasanoff, *The Fifth Branch,* 17, for a more complete discussion.

10.   Daniel Sarewitz, "Science and environmental policy: an excess of objectivity," in R. Frodeman, ed., *Earth Matters: The Earth Sciences, Philosophy, and the Claims of Community* (New York: Prentice Hall, 2000), 80. A useful introduction to the debate can be found in Keith Parsons, ed., *The Science Wars: Debating Scientific Knowledge and Technology* (Amherst, NY: Prometheus Books, 2003).

11.   Sarewitz, "Science and environmental policy," 90.

Science, however, has much to offer. We would not enjoy the many material benefits of modern life without the contributions of dedicated scientists, but the public should nevertheless be wary of sensational claims, particularly those arising from such soft sciences as ecology or epidemiology or promoted by advocacy groups such as Friends of the Earth or the Union of Concerned Scientists. Conscientious scientists are always on the lookout for the pitfalls that come from premature conclusions or from ignoring confounding observations. What may be reported in the media as a breakthrough one day is often contradicted by conflicting observations later. Uncertainty and scepticism are thus at the heart of good science. As the great theoretical physicist Richard Feynman (1918-88) cautioned: "it is imperative in science to doubt; it is absolutely necessary, for progress in science, to have uncertainty as a fundamental part of your inner nature. ... Nothing is certain or proved beyond all doubt."[12]

When it comes to public policy, however, uncertainty is a big problem, particularly for governments under pressure to make policy on the basis of uncertain science. Governments do not like to take decisions based on speculative reasoning. Even more than their preference for one-handed economists, politicians look for one-handed scientists. But the best that scientists can do as scientists is to discuss probabilities. Governments, nevertheless, seek certainty and thus we get "official" science, which is often the product of an advisory process that, in economist David Henderson's words, is "marred by chronic and pervasive bias"[13] as activist scientists turn into stealth policy advocates, often aided by officials and policy advisers with their own agendas.

Guelph University economist Ross McKitrick argues that in public policy discussions, science occupies a privileged position; governments generally accept research standards acceptable to academia that would never be accepted in a court of law or in a securities filing. He explains that "this is not necessarily a problem for the academic purpose being served, since researchers have to have considerable leeway to make their mistakes in public in order to ensure scholarly communication remains open and important topics are probed thoroughly. The problem arises when governments assume journal peer review amounts to a standard of verification similar to

---

12. Richard Feynman, "The Relation of Science and Religion," transcript of a talk at the Caltech YMCA Lunch Forum, May 2, 1956.

13. David Henderson, "Economists and Climate Science: A Critique," *World Economics* 10:1 (January-March 2009), 75.

what would be applied in a business setting or a trial procedure. This is a disastrous assumption."[14]

Activist support for certain scientific conclusions, however, should not be taken as unqualified support for all of science. Many of the most vigorous proponents of a green agenda, for example, tend to be suspicious and fearful of many scientific breakthroughs of the past half century such as genetic modification or vaccination against diseases. Activists frequently blame science and technology for contributing to environmental degradation and to global warming but then seize on science when it can be used to reinforce their views and beat back their opponents. Thus the people who were at the forefront in opposing all things nuclear are the same people who are convinced that the word of climate scientists is of the highest order because it is science, and peer-reviewed at that. What their views often have in common is that they are progressive and often even anti-human.[15]

The confusion is evident in many recent policy controversies. Once governments have pronounced on a matter, often as a result of lobbying by various interests and with only a cursory examination of the science, real science goes out the window and official science takes over. As McKitrick argues, official science lacks three critical safeguards: balance, due diligence, and full disclosure.[16] One can add a fourth: an ability to adjust to new evidence and insights. In the absence of these safeguards, science can easily be captured by vested interests that can then use the authority and resources of government to marginalize critics and advance their own perspectives. Within the halls of government, policy then directs science, dissent is stifled, and healthy discussion of alternative perspectives is discouraged.

The UN's Intergovernmental Panel on Climate Change (IPCC) has taken official science to a whole new level, corrupting many of

---

14. Ross McKitrick, "Bringing Balance, Disclosure and Due Diligence into Science-based Policymaking," in Jene M. Porter and Peter W.B. Phillips, eds., *Public Science in Liberal Democracy* (Toronto: University of Toronto Press, 2007), 260-1.

15. See Robert Zubrin, *Merchants of Despair: Radical Environmentalists, Criminal Pseudo-Scientists, and the Fatal Cult of Antihumanism* (New York: New Atlantis Books, 2012) and Wesley J. Smith, *The War on Humans* (Seattle: Discovery Institute, 2014) for full explorations of this theme. Jeffrey Foss similarly explores the anti-human dimension of modern environmentalism in *Beyond Environmentalism: A Philosophy of Nature* (Hoboken, NJ: John Wiley & Sons, 2009), 57-75.

16. McKitrick, "Bringing Balance, Disclosure and Due Diligence into Science-based Policymaking."

the normal safeguards built into the scientific process in order to provide governments with certainty and a basis for action. Its 2007 *Summary for Policy Makers,* for example, posits a 95 percent confidence level that global warming in the 20th century was largely anthropogenic, but the underlying scientific reports from the three Working Groups use the words "uncertain" and "uncertainties" 1,300 times.[17] The IPCC's mandate was to find the *human* impact on climate change, which disposed it to ignore or underestimate the many other factors that influence the highly complex processes of ever-changing climate. In its 2013-14 Report, despite a 15-year lull in global warming, the IPCC stuck to its mantra and insisted that "it is *extremely likely* that human influence has been the dominant cause of the observed warming since the mid-20th century."[18]

Retired MIT atmospheric physicist Richard Lindzen suggests that "when an issue becomes a vital part of a political agenda, as is the case with climate, then the politically desired position becomes a *goal* rather than a *consequence* of scientific research." [emphasis added] He notes further: "The temptation to politicize science has always been high, and political organizations have long sought to improve their own credibility by associating their goals with 'science' – even if this involves misrepresenting the science."[19] Regarding the latest 2013 assessment report of the IPCC, Lindzen told the *Climate Depot* blog: "I think that the latest IPCC report has truly

---

17. S.F. Hayward, K.P. Green, & J.M. Schwartz, "Politics Posing as Science: a Preliminary Assessment of the IPCC's Latest Climate Change Report," *AEI Outlook,* December 2007. Rachael Jonassen and Roger Pielke, Jr. similarly found that the authors of the *Summary for Policymakers* posited a much higher level of certainty than the authors of the underlying scientific reports. "Improving conveyance of uncertainties in the findings of the IPCC," *Climate Change* 69:1 (March 2005).

18. UN IPCC AR5 WG1, *Summary for Policymakers,* 17. David Legates et al. indicate that only 0.3 per cent of the abstracts of 11,944 scientific papers on climate-related topics from 1991-2011 explicitly stated an opinion that more than half of the global warming since 1950 had been caused by human emissions of CO2 and other greenhouse gases. "Climate Consensus and 'Misinformation': A Rejoinder to Agnotology, 'Scientific Consensus and the Teaching and Learning of Climate Change,'" *Science & Education,* August 2013. The claim that 97 per cent of scientists agree with the IPCC is a contorted example of the misapplication of agnotology, the study of culturally induced ignorance or doubt, particularly the publication of inaccurate or misleading scientific data. Legates et al. thoroughly demolish the frequently advanced 97 per cent consensus claim. See Box 7-1.

19. Richard Lindzen, "Climate Science: Is It Currently Designed to Answer Questions?" Paper prepared for presentation to a conference in San Marino, Italy, September 27, 2008, at arxiv.org/pdf/0809.3762v4.pdf. Lindzen held the Alfred P. Sloan chair at MIT until his retirement in 2013.

sunk to [the] level of hilarious incoherence. ... It is quite amazing to see the contortions the IPCC has to go through in order to keep the international climate agenda going."[20]

The stifling impact of official science is compounded by the tyranny of highly specialized experts prepared to speak with confidence from a narrow base on a broad subject. Feynman observed: "In this age of specialization men who thoroughly know one field are often incompetent to discuss another."[21] Over the past half century, knowledge and research have become ever more narrow and specialized and cross-cutting interdisciplinary work ever rarer. In both research laboratories and academia, money and prestige flow increasingly to the narrowly focused. As a result, many investigators need to accept on faith the conclusions of various other experts in order to push the boundaries of their own areas of specialization.

In these circumstances, it becomes possible for an environmental economist like Simon Fraser University's Marc Jaccard to model how best to use carbon taxes for reducing carbon emissions to more politically acceptable levels without any attempt to examine whether the scientific case for reducing CO_2 has any merit; he relies on others to make that judgment. By this process, questionable ideas are disseminated and fixed in the "paradigm" of the moment, and the point is reached at which failure to conform becomes a liability. British philosopher Martin Cohen observes, "today, global-warming 'deniers' have all been told they must fall into line with 'the science'. But this is not science; this is propaganda. And we are not being asked to be more rational but to suspend our own judgment completely. That, not 'runaway climate change', is the most dangerous threat to the world today."[22] After centuries of progress in unraveling the mysteries of nature, science is in danger of descending into increasing disrepute, particularly in areas that are policy-relevant or that satisfy political or other non-scientific goals. Australian climate scientist Garth Paltridge worries that we "have to consider the possibility that the scientific establishment behind the global warming issue has been drawn into the trap of seriously overstating the climate problem – or, what is much the same thing, of seriously understating the uncertainties associated with the climate problem – in

20.   "MIT Climate Scientist Dr. Richard Lindzen Rips UN IPCC Report," *Climate Depot,* September 27, 2013.

21.   Feynman, "The Relation of Science and Religion."

22.   Martin Cohen, "Beyond Debate?" *Times Higher Education Supplement,* December 10, 2009.

its effort to promote the cause. It is a particularly nasty trap in the context of science, because it risks destroying, perhaps for centuries to come, the unique and hard-won reputation for honesty which is the basis for society's respect for scientific endeavour."[23]

Economists distinguish between positive and normative economics. The former *describes* how economic phenomena function; the latter *prescribes* how to use solutions grounded in economics to make the world a better place. A similar distinction can be useful in looking at science. Positive science seeks to understand how particular natural phenomena function by asking, as described by Feynman, "If I do this, what will happen?"[24] Normative science focuses on how to use science to improve life and solve social and other problems. The former is the proper domain of scientists; the latter should be informed by positive science and the insight of scientists but should encompasses much more expertise and may ultimately involve political decisions and moral values. Official science confuses the boundaries between the two and ushers in a cascade of potential problems.

Most scientists with experience working in the policy realm caution against falling into the traps that emerge when they exceed the boundaries of their expertise. Under these circumstances, it is critical that policy makers and analysts gain a much better understanding of the limits and reliability of modern science and appreciate the extent to which funding has become a critical influence on scientific outcomes.

Is modern science reliable?

Thirty years ago, Richard Roberts, then at the US National Bureau of Standards, estimated that at least half of all published scientific papers were either unusable or unreliable. It was a provocative claim but garnered little attention.[25] In 2005, however, in a widely read article, medical researcher John Ioannidis similarly argued that "in modern research, false findings may be the majority or even the vast majority of published research claims." He argued that "the

---

23. Garth W. Paltridge, "Uncertainty, scepticism and the climate issue," in Alan Moran, ed., *Climate Change: The Facts* (Melbourne: Institute of Public Affairs, 2015), kindle edition.

24. Feynman, "The Relation of Science and Religion."

25. Richard R. Roberts, "An unscientific phenomenon: Fraud grows in laboratories," *Science Digest* (June, 1977), 38, cited in Alexander Kohn, *False Prophets* (Oxford: Basil Blackwell, 1986), 2.

combination of various design, data, analysis, and presentation factors ... tend to produce research findings when they should not be produced." Particularly troubling was his conclusion that "the hotter a scientific field (with more scientific teams involved), the less likely the research findings are to be true."[26] Ioannidis' findings also point to the prevalence of scientists publishing results that confirm their initial findings while ignoring negative results. Selective reporting has become a major problem in science.

Ioannidis' findings were largely based on his knowledge of research in biomedicine. In 2011, however, he made an even wider claim: "False positives and exaggerated results in peer-reviewed scientific studies have reached epidemic proportions in recent years. The problem is rampant in economics, the social sciences, and even the natural sciences, but it is particularly egregious in biomedicine. Many studies that claim some drug or treatment is beneficial have turned out not to be true."[27] In a reply to a comment on his original article, he pointed to the bottom line: "confidence in the research enterprise is probably undermined primarily when we claim that discoveries are more certain than they really are, and then the public, scientists, and patients suffer the painful refutations."[28]

In neither his original article nor in its echo six years later did Ioannidis suggest fraud. As discussed further below, increasing incidence of misconduct presents its own problems. Ioannidis' claim was much more serious: much of what is published in scientific journals is premature and implies more confidence than the research warrants, misleading other researchers, the public, and policy makers. He points to the reluctance of researchers to indicate the limits of their studies, the failure to report both negative and positive evidence, the temptation to over-interpret the data, and the extent to which they torture the data to find statistical significance when the data at best show weak correlations. In short, scientists, being human, "are tempted to show that they know more than they do." Nevertheless, he concludes, "the crisis should not shake confi-

---

26. John P. Ioannidis, "Why Most Published Research Findings Are False," *Public Library of Science (PLoS): Medicine* 2:8 (August 30, 2005), 124. By early July 2015, Ioannidis' article had been cited 1,673 times and viewed 1,335,602 times online.

27. Ioannidis, "An Epidemic of False Claims," *Scientific American,* June 2011, 16. See also See Jonah Lehrer, "The Truth Wears Off," *The New Yorker,* December 13, 2010 for a broader discussion of this phenomenon.

28. Ioannidis, "Author's Reply to Goodman and Greenland," *PLoS Medicine* 4:6e (June 26, 2007), 215.

dence in the scientific method. The ability to prove something false continues to be a hallmark of science. But scientists need to improve the way they do their research and how they disseminate evidence."[29]

Taubes puts the issue of studies that have been proven wrong into a useful perspective: "Many explanations have been offered to make sense of the here-today-gone-tomorrow nature of medical wisdom – what we are advised with confidence one year is reversed the next – but the simplest one is that it is the natural rhythm of science. An observation leads to an hypothesis. The hypothesis (last year's advice) is tested, and it fails this year's test, which is always the most likely outcome in any scientific endeavour. There are, after all, an infinite number of wrong hypotheses for every right one, and so the odds are always against any particular hypothesis being true, no matter how obvious or vitally important it might seem."[30] What applies to medical research applies equally to other areas of science, particularly science that plays a critical role in the development of public policy.

Properly done, science is about advancing hypotheses based on initial observations and then rigorously testing them to determine their strengths. Studies done to test, verify, and replicate an original study are an essential part of science. The problem we have is cultural. The modern culture of science, in which funding, careers, prestige, and enterprise depend on getting it right, makes it difficult to accept that an hypothesis has been falsified. It also devalues replication work and overvalues novelty. More troubling is that in today's media culture the initial study is more likely to gain attention and stick in the popular imagination, while the many later confounding studies are ignored. The University of Montana's Daniel Kemmis concludes: "So why would anyone continue to speak and act as if good science by itself could get to the bottom of these bottomless phenomena and in the process give us 'the answer' to difficult ... issues? In large part this is simply a holdover of an anachronistic view of how the world works and of what science can tell us about that world. In this sense, the repeated invocation of good science as the key to resolving complex ecosystem problems has itself

---

29. Ioannidis, "An Epidemic of False Claims."
30. Gary Taubes, "Do We Really Know What Makes Us Healthy?" *New York Times Magazine,* September 16, 2007.

become bad science. What is infinitely worse is that this bad science is all too readily made the servant of bad government."[31]

Ian Boyd, chief scientific adviser at the UK Department of Environment, Food and Rural Affairs, drew the obvious conclusion from the disturbing phenomenon of science gone wrong: "Unreliability in scientific literature is a problem for people like me .... To counsel politicians, I must recognize systematic bias in research. Bias is cryptic enough in individual studies, let alone in whole bodies of literature that contain important inaccuracies. ... It could stem from the combined effects of how science is commissioned, conducted, reported, and used, and also from how scientists themselves are incentivized to conduct certain research" and report its results.[32]

Boyd is not alone; other scientists engaged in the interface between science and its engineering, health, policy, and other applications have begun to demand recognition of the problem and to propose steps to address it. Sarewitz, whose work focuses on the social benefits of scientific research, writes: "alarming cracks are starting to penetrate deep into the scientific edifice. They threaten the status of science and its value to society. ... Their cause is bias, and the threat they pose goes to the heart of research. ... Science's internal controls on bias were failing, and bias and error were trending in the same direction – towards the pervasive over-selection and over-reporting of false positive results. ... A biased scientific result is no different from a useless one. Neither can be turned into a real-world application. ... It is likely to be prevalent in any field that seeks to predict the behaviour of complex systems."[33] More problematic is the extent to which bias originates as a result of public funding and ends up as the basis for flawed public policies. Butos and McQuade warn, "there are serious reasons ... for thinking that the liaison between government and science carries with it unrecognized dangers for the functioning and integrity of science as a reliable generator of knowledge,"[34] let alone as input to public policy.

Henry Bauer, a retired chemist at Virginia Polytechnic University, worries that increasing dogmatism in some fields is leading to

---

31. Daniel Kemmis, "Science's role in natural resource decisions," *Issues in Science and Technology* (Summer 2002).

32. Ian Boyd, "A standard for policy-relevant science," *Nature* 501 (September 12, 2013), 159.

33. Daniel Sarewitz, "Beware the creeping cracks of bias," *Nature* 485 (May 10, 2012), 149.

34. Butos and McQuade, "Government and Science," 177.

what he characterizes as "dominant knowledge monopolies" committed to protecting their theories while denigrating nonconforming research and stifling the search for scientific truth. One of the founders of the science studies movement, Bauer believes that the issue is less a matter of some researchers being right and others wrong and more a matter of maintaining a perpetually sceptical approach to evolving issues.[35] Often, it can be less a matter of a critic's being right or wrong than of a fellow scientist's having identified a flaw in a mainstream argument that needs to be addressed if an hypothesis is not to fail. Climate science, for example, is replete with problems involving both data and explanations.

The problem is compounded by what science blogger Eric Raymond calls an error cascade, i.e., when researchers begin to trim their observations to fit within a perceived or prevailing consensus, particularly one that is policy-relevant or satisfies political or other non-scientific goals. Investigators may privately consider the consensus wrong and incapable of explaining the phenomenon being studied, but peer pressure keeps them from speaking out. Raymond observes: "When politics co-opts a field that is in the grip of an error cascade, the effect is to tighten that grip to the strangling point. ... Consequently, scientific fields that have become entangled with public-policy debates are far more likely to pathologize – that is, to develop inner circles that collude in actual misconduct and suppression of refuting data rather than innocently perpetuating a mistake. ... When anyone attempts to end debate by insisting that a majority of scientists believe some specified position, this is the social mechanism of error cascades coming into the open and swinging a wrecking ball at actual scientific method right out where everyone can watch it happening."[36]

The impact of scientific misconduct

The increasing extent to which research is both biased and less reliable than claimed is compounded by the rising incidence of scientific misconduct. The most egregious instances often lead to exposure and retraction of the implicated research, but lesser examples continue to taint the literature for years and can even become part of established paradigms. One barometer is the extent of retractions

---

35. See Henry Bauer, *Dogmatism in Science and Medicine* (Jefferson, NC: McFarland and Company, 2012).

36. Eric S. Raymond, "Error cascade: a definition and examples," *Armed and Dangerous*, February 4, 2010.

in scientific journals. A relatively rare phenomenon a generation ago, retractions have become a flood. An assessment in the *Proceedings of the National Academy of Sciences* limited to medical literature found that 2,407 articles had been retracted by the middle of 2012. "Fraud or suspected fraud was responsible for 43 percent of the retractions. Other types of misconduct – duplicate publication and plagiarism – accounted for 14 per cent and 10 per cent of retractions, respectively. Only 21 percent of the papers were retracted because of error."[37]

Researchers at the US Office of Research Integrity (ORI) found a disturbing disconnect between the number of reported cases of fraud and the suspected number based on surveys of holders of US government research grants. Their findings suggested that there should be about 2,335 cases reported annually, while in fact the number of cases investigated by the ORI only added up to 24 annually, i.e., one percent of possible cases. Whatever the reasons – institutional or personal – their findings point to a "failure to foster a culture of integrity" in the scientific research community.[38]

An earlier study of scientific integrity conducted by investigators at the University of Minnesota found similar evidence of willingness to skirt the edges of truthfulness (See Table 2-1). Given that the survey was mail-in and voluntary with a participation rate of 52 percent for mid-career scientists and 43 percent for early career scientists, it was not unreasonable for the authors to conclude that "our approach certainly leaves room for potential non-response bias; misbehaving scientists may have been less likely than others to respond to our survey, perhaps for fear of discovery and potential sanction."[39] One-third of the respondents admitted to at least one of the top ten listed failings.

In 2009, Daniele Fanelli, a researcher at the University of Edinburgh interested in the sociology of science, carried out a meta-analysis of 21 surveys of scientists' attitudes towards the integrity of research. He found that "on average, about 2 percent of scientists admitted to having fabricated, falsified or modified data or results

37. Zoë Corbyn, "Misconduct is the main cause of life-science retractions," *Nature* 490 (October 4, 2012), 21.

38. Sandra L. Titus, James A. Wells and Lawrence J. Rhoades, "Repairing research integrity," *Nature* 453 (June 19, 2008), 980-982.

39. Brian C. Martinson, Melissa S. Anderson and Raymond de Vries, "Scientists behaving badly," *Nature* 435 (June 9, 2005), 737.

Table 2-1: Scientists behaving badly

Percentage of scientists who say that they engaged in the behaviour listed within the previous three years (n=3,247)	All	Mid-career	Early career
• Inadequate record keeping related to research projects	27.5	27.7	27.3
• Changing the design, methodology or results of a study in response to pressure from a funding source	15.5	20.6	9.5***
• Dropping observations or data points from analysis based on a gut feeling that they were inaccurate	15.3	14.3	16.5
• Using inadequate or inappropriate research designs	13.5	14.6	14.2
• Overlooking others' use of flawed data or questionable interpretation of data	12.5	12.2	12.8
• Withholding details of methodology or results in papers or proposals	10.8	12.4	8.9**
• Inappropriately assigning authorship credit	10.0	12.3	7.4***
• Circumventing certain minor aspects of human-subject requirements	7.6	9.0	6.0**
• Failing to present data that contradict one's own previous research	6.0	6.5	5.3
• Publishing the same data or results in two or more publications	4.7	5.9	3.4**
• Unauthorized use of confidential information in connection with one's own research	1.7	2.4	0.8***
• Relationships with students, research subjects or clients that may be interpreted as questionable	1.4	1.3	1.4
• Using another's ideas without obtaining permission or giving due credit	1.4	1.7	1.0
• Not properly disclosing involvement in firms whose products are based on one's own research	0.3	0.4	0.3
• Falsifying or 'cooking' research data	0.3	0.2	0.5
• Ignoring major aspects of human-subject requirements	0.3	0.3	0.4

Note: significance of X^2 tests of differences between mid- and early career scientists are noted by** (P<0.01) and *** (P<0.001)

Source: Adapted from Martinson, Anderson, and de Vries, "Scientists behaving badly," *Nature* 435 (June 9, 2005), 737.

at least once – a serious form of misconduct by any standard ... and up to one third admitted a variety of other questionable research practices including 'dropping data points based on a gut feeling', and 'changing the design, methodology or results of a study in response to pressures from a funding source'. In surveys asking about the behaviour of colleagues, fabrication, falsification, and modification had been observed, on average, by over 14 percent of respondents, and other questionable practices by up to 72 percent."[40]

40. Daniele Fanelli, "How Many Scientists Fabricate and Falsify Research? A Systematic Review and Meta-Analysis of Survey Data." *Public Library of Science One* 4:5 (May 2009), 5738.

Science is built on trust. In any field of science, individual scientists are working on a small part of the whole and rely on other scientists to help fill in the details leading to a more complete understanding. Sociologist Harriet Zuckerman explains: "the institution of science involves an implicit social contract between scientists so that each can depend on the trustworthiness of the rest. ... The entire cognitive system of science is rooted in the moral integrity of aggregates of individual scientists."[41] Trust is an integral part of the scientific method first advanced by Francis Bacon in the 17th century and pursued by generations of scientists ever since, as encapsulated in the opening paragraph of the latest edition of the US National Academies' *On Being a Scientist: A Guide to Responsible Conduct in Research*:

> The scientific enterprise is built on a foundation of trust. Society trusts that scientific research results are an honest and accurate reflection of a researcher's work. Researchers equally trust that their colleagues have gathered data carefully, have used appropriate analytic and statistical techniques, have reported their results accurately, and have treated the work of other researchers with respect. When this trust is misplaced and the professional standards of science are violated, researchers are not just personally affronted – they feel that the base of their profession has been undermined. This would impact the relationship between science and society.[42]

The scientific method

Modern science came into its own with the recognition that human observation of natural phenomena could be organized systematically, thus allowing scientists to develop hypotheses and theories about the natural world not from first principles but from detailed experimentation and observation. While philosophers of science may continue to pay attention to the work of Aristotle, Plato, Galen, and other pre-modern thinkers, the reality is that they set science off on the wrong track for nearly two millennia. It was not until the 17th century with the influence of thinkers such as Bacon that science began to adopt what has become known as the *scientific method*. Bacon believed that science should be grounded in carefully verified

---

41. Harriet Zuckerman, "Deviant behaviour and social control in science," in Edward Sagarin, ed., *Deviance and Social Change* (Beverley Hills, CA: Sage, 1977), 113.

42. Preface to the new 3rd edition of *On Being a Scientist: A Guide to Responsible Conduct in Research* (Washington: National Academies Press, 2009).

observations of the natural world. As confidence in the new method evolved, scientists increasingly focused their efforts on systematic observation, precise measurement, controlled experimentation, and the development of data leading to the formulation, testing, and modification of hypotheses with predictive power.

As scientific knowledge progressed, scientists accepted that any theory or hypothesis was only as good as the observations – or evidence – on which it rested; any evidence that contradicted a hypothesis immediately brought it into question. Scientific enquiry thus involves learning about the natural world on the basis of *testable* explanations capable of being *replicated* by others. Replication ensures that initial results are not atypical and strengthens regression to the mean by reducing the statistical impact of outliers. The extent to which scientists no longer pursue replication studies – due to both cost and lack of incentives – may be one of the reasons that so many findings turn out to be premature or false. Novelty is much higher on the incentive scale than confirmation.

More sophisticated technologies and measurement techniques have remained key to scientific progress, allowing scientists to observe natural phenomena more finely and consistently. At the same time, as Stephen Jay Gould reminds us, "science, as done by human beings, could only be envisaged and practiced within a constraining and potentiating set of social, cultural, and historical circumstances – a variegated and changing context that, by the way, makes the history of science so much more interesting, and so much more passionate, than the cardboard Whiggery of conventional marches to truth over social impediments."[43] The extent to which the product of science is objective or subjective (i.e., is the pure result of what scientists do based on the scientific method or is it shaped by social, cultural, and historical circumstances?) is the stuff of much debate, less among scientists than among philosophers of science. Labels may have changed and arguments may have become more complex and sophisticated, but the fundamental divide remains.

Most working scientists assume that in their quest for the best explanations of natural phenomena they seek to explain the operation of nature as truthfully as they can, i.e., by developing objective knowledge. The results of their efforts are not expressed in terms of truth or proof but as matters of probability, based on well-

---

43. Stephen Jay Gould, "Deconstructing the 'Science Wars' by Reconstructing an Old Mold," *Science* 287:5451 (January 14, 2000), 254.

established quantitative measures, such as statistical significance.[44] To add credibility, much of science includes information on deviations from the mean or error bars. Many would agree with biologist Austin Hughes that if science has any claims to authority, it does not lie in its practitioners and their credentials but in rigorous adherence to high methodological standards.[45] Philosophers of science, on the other hand, are interested in understanding the viability and value of the scientific method in developing human knowledge and understanding of the natural world. For them, the march of science may not be as objective as many scientists believe. The two best known modern expressions of the philosophy of science were advanced by Karl Popper (1902-94) and Thomas Kuhn (1922-96).[46] Their competing views of the scientific enterprise will prove helpful in unravelling the nature of competing perspectives on climate change and the need, if any, for a public policy response.

---

44. In an interesting sleight of hand rarely appreciated by policy makers, the UN's IPCC expresses its views in terms of *quantitative* probabilities (e.g., very likely = 90 per cent certainty), based on the informed *opinion* of those participating in its assessment but excluding those who do not form part of the in-group. Expressing opinions in quantitative terms for issues that essentially cannot be quantified may give the appearance of being scientific but smacks of charlatanism. For an explanation of how this developed, see Stephen H. Schneider, *Science as a Contact Sport: Inside the Battle to Save Earth's Climate* (Washington: National Geographic, 2009), 148-54. See also Arthur Rörsch, "Post-Modern Science and the Scientific Legitimacy of the IPCC's WG1 AR5 Report."

45. Austin L. Hughes, "The Folly of Scientism," *The New Atlantis* (Fall 2012), 32-50. Hughes is Carolina Distinguished Professor of Biological Sciences at the University of South Carolina.

46. Popper's principal work on the philosophy of science can be found in *The Logic of Scientific Discovery*, first published in German in 1934 and reissued with additional material in English in 1959. A second English edition with additional material was published in 1968, available as a Routledge Classic (London 2002). Kuhn's controversial and widely read book, *The Structure of Scientific Revolutions* (Chicago: University of Chicago Press, 1962), has been reissued many times. A good introduction to what the two philosophers of science had in common and where they disagreed can be found in Steve Fuller, *Kuhn vs. Popper: The Struggle for the Soul of Science* (New York: Columbia University Press, 2004). The book is based superficially on an encounter between the two at Bedford College at the University of London on July 13, 1965, and provides a sympathetic view of their respective approaches to the scientific process as well as an intellectual history of the sources and results of their ideas. There are, of course, many other views among philosophers of science. See, for example, Peter Godfrey-Smith, *Theory and Reality: An Introduction to the Philosophy of Science* (Chicago: University of Chicago Press, 2003). The following pages draw extensively on Fuller's work.

For policy makers and society at large, the issue at stake is the *authority* of science. To thinkers like Kuhn and his followers, much of science is a subjective social construction subject to revision. Followers of Popper, on the other hand, insist that science involves an objective quest for truth and that its authority depends on the ability to marshal real-world observational evidence that validates scientific hypotheses and conclusions. The late American pragmatist Richard Rorty characterized these competing perspectives to be "between those who believe in truth and rationality and those who do not. [Realists]... believe that science tells us the way things really are; they take the paradigm of rationality to be scientific inquiry, just as the paradigm of truth is the result of that inquiry. ... [They insist] that natural science enjoys a special relationship to reality. ... [For relativists], the very idea of scientific objectivity ... is self-deceptive and fraudulent. ... A third group ... believe neither that science has a special relationship to reality nor that its pretensions need to be unmasked. The community of natural scientists is, they think, a model of intellectual rectitude, and yet its virtues – willingness to hear the other side, to think through the issues, to examine the evidence – have nothing to do with the fact that the objects natural scientists investigate are found rather than made."[47] All three sets of arguments provide some insight, and policy makers need to be wary of that reality. Authority, certainty, and consensus in science are not always what they are claimed to be.

Popper did not think that belief had a place in science. To him the core issue for science was the extent to which a scientific proposition could be *falsified* and thus subject to potential revision and refinement. As Hughes explains, "A falsifiable theory is one that makes a specific prediction about what results are supposed to occur under a set of experimental conditions, so that the theory might be falsified by performing the experiment and comparing predicted to actual results. A theory or explanation that cannot be falsified falls outside the domain of science."[48] A central tenet of Popper's philosophy was the critical importance of an open society, a society in which people were free to confront their own decisions and take responsibility for their moral and other choices.[49] Insisting that only

---

47. Richard Rorty, "Phony Science Wars," *The Atlantic Monthly* 284:5 (November 1999), 122.

48. Hughes, "The Folly of Scientism," 34.

49. This proposition was developed in his best-known work, *The Open Society and its Enemies* (London: Routledge, 1945). While more focused on epistemology

falsifiable propositions were scientific was wholly consistent with his view that an open society must be open to alternative views. He maintained that the right to question a scientific proposition is open to anyone willing to marshal the evidence and take responsibility for any resulting alternative hypothesis. An open approach to science is also key to its *replicability*, one of the keys to maintaining science's integrity. In an open society, scientists need to ensure access not only to their findings but also to their methodology and data.

Popper accepted that science was a collective enterprise involving many contributors and that such enterprises tended to gravitate towards the mean. Nevertheless, he maintained that the best science challenges conventional wisdom in order to improve scientific knowledge. Science progresses on the basis of testing by independent and competing minds and as such seeks, but rarely reaches, universal truths. Kuhn described science as a process of indoctrination leading to dogma rather than to falsifiable propositions. Popper also recognized that knowledge can be a basis for accumulating power and that dominant theories can exercise a cumulative advantage over competing theories: he found neither characteristic to be a scientific virtue but rather evidence of human frailty. Scientific pluralism characterized by competing hypotheses was much more likely to advance scientific knowledge.

Kuhn, conversely, was primarily interested in how scientific knowledge was produced, and he derived his perspective from the history of science rather than from first principles. He believed that the scientific process was much messier than widely assumed and that much of science was conditioned by the context in which it was developed and was not the result of the linear accumulation of knowledge. He saw the evolution of science as the result of a series of periodic revolutions, each of which overturned an established scientific order and replaced it with a new one. He called these periods of stability "paradigms." A paradigm represents the consensus among the community of scientists working in a field and their choice of a satisfactory theory; rival paradigms are unhelpful and point to a field in crisis.[50] Once a paradigm is established, research

---

than political philosophy, the book's main influence lay in political theory as a defence of democracy.

50. In response to his critics, he indicated that choice of theory was important and that different scientists looking at the same evidence could choose different theories. A good theory satisfied five criteria: accuracy, consistency, scope, simplicity, and fruitfulness. "Objectivity, Value Judgment, and Theory Choice," in

within that field is guided by it, and scientists are institutionally predisposed to validate it in their daily work. As he wrote in *The Structure of Scientific Revolutions:*

> 'Normal science' means research firmly based upon one or more past scientific achievements, achievements that some particular scientific community acknowledges for a time as supplying the foundation for its further practice. Today such achievements are recounted, though seldom in their original form, by science textbooks, elementary and advanced. These textbooks expound the body of accepted theory, illustrate many or all of its successful applications and compare these applications with exemplary observations and experiments.'[51]

Kuhn characterized "normal" scientific research to be a matter of puzzle-solving within the stable framework of the dominant paradigm. He characterized solutions that did not fit within the paradigm to be mistakes that would be resolved with new data and experiments. When, however, more and more puzzles emerged that could not be solved within the paradigm, the paradigm was said to be in crisis, leading to a period of uncertainty until a new paradigm emerged and normal science continued. This process was more a matter of the interactions and strategies of the scientists themselves than of its own innate logical structure. Unlike Popper, who viewed science as the *result* of rigorous enquiry, Kuhn saw it as a *process* that describes what scientists do at any point in time. Truth and objectivity are irrelevant. In such circumstances, the pathologies described in the next chapter can easily take hold, particularly the malign impact of science dominated by government funding and catering to political objectives. As Darwall observes, scientists tend to have "a cultural aversion to learning from the past. For them, history is not so much a closed book as irrelevant to the problems of the future,"[52] a perspective that disposes scientists to become easily captive of moralistic and progressive causes and to assume a highly expansive view of their role in solving global problems.

It is not difficult to find examples of paradigm shifts in the history of science. Arthur Holmes, for example, by applying radiometric dating to rocks, gradually pushed back scientific estimates of

---

Kuhn, *The Essential Tension: Selected Studies in Scientific Tradition and Change* (Chicago: University of Chicago Press, 1977), 320–39.

51. Thomas S. Kuhn, *The Structure of Scientific Revolutions*, 3rd edition (Chicago: University of Chicago Press, 1996), 10.

52. Rupert Darwall, *The Age of Global Warming: A History* (London: Quartet, 2013), 349.

the age of the earth from Lord Kelvin's long accepted estimate of 100 million years. Over a period of some thirty years, as radiometric dating became more sophisticated, Holmes was able to determine the now widely accepted age of about 4.5 billion years and opened up earth science to whole new areas of research. His view was long resisted but had become the new paradigm by the 1940s. Similarly, Alfred Wegener proposed as early as 1912 that the continents had at one time formed one giant land mass that had gradually split up with the parts drifting to their current locations. His idea originated in the contours of the continents, which suggested that their shapes resembled pieces of a puzzle, as well as in the similarities of fossils found on different continents. It was not until the 1950s, however, that advances in paleo-magnetism confirmed the essential contours of Wegener's theory of continental drift. Modern geology is deeply indebted to Holmes' and Wegener's insistence that the physical evidence did not support the prevailing paradigm and required a new and more convincing explanation.

A well-known modern example of the stickiness of an established paradigm comes from the medical profession and how it formerly dealt with stomach ulcers. For many years, doctors were convinced that stomach ulcers were brought on by stress. In 1982, however, two Australian scientists, Robin Warren and Barry Marshall, indicated that the real cause of these ulcers was the *helicobacter pylori* bacterium. Their hypothesis was dismissed until Marshall drank a petri dish of the stuff, gave himself an ulcer, and treated it successfully with antibiotics. They were awarded the Nobel prize for Medicine in 2005.[53] Based on increasing literature questioning the long-held view of the relationship between dietary fat and cardiovascular disease, this aspect of bio-medicine suggests a paradigm in crisis, as are a number of related areas of medical conventional wisdom, including the benefits of statin therapy, the need to medicate modest increases in blood pressure with age, and the definition of late onset diabetes.

Kuhn's view has been characterized as institutional, i.e., that the results of science are determined within the institutions and conventions of practicing scientists. It is limited to those who have the recognized expertise established through specialized training, institutional practice, and recognition within the community of scientists. Defenders of the institutional approach, such as the community of

---

53. Tom Quirk, "Of climate science and stomach bugs," *Quadrant Online,* January 21, 2013.

climate scientists associated with the UN's IPCC, like to point to its self-correcting nature by the community of "experts." The findings of Ioannidis and others, however, provide little comfort for this smug attitude. Hughes points out that "the history of science provides examples of the eventual discarding of erroneous theories. We should not, however, be overly confident that such self-correction will inevitably occur nor that the institutional mechanisms of science will be so robust as to preclude the occurrence of long dark ages in which false theories hold sway. The fundamental problem raised by the identification of 'good science' with 'institutional science' is that it assumes the practitioners of science to be inherently exempt, at least in the long term, from the corrupting influences that affect all other human practices and institutions."[54]

Kuhn's critics have argued that his view provides a rationale for Big Science – the science practiced in the dominant institutions characterized in 1961 by President Eisenhower as the military-industrial complex staffed by a scientific-technological elite. Imre Lakatos, the organizer of the 1965 Kuhn-Popper debate, similarly warned of the dangers to science of becoming embroiled in political goals: "In my view, science as such, has no social responsibility. In my view it is society that has a responsibility – that of maintaining the apolitical, detached scientific tradition and allowing science to search for truth in the way determined purely by its inner life. Of course, scientists, as citizens, have a responsibility, like all other citizens, to see that science is applied to the right social and political ends. This is a different, independent question."[55]

Kuhn's view lends itself to an elitist view of science and to governmental and industrial support for scientists' work, as has become more evident in the intervening years, and as documented by Butos and McQuade and others. Both sources of support tend to be committed to dominant paradigms and to the view that only some scientific enterprises should be supported, often with specific applications in mind. Dutch science blogger Jaap Hanekamp points out that "the prevailing epistemic community could well hinder freedom of research, which could result in impeding certain research themes that are not regarded as in line with the dominating para-

---

54. Hughes, "The Folly of Scientism," 37.
55. Imre Lakatos, *Mathematics, Science and Epistemology: Philosophical Papers*, v. 2 (Cambridge: Cambridge University Press, 1978), 258.

digm and thereby ignored for less than charitable reasons."[56] It also appeals to those scientists who put great stock in peer review and publishing in the right journals. It encourages gate-keeping by the anointed to keep unqualified critics – those that disagree with the dominant paradigm – from muddying the waters of normal science. It also leads to efforts to ensure that those working outside the dominant paradigm are denied funding for their work. Popper took a much more democratic view, emphasizing the irrelevance of the provenance of an idea and focusing instead on the quality of the evidence and the extent to which an hypothesis could be replicated or falsified. Jerome Ravetz, one of the founders of "post-normal" science discussed below, observes that "Kuhn's disenchanted picture of science was so troubling to the idealists (e.g., Popper) because in his 'normal' science, criticism had hardly any role. For Kuhn, even the Mertonian principles of ethical behaviour were effectively dismissed as irrelevant."[57]

Over the years, Kuhn's view has captured well the changing fashion in the humanities and social sciences, where paradigm shifts are frequent and often traumatic. It has fared less well in describing the scientific process in the natural sciences. Most scientific breakthroughs have been the result not of Kuhnian paradigm shifts but of dedicated scientists finding fault with the work of others and thus refining human understanding of physical processes, a quest that never ends, even as scientific understanding improves. Kuhn may have accurately described how the scientific enterprise proceeds, but, as Feynman and other major figures in modern science have sought to inculcate in their students, a good scientist accepts the strictures emphasized by Popper. If not, the results are the problems identified by Ioannidis and others that are painfully evident in some politically sensitive areas of scientific investigation. In such circumstances, not only does science lose, but so also does society at large

---

56. Hanekamp, "Post-normal failings," at *Climategate.nl,* March 15, 2010. Originally trained as a chemist, Hanekamp is now focused on the science and philosophy of risk analysis, particularly the misapplication of the precautionary principle.

57. Jerome Ravetz, "Climategate: Plausibility and the blogosphere in the post-normal age," *WattsUpWithThat,* February 9, 2009. Robert Merton was an influential mid-20th century sociologist of science who emphasized that science should be governed by four principles that make up the ethos of science: universalism, communalism, disinterestedness, and organized scepticism. Robert K. Merton, "The Normative Structure of Science," in Merton, *The Sociology of Science: Theoretical and Empirical Investigations* (Chicago: University of Chicago Press, 1942).

as the result of a misbegotten application of science that fails to meet the highest standards.

Kuhn's perspective has also fostered the development of scientism, the view that empirical science constitutes the most authoritative worldview or most authentic part of human learning to the exclusion of other viewpoints. This has led, quite naturally, to the belief that all reality can be explained fully by understanding physical processes. This perspective has been reinforced by the emergence of Neo-Darwinism, the application of Darwin's theory of natural selection to human behaviour, including in areas of life once assumed to be nonmaterial: emotions, thoughts, habits, and perceptions. To neo-Darwinians such as Richard Dawkins or Daniel Dennett, there is no divide between mind and matter. Everything can be explained on the basis of physical processes. The geneticist Francis Crick put it in somewhat stark terms: "You, your joys and your sorrows, your memories and your ambitions, your sense of personal identity and free will, are in fact no more than the behaviour of a vast assembly of nerve cells and their associated molecules." Who you are is "nothing but a pack of neurons."[58] In materialism, there is no room for a divine being or for free will or for any force other than randomness. Huston Smith, one of the leading scholars of the world's religions, dismisses scientism as the world's "littlest religion," because its reductionist assumptions make everything it touches little and the world it describes is "too small for the human spirit."[59]

Not all philosophers of science, however, agree that scientism provides a satisfying description of modern science. Thomas Nagel, in a controversial book, explained that "materialism is the view that only the physical world is irreducibly real, and that a place must be found in it for mind, if there is such a thing. This would continue the onward march of physical science, through molecular biology, to full closure by swallowing up the mind in the objective physical reality from which it was initially excluded."[60] In materialism, consciousness is a purely biological phenomenon that can be explained on the basis of physical processes in the brain. Nagel rejects this extreme materialism and, while remaining an atheist who sees no

58. Francis Crick, *The Astonishing Hypothesis* (New York: Touchstone, 1995), 3.

59. Huston Smith, "Scientism: The World's Littlest Religion – How Theology Must Confront the New Global Religion of Scientism," *Touchstone Magazine* 10:3 (Summer 1997).

60. Thomas Nagel, *Mind and Cosmos: Why the Materialist Neo-Darwinian Conception of Nature is Almost Certainly False* (New York: Oxford, 2012), 37.

need for a divine being, posits that there is more than a purely random universe without purpose, free will, or spiritual life.

The reaction to Nagel's book was fierce. Many in the intellectual community took offense that Nagel dared to question Darwinism, which, for many modern intellectuals, is the touchstone of scientific thinking. Anything that questions Darwin is heresy but, as the American writer Leon Wieseltier reminds us, "the problem of the limits of science is not a scientific problem. It is also pertinent to note that the history of science is a history of mistakes, and so the dogmatism of scientists is especially rich."[61] In an amusing article, Andrew Ferguson recounts how the *bien pensants* of contemporary intellectual life circled the wagons and fired fusillade after fusillade at their erring colleague. He characterized Nagel's book as "a work of philosophical populism, defending our everyday understanding from the highly implausible worldview of a secular clerisy."[62]

The Nagel controversy may at first blush seem a far stretch from concern about the reliability of modern science, but it helps to explain why academic and intellectual elites have, in great droves, abandoned religion and philosophy and embraced scientism, granting more authority to science than many scientists would claim for themselves.[63] Scientism leads to caricatures of science. In an hysterical comment, Harvard's Steven Pinker insisted that Nagel's flight from received opinion "will give ammunition to disturbing anti-science, anti-reason forces in the contemporary political power structure" and will provide "comfort to a powerful and well-funded lobby in this country that is trying to discredit the entire institution of science as a close-minded, ideological propaganda front which is determined to promote a secular, materialistic, anti-Judaeo-Christian liberalism. This emboldens them to blow off the scientific consensus about man-made climate change, corrupt science education, suppress research on gun violence, and criminalize lifesaving medical research. ... This is about the future of the Planet."[64]

---

61. Leon Wieseltier, "A Darwinist Mob Goes After a Serious Philosopher," *New Republic,* March 8, 2013.

62. Andrew Ferguson, "The Heretic: Who is Thomas Nagel and why are so many of his fellow academics condemning him?" *Weekly Standard* 18:27(March 25, 2013).

63. See Hughes, "The Folly of Scientism." He believes that "continued insistence on the universal competence of science will serve only to undermine the credibility of science as a whole," 50.

64. Steven Pinker, Comment on Leon Wieseltier, "A Darwinist Mob Goes After a Serious Philosopher," *New Republic Online,* March 8, 2013 at 5:36pm.

Ferguson points out that for intellectuals like Pinker, Dennett, and Dawkins, materialism is a critical tenet of their secular faith rather than the working assumption of most practicing scientists. "Scientists do their work by assuming that every phenomenon can be reduced to a material, mechanistic cause and by excluding any possibility of nonmaterial explanations. And the materialist assumption works really, really well – in detecting and quantifying things that have a material or mechanistic explanation." The materialists, however, go much farther, insisting that "if science can't quantify something, it doesn't exist, and so the subjective, unquantifiable, immaterial manifest image of our mental life is proved to be an illusion."[65] Christian philosopher Alvin Plantinga adds: "those who champion [materialism] tend to wrap themselves in science like a politician in the flag."[66]

Enter post-normal science

In public policy areas that rely on scientific input, some philosophers have developed a new assessment of the scientific enterprise which they call *post-normal science*. They accept that most science is pursued as normal science in the sense described by Kuhn but that there are circumstances when normal science is inadequate to society's needs. First enunciated by Silvio Funtowicz and Jerome Ravetz, this perspective has infected much of advocacy science and provides a rationale for noble-cause corruption, both discussed in the next chapter. From their perspective, when "facts are uncertain, values in dispute, stakes high, and decisions urgent," such traditional values as objectivity and certainty are inappropriate, as is the puzzle-solving of normal science. Rather, the necessary science requires input from an extended peer community as well as a willingness to engage in value-laden judgment. When science is in this post-normal stage, scholarly activities are dominated by goals influenced by political and societal actors and involve a strong two-way dialogue with society. The need for this new approach arises from the fact that "after centuries of triumph and optimism, science is now called on to remedy the pathologies of the global industrial system of which it forms the basis. ... The reductionist, analytical worldview which divides systems into ever smaller elements, studied by ever more esoteric specialism, is being replaced by a sys-

---

65. Andrew Ferguson, "The Heretic."

66. Alvin Plantinga, "Why Darwinist Materialism is Wrong," *The New Republic* (November 16, 2012).

temic, synthetic and humanistic approach. ... The old dichotomies of facts and values, and of knowledge and ignorance, are being transcended."[67] Funtowicz and Ravetz argue that post-normal science is necessary to address issues in health and environmental science in general and to address climate science in particular. They contend that when the normal scientific process of *validation-by-evidence* no longer works, there is still science, but it is now in a post-normal state, where it is propelled into the political domain; despite uncertainties, it can still address urgent policy decisions. Ravetz explained further at a popular science blog:

> As Thomas Kuhn described 'normal science', which (as he said) nearly all scientists do all the time, it is puzzle-solving within an unquestioned framework or 'paradigm'. Issues of uncertainty and quality are not prominent in 'normal' scientific training, and so they are less easily conceived and managed by its practitioners.
>
> Now, as Kuhn saw, this 'normal' science has been enormously successful in enabling our unprecedented understanding and control of the world around us. But his analysis related to the sciences of the laboratory, and by extension the technologies that could reproduce stable and controllable external conditions for their working. Where the systems under study are complicated, complex, or poorly understood, that 'textbook' style of investigation becomes less, sometimes much less, effective.[68]

Ravetz came relatively late to his idea of post-normal science, but his earlier work showed him to be a post-modern philosopher steeped in the ideas and language of Neo-Marxism. Marx had insisted that the purpose of philosophy was not to *interpret* but to *change* knowledge. The idea that the validation of knowledge is political became a central thesis of Neo-Marxists and took serious hold in the modern secular university in the 1970s and 1980s, particularly in the social sciences and humanities. With their concept of post-normal science, Ravetz and Funtowicz sought to extend that thesis to science itself.

Areas that lend themselves particularly well to the claims of post-normal science are those for which direct observation and experimentation are not possible but for which computer models can be used as a virtual substitute. Australian scholar Aynsley Kellow

---

67. Funtowicz and Ravetz, "Science for the Post-Normal Age," *Futures*, 25 (1993), 739-40.

68. Jerome Ravetz, "Climategate: Plausibility and the blogosphere in the post-normal age," *WattsUpWithThat*, February 2, 2009.

gives it the apt name of *virtual science* and points to climate science as a prime example.[69] The global climate cannot be subjected to controlled experiments, and much of the science is focused on projections, i.e., the unknowable. With the development of powerful computers, models can now provide simulated experiments. Calling the results scientific, however, is a stretch. English science writer Matt Ridley astutely observes: "None of this would matter if it was just scientific inquiry, though that rarely comes cheap in itself. The big difference is that these scientists who insist that we take their word for it, and who get cross if we don't, are also asking us to make huge, expensive and risky changes to the world economy and to people's livelihoods."[70]

From this perspective, the UN's IPCC is practicing an emerging but contested form of science, bringing scientists, governments, and NGOs together in a common enterprise.[71] Trying to treat science that is in a post-normal condition as normal science is, according to its defenders, self-defeating. Involving the extended peer community, conversely, strengthens the science, so long as all accept that the science is in a post-normal condition. Attacking and discrediting sceptics and critics are not only legitimate activities, but necessary, well-illustrated by the University of London's Mike Hulme in his review of Singer's and Avery's *Unstoppable Global Warming*, a book he bluntly calls wrong because it lies outside of "settled" science and is thus best characterized as pseudo-science. In Hulme's view, the authors' disagreement with climate science is not a matter of science but of the political values embedded in the settled science. He writes: "Scientific knowledge is always provisional knowledge, [in] that it can be modified through its interaction with society. ... [but] climate change seems to fall in [the 'post-normal'] category. ... The danger of a 'normal' reading of science is that it assumes science can first find truth, then speak truth to power, and that truth-based policy will then follow. ... If scientists want to remain listened to, to bear influence on policy, they must recognize the social limits

---

69. Aynsley Kellow, *Science and Public Policy* (Cheltenham, UK: Edward Elgar, 2007).

70. Matt Ridley, "The Climate Wars' Damage to Science," *Quadrant Online*, June 19, 2015.

71. Paltridge observes that "postmodern science envisages a sort of political nirvana in which scientific theory and results can be manipulated to suit either the dictates of political correctness or the policies of the government of the day." "Uncertainty, scepticism, and the climate issue," kindle edition.

of their truth seeking and reveal fully the values and beliefs they bring to their scientific activity. ... Climate change is too important to be left to scientists – least of all the normal ones."[72]

Evidently Hulme does not believe that the extended peer community includes normal scientists such as Singer, Avery, and other critics of science as practiced by the IPCC community, an attitude that is evident among many members of the official climate science community: well-credentialed scientists who do not accept that the science is settled are beyond the pale. Nor does Hulme appear to accept the long-standing virtue of scepticism. The idea of a non-sceptical scientist is in itself bizarre. Doing science, as Popper, Feynman, Merton, and others indicated, requires a perpetual attitude of scepticism. To the committed on any issue, however, scepticism undermines the authority that is being sought and must be rooted out. As sociologist Frank Furedi points out, "Scepticism today, as in the past, has a bad name because for the dogmatic believer any sign of doubt, hesitation, uncertainty, questioning and even indifference is interpreted as disbelief."[73]

Ravetz was much influenced by concern among environmentalists that science was proving inadequate to address environmental policy issues. In his view, the *irreducible uncertainties* of environmental science should not prevent scientists from making positive claims, particularly by involving *the extended peer community,* i.e., other stakeholders and interest groups. In post-normal science, it is critical that they participate in problem-solving strategies and decision-making. Political interactions can thus be used to generate knowledge and develop the conclusions upon which government policy can subsequently be based.[74] Dennis Bray and Hans von Storch explored the dilemma for many environmental and health activists of determining who is competent to engage in risk-based

---

72. Mike Hulme, "The Appliance of Science," *The Guardian,* March 14, 2007. The book being reviewed was Fred Singer and Dennis Avery, *Unstoppable Global Warming: Every 1500 years* (Lanham, MD: Rowman and Littlefield, 2007). An extended version of Hulme's views can be found in *Why We Disagree About Climate Change: Understanding Controversy, Inaction, and Opportunity* (New York: Cambridge, 2009).

73. Frank Furedi, "Why scepticism is still 'the highest duty'," *Spiked,* April 26, 2010.

74. See Bruna De Marchi and Jerome R. Ravetz, "Participatory Approaches to Environmental Policy," *Environmental Valuation in Europe,* Policy Research Brief Number 10, Cambridge Research for the Environment. See also his earlier work, Jerome R. Ravetz, *Scientific Knowledge and Its Social Problems* (Oxford: Oxford University Press, 1971).

analysis and policy advice: "One widely held view in this regard is that the public should be excluded from the policy process associated with risks since the public are generally too ill informed to make rational choices. ... [At the same time,] scientific credentials, whether relevant or not to the topic at hand, are often deemed sufficient to make comment well beyond the area of scientific expertise."[75] Expert advising, however, was considered by many activists to be geared to the needs of industry and other narrow interests rather than to the public interest. Post-normal science was the answer.

Hanekamp astutely points to the principal problem with post-normal science: "Facts are never in dispute, otherwise they would not be called facts, ... and if we have not arrived at the facts of that particular slice of reality (which is hardly unusual in science)... how can we know that the stakes are, in fact, high or that decisions are required urgently?"[76] University of Windsor philosopher Christopher Tindale points to another problem: "Such recourse to audiences and to their own standards of acceptance raises not only the specter of relativism ... but the more serious problem of allowing what intuitively seems impermissible when we look beyond the restricted interests of specific audiences. ... When an audience does not see the sleight of hand involved, or raises no objections, should we allow the questionable reasoning of an arguer?... If we are prepared to extend to individual audiences carte blanche authority to set the standards of acceptability, then we fall prey to the vicissitudes of popularity that have plagued argumentation theory, primarily in the form of *ad populum* arguments."[77]

The idea of "acceptability" results in a pernicious relativism that renders it duplicitous and subverts the long-held aim of science to secure objective knowledge. Within the context of contested issues such as health, safety, and the environment, it becomes possible for activists to assume a particular risk and assert that it requires stringent policies, challenging other societal interests to prove that there is an absence of risk. The result is an ever-increasing regulatory

75. Dennis Bray and Hans von Storch, "Climate Science: An Empirical Example of Post-Normal Science," *Bulletin of the American Meteorological Society* 80:3 (March 1999), 453. Lewis, *Technological Risk*, explains this dilemma in considerable detail, exploring many examples of the gulf between expert and popular opinion.

76. Hanekamp, "Post-normal failings," at climategate.nl/2010/03/15/post-normal-failings.

77. Christopher W. Tindale, *Acts of Arguing: A Rhetorical Model of Argument* (Albany: State University of New York, 1999), 114-15.

state resulting not only from climate science but involving a whole panoply of health, safety, and environmental controls, with minimal regard for personal choice, responsibility, costs, and benefits. In Europe in particular, post-normal science has proven a perfect handmaiden for the precautionary principle, leading to a spiral of risk-avoiding regulations and corruption of the peer review process. Hanekamp notes that the activist claim of "saving the planet or protecting the vulnerable do not leave time for science to develop objective knowledge."[78]

Von Storch and Bray remark that one of the things that they learned from a survey of attitudes among active researchers in climate science is that some are eager to engage in the policy debate, while others are more reticent. They write: "As scientists move from their specific areas of expertise, as would be expected, the diversity of opinion widens. Unfortunately, however, it is these opinions that … are the most sought after in the policy realm, and it is these opinions that the climate scientists can only present at the level of the lay perspective since they are not formally trained (at least in most cases) to assess social or economic matters in a formal manner. … Just as a sociologist or economist could not provide a very enlightening diatribe on atmospheric physics, so too should a climate scientist be cautious of making social and economic commentary."[79] Their solution is not post-normal science but greater cooperation across disciplines and the development of interdisciplinary competence.

Already in mid-century Popper, scarred by the world's experience with the claims of communism and fascism, had warned of the dangers of losing science's goal of pursuing objective knowledge and becoming embroiled in political and social goals:

> Utopian aims are designed to serve as a basis for rational political action and discussion, and such action appears to be possible only if the aim is definitely decided upon. Thus the Utopianist must win over, or else crush, his Utopianist competitors who do not share his own Utopian aims and who do not profess his own Utopianist religion. But he has to do more. He has to be very thorough in eliminating and stamping out all heretical competing views. For the way to the Utopian goal is long. Thus the rationality of his political action demands constancy of aim for a long time ahead; and this can only be achieved if he not

78. Hanekamp, "Post-normal failings."

79. Hans von Storch and Dennis Bray, "Climate Science: An Empirical Example of Post-Normal Science," *Bulletin of the American Meteorological Society* 80:3 (March 1999), 454.

merely crushes competing Utopian religions, but as far as possible stamps out all memory of them.[80]

Ioannidis, Sarewitz, Fanelli, and other investigators pointing to the disturbing rise in questionable science as well as scientific misconduct would all consider themselves to be working from a largely Popperian perspective. While not dismissing the value of Kuhn's insights into the everyday practice of science, they would insist that science's value to society depends on the integrity of its practitioners, and integrity involves trust, objectivity, and a search for truth. For them, the claims of post-normal science are a caricature of how science should work, particularly in areas of contested science. Nevertheless, policy making in such contested areas as climate science but extending to health, safety, and environmental issues more generally, is rife with claims that only make sense as post-normal science, and, increasingly, post-normal science has become all too prevalent in official science.

Philosopher Susan Haack, a pragmatist at the University of Miami, has provided what many practicing scientists would consider a more balanced assessment of the value and reliability of their enterprise. She writes: "Science is not sacred: like all human enterprise, it is thoroughly fallible, imperfect, uneven in its achievements, often fumbling, sometimes corrupt, and of course, incomplete. ... What we need is an understanding of inquiry in the sciences which is, in the ordinary, non-technical sense of the word, realistic, neither overestimating nor underestimating what the sciences can do."[81] She calls for neither uncritical admiration nor denigration or hostility. Rather, understanding science requires an informed and critical eye. In an era of increasing specialization and complexity in science and of scientific illiteracy among most lay people, including particularly those engaged in making public policy, this is a major challenge to overcome. On a widening range of public policy issues, it remains an unmet challenge and poses dire consequences for society and public policy.

*****

---

80. Popper, "Utopia and Violence," first published in 1948 and reprinted in *Conjectures and Refutations: The Growth of Scientific Knowledge* (London: Routledge Classics, 2002), 483.

81. Susan Haack, *Defending Science – Within Reason: Between Scientism and Cynicism* (New York: Prometheus, 2003),19.

The ideal and the reality

It would not be unfair to conclude that Popper, the philosopher, focused on how science *ought* to be conducted while Kuhn, the historian, described how it *is* conducted. In emphasizing the subjective nature of science, Kuhn's followers have gone much farther, leading to such rationalizations for questionable conduct as virtual science and post-normal science. For Popper, the only serious question was whether a proposition could be replicated and falsified which, in turn, required that scientists pursue the scientific method as traditionally understood while recognizing that few propositions would ever attain the status of "settled" science. Kuhn, on the other hand, could easily understand what scientists mean by a consensus: the paradigm within which they are operating is settled and much of their work is a matter of refinement and detail. At some future point those details may accumulate to the point that the paradigm is no longer tenable, leading to a crisis and a revolution, but most scientists will never experience such a situation. An assessment of science as practiced, however, will suggest that there remains much in science that adheres to the high standards championed by Popper but that a disturbing number of ideological and institutional factors are providing room for questionable science and less-than-honest scientists, both eroding society's confidence in science and compromising science-based public policy.

Good science is iterative and incremental, i.e., scientists repeat their experiments many times over, often with small variations, in order to strengthen their hypotheses. Failure to pursue this tedious work is often a recipe for problems, as documented by Roberts, Ioannidis, Fanelli, and others. A good example of failure to pursue this resource-intensive, time-consuming task was documented by two Australian drug researchers, Glenn Begley and Lee Ellis. They sought to replicate published findings in their own laboratory and found that they could do so in only 6 of 53 studies. They cite a number of other researchers who identified similar problems and conclude: "the inability of industry and clinical trials to validate results from the majority of publications ... suggests a general systemic problem. ... Responsibility for design, analysis and presentation of data rests with investigators, the laboratory and the host institutions. All are accountable for poor experimental design, a lack of robust supportive data or selective data presentation. The scientific process demands the highest standards of quality, ethics and rigour."[82] In a later article, Begley notes that in many of the cases he examined, 1) experiments had not been performed blinded, 2) experiments were not repeated, 3) experiments were only partially reported, 4) control experiments were not done or, if done, not reported, or 5) experiments were poorly designed from the outset. In properly conducted studies, "it is now the gold standard to blind investigators, include concurrent controls, rigorously apply statistical tests and analyze all patients – we cannot exclude patients because we do not like their outcomes." The result is "a plethora of studies that don't stand up to scrutiny."[83]

In good science, each iteration of an experiment allows researchers to refine their observations and strengthen their understanding of the phenomenon being studied. Precise measurements expressed in mathematical terms are essential to this process and underline the critical role of mathematics and statistics in modern scientific praxis. Science is never settled. Modern science is not so much about finding proof as it is about determining probability. Proof suggests certainty whereas scientists must always remain open to the idea that more detailed observations, better data, and new insights may indicate flaws in an hypothesis and confirm the need for a new, improved one. As a result, scientists indicate levels of confidence in the value of an hypothesis in terms of probability, accompanied by error bars and other statistical data useful to the informed reader.

Science is as much a matter of asking questions as it is of finding answers. It would surprise most laymen to learn that in science many answers have a very short shelf life before they are replaced by better answers that result from further probing and questioning. Most science papers are now cited in the literature for no more than five years. Neuroscientist Stuart Firestein maintains that "the contemporary view of science puts too much emphasis on answers. What leads to good science is uncertainty. That doesn't mean scientists shouldn't be certain about their findings. It means they should be comfortable that their findings are not the final answer. ... Being a scientist requires having faith in uncertainty, finding pleasure in mystery, and learning to cultivate doubt. ... Facts change, revisions are made, but it adds up to progress. In science, revision is a victory.

---

82. C. Glenn Begley and Lee M. Ellis, "Raise standards for preclinical cancer research," *Nature* 483 (March 29, 2012), 532-33.

83. C. Glenn Begley, "Six red flags for suspect work," *Nature* 497 (May 23, 2013), 433-4.

And that process of revision has accelerated significantly in the last few decades. ...Unsettled science is not unsound science." [84] Firestein's views stand in sharp contrast to the ideas displayed by climate scientists such as Kevin Trenberth and Michael Mann, for whom the science of climate change is "settled" and the remaining issues are a matter of appropriate application of that science to solve the climate "crisis."

The stickiness of Kuhnian paradigms reflects the human reality that scientists, having invested considerable psychological and other resources into a paradigm, find it difficult to accept that there is a different route to a better understanding of the phenomena being studied. In the face of anomalies and contradictory evidence they will stick to that paradigm until contradictory evidence is overwhelming. Economists refer to this as sunk costs. As a matter of psychology, investigators are more likely to look for evidence that confirms their hypotheses than the other way around. Science, properly done, requires researchers to behave against their natural inclinations. Failure to do so explains the epidemic of published papers that have been demonstrated to be non-replicable or otherwise unreliable.

Science is a social process – scientists want their work to be accepted by the larger community of scientists and incorporated into their work. Most scientific progress is the result of contributions by hundreds if not thousands of scientists who provide the basis for a new insight or development that may be advanced by a single scientist or group of scientists.[85] An open invitation to replicate the work of others is critical to validating and confirming scientific breakthroughs. The success of this process also requires that scientists be open to sharing their data and workings, an issue that has gained increasing prominence and discussion in scientific circles.

Science is not democratic, evaluating the validity of an hypothesis by vote. No matter how many scientists are agreed that a particular hypothesis offers a satisfying explanation of a particular phenomenon, any observation contrary to the hypothesis will bring it into question. Indeed, many of the greatest breakthroughs in science,

---

84. Stuart Firestein, "Certainly Not! Philosophy: Good science requires cultivating doubt and finding pleasure in mystery," *Nautilus* 2, adapted from his book, *Ignorance: How it Drives Science* (Oxford: Oxford University Press, 2012).

85. In the science literature, it is common to see articles, even short articles of a page or two, authored by as many as ten, twenty, or even thirty scientists, each of whom contributed at some point to the reported findings.

from Newton to Einstein, have been realized by a contrarian questioning the prevailing view. In areas of scientific research that wear the mantle of post-normal or virtual science, however, claims to consensus are a critical part of concerted public relations campaigns designed to strengthen claims to scientific status and authority. Over the years, for example, learned societies have scrupulously avoided staking out positions on controversial issues by taking their cue from one of the oldest, the Royal Society in the UK. Founded in 1660 with a charter from King Charles II with the motto *nullius in verba* – take no one's word for it. It maintained for more than 300 years that controversy was an issue for the members, not for the Society. In a 1753 statement, the Society maintained that "…it is an established rule of the Society, to which they will always adhere, never to give their opinion as a Body upon any subject either of Nature or Art, that comes before them." Two hundred years later, Edgar Lord Adrian, in his farewell address as outgoing president, could still state that: "It is neither necessary nor desirable for the Society to give an official ruling on scientific issues, for these are settled far more conclusively in the laboratory than in the committee room."[86]

That firm policy came to an end in the opening decade of the 21st century as a series of activist presidents – Robert May, Martin Rees, and Paul Nurse – began speaking on behalf of the Society about the dangers of global warming. The Society also published a series of pamphlets endorsing the IPCC view of the science and warned journalists about reporting sceptical views. May even led an effort to organize other national associations to speak about the danger of questioning the settled science of the IPCC. By 2009, this activism had reached the point that a number of fellows sought to rein it in and return the Society to its neutral roots. The best that they could achieve was the release of a more balanced statement. The most recent president – Nurse (2010-14) – broadened the debate by insisting that the Society has a duty to provide society with its advice on controversial issues, the very opposite of what its founders had established as its reigning philosophy.[87] The Royal Society is not alone. Many other science organizations have experienced the same tensions, including the American Physical Society, the Ameri-

---

86. Edgar Adrian, "On the Functions of the Royal Society," *British Medical Journal* 2(4953), December 10, 1955, 1444.

87. The story of the Society's descent into partisan politics can be found in Andrew Montford, *Nullius in Verba: On the Word of No One*, Global Warming Policy Foundation, Report No. 6 (London, 2012).

can Meteorological Society, the American Geophysical Union, and the American Association for the Advancement of Science.[88] For activist scientists, being able to express their views through such clearly political groups as the Union of Concerned Scientists is not enough. In the era of post-normal science, balance and neutrality are no longer scientific virtues. All of science must be coerced into accepting the dogma. As Richard Lindzen laments, the politicization of science resulting from these efforts has led to "the legitimate role of science as a powerful mode of inquiry [being] replaced by the pretence of science to a position of political authority."[89]

Given the complexity of modern science, most scientists are not inclined to politicize the scientific enterprise in which they are personally engaged. They are, in Isaiah Berlin's memorable metaphor, hedgehogs: they know a tremendous amount about their small area of specialization; they may know it better than anyone who has ever lived; but they tend to know little about science in general and are often naïve about – and not engaged in – broader political and policy controversies. There are few foxes in today's scientific enterprise, scientists who understand something about a lot of things, but few of them in depth. This phenomenon has disposed some scientists to become overconfident in their knowledge and to make prognostications beyond their competence. Two recent popular books explore this disturbing phenomenon: David Freedman, *Wrong* and Dan Gardner, *Future Babble*.[90] Gardner's book relies heavily on the work of psychologist Philip Tetlock, *Expert Political Judgment: How Good Is It? How Can We Know?* Tetlock had focused on two key issues: the relative strengths and weaknesses of human versus model forecasting and the relative strengths and weaknesses of generalists versus specialists. He found that the best experts in making political estimates and forecasts are no more accurate than fairly simple mathematical models. Human experts have failed to out-perform simple linear models in over one hundred fields of expertise covering fifty years of research.

---

88. The AAAS wrote a letter in 2009 to US senators urging action on climate change and endorsing the "consensus" position on behalf of 18 scientific organizations and associations. See Letter from the AAAS to the US Senate at www.climate-shifts.org/?p=3394.

89. Richard Lindzen, "Foreword," to Montford, *Nullius in Verba*.

90. David Freedman, *Wrong: Why Experts Keep Failing Us and How to Know When Not to Trust Them* (New York: Little Brown, 2010) and Dan Gardner, *Future Babble: Why Expert Predictions Fail – and Why We Believe Them Anyway* (Toronto: McClelland and Stewart, 2010).

Even more interestingly, however, Tetlock discovered that pundits and academics with a broad knowledge base (foxes) have a much better track record than those with a deep grasp of a single area or those committed to a single grand vision (hedgehogs).[91]

Paul Ehrlich and David Suzuki are good examples of experts with deep knowledge of a single subject who are prepared to expound at length about future catastrophes. Both were trained as entomologists: Ehrlich's area of expertise is in butterflies and Suzuki's in fruit flies. Both fall into Tetlock's category of prognosticators with well-deserved reputations for getting things wrong, often spectacularly so, never learning from their mistakes, and dismissive of any information that does not fit within their preconceived notions. Both have traded on their credentials as scientists to command respect in areas far from their scientific expertise. Both are veteran doomsayers, much lionized by the media who mistake their media popularity for confirmation of their status as experts. Ehrlich extrapolated population behaviour in insect colonies to that of humans and concluded that we are doomed because we have outgrown our ecological niche. In book after book he has predicted imminent mass starvation and related mass disasters.[92] Suzuki has spent little time doing research, preferring to work with television as a medium to spread his message about the negative impact of humans on global ecology.[93]

The popular authority of scientists such as Ehrlich and Suzuki rests, of course, on the widely held view that science is objective and value-free; scientists are not above feeding this perception. Scientists, however, are human, and the advance of science is as subject to hu-

---

91. Philip Tetlock, *Expert Political Judgment: How Good Is It? How Can We Know?* (Princeton: Princeton University Press, 2006).

92. Together with his wife Anne, also a biologist, and sometimes with others, such as Obama science adviser John Holdren, Ehrlich has written many books and articles, all of which can be classified as jeremiads and all of which have proven to be wrong. Among his books are Paul Ehrlich and Anne Ehrlich, *The Population Bomb* (New York: Ballantine Books, 1968); *The Population Explosion* (New York: Simon and Schuster, 1990); and *Betrayal of Science and Reason: How Anti-Environmental Rhetoric Threatens Our Future* (Washington: Island Press, 1997).

93. Suzuki's books include Anita Gordon and David Suzuki, *It's a Matter of Survival* (Cambridge, MA: Harvard University Press, 1991); Suzuki and Holly Dressel, *Good News for a Change: How Everyday People Are Helping the Planet* (Vancouver: Greystone Books, 2003); and Suzuki and David Robert Taylor, *The Big Picture: Reflections on Science, Humanity, and a Quickly Changing Planet* (Vancouver: Greystone Books, 2003). All are best characterized as religious tracts written to encourage and reinforce environmentalism as a modern, secular faith.

man frailties as any other field of human endeavour. Scientists have agendas and opinions, and they are interested in careers. Cognitive biases, confirmation bias, and logical fallacies abound as much in scientific work as in that of economists and other social scientists.

Most advances in science are reported in articles in the thousands of science journals, many of them quite short and many of them read only by a few researchers working in the same specialized field. The large number of journals and published articles points to the increasing complexity and specialization of scientific research and the sheer volume of data and hypotheses being generated by researchers. Technology guru David Weinberger observes that "with the new database-based science, there is often no moment when the complex becomes simple enough for us to understand it. The model does not reduce to an equation that lets us then throw away the model. You have to run the simulation to see what emerges. ... [The process of generating scientific knowledge has become] so complex that only our artificial brains can manage the amount of data and the number of interactions involved." The rapid growth in the generation of knowledge, he explains, has become possible because the cost of sharing knowledge has greatly decreased while computers capable of handling huge amounts of data have become exponentially smarter. Nevertheless, "models this complex – whether of cellular biology, the weather, the economy, even highway traffic – often fail us, because the world is more complex than our models can capture."[94] As a result, the scope for error and for fooling ourselves has become immense, as indicated by the findings of Ioannidis and other researchers.

Few practicing scientists succeed in publishing in the prestige journals, such as *Nature* or *Science*. Rather, the vast web of science, growing at a rate of 27,000 new journal articles per week,[95] consists of lower tier and highly specialized journals. Of those 27,000 articles, only a very small, but rising, number are eventually retracted. Only the most sensational of these retractions are widely noticed. Most retracted articles live on as part of a field's literature. One researcher looked at 1,112 retracted papers from 1997–2009 and found them widely cited, with the retractions mentioned in only four per cent of

94. David Weinberger, "To Know but Not Understand," *The Atlantic*, January 3, 2012. The article provides an edited excerpt from his new book, *Too Big to Know* (New York: Basic Books, 2012).

95. Richard Van Noorden, "The trouble with retractions," *Nature* 478 (October 6, 2011), 26-8.

the citations.[96] An earlier study examined 235 articles retracted during 1966-96; they were cited 2,034 times after their withdrawal, with fewer than eight per cent acknowledging the retraction. The authors concluded: "retracted articles continue to be cited as valid work in the biomedical literature after publication of the retraction; these citations signal potential problems for biomedical science."[97]

Most researchers are unaware that a paper has been retracted and assume the results remain valid. As science journalist Richard Van Noorden explains, problems "include opaque retraction notices that don't explain why a paper has been withdrawn, a tendency for authors to keep citing retracted papers long after they've been red-flagged ... and the fact that many scientists hear 'retraction' and immediately think 'misconduct' – a stigma that may keep researchers from coming forward to admit honest errors."[98] Not all retractions are the result of misconduct. In Van Noorden's survey, 28 percent were the result of researchers reporting an honest error and retracting the resulting research paper.

And what about published papers that may not have been the result of misconduct or error and then retracted but have been thoroughly rebutted or even debunked by subsequent researchers? Do such papers live on or does the community of scholars engaged in that area reject their conclusions? Jeannette Banobi and her colleagues at the University of Washington concluded that the original paper continues to be cited, rarely with an acknowledgment of the rebuttals. They ask: "How does science progress? A naïve view is that scientists propose new ideas and hypotheses and these are either accepted or rejected according to the evidence at hand. In practice it takes considerable evidence to cause the scientific community to abandon an established idea." Their research showed "strong evidence that rebuttals scarcely alter scientific perceptions about the original papers. ... For every article that cited the rebuttal, there were 17 that ignored the rebuttal and cited only the original, and among this silent majority, 95 percent uncritically accepted the findings of the original article."[99]

96. Van Noorden, "The trouble with retractions," 26.

97. John M. Budd, Mary Ellen Sievert, and Tom R. Schultz, "Phenomena of Retraction: Reasons for Retraction and Citations to the Publications." *Journal of the American Medical Association*, 280:3 (July 15, 1998), 296-98.

98. Van Noorden, "The trouble with retractions," 26.

99. Jeannette A. Banobi, Trevor A. Branch, and Ray Hilborn, "Do rebuttals affect future science?" *Ecosphere* 2:3 (March 2011), Article 37, 1.

Formal rebuttals in the scientific literature remain relatively rare. Discussion, yes; rebuttals, not so much. On the other hand, the emergence of the internet has led to much more discussion of questionable science than ever before, particularly of science that is in the public eye, from conventional medicine to climate change. For some scientists, this has become an annoying new factor; others welcome it and happily join in the discussion. Embattled scientists insist that science should only be debated in the peer-reviewed literature; those with a more open mind recognize that error is error, whether pointed out by a credentialed scientist or a rank amateur in the field.

The problem, however, may not be that there is a reluctance to report misconduct and errors leading to retractions or to acknowledge that a paper has been rebutted but that too many papers initially make it through the review process and get published. It is difficult to imagine that researchers, even in the widely diverse range of science covered by the Web of Science, have 27,000 new ideas, insights, and findings to report every week. The pressure to publish – as a matter of prestige, reputation, funding, promotion, and other considerations – is creating a monster that is leading to the publication of too much marginal and questionable science. The current system encourages scientists to milk the maximum number of papers out of a single piece of research.[100] The rising number of retractions – .02 percent of what is published – is, in Ioannidis' view, "the tip of the iceberg, too small and fragmentary for any useful conclusions to be drawn about the overall rates of sloppiness or misconduct."[101]

For all practical purposes, we can conclude that science is practiced today by "scientists" working within a hierarchically organized community of practitioners within the same field or sub-field of science. It is no longer the realm of amateurs but of professionals who have been rigorously trained in their areas of specialization, organized into laboratories, universities, and other specialized institutions and communicating in a specialized language in journals and at conferences organized for the benefit of other specialists. Peer review is used to ensure that only initiated members of a specialized community may participate. There is little room for outsid-

---

100. See the discussion in William Broad and Nicholas Wade, *Betrayers of the Truth: Fraud and Deceit in the Halls of Science* (New York: Simon and Schuster, 1983), 212ff.

101. Quoted in Van Noorden, "The trouble with retractions," Box, 28.

ers with different interests and assumptions. Communication with the broader public is the preserve of another group of specialists. There is nothing democratic about the practice of science. Australian sociologist of science Brian Martin explains:

> One cannot do scientific research, or conceive of current scientific knowledge or institutions, without at the same time accepting many assumptions currently built into the structure and organization of the scientific enterprise…. At an individual or personal level, individual scientists can be and are dogmatic and biased; they make assumptions about the directions, uses and conclusions of their work. At the level of research organizations, the organizations for which scientists work have vested interests in certain types of research and in obtaining certain types of results. At the level of scientific disciplines, it is implicitly assumed that it is useful to single out particular aspects of the universe for study, and advantageous to study them in special ways. At the level of the material organization of society, pressures for the selective development of science and for specific applications of scientific knowledge lead to the development of tools useful primarily to select groups. At an ideological level, scientific theories can provide justification for policies and practices or provide the authority and exclusive plane for discussion.[102]

Despite the increasing gulf between science and the public as a result of complexity and specialization, the broader public does have a substantial interest in the results of science. The application of science has an important bearing on the way we live our lives, usually for the better, but not always. More importantly, the public has a critical stake in the scientific endeavour because, increasingly, public policy is grounded in scientific or expert knowledge, and much of science is funded from public resources. The pursuit of most research is resource-intensive and thus requires funding, often massive funding. The days of gentlemen researchers are long past, as are the days of disinterested researchers working in a university lab as part of their broader university responsibilities to teach and advance knowledge. Today, there are three principal sources of funding for significant research: government, industry, and foundations. All three come with strings and values attached, some overtly and others more subtly. All three sources of funding introduce sometimes unwelcome complications. The need for funding explains, at least in part, the rise in questionable results, from outright fraud to more subtle forms of corruption. The public thus has a le-

---

102. Brian Martin, *The Bias of Science* (Canberra: Society for Social Responsibility in Science, 1979), 6. Martin is a professor of social sciences at the University of Wollongong, with a PhD in physics.

gitimate interest in ensuring the accountability of science, as Popper insisted. Claims of post-normal science can only complicate the need to demand accountability and to reject the special claims of those working on such policy-sensitive issues as climate change and species diversity.

In the face of some of the difficulties that have undermined the trust that science once enjoyed and the authority that some still claim, wary public policy makers, advisers, and analysts need to adopt one of the long-held mantras in arms control negotiations: trust but verify. They also need to accept the advice of neuroscientist Stuart Firestein: accept uncertainty and be wary of "the too-well-crafted explanation, the one that explains everything; [it] should set off red flags, warning us that we are likely being deceived, misled, or outright duped."[103] Science continues to have much to offer, and grounding public policy in good science has more to recommend it than policy that is purely a matter of who can enlist the loudest voices. Nevertheless, conscientious policy makers need to develop keener antennae for some of the less admirable aspects of modern science in order to separate the wheat from the chaff.

＊＊＊＊＊

---

103. Stuart Firestein, "Certainly Not! Philosophy: Good science requires cultivating doubt and finding pleasure in mystery."

3 | Science and Its Pathologies

Historically, the claim of consensus has been the first refuge of scoundrels; it is a way to avoid debate by claiming that the matter is already settled.

Michael Crichton, novelist-physician, 2003[1]

Scientists will only be able to command trust in society if they follow basic professional standards. Prime among them is to publish the results of their research, no matter if they support a desirable storyline or not.

Reiner Grundmann, German climate scientist, 2013[2]

For more than four centuries, scientists have made giant strides in understanding the natural world. This progress has provided the basis for major advances in medicine, technology, and many other endeavours that have vastly enriched the human experience.[3] As the role of science in many areas of life has grown, the scientific enterprise has also developed some pathologies reflective of the darker side of human nature. Some of the problems are institutional and ideological, but it would not be fair to tar all of science with the same brush. There are many fields of science that continue on the path pioneered by the great scientists of the past. Problems are most acute in those scientific disciplines that are key to addressing controversial areas of health and environmental policy. It is also important to distinguish between areas of science that are controver-

---

1. Michael Crichton, "Aliens cause Global Warming," speech at the California Institute of Technology, January 17, 2003.

2. Reiner Grundmann, "Science for a good cause," *Die Klimazwiebel*, August, 2013.

3. See Steven Johnson, *How We Got to Now* (New York: Riverhead Books, 2014) for a fascinating account of the way in which advances in one area of science or technology can transform other fields and advance human knowledge and welfare in unexpected ways.

sial because some members of the public are uncomfortable with their application, for example, reproductive medicine or genetic modification, and those areas that have been politicized in order to drive a particular political agenda as is the case with climate change. The latter presents a set of issues very different from the former, although both can lead to difficult public policy decisions.

Science and peer review

In some of the more contentious areas of science, apologists insist that the authority of science should only be extended to projects and papers that have passed the bar of peer review, an idea that might have appealed to Kuhn but not to Popper or to many of the great scientists of the past. Current academics often insist that peer review acts as a guarantor of quality and provides an important check on the initiation and publication of errant and irresponsible scholarship. Their insistence cannot withstand serious scrutiny. Peer review tends to reinforce conventional wisdom and the dominant paradigm and to bar access to non-conforming scholars. Veteran British geneticist Sydney Brenner believes peer review does little more than drive research to the mediocre mean.[4] Climate scientist Judy Curry adds that peer review produces "research that dots i's and crosses t's and that promotes conformity of thought."[5] The extent to which questionable and marginal science is published stands as an indictment of the process and of the lack of due diligence exercised by journal editors and reviewers. Equally indictable is the extent to which solid and respectable research may be blocked from gaining financial support or publication because it disagrees with prevailing wisdom. The rising role of the internet in disseminating and discussing science is further undermining peer review's claim to authority.

Scientific societies such as the Royal Society in England pioneered peer review in the 17th century as a way of encouraging discussion and dissemination of scientific discoveries. William Boyle, one of the founders of the Royal Society, established the convention that date of publication rather than discovery should privilege scientists to claim status as originators of an idea. Rigorous debate

---

4.   Elizabeth Dzeng, "How Academia and Publishing are Destroying Scientific Innovation: A Conversation with Sydney Brenner," *Kingsreview*, February 23, 2013. Brenner was honoured with the Nobel prize in physiology in 2002.

5.   Judith Curry, "Are academia and publishing destroying scientific innovation?" *Climate Etc.*, April 8, 2014.

among peers then followed rather than preceded publication. By the second half of the 20th century, however, peer review had evolved into the current practice prevalent in almost all fields of scholarly enquiry: a process geared to ensuring a level of quality and integrity in both proposed and published research. As Harvard psycho-biologist William Anderson writes: "Peer review of scientists' work is necessary to the scientific enterprise. It requires open sharing of original data and recognition that colleagues, including hostile ones, may detect errors that confound our fondest hopes. Peer review is no guarantee of sound science, but it is one indispensable safeguard against avoidable error. The essential condition of peer review is that the peers not be deliberately selected by journal editors to be predisposed to agree with or condemn the work of others. This is a serious hazard, especially in disciplines that have only a few real experts all known to one another. It depends entirely on the integrity of the editors and the peer referees."[6]

Despite the claims of some apologists, peer review was never intended to constitute either endorsement or authorization of the research in question; rather, it was meant to inform readers that the community of researchers in the field considered the research to be worthwhile and consistent with the field's research standards. The Hoover Institution's Peter Berkowitz notes, however, that it was always based on a false belief that peer review ensured fairness: "the peer review process violates a fundamental principle of fairness. We don't allow judges to be parties to a controversy they are adjudicating, and don't permit athletes to umpire games in which they are playing. In both cases the concern is that their interest in the outcome will bias their judgment and corrupt their integrity. So why should we expect scholars, especially operating under the cloak of anonymity, to fairly and honourably evaluate the work of allies and rivals?"[7]

In reviewing submissions, funding agencies, editors of journals, and university presses first assure themselves that the research in question is worth funding or publishing; they then send it to three to five researchers in the field and seek their guidance. Assessment is often discipline-specific and interdisciplinary review is rare. Requests to review a manuscript or research application are an inte-

---

6.   William Anderson, "Some Like It Warm," *First Things,* February 2010.

7.   Peter Berkowitz, "Climategate Was an Academic Disaster Waiting to Happen," *Wall Street Journal,* March 13, 2010.

gral part of a scholar's contribution to a discipline and assume a level of competence and integrity on the part of reviewers. Referees are offered a range of options, from acceptance to rejection. A common option is to recommend funding or publication after suggested revisions. If the reviewers agree that the research is worth funding or publishing but needs revisions, the originator is given the opportunity to make the revisions and resubmit the revised proposal or paper. In some fields, funding of a research proposal or publication of an article may involve extensive comment and recommendations for revision. Different funding agencies, journals, and presses have their own rules on what to do with split reviews, from rejection to a further round of reviews.

The best peer review is double blind: the author does not know the referees and the referees do not know the author. Anonymity promotes integrity in the review process and helps to reduce both pal review and efforts to denigrate the work of a competitor by blocking funding or publication. Unfortunately, researchers who constitute a "community of experts" may put personal ambitions ahead of integrity. As a result, the process is not always as disinterested as it claims to be. Political economist Robert Higgs, reflecting back on a long academic career, notes that "peer review, on which lay people place great weight, varies from being an important control, where the editors and the referees are competent and responsible, to being a complete farce where they are not."[8]

Vested interests are as common among scholars as among participants in any other field of human endeavour. Sociologist Frank Furedi notes: "the contradiction between working as a member of an expert community and one's own personal interests cannot always be satisfactorily resolved."[9] Even more ominously, in too many subfields, peer review has descended into pal review and gate keeping, ensuring that only those who share the dominant researchers' perspectives are published in the lead journals. This is particularly common in fields that have become politically sensitive. Post-normal science and gate keeping are often highly correlated. In such cases, claims of consensus in the field have led to extraordinary measures to reinforce current dogma. Princeton physicist William Happer, for example, in his testimony before the US Senate

8.   Robert Higgs, "Peer Review, Publication in Top Journals, Scientific Consensus, and So Forth," *The Independent Institute,* May 7, 2007.

9.   Frank Furedi, "Turning peer review into modern-day holy scripture," *Spiked,* February 23, 2010.

Environment and Public Works Committee, pointed out that "many distinguished scientific journals now have editors who further the agenda of climate-change alarmism. Research papers with scientific findings contrary to the dogma of climate calamity are rejected by reviewers, many of whom fear that their research funding will be cut if any doubt is cast on the coming climate catastrophe."[10] Sceptical climate scientists have found their work rejected for frivolous reasons. In such cases, peer review has become wholly subordinated to a political agenda. British philosopher of science Donald Gillies concludes that too often "research assessment based on peer review is likely to concentrate funding on the most popular, or mainstream, research programs, while withdrawing funding from, and sometimes closing down altogether, minority research programs on which few researchers are working."[11]

As the stakes in science have grown, both financially and otherwise, the so-called prestige journals increasingly favour publishing what Patrick Michael calls "flashy" research. He points out: "I reviewed 13 months of both *Science* and *Nature*, and sorted every article or story about climate change or its impact into three piles: worse, better, or neutral compared with previous studies. Of the 115 entries, 23 made the 'neutral' pile, 83 were in the 'worse' stack, and nine were in the 'better.' The probability of the journals not having a bias is as likely as a coin being flipped 92 times and showing heads or tails fewer than nine times."[12]

---

10. William Happer, "Climate Change," Statement before the US Senate Environment and Public Works Committee, February 25, 2009. Ross McKitrick provides an excellent overview of the extent to which peer review has descended into gate keeping in "Science and Environmental Policy-Making: Bias-Proofing the Assessment Process," *Canadian Journal of Agricultural Economics* 53 (2005), 275–290. The Science and Technology Committee of the UK Parliament initiated hearings in January 2011 on peer review. Much of the testimony by experienced scholars is disturbing, demonstrating the extent to which peer review has become a gate-keeping process protecting dogma and paradigms and is now wholly at odds with the spirit of innovation and constructive criticism that should mark scholarship.

11. Donald Gillies, written evidence submitted to the UK Parliamentary Committee on Science and Technology, March 3, 2011. Gillies, a student of Karl Popper and former editor of *The British Journal for the Philosophy of Science*, is Emeritus Professor, University College London. In *How Should Research be Organised?* (London: College Publications, 2008), he provides a much more general indictment of the conduct of modern scholarship.

12. Patrick J. Michaels, "Putting Headlines Ahead of Science," *Orange County Register*, January 2, 2014. Garth Paltridge concludes that it is "vastly more difficult to publish results in climate research journals if they run against the tide of politi-

Peer review can be capricious. A manuscript that has been found seriously deficient and rejected by one group of reviewers can be enthusiastically endorsed by another. Manuscripts rejected one year by a journal have been eagerly published a few years later by the same journal, without any indication that editors or referees were aware of the paper's earlier history. Even with a surfeit of journals available in any field, some papers can still make the rounds for years looking for a home. Some academics have adopted the strategy of submitting simultaneously to more than one journal in the hope that at least one will be interested. Physicist Frank Tipler recounts how referees routinely reject papers with new ideas. "Prior to the Second World War, the refereeing process worked primarily to eliminate crackpot papers. Today, the refereeing process works primarily to enforce orthodoxy. ... If one reads memoirs or biographies of physicists who made their great breakthroughs after, say, 1950, one is struck by how often one reads that 'the referees rejected for publication the paper that later won me the Nobel Prize.' ... We have a scientific social system in which intellectual pygmies are standing in judgment of giants."[13]

The healthy aspect of the peer-review process is that it can strengthen arguments and the supporting evidence. Less healthily, however, it stifles innovation and can become incestuous. Climate science and a number of its principal organs, including *Science* and *Nature*, have become fully captive of the most disturbing elements of peer review. Some of the most important and innovative work in a variety of fields may fall afoul of peer review and have difficulty finding a place in the most prestigious, widely read journals.[14] The

---

cally correct opinion, which is why most of the sceptic literature on the subject has been forced onto the web, and particularly onto weblogs devoted to the sceptic view of things. ... Their output ... is well on the way to becoming a practical and stringent substitute for peer review." "Uncertainty, sceptics and the climate issue," in Alan Moran, ed., *Climate Change: The Facts* (Melbourne, Australia: Institute of Public Affairs, 2015), kindle edition.

13. Frank J. Tipler, "Refereed Journals: Do They Insure Quality or Enforce Orthodoxy?" *International Society for Complexity, Information, and Design (ISCID) Archive,* June 30, 2003, 1-2.

14. Fascinating insight into the furore generated by Cambridge University Press's decision to publish Bjørn Lomborg's *The Skeptical Environmentalist* in 2001 is provided by one of the editors at the Press, Chris Harrison, in "Peer review, politics, and pluralism," *Environmental Science & Policy* 7 (2004), 357–368. Lomborg's manuscript had been subjected to a thorough peer review, but the referees had not included the self-appointed keepers of environmental science orthodoxy. The result was a vicious campaign to punish the Press and to force

late medical researcher, David Horrobin, noted that "public support can only erode further if science does not put its house in order and begin a real attempt to develop validated processes for the distribution of publication rights, credit for completed work, and funds for new work. Funding is the most important issue that most urgently requires opening up to rigorous research and objective evaluation."[15] The extent of flaws in the peer review process prompted the US Supreme Court to rule in Daubert v. Merrell Dow Pharmaceuticals (1993) that peer review is not in and of itself evidence of the validity of expert testimony.[16]

Progress in scientific understanding is based on replication, confirmation and/or falsification; whether these goals are met by the work of credentialed experts or by inspired amateurs is irrelevant. An hypothesis remains an hypothesis until it has been verified by real-world observation and replicated by other researchers. Verification remains subject to the caveat that future observations falsifying an hypothesis can bring the whole theory into question. The descent of some fields of scientific research into dogma upheld by pal review may look superficially authoritative, but, in fact, the dogma may shield a cascade of questionable science. Computer models, for example, do not verify, but they can falsify. In the case of greenhouse-gas induced global warming, there has to date been a lot of theorizing, a great deal of computer modelling, little verification, and much falsification. The leaked e-mails from the Climatic Research Unit (CRU) at the University of East Anglia, the subject of much controversy in the weeks leading up to the 2009 Copenhagen climate conference, exposed peer review for what it often is: the defense of entrenched interests.

The desire to ensure the "authority" of science in underwriting public policy or strengthening political campaigns has made funding agencies, journal editors, and referees more sensitive as to whether a particular proposal or finding will reinforce or under-

withdrawal of Lomborg's book. The Press stood by its decision and continued to publish many important books in the field, including all the official reports of the UN's Intergovernmental Panel on Climate Change. *The Skeptical Environmentalist* was a huge commercial success and launched Lomborg's global reputation.

15. David F. Horrobin, "Something Rotten at the Core of Science?" *Trends in Pharmacological Sciences,* 22: 2 (February 2001). Horrobin was editor of *Medical Hypotheses,* a journal devoted to publishing innovative and contrarian views.

16. Daubert v. Merrell Dow Pharmaceuticals, *Judgment* (1993), US Supreme Court (92–102), 509, 579.

mine the prevailing orthodoxy. Peer review from this perspective leads to claims that a paper may be "irresponsible," "misleading," or even "wrong." In such cases, the review becomes less a matter of determining whether the paper or proposal in question is a worthwhile contribution to the scientific quest for knowledge and more a matter of determining whether it strays outside the accepted boundaries of the dominant paradigm. In such cases, politics trumps science and peer review becomes censorship. The reliability of the science that emerges from this kind of peer review does not serve the public interest as well as science that provides room for competitive perspectives, no matter how challenging this may be to those attached to a particular point of view. In the long run, questionable peer review practices are likely to undermine confidence in the reliability of science as a whole.

In science, there are many issues for which the same evidence will lead to different conjectures and conclusions, particularly in areas of science in which hypotheses remain tentative. Neither the progress of science nor the making of public policy is well served if one group finds it expedient to suppress opposing perspectives. Often, in such cases, both sides need to be heard and final judgment suspended until further evidence tilts the balance in one direction or the other. Even then, science must always remain open to the possibility that later evidence may change the balance once again. Anything other than a peer review process that accepts competing perspectives does a disservice to both science and its users.

More broadly, the descent from peer review to pal review is symptomatic of a broader problem: institutional pressure to publish and keep the lid on misconduct in order not to kill the goose that lays the golden egg of research funding. Every university president and laboratory administrator is under constant pressure to increase funding, particularly for research, which in turn keeps the pressure on deans, chairs, and senior researchers to seek research funding and pump out the results. Research funding and publication are the principal measures of scientific merit, whether in academia or research institutions, and neither researchers nor their supervisors are interested in cleaning up what is obviously a growing problem. Major scandals, of course, get enough publicity so that others are unlikely to make use of tainted work. But smaller admissions of error or discoveries of misconduct as well as rebuttals rarely get that kind of notice; as a consequence, the resulting retractions or rebuttals have little impact on the continued use of the discredited work.

Berkowitz sadly concludes that "our universities, which above all should be cultivating intellectual virtue, are in their day-to-day operations fostering the opposite. Fashionable ideas, the convenience of professors, and the bureaucratic structures of academic life combine to encourage students and faculty alike to defend arguments for which they lack vital information. They pretend to knowledge they don't possess and invoke the authority of rank and status instead of reasoned debate."[17]

Science, money, and politics

Contemporary science has become a multi-billion dollar enterprise, requiring extensive laboratories and sophisticated equipment for research purposes. Whether conducted at private or government laboratories or at universities, there is never enough money to satisfy the demand for new equipment, assistants, field trips, and conference attendance. But, as Richard Lindzen warns, "expanded funding is eagerly sought, but the expansion of funding inevitably invites rent-seeking by scientists, university administrations, and government bureaucracies."[18] There are also more and more scientists pursuing the same funds, whether provided by government, the private sector, or foundations. Daniel Greenberg notes that the scientific enterprise now provides its own version of Parkinson's law: "research expands to absorb the money available for its conduct."[19] And there is never enough money. Only about 20 per cent of research applications for US federal funding succeed, with the rest scrambling for private and foundation money or seed money from a university's scarce research funds in the hope that it will pave the way to a more successful application next time around.

Governments have been in the funding business since the 1940s when politicians and officials took up the mantra that basic science underpins innovation and thus fuels economic growth. The direct returns to society are admittedly quite small, but the indirect returns are imagined to be very large. In the United States, a 1945 report to the president by the head of the US Office of Scientific Research and Development, Vannevar Bush, set a tone that has been

---

17. Berkowitz, "Climategate Was an Academic Disaster Waiting to Happen."

18. Richard Lindzen, "Science in the Public Square: Global Climate Alarmism and Historical Precedents," *Journal of American Physicians and Surgeons* 18:3 (Fall 2013), 70.

19. Daniel S. Greenberg, *Science for Sale: The Perils, Rewards, and Delusions of Campus Capitalism* (Chicago: Chicago University Press, 2007), 29.

followed ever since in industrialized countries.[20] Bush believed that
basic research was critical in advancing the national interest for
both security and commercial reasons and that continued government support for science and technology would ensure US leadership. Other governments followed the US lead, although with much
more limited resources than the US was able to marshal. A 2012 report by the Council of Canadian Academies provides a good modern example of this basic article of faith justifying strong public
support for scientific research: "Discovery research in the natural
sciences and engineering (NSE) is a key driver in the creation of
many public goods. Scientific advances help catalyze innovation,
create new knowledge, foster economic prosperity, improve public
health, enable better protection of the environment, strengthen national security and defence, and contribute in myriad other ways to
national and sub-national policy objectives. For all of these reasons,
most governments around the world wisely invest substantial public resources in supporting discovery research in the NSE."[21]

In addition to funding university-based research, many governments maintain their own research facilities. Canada, for example, has long conducted research at the labs of the National Research Council (NRC) – established in 1916 – often in cooperation
with university-based scientists. The Departments of Fisheries, the
Environment, and Energy and Resources have mandates to conduct
issue-specific research, sometimes at their own facilities and sometimes in cooperation with university-based scientists. The trend in
Canada, however, is towards funding research by university-based
researchers rather than funding government-based research. In the
United States, scientists at NASA, NOAA, UCAR, NCAR, the Energy Department, and other agencies are major contributors to scientific research and compete with university-based scientists for space
in the scientific journals. The same situation exists in Japan, Europe,
Australia, and other developed countries, as well as in emerging
economies such as China, India, and Brazil, which increasingly see
the value of sponsoring research.

Depending on one's perspective, the amount of money spent by
governments on research can be characterized as vast to insufficient.

20. "Science, The Endless Frontier," A Report to the President by Vannevar Bush,
     July 1945.

21. Council of Canadian Academies, Expert Panel on Science Performance and Re-
     search Funding, *Informing Research Choices: Indicators and Judgment* (Ottawa,
     2012), xi.

In 2009, the US federal government devoted $133.3 billion to R&D, representing a little less than a third of total US expenditures from all sources. Of the federal effort, half was for defence-related R&D and the rest for all other research. Of that $65.2 billion, less than half was split between applied and basic research; the rest went to development. Thus basic research benefited from about $14 billion in federal support.[22] About 60 per cent of that amount was spent on university-based research. It is a princely sum and more than any other government spends, but it responds to only about a fifth of the demand and provides grist for the steady drumbeat from the science community that it is far from enough. In absolute terms, US spending on all R&D far exceeds that of any other country: in current dollars, US-based funders spent $401.6 billion in 2009, more than twice as much as the second-highest R&D investor, China, which spent $154.1 billion. American spending on R&D also outweighed the total R&D spending of the European Union, which came to $297.9 billion. Canadian numbers are significantly smaller, but in terms of money spent on university-based research, the share of GDP is slightly higher.[23]

Despite the billions of dollars devoted to scientific research, there is never enough. Few scientists would ever suggest that perhaps society should spend some of this money on other priorities. The media can always be relied on to print articles that bemoan the lack of funding for science, claiming that a country or university is falling behind in the research race, a race characterized as being similar to the arms race. Bruce Alberts, editor-in-chief of *Nature*, complained that "the declining opportunities for research funding have made survival for some of the most able researchers resemble a lottery – or perhaps Russian roulette is a better analogy. The effect on the US research system seems devastating."[24] A related worry involves threats of impending brain drains to more generous countries/universities if the government does not step up to the plate and increase funding.[25]

---

22. Numbers derived from Joseph V. Kennedy, "The Sources and Uses of US Science and Funding," *The New Atlantis,* Summer 2012, 3-22.

23. Not all this money is well spent. See, for example, the critical assessment of the amount of money spent on frivolous research by Henry I. Miller, "Investing in Bad Science," *Policy Review* 177 (February 1, 2013).

24. Bruce Alberts, "Am I wrong?" *Nature* 339 (March 15, 2013), 1252.

25. See, for example, Ivan Semeniuk, "Canada losing ground in global science: report," *Globe and Mail,* May 22, 2013; Mark Stokes, "The folly of science on a

Presidents, vice-presidents, and deans at the major research universities, with the help of specialized staff, devote almost all their time to raising funds for new buildings, centers, laboratories, and other accoutrements of modern research. The bulk of the money is swallowed up by research in the bio-sciences and military-related fields, but significant sums remain available for basic physics, chemistry, climate science, and other endeavours. Politicians find it difficult to resist calls for research money to cure cancer and other medical scourges, and climate science has recently climbed into a similarly irresistible niche, with researchers learning that almost any research that they can tie to climate change improves their chances for success.[26]

At the major universities – both private and public – senior administrators successful at gaining major grants as well as attracting star researchers and the money they command now receive salaries and fringe benefits rivalling those of winning football and basketball coaches. Their roles are not dissimilar: both raise prestige and bring in money from alumni, an important source of income, particularly for capital projects. A presidential box in a university football stadium is often a key asset in raising funds from alumni and corporate donors, including those investing in the university's re-

---

shoestring," *The Guardian,* April 16, 2013; and Virginia Gewin, "Science Funding: Flirting with Disaster," *Nature* 498 (June 26, 2013), 527-8. Almost every science advocacy group can tell of heart-rending stories of life-saving, wealth-enhancing, planet-saving scientific work that has been thwarted by lack of funding. Most scientific journals carry periodic editorials lamenting the perilous state of funding. Most believe that an appropriate sum would be 3 per cent of GDP for basic science research, a sum vastly larger than any society has ever achieved and one that would inevitably require a re-ordering of other societal priorities. In a period of fiscal constraint, when virtually every OECD government is facing mounting debts and deficits, such a sum is unlikely to be obtained in the foreseeable future.

26. Richard Lindzen provides two telling examples: "A $197,000 grant went to a psychologist who wrote: 'Climate change represents a moral challenge to humanity, and one that elicits high levels of emotion. This project examines how emotions and morality influence how people send and receive messages about climate change, and does so with an eye to developing concrete and do-able strategies for positive change.' A grant for more than $400,000 went to a political scientist who wrote: 'Common sense says that claims about how social and political life ought to be arranged must not make infeasible demands. This project will investigate this piece of common sense and explore its implications for a number of pressing issues, such as climate change, multiculturalism, political participation, inequality, historical justice, and the rules of war.'" "Science in the Public Square: Global Climate Alarmism and Historical Precedents," 70.

search activities. Star researchers with proven track records are often no longer required to do much teaching; their sole educational role is often confined to supervising graduate students and post-doctoral fellows working in their labs on projects largely related to the supervisor's research interests. Undergraduates, saddled with rising tuition fees to help pay for expensive faculty and their research requirements, are now rarely taught by senior full-time faculty, finding their classes and labs staffed by post-docs, adjunct faculty, or graduate students. In most fields, tenure and promotion are directly tied to a faculty member's publication and fund-raising record. The publish-or-perish creed is deeply engrained in academic cultures, explaining why so much marginal and misleading research is published. The explosion in the number of academic journals reflects the need to find outlets for much of this research. Enterprising faculty have learned that their value to the university is enhanced if they edit or sit on the editorial boards of journals. Many research universities now hire scientists who only do research. Even more than faculty, their value to the university is determined by their ability to attract funding for their research projects, which, in turn, is often dependent on their publishing track record.

One of the incentives for universities to attract research is that funders will often agree that a certain percentage of the grant may be spent on overhead, loosely defined, but meant to defray the costs of administrators, facilities, and other expenses of running a research university. The major return to the university is two-fold: prestige and income from projects that result in valuable patents and licensing arrangements. Universities now require that any intellectual property resulting from university-based research belongs to the university rather than to the researcher or the funding agency. Law suits to defend this claim have become a normal part of the US academic scene. Universities, which at one time took pride in sharing their research with any one interested, have morphed into institutions prepared to defend secrecy and proprietary knowledge.[27] In climate science and bio-medical research, questionable findings of-

---

27. The lengthy dispute between the American Tradition Institute and the University of Virginia involving access to the publicly funded research data and e-mails of climate scientist Michael Mann provides a case in point. Former Virginia attorney general Ken Cuccinelli joined the quest but was equally rebuffed by the University, a public institution. The hyper-sensitive Mann has also lashed out and sued a number of critics for defamation, including Canadian climate scientist Tim Ball and journalist Mark Steyn. See Darren Jonescu, "Mann vs. Steyn: Heresy Shall Be Crushed," *American Thinker,* July 26, 2013.

ten prove difficult to audit or replicate because researchers, with the full backing of their universities, argue that their data and methodology are proprietary and not available without their permission, even if the underlying research has been funded with public money.

While public funds remain critical to most research universities, private firms have also made significant contributions, principally in areas with potential for profitable applications, e.g. in biomedicine or electronics. The basic chemistry behind many new drugs, for example, frequently originates in a university lab with government funding, but the clinical trials and other parts of the translation of the basic science into a usable drug are often assumed by private firms. In gross terms, private funding adds up to about twice the public effort, because it includes the enormous costs of development as well as the marketing of the product.[28] In basic research, the private contribution is less than that of government. The private sector funds both university-based research and non-academic research at contract research organizations (CROs). Private firms may prefer contract research over academic research in sensitive areas involving potential conflicts of interest or disputes about ownership of intellectual property. Some university-based scientists may also be affiliated with CROs.

Finally, a significant part of research is funded by foundations and non-profit organizations. Virtually every major disease, for example, has an organization devoted to, among other activities, raising funds for research to cure the disease, including cancer, heart and stroke, and diabetes. Many non-governmental organizations (NGOs) also fund research. Environmental NGOs (ENGOs) in particular sponsor research to promote their cause. Many ENGOs now command budgets in the millions of dollars, and while they spend a good deal of their budgets on mutually reinforcing fund-raising and awareness campaigns, they also devote some of their resources to sponsored research, for example, on climate change. Much of this research supports advocacy science, but some may also advance scientific knowledge as input into more disinterested research.

The syndrome of research-heavy universities is at its most acute in the United States, but it is not unknown at Canadian, European, and other universities, all of which are in the hunt for money. In

---

28. See Marcia Angell, *The Truth About the Drug Companies: How They Deceive US and What to do About It* (New York: Random House, 2004). Angell is a former editor-in-chief of the *New England Medical Journal* and for many years had a front-row seat from which to observe how the drug companies operate.

universities across the OECD, faculty are under constant pressure from senior administrators to apply for research grants and to raise the university's research profile. Faculty members who perform useful, published but unfunded research are inherently less valuable to a university than those who bring in research money. Universities will often protect researchers whose work has attracted criticism so long as they bring in the money.[29] University ratings agencies look closely at the amount of research money a university attracts in determining rankings. In Canada, the *MacLean's* annual survey puts great stock in research funding that gives universities with a large bio-medical research complex associated with a medical school a huge advantage over universities that lack medical schools.

In a perfect world of disinterested scientists advancing science on a purely objective basis, the sources of research funding would be irrelevant. In the real world the source of funding is consequential. British science researcher Sonja Boehmer-Christiansen, for example, told the UK House of Commons:

> Most climate change [research] since the late 1980s has been government- and grant- funded with the clearly stated objective that it must support a decarbonisation agenda for the energy sector. Scientific research as advocacy for an agenda (a coalition of interests, not a conspiracy), was presented to the public and governments as protection of the planet. This cause of environmental protection had from the start natural allies in the EU Commission, United Nations, and the World Bank. [The Climatic Research Unit at the University of East Anglia], working for the UK government and hence the IPCC, was expected to support the hypothesis of man-made, dangerous warming caused by carbon dioxide, a hypothesis it had helped to formulate in the late 1980s and which became 'true' in international law with the adoption of the 1992 Framework Convention on Climate Change.[30]

There is very little money available for pure, disinterested, independent research. Australian science blogger Jo Nova points out, "when grants, careers, junkets, book sales, and offers to sit on golden commissions are on the line, it doesn't take much motivated rea-

---

29. As noted, Michael Mann, a central figure in the Climategate and Hockey Stick scandals discussed in chapters four and five, provides a good example. Both his former university, the University of Virginia, and his current employer, Pennsylvania State University, have gone to great lengths to protect him.

30. "Memorandum submitted by Dr. Sonja Boehmer-Christiansen," for the UK House of Commons Science and Technology Committee Hearings into the CRU in 2009-2010.

soning to find excuses to believe your work is 'science' even as you ignore opportunities to follow data that don't quite fit, or delay publications of inconvenient graphs, while you double check, triple check, and invite like-minded colleagues to help find reasons the graphs are not important."[31] From governments to foundations and private firms, funding organizations have values and objectives that influence decisions about who and what receive funding, and researchers are well aware of those values and objectives as they formulate their research. There are some celebrated cases of scientists who have wilfully ignored the objectives of their funders and have paid the price, for example, by disclosing the adverse results of a drug when their research contract specifically stipulates that they may not do so without permission from the sponsor. Such contrarians are more admired than emulated. Whether it is the Department of Defense, the NIH, NSERC, the British Council, IBM, Pfizer, the Worldwide Fund for Nature, or the Pew Charitable Trust, the science that each funds can hardly be characterized as a disinterested search for the truth.[32]

Psychologists who specialize in understanding how the human brain rationalizes conflicting evidence have identified a number of unconscious strategies, including confirmation bias and motivated reasoning. The latter is usually taken to be a more severe manifestation of the former. Both lead people to ignore or discount contrary evidence that conflicts with a prior belief. People will respond defensively to contrary evidence and try to discredit it. When it comes to funding, both confirmation bias and motivated reasoning are easily explained. Acceptance of contrary evidence can handicap the ability to raise funds as easily as in the past. In highly controversial areas of science, cases of confirmation bias and motivated reasoning are frequent, particularly when the funding stakes are high. Yale's Dan Kahan writes: "Examples of the goals or needs that can motivate cognition are diverse. They include fairly straightforward things, like a person's financial or related interests. But they reach more intangible stakes, too, such as the need to sustain a positive

---

31.　Jo Nova, "Consensus Police: 101 'motivated' reasons not to be a sceptic," *Joannenova.com*, August 23, 2013.

32.　For the record and in the interest of full disclosure, the research for this book has received no funding from any source other than Carleton University, which provided me with a sabbatical for the 2013-14 academic year to complete a project which occupied seven years of my research time as part of my normal academic duties at Carleton.

self-image or protect connections to others with whom someone is intimately connected and on whom someone might well depend for support, emotional or material."[33] Climate scientist Judith Curry argues that "motivated reasoning by climate scientists is adversely impacting the public trust in climate science and provides a reason for people to reject the consensus on climate change science."[34] When motivated reasoning becomes systemic, it leads to noble cause corruption and advocacy science.

Noble cause corruption and advocacy science

On February 28, 2012, motivated reasoning was on full display in a London lecture hall. Sir Paul Nurse, President of the Royal Society, was giving the annual Dimbleby Lecture. He asked: "So, what is special about science that means we should trust it? What makes it so good at generating reliable knowledge about the natural world?" Good question, to which he gave a generally excellent answer. He explained that at the centre of science lies "the ability to prove that something is not true. ... This distinguishes it from beliefs based on religion and ideology, which place much more emphasis on faith, tradition, and opinion. As a scientist I have to come up with ideas that can be tested. Then I think of experiments to test the idea further. If the result of the experiment does not support the idea then I reject it, or modify it, and test it again. If the results of the experiments always support the idea, then it becomes more acceptable as an explanation of the natural phenomenon. ... Early on in a scientific study knowledge is often tentative, and it is only after repeated testing that it becomes increasingly secure. It is this process that makes science reliable, but it takes time."[35] Popper would have agreed. If he had stopped at that point, all would have been well. But he did not. He went on: "It is impossible to achieve complete certainty on many complex scientific problems, yet sometimes we still need to take action. The sensible course is to turn to the expert scientists for their *consensus* view."[36] [emphasis added]

---

33. Dan Kahan, "What is Motivated Reasoning and How Does It Work?" *Science+Religion Today*, May 4, 2011.

34. Judith Curry, "Scientists and Motivated Reasoning," *Climate Etc.*, August 20, 2013.

35. Sir Paul Nurse, "The New Enlightenment," The Richard Dimbleby Lecture 2012, February 28, 5, 6.

36. Nurse, "The New Enlightenment," 7.

Nurse's lecture prompted some heated discussion, particularly among climate scientists. Georgia Tech's Curry observed that "the perceived need for a scientific consensus leads to the situation whereby any disagreement with the consensus is mistakenly viewed as arising from ideology, politics, and consensus. In reality, politics comes in when solutions are discussed, and a scientific consensus that is married to a specific policy option precludes having a mature discussion about these issues." She believes that in controversial areas of public policy, scientists are better advised to "clearly explain the levels and types of uncertainty and areas of ignorance" than to try to create the false impression of a consensus. She concludes that "Nurse gets credit for grappling with this important and difficult issue, but his talk exposes many issues that have contributed to the growing dysfunction at the climate science-policy interface."[37]

Nurse is not alone. Many scientists believe that public policy makers need help to interpret science and choose the right policy: without scientists providing both an expert analysis of the problem and an appropriate prescription, the problem will not be resolved. Critics of this perspective point out that, in such cases, scientists are straying beyond their competence and are trying to use their scientific expertise as a way to frame the problem and shape solutions that suit their political preferences.

When the boundary between advocacy and science becomes blurred, society loses. Claims made by an advocacy community can be, and often are, discounted due to their source. Claims made by a scientific community are often accepted at face value, perhaps deservedly so, if they reflect the results of scientific effort. They are problematic, however, when the two communities are no longer separated by clear boundaries. UK science journalist Matt Ridley provides a disturbing catalogue of recent cases of advocacy science and concludes: "It's hard for champions of science like me to make our case against creationists, homeopaths and other merchants of mysticism if some of those within science also practise pseudoscience."[38] A report on a seminar on science advocacy similarly points out: "when scientists themselves are perceived as advocates,

---

37. Curry, "Sir Paul Nurse on the science-society relationship," *Climate Etc.,* March 14, 2012.

38. "Policy-based evidence-making," *The Times,* December 9, 2014. An expanded version appeared on his blog the next day with further examples of science in the service of policy.

their views are often discounted, even if they are being objective. Further, that discount can threaten the perceived legitimacy of the advocate's field. ... [Advocacy] poses the temptation of distorting or tainting the science or substituting a personal opinion for a scientific one. Most seriously, if scientists become advocates they risk losing their good name as scientists. ... If the hallmarks of science are accountability, fairness, and honesty, then those traits may be incompatible with effective advocacy."[39]

Marine biologist Jake Rice captures well what troubles some scientists and members of the public: "Society benefits from well-informed experts participating in public dialogue on policy issues, and providing information on how consistent policy alternatives are with the scientific information in their area of expertise. However, when those experts place their desired policy outcomes ahead of the basic principles of sound, objective science, an important boundary is crossed. ... Partisan groups lobbying for preferred outcomes have a long history of the selective use of information to support predetermined conclusions."[40] Rice may be right, but he fails to grasp the moral righteousness that motivates the scientist as advocate. Activists in all areas have two things in common: they are typically progressives, leading them to perceive themselves to be morally superior to their critics, and they insist that those in power fail to pay them the attention they deserve. Failure in getting their policy preferences implemented is rarely interpreted as a rejection of their preferences in favour of other societal interests and priorities but rather as a failure of policy makers to listen and understand the moral urgency of their cause. Advocacy thus arises out of a deep conviction of righteousness and a myopia about competing societal interests. From an activist perspective, the science is so clear and compelling that it screams for immediate and unreserved policy measures.

Activist scientists often confuse criticism of their advocacy with attacks on their science, refusing to accept that there may be genuine differences in the science itself. David Wallace, for example, vice-president of the Royal Society, appealed to the media for vigilance against "individuals on the fringes, sometimes with financial support from the oil industry, who have been attempting to cast

39. Deborah Runkle and Mark S. Frankel, "Advocacy in Science," Summary of a Workshop convened by the American Association for the Advancement of Science, Washington, DC, October 17-18, 2011.

40. Jake Rice, "Food for thought: Advocacy science and fisheries decision-making," *ICES Journal of Marine Science,* 68:10 (2011), 2007.

doubt on the scientific consensus on climate change."[41] His boss,
Nurse, took to the airwaves of the BBC to explain "why public trust
in key scientific theories has been eroded – from the theory that
man-made climate change is warming our planet, to the safety of
GM food, or that HIV causes AIDS."[42] Hard as he tries, Nurse never
seems to grasp the problem: the corrosive impact of overreaching
advocacy science on public confidence. As climate modeller Tamsin
Edwards puts it: "We risk our credibility, our reputation for objectivity, if we are not absolutely neutral. At the very least, it leaves us
open to criticism. ... I care more about restoring trust in science than
about calling people to action; more about improving public understanding of science so society can make better informed decisions,
than about making people's decisions for them. Science doesn't tell
us the answer to our problems. Neither should scientists."[43]

Dorothy Nelkin was a pioneer in analyzing the relationship between science and society, including the emergence of advocacy science. She provided a detailed assessment of the role of scientists in
addressing controversial issues such as the pollution of the Cayuga
River in Cleveland and the location of nuclear power plants. She
observed:

> A striking feature of the new scientific activism is the public nature of
> its activities and the willingness of activists to engage in and, indeed, to
> abet political controversy. Disputes among scientists are normally re-
> solved within the scientific community using well-established provi-
> sions of collegial review. However, recently, scientists appear willing
> to air grievances in a political forum – through mass media, litigation,
> or appeals to citizens' groups or political representatives. Citizen par-
> ticipation is sought today for a different reason – as a means to increase
> the political accountability of science. While activists in the 1940s
> fought against political control over research, their recent counterparts
> – by calling public attention to conflicts of interest within the scientific
> community – seek to increase political control. Such actions have polar-
> ized the scientific community, as less radical scientists seek to maintain
> intact the principles of autonomy and self-regulation that were fought
> for by activists nearly 30 years ago.[44]

---

41. Neil Collins, "Global warming generates hot air," *The Telegraph,* May 16, 2005.

42. BBC, Program Notes for "Science Under Attack," aired January 24, 2011.

43. Tamsin Edwards, "Climate Scientists must not advocate particular policies," *The
     Guardian,* October 8, 2013. Trained in physics, Edwards is currently engaged in
     modelling research at the University of Bristol.

44. Dorothy Nelkin, "Scientists in an Adversary Culture," Paper presented to the
     Organization of American Historians, April, 1978, quoted in Mary Douglas and

Advocacy today is most firmly entrenched among environmental scientists. Its origins, however, predate the birth of modern environmentalism in the 1960s and can be found in the anti-nuclear movement of the 1950s with its fears of a nuclear winter. Crichton pointed to the wave of contemporary fear-mongers as "demons ... invented by scientists."[45] He went on to encapsulate in a few words the pathology of advocacy science: "Evidentiary uncertainties are glossed over in the unseemly rush for an overarching policy, and for grants to support the policy by delivering findings that are desired by the patron. Next, the isolation of those scientists who won't get with the program, and the characterization of those scientists as outsiders and 'sceptics' – suspect individuals with suspect motives, industry flunkies, reactionaries, or simply anti-environmental nutcases. In short order, debate ends, even though prominent scientists are uncomfortable about how things are being done."[46]

Science's value to policy makers lies in its ability to *understand* and *explain* how complex natural phenomena function and to *predict* how these phenomena will respond to changing conditions. Science can play a critical role not only in explaining natural phenomena but also in setting the limits of policy measures. Policy making is future-oriented, focused on alternative policy measures that may be appropriate to address emerging conditions. Predictions that qualify as scientific are those that are consistent and independent of time and place. Some phenomena, however, cannot be reduced to the point of being predictable. The best that scientists can do is to provide a range of probabilities based on past performance. Weather is a good example as are other earth-system phenomena. Predicting the future of complex systems is wholly different from the ability of scientists to reduce natural phenomena to their constituent components and then to predict the behaviour of each. The interaction of more and more components makes predictions less and less reliable. Scientists specializing in environmental and bio-medical phenomena are thus often handicapped in providing policy makers with the

---

Aaron Wildavsky, *Risk and Culture: An Essay on the Selection of Technological and Environmental Dangers* (Los Angeles: University of California Press, 1983), 64. For a discussion of advocacy's perverse impact on research and policy, see Stanley Trimble, "The Double Standard in Environmental Science," *Regulation*, Summer 2007, 16-22.

45. Crichton, "Aliens Cause Global Warming," speech at the California Institute of Technology, January 17, 2003, 2.

46. Crichton, "Aliens Cause Global Warming," 9.

precision and certainty they would like. White-coated propagandists,[47] trying to reach beyond these inherent uncertainties to provide a level of certainty that is rarely warranted, thus mislead policy makers. Doing so typically involves bringing in values and preferences that add urgency to the issues and reduce the uncertainties, invoking claims of consensus, as advocated by science historian Naomi Oreskes: "The criteria that are typically invoked in defense of the reliability of scientific knowledge – quantification, replicability, falsifiability – have proved no guarantee. ... In all but the most trivial cases, science does not produce logically indisputable proofs about the natural world. At best it produces a robust consensus based on a process of inquiry that allows for continued scrutiny, reexamination, and revision."[48] Whether this is called post-normal science, advocacy, virtuous corruption, or noble cause corruption, it is no longer science.

Roger Pielke, Jr. offers an interesting twist. He suggests that scientists engaged in policy-relevant science need to be much more sensitive to the difference between *policy* and *politics*. Good policy analysis expands the choices available to decision makers by offering alternative ways to resolve an issue. Politics, on the other hand, seeks to narrow the choices on the basis of predetermined political values and preferences. Pielke writes: "Because scientific results always have some degree of uncertainty and a range of means is typically available to achieve particular objectives, the task of political advocacy necessarily involves considerations that go well beyond science. ... Science never compels just one political outcome ... In thinking about how things might be different it is absolutely critical to differentiate *scientific results* from their *policy significance*."[49]

Advocacy science and its various cognates are unlikely to disappear. As a result policy makers and advisers need to develop much more sensitive antennae about the science in question in order to distinguish between scientific advice and advocacy. Failure to do so will lead to the many traps that have been laid by advocacy science: incomplete and selective information, more certainty than is

47.   See Charles Krauthammer in "Obama plays high priest to climate change religion," *National Post*, February 22, 2014.

48.   Naomi Oreskes, "Science and public policy: what's proof got to do with it?" *Environmental Science and Policy* 7 (2004), 380, 370-71.

49.   Roger A. Pielke, Jr., "When scientists politicize science: making sense of controversy over *The Skeptical Environmentalist*," *Environmental Science and Policy* 7 (2004), 406, 414.

warranted, claims of consensus, intolerance of alternative views, and other pathologies. Philip Handler, long-time president of the US National Academy of Sciences, concluded in 1980 that "scientists best serve public policy by living within the ethics of science, not those of politics. If the scientific community will not unfrock the charlatans, the public will not discern the difference – science and the nation will suffer."[50] Unfortunately, Handler's conclusions were less and less honoured in the years following his death (1982), and science has been the poorer for it. The current president, Ralph Cicerone, is among the most active global warming alarmists.

Most scientific enquiry continues to be conceived in realist terms with scientists trained to develop objective representations of the natural world. There are areas in which the painstaking work of verification, replication, and falsification is difficult and the demand for policy is pressing – for example, in public health and environmental science. As a result, maintaining a dividing line between expert advice and advocacy has proven particularly difficult. These are also areas in which sound science too often degenerates into junk science.

Junk science and epidemiology

In 1953, Nobel prize-winning chemist Irving Langmuir gave a talk at a colloquium at Princeton University in which he described a number of aberrations in the practice of science, one of which he called "pathological" science. He told his Princeton audience of cases,

> ... where there is no dishonesty involved but where people are tricked into false results by a lack of understanding about what human beings can do to themselves in the way of being led astray by subjective effects, wishful thinking or threshold interactions. These are examples of pathological science. These are things that attracted a great deal of attention. Usually hundreds of papers have been published upon them. Sometimes they have lasted for fifteen or twenty years and then they gradually die away.

He also set out the six basic symptoms of pathological science that have since become know as Langmuir's Laws of Bad Science:

> 1) The maximum effect that is observed is produced by a causative agent of barely detectable intensity, and the magnitude of the effect is

---

50. Philip Handler, "Public Doubts About Science," Editorial, *Science* 208 (June 6, 1980), 1093.

substantially independent of the intensity of the cause; 2) the effect is of a magnitude that remains close to the limit of detectability; or, many measurements are necessary because of the very low statistical significance of the results; 3) claims of great accuracy; 4) fantastic theories contrary to experience; 5) criticisms are met by *ad hoc* excuses thought up on the spur of the moment; and 6) the ratio of supporters to critics rises up to somewhere near 50 per cent and then falls gradually to oblivion.[51]

Langmuir raised a difficult issue: how do scientists determine the boundary between a finding that lies outside the mainstream (Kuhn's concept of normal science) and important new findings? Is the finding a revolutionary new insight or an example of pathological or junk science? Popper's answer would have been that a pathological finding would not pass the test of falsifiability. More generally, the inability of others to verify or replicate the results is a sure sign of trouble. Langmuir's answer was more complex, identifying a series of salient characteristics.

Since Langmuir's time, there has been no shortage of candidates who have contributed to the growth of pathological science. Dozens of these claims live on, either because the original scientists stick with what they have discovered or, more ominously, because the bad or junk science gains a popular following. And sometimes, what may at first seem to be pathological turns out to be a revolutionary new insight. Warren and Marshall's conjecture that ulcers were caused by bacteria rather than stress was ridiculed by the biomedical community until the two Australians demonstrated that they were right and that others were able to replicate their findings. Their hypothesis obviously was not a matter of pathological science. At the same time, there are dozens of claims every year that quickly gain a popular following but that cannot be replicated and that subsequently demonstrate all the pathologies identified by Langmuir. One famous example is the claim made in 1989 by electrochemists Stanley Pons and Martin Fleischmann that they had successfully generated a nuclear reaction at low temperatures known as cold fusion. Their claim attracted a tremendous amount of interest, and there were efforts to replicate their breakthrough. Cold fusion remains one of the most sought-after unproven hypotheses in science, but the Pons-Fleischmann claim turned out to be a mirage. A small

---

51. Irving Langmuir, "Colloquium at The Knolls Research Laboratory," December 18, 1953, transcribed and edited by R. N. Hall.

band of enthusiasts, however, meets every year to discuss the prospects, even though few continue to take them seriously.[52]

The Pons-Fleischmann episode was soon resolved in favour of sound science, but many other instances have a longer shelf life than they should. The late Columbia University chemist Nicholas Turro sums up the perverse incentive structures that have led to a society bombarded with junk science:

> Extra-scientific considerations such as media attention, professional standing, promises of monetary gain, ideological predilections, *hubris Nobelicus*, and pressures from interested parties outside the scientific community all can contribute to self-delusion. The exigencies of funding tempt even the most scrupulous basic researcher to overstate practical benefits when describing new work to potential supporters. Today's academic environment – which can appear more like a media fishbowl than an ivory tower – also presents the scientist with ample channels to speak to the general public, with a considerable risk of misrepresenting the content, purpose, and potential of a scientific discovery, either in an effort to simplify professional jargon (the ever-present problem of 'dumbing it down') or in the highly contagious enthusiasm over an untested idea.[53]

People now make a living either promoting or debunking junk science. Steve Milloy runs a popular weblog, *Junkscience.com*, devoted to exposing many of the claims of the devotees of questionable but popular science, particularly claims related to health, environmental, and safety issues, the focus of many activist campaigns.[54] As his and similar websites make clear, once a scientific claim is in the public domain and attracts a following, it will continue to be touted as a matter of science rather than opinion. Milloy defines junk science as "faulty scientific data and analysis used to advance special and, often, hidden agendas." He finds its practice to be ubiquitous, involving a wide array of societal actors. For example:

- The media may use junk science for sensational headlines and programming;
- Personal injury lawyers may use junk science to convince juries to award huge verdicts;

---

52. The Pons-Fleischmann saga is discussed in Robert Park, *Voodoo Science: The Road from Foolishness to Fraud* (New York: Oxford University Press, 2000), 10-27.

53. Nicholas J. Turro, "Toward a general theory of pathological science," at columbia.edu/cu/21stC/issue-3.4/turro.html.

54. See also Steve Milloy, *Green Hell: How Environmentalists Plan to Control Your Life and What You Can Do to Stop Them* (Washington, DC: Regnery, 2009).

- Social and environmental activists may use junk science to achieve social and political change;
- Government regulators may use junk science to expand their authority and increase their budgets;
- Businesses may use junk science to bad-mouth competitors' products or to make bogus claims about their own;
- Politicians may use junk science to curry favor with special interest groups or to be politically correct;
- Individual scientists may use junk science to achieve fame and fortune; and
- Individuals who are ill (real or imagined) may use junk science to blame others for causing their illness.[55]

The Hoover Institution's resident scientific sceptic, Henry Miller, notes that today we see many "examples of radical activists exploiting widespread ignorance of science, pushing a kind of New Age, anti-technology, anti-business, 'return-to-unspoiled-nature' ideology. Purveyors of superstition and darkness now spur concerns about many products and technologies, including vaccines, nuclear power, pesticides, genetically engineered foods, and chemicals found in an array of consumer products."[56] Actress Jenny McCarthy and her army of supporters, for example, continue their campaign to convince parents that the MMR (measles, mumps, rubella) vaccine causes autism. Originally, this claim had a basis in science, but the article in the *Lancet* by British researcher Andrew Wakefield has long since been debunked and the author disciplined by the British Medical Association.[57] Suzanne Somers is another actress with a large following convinced by her junk science crusades opposing hormone replacement therapy and fluoridation, and supporting alternatives to chemotherapy. Each of these is based on promising claims by one scientist or another that have since been thoroughly discredited. The internet abounds with similar claims by determined activist groups that may once have had a basis in a peer-reviewed published paper but have since been discredited. British doctors Ben Goldacre and James Le Fanu have carved out reward-

---

55. Milloy, "Junk Science?" at junksciencearchive.com/define.html.

56. Henry Miller, "Junk Science Attacks On Important Products And Technologies Diminish Us All," *Forbes*, October 23, 2013.

57. See Ben Goldacre, *Bad Science: Quacks, Hacks, and Big Pharma Flacks* (Toronto: McClelland and Stewart, 2010), 208-52 for a full discussion of this medical hoax and its tragic consequences.

ing careers debunking health and nutrition-related junk science.[58] Milloy and Miller pursue a wider array of questionable claims.

One of the more common pathways for the growth of junk science is the media's anxiety of the week. As Miller points out, "every editor knows that a headline that elicits panic attracts more readers than one explaining that everything is just fine, so explication too often takes a back seat to sensation."[59] It starts with a sensational article indicating that scientists have linked a factor – environmental, life-style, or nutritional – to an increased risk of a dreaded disease such as cancer or Alzheimer's. The article frequently regurgitates a press release from a journal or from a scientist's public relations department touting a particular research finding. Often these are little more than epidemiological, i.e., observational, studies that are not ready for prime time. A high proportion of the studies identified by Ioannidis and other researchers as being proved wrong or misleading within a short period of time are typically epidemiological. Stanley Young, a statistician with the US National Institute of Statistical Sciences, concludes that the "empirical evidence is that 80-90 per cent of the claims made by epidemiologists are false; these claims do not replicate when retested under rigorous conditions," i.e., they are falsified by randomized, controlled placebo-based trials. Epidemiologists rationalize this low success rate by insisting that "it is better to miss nothing real than to control the number of false claims they make."[60] For example, a 1981 study indicated that drinking two or three cups of coffee per day can triple the risk of pancreatic cancer; a larger follow-up study published in 2001 concluded that there was no correlation between coffee and pancreatic cancer. A 1986 study found that drinking coffee reduces the risk of colorectal cancer, but a 2005 study found no basis for this conclusion. The epidemiological literature abounds with such discredited studies, and many of them

---

58. Goldacre writes a regular column for *The Guardian*, has written the best sellers *Bad Science: Quacks, Hacks, and Big Pharma Flacks*, and *Bad Pharma: How Drug Companies Mislead Doctors and Harm Patients*, and maintains a successful blog. Le Fanu writes a column in the *Daily Telegraph*, "Doctor's Diary," and is the author of *The Rise and Fall of Modern Medicine* and *Why Us: How Science Rediscovered the Mystery of Ourselves*, and many other books on health and medicine.

59. Henry I. Miller, "The Controversy Around BPA: Bad Reporting On Bad Science," *Forbes*, October 10, 2012.

60. S. Stanley Young, "Everything is Dangerous: A Controversy," June, 2008. Similar to the concerns expressed by statisticians such as William Briggs about the abuse of statistics by climate scientists, Young's presentation indicates the extent to which statisticians find the claims of epidemiologists to be fanciful at best.

continue to influence public attitudes and even public health policy.[61]

Epidemiology is the science of studying the prevalence of health and disease in populations and relating them to the presence of nutritional, environmental, and other agents, exposures, and risk factors. It provides "the basic science for public health, as the clinical disciplines do for medicine; epidemiology [tries] to explain, and thereby suggest ways to improve, the experience of disease and health in populations as distinct from individual patients."[62] Epidemiology relies on gathering detailed data on particular populations and then subjecting that data to analysis to determine possible relationships between risk factors and the presence – or absence – of disease.

Epidemiological studies can be pursued in a number of ways, all of which rely on finding an association or correlation between factor A and result B. Cohort studies follow a healthy group of people over time (with different intakes of, say, coffee) and look at who gets a disease. Case-control or retrospective studies examine people with and without a certain disease and compare their prior life based, for example, on how much coffee they drank, in order to determine whether people who got the disease drank more coffee in their past than those who did not. Cross-sectional studies compare a group's present lifestyle, i.e., how much coffee they drink now, with their present health status. All three kinds of studies are relatively inexpensive and present few ethical problems. They do require considerable statistical skill in analyzing the data as well as a good grasp of the underlying science in order to frame the study's questions properly. They often need to rely on people's memories and their willingness to fill in detailed questionnaires. They may also look for causal relationships where none exist. For example, B may not be caused by A, but by C and D acting on E, factors not considered by the researchers.

A positive association between A and B may point to causality and to a testable hypothesis, but only if the association is strong and

---

61. Andreas von Bubnoff, "Numbers Can Lie," *Los Angeles Times*, September 17, 2007.

62. Steve Wing, "The Limits of Epidemiology," *Medicine & Global Survival* 1: 2(1994), 74. Wing catalogues many of the advances in public health that have resulted from the work of epidemiologists. Gary Taubes, on the other hand, in a much debated article, surveys the many problems and limitations of epidemiology as currently practiced. "Epidemiology Faces Its Limits," *Science* 269 (July 1995), 164-9.

not random, if there appears to be a plausible causal mechanism, or if there is no more likely explanation on offer. In such cases, the challenge is to refine the hypothesis by considering and testing for bias, confounding factors, and other explanations. In the modern rush to publish and apply for more funding, the weakest link is often the first problem, i.e., lack of a strong correlation, but the next two are not far behind. There is no shortage of journal articles based on a weak association, no plausible underlying mechanism, and no effort to explore alternative explanations. Even when a study admits its limitations, press releases and media stories do *not* and often imply claims that are not warranted by the study. The result is a confused public presented with conflicting claims about health and nutrition or health and environmental factors. The credibility of science as a whole suffers from these simplistic, premature rushes to judgment on complicated issues, and the discredited results often join the ranks of junk science, to be touted as "scientific proof" of one spurious claim or another.

Epidemiology is the basis for much public policy in the health field, regardless of its questionable value: for example, correlations between salt intake and risk of stroke, dietary fat and risk of cardiovascular disease, and recommended diets to lose weight and improve well-being. These and other public health pronouncements have in common the fact that they are based on little more than epidemiology. Few are based on follow-up studies that try to establish plausible chemical, biological, or other mechanisms that would explain the association initially observed. Many positive studies are balanced by negative studies pointing to no correlation. In many cases, randomized controlled clinical trials have dismissed the claims of an earlier epidemiological study. Epidemiologists Roberta Ness and Richard Rothenberg admit that "only a body of evidence, fully accumulated over time, will inform causal thinking and prevention strategies. While this is happening, we must work to change the way in which the media report scientific findings and the public absorbs them."[63] That epidemiologists continue to have an audience leads to claims that attract press attention and popularisers, but more sober scientists point to the limitations of most epidemiological studies, including confounding studies. Epidemiological studies often appeal to those who are moved by "the need to act" but are dismissed by the wider community of scientists who insist that a

---

63. Roberta B. Ness and Richard Rothenberg, "Critique of Epidemiology: Changing the Terms of the Debate," Editorial, *Annals of Epidemiology* 17 (2007), 1011.

more sceptical attitude is fundamental to building a reliable body of knowledge.

Epidemiology proved its value in determining the causes of infectious diseases in the 19th and early 20th centuries. The classic study was one in 1854 by John Snow, who identified the Broad Street water pump in London, England as the source of an outbreak of cholera. Today, epidemiology can contribute importantly to public health by, for example, identifying possible side effects resulting from the use of specific therapeutic drugs. More generally, however, epidemiologists need to accept the limits that are inherent in their discipline, i.e., biases, uncertainties, and methodological weaknesses that can do no more than suggest possible associations. Epidemiologists can identify associations or correlations which can usefully be examined further with other tools, such as randomized controlled clinical trials, but they cannot establish cause and effect. In many cases, the testing required to establish more than correlation can be ruinously expensive or even unethical. Even the gold standard of clinical trials, i.e. a large, random, double blind, placebo-based trial, has its own limitations imposed by the reality of the human condition. Researchers rely, for example, on motivated volunteers rather than on a cross section of ordinary people. Typical is the so-called healthy user bias (i.e., trials involving faithful adherents to a prescribed regime who are also likely to be conscientious about other factors that lead to good health), a phenomenon that has confounded many epidemiological studies. Poverty is another confounding factor: poor people are less educated than wealthy people, smoke more and weigh more, eat what's cheap rather than what is healthy, exercise less, are likely to have higher blood pressure and other cardiovascular risk factors, and live in more stressful environments. The pressure to publish and the media's need for drama, however, rarely lead to the kind of restrained public reporting that would indicate the limitations of a study.

Many epidemiological studies constitute little more than what British blogger John Brignell calls data drags.[64] Le Fanu even suggests that closing all departments of epidemiology would lead to a singular improvement in public health.[65] The availability of large data bases from such studies as the Harvard Nurses Health Study

---

64. Brignell, trained in statistics and in the calibration of machinery, is the author of a highly critical assessment of epidemiology: *The Epidemiologist: Have They Got Scares for You* (self-published, ISBN 0-9539108-2-2, July 1, 2004).

65. James Le Fanu, *The Rise and Fall of Modern Medicine* (London: Basic Books, revised edition, 2012).

invites researchers to tease out more and more results from data that were never collected for that kind of analysis. A study of this kind, based on hundreds of questions repeated over many years, will yield hundreds of papers. The Harvard study, which has been ongoing since 1976, constitutes what the late Irish public health specialist Petr Skrabanek termed a "risk-factorology" study, i.e., a large data base analyzed to match dozens of factors to specific diseases. Positive associations are then characterized as risk factors and negative ones as protective factors. In too many cases different studies reach contradictory conclusions and report associations that are not strong enough to carry any evidentiary weight.[66]

The numbers used in public discussions of epidemiological findings add to the possibilities of misleading science. A study ostensibly indicating that a particular factor may increase *relative risk* in certain populations by 30 per cent sounds alarming to most people but becomes much less so when the numbers are explained. If, in the general population, the *absolute risk* is one in ten thousand, a thirty per cent increase is still only a 1.3 chance in ten thousand. In a large population, the numbers may add up, but the reality remains that the risk is very small, particularly when compared with the many other risks that are part of the human condition. In public health, concern about the many relative risk factors that epidemiologists identify and that lead to media alarm stories should be evaluated against the background of a population that has seen steady increases in both health and longevity. The general increase in well-being stands as a huge confounding factor for the many studies that claim increased risks from nutritional and environmental factors.

All too often, published reports focus on positive associations and ignore negative ones. Other findings are no more than artefacts of the researcher's methodology. Failure to frame an issue properly and to follow positive findings with more reliable controlled clinical studies leaves the initial results out there to feed the world of activism and junk science. Young cautions that researchers "think it is fine to ask many questions of the same data set ... the more things you check, the more likely it becomes that you'll find something that's statistically significant just by chance, luck, nothing more."[67]

---

66. Petr Skrabanek, "The poverty of epidemiology," *Perspectives in Biology and Medicine* 35:2 (Winter 1992), 182–5.

67. Quoted in Andreas von Bubnoff, "Numbers Can Lie," *Los Angeles Times,* September 17, 2007.

For some analysts, particularly those committed to a constructivist view of science, i.e., that scientific knowledge is the result of
human experience in a socially constructed setting, all scientific
knowledge has validity; from this perspective, distinguishing between sound science and junk science is quite unnecessary. What
matters is the degree to which a scientific proposition commands "a
robust consensus." Most practicing scientists, however, are keenly
aware that invoking the authority of science for findings that cannot
meet the tests of verification, replication, and falsification can cause
serious problems and should properly be dismissed as junk science.

While junk science is particularly prevalent in health and nutrition studies, it also creates confusion and more confidence than
warranted in other areas, including ecological, environmental, and
climate science. Similar to the problem with epidemiological work
on nutrition and health, most ecological, environmental, and climatological studies are based on observation and are either difficult or
expensive to replicate, verify, or falsify on the basis of experimental
science, a fact underlining the need for scientists in these fields to be
particularly careful in reaching conclusions and prescribing solutions. Unfortunately, as climate scientist Claire Parkinson observes,
"the majority of the scientists I have worked with over the years do
not exercise a level of care that would lead me to be certain of the
results they present in their research papers. There are far too many
instances where the text they write says something that is contradicted by the figures or by the numbers or is in error in other
ways."[68] Not surprisingly, after health and nutrition, the trio of ecological, environmental, and climate science take pride of place in influencing public policy and in attracting popular movements.

Official science and politics

The cognitive dissonance that fuels advocacy and junk science is but
a prelude to the self-deception required to *maintain* official science.
Scientific advice that underpins public policy may place both scientists and policy makers in a difficult position. Scientists must ensure
that policy makers have a grasp of the uncertainties and complexity
inherent in much of science, even as they try to provide the policy
maker with advice that reduces the issue to its essentials so that decisions can be made and policies, if appropriate, adopted and im-

---

68. Claire L. Parkinson, *Coming Climate Crisis? Consider the Past, Beware the Big Fix*
     (Lanham, MD: Rowman & Littlefield, 2010), 279.

plemented. The policy maker, on the other hand, who is looking for certainty and simplicity, must recognize the limitations of good scientific advice and accept that clear, unambiguous policy options may not be possible. In these circumstances, Sylvia Jasanoff concluded from her study of policy making by both the US Food and Drug Administration (FDA) and the US Environmental Protection Agency (EPA) that "repeated rounds of analysis and review may be required before an agency reaches a conclusion that is acceptable at once to science and to the lay interests concerned with regulation."[69] The problem becomes even more complicated in explaining the adopted policy to the public. As McKitrick indicates, "it is understandable that scientific input to a democratic decision-making process will at times cause tension between the desire for the best information – however complex – and the desire for informed, voluntary consent by those being governed."[70]

Policy on science-based issues is often developed in the context of public discussion dominated by committed activist lobbies with a strong bias towards the preferred policy, a media looking for drama and industry or other groups counselling caution. The issue may be technically very demanding, leading to fierce debate even among experts, but activists will succeed in painting it in simplistic terms, and the media will frame it along similar lines. Typically, there is little room for the sober and nuanced debate that would lead to a measured response. If the issue is significant enough, the resources devoted to convincing governments to pursue a particular course of action can be enormous. There is rarely opportunity for delay until the science becomes less ambiguous and the policy pathway more obvious. Whether it is a municipal decision to ban pesticide use on lawns, a provincial decision to build a nuclear power plant, or a national decision to extend protection to an alleged endangered bird, both scientists and decision-makers are under pressure to come up with a solution. In these circumstances, it is not unusual for politicians to invoke the precautionary principle and opt for the politically popular rather than the scientifically responsible decision.

Once a course of action is adopted and implemented, the tensions between scientists and policy makers can reach new heights. Science

---

69. Sheila Jasanoff, *The Fifth Branch: Science Advisers as Policymakers* (Cambridge, MA: Harvard University Press, 1990), 250.

70. Ross McKitrick, "Bringing Balance, Disclosure and Due Process into Science-Based Policy-Making," in Jene W.B. Porter, ed., *Public Science in Liberal Democracy* (Toronto: University of Toronto Press, 2007), 240.

evolves, and advice that may have made sense earlier can become obsolete or even wrong. Such a turn of events may arrive more rapidly
than desired or anticipated. The underlying science may have been
based on the kinds of studies discussed earlier that seemed promising
at the time but proved to be a mirage or premised on the work of scientists with an agenda. Still other factors may turn what seemed a
good idea into a questionable one. The wise approach, of course, is to
admit that the policy is no longer appropriate and to change course.
Unlike private businesses, however, governments are not very good
at cutting their losses and admitting the need for re-evaluation. The
result is official science: research and advice pursued in order to vin-
dicate and continue to justify public policy. Policy is no longer a consequence of science; rather, science becomes a consequence of policy
or, as Mary Douglas concluded: "When science is used to arbitrate ...
it eventually loses its independent status, and like other high priests
who mix politics with ritual, finally disqualifies itself."[71]

Ironically, it is government scientists and senior officials who often adopt this approach rather than ministers or other political-level
officials. Once a policy has been adopted, other scientists are recruited to become part of the advisory process, including sympathetic scientists outside of government, editors of influential journals, and
leaders of scientific associations. These recruits reinforce the tenets of
the officially adopted scientific position with papers, articles, and
pronouncements that assure the public that the science is settled except for the claims of a few cranks and shills. As retired Penn State
physicist Craig Bohrens laments, in this process "incompetent, dishonest, opportunistic, porch-climbing scientists will provide certainty
where none exists, thereby driving out of circulation those scientists
who can only confess to honest ignorance and uncertainty."[72] As a result, scientific advice to policy makers becomes ever more one-sided.
The problem is even more acute when the policy advisory process is
infected from the start by activist scientists and officials with their
own agendas. In these circumstances, the internal policy advisory
process becomes captive of advocates who then tilt that advice on a
single axis, distorting the science.

Politicians may sometimes be more comfortable with this kind
of advice because it allows them to use "scientists tell us" as an ex-

---

71. Mary Douglas, *Risk and Blame* (London: Routledge, 1992), 33.

72. As quoted in Ross McKitrick, "Bringing Balance, Disclosure and Due Process in-
     to Science-Based Policy-Making," 248-9.

cuse for taking an unpopular position. In official science, politicians are more comfortable with a difficult position if they can appeal to the authority of science, an authority that is undermined if there is a perception that the policy option lacks consensus. There is a vested interest in forming a consensus and in marginalizing and demonizing those who disagree. Government scientists and officials who maintain that the science does not support the chosen option are re-assigned or dismissed. Governments speak with one voice, and it is understandable that, once a policy has been adopted, officials and scientists in the government's employ should not be speaking in public and undermining the government's decision. It takes a brave official who is prepared to sacrifice a career in order to rescue a government saddled with a wrong-headed policy. The inertia to stay the course usually proves irresistible and the strength of will needed to change direction herculean. The negative repercussions of wrongly adopted science-based policy may have been a relatively minor problem for earlier generations, but in an age of activist governments prepared to step in and regulate almost any issue, many of them science-based, the impact can be significant.

Some analysts have tried to suggest that the failure of some governments to pursue the policy choices dictated by science is a partisan matter. Environmental activist and journalist Chris Mooney, for example, gained a wide audience during the second Bush administration in the United States with his polemical *The Republican War on Science,* arguing in particular that the administration's view of climate change made a mockery of the "settled science."[73] The idea was not original with Mooney, but it attracted attention because it fit with the broader view of the Bush administration among progressives and many in the media. The Bush administration was seen as wilfully ignorant, rejecting the advice of scientists and pursuing policies at odds with science, from stem-cell research to climate change. In Canada, activist Chris Turner reached similar conclusions in *The War on Science: Muzzled Scientists and Wil-*

---

73. Chris Mooney, *The Republican War on Science* (New York: Basic Books, 2005). He followed up in 2012 with *The Republican Brain: The Science of Why They Deny Science - and Reality* (New York: Wiley). Labelling Republicans as anti-science remains a mainstay of left-wing journalism in the United States. See, for example, Sean McElwee, "GOP is an anti-science party of nuts (sorry, Atlantic!)," *Salon,* November 13, 2013, written in response to Mischa Fischler, "The Republican Party Isn't Really the Anti-Science Party," *The Atlantic,* November 11, 2013. Fischler has by far the more credible argument.

ful Blindness in Stephen Harper's Canada.[74] He found that government scientists were not allowed to speak freely and that funding for government science was being reduced.

Science writers Alex Berezow and Hank Campbell provide an effective antidote with *Science Left Behind,* persuasively making the case that, for most politicians, the usefulness of funding is determined by its ability to support existing policy preferences, whether they are left, right, or center. In order to set the record straight, they furnish their own catalogue of issues demonstrating what they call the progressive war on science, from progressives' affectation for organic food and fear of genetically modified products to their attacks on vaccines and other life-saving drugs.[75] As documented by Goldacre, Le Fanu, Milloy, and Miller, junk science attracts followers of all political persuasions, although there may well be differences in the delusions favoured by one group over another. At the same time, as Roger Pielke, Jr. warns, "invoking the phrase 'junk science' means that one believes that political agendas following from that science must be ill conceived and not deserving of support. Invoking the phrase 'sound science' means that one believes that political agendas following from that science are right, just, and deserving of support. Battles take place over whether science is sound or junk instead of debating the value or practicality of specific policy alternatives."[76]

In *Science Left Behind,* Berezow and Campbell explore the theme that activists have abused and misused science to advance their agendas and that governments have been willing to regulate in order to appease them, often doing more harm than good. The US-based Center for Science in the Public Interest, for example, often promotes science that supports the interests of misinformed and misleading activists.[77] In each instance, there is a reliance on media drama and ignorance to advance their causes to the detriment of the

---

74. Chris Turner, *The War on Science: Muzzled Scientists and Wilful Blindness in Stephen Harper's Canada* (Vancouver: Greystone Books, 2013). Labelling Conservatives as anti-science is a staple of Canadian left-wing commentary.

75. Alex B. Berezow and Hank Campbell, *Science Left Behind: Feel-Good Fallacies and the Rise of the Anti-Scientific Left* (New York: PublicAffairs, 2012). Unlike Mooney, Berezow is a trained scientist, with a PhD in microbiology.

76. Roger A. Pielke, Jr., "When scientists politicize science: making sense of controversy over *The Skeptical Environmentalist*," *Environmental Science and Policy* 7 (2004), 409.

77. See Henry Miller, "The Human Cost of Anti-Science Activism," *Policy Review* 154 (2010).

public interest. An industry that is complicit in undermining the public interest in order to advance its own set of priorities is also an important part of the story. Occasionally, politicians may be moved by moral, religious, or ideological convictions to oppose or advance a position that is favoured by scientists, but the most important factor in political decision-making is electoral politics. Steve Fuller argues that "politicians don't ask scientists for advice because they want the scientists to rule on their behalf. Scientists are asked more in the spirit of a special interest group, albeit one with considerable mystique, rather like the church. Just as politicians would ideally like to have the church on their side, so too they would like to have the scientific community. However, politicians need to keep a lot of interests and prospects in balance, since in the end it is all about winning elections. And neither the clerics nor the scientists need to face the electorate. It's as simple as that."[78]

In a democracy, people expect the officials they deal with to be non-partisan servants of the government. They expect that programs are delivered without political bias, and they believe that civil servants provide non-partisan advice. Like the rest of us, civil servants have values, interests, and preferences that inform their work, but we still expect them to serve their political masters faithfully without reference to their own preferences. To the extent that this is not true, the policy and regulatory process is the poorer, serving neither policy makers nor the people who elect them and possibly deceiving both. In the final analysis, political authorities will decide on a course of action that suits their political needs, always aware that their choices are ultimately subject to validation or rejection by the voters. Citizens should, nevertheless, be able to rely on the integrity of the advisory process that led to a particular decision. As should have become clear by now, the standard that the science in question was published in a peer-reviewed journal hardly meets this test. For McKitrick, the decision-making process should be held to a minimum standard of balance, disclosure and due diligence. Instead, the process is often compromised by reliance on science that is tainted by activism, error, dishonesty, or some of the other problems discussed above. Meanwhile, the public remains unaware of the extent to which the process lacks integrity. McKitrick concludes: "if we want sound policy we have to have a mechanism for communicating honest, complex, deep science into the policymaking

---

78. Steve Fuller, "When Scientists Lose Touch … The Case of David Nutt," *War-wickblogs,* November 3, 2009.

process, without distorting or stripping down the content along the way."[79]

Essential to maintaining the integrity of scientific advice is the recognition that science can rarely deliver the certainty and consensus that politicians prefer. Judith Curry is one of the few climate scientists who has spent the years since the scandal of Climategate and the disaster of the Copenhagen Climate Conference (2009) coming to terms with what she characterizes as the "uncertainty monster." She argues that "the consensus approach being used by the IPCC has failed to produce a thorough portrayal of the complexities of the problem and the associated uncertainties in our understanding. ... Improved understanding and characterization of uncertainty [are] critical information for the development of robust policy options. When working with policy makers and communicators, it is essential not to fall into the trap of acceding to inappropriate demands for certainty; the intrinsic limitations of the knowledge base need to be properly assessed and presented to decision makers."[80] Sarewitz makes essentially the same point when he concludes that "the very idea that science best expresses its authority through consensus statements is at odds with a vibrant scientific enterprise. Consensus is for textbooks; real science depends for its progress on continual challenges to the current state of always-imperfect knowledge. Science would provide better value to politics if it articulated the broadest set of plausible interpretations, options and perspectives, imagined by the best experts, rather than forcing convergence to an allegedly unified voice."[81]

The fact that some scientists insist that there is a consensus on an issue often means that there is *not* and that scientists who insist that there *is* have a political agenda. Certainly, no claims need to be made in areas for which there genuinely *is* consensus. Curry concludes that "scientists do not need to be consensual to be authoritative. Authority rests in the credibility of the arguments, which must

79.   Ross McKitrick, "Bringing Balance, Disclosure and Due Process into Science-Based Policy-Making," 261. See also B.D. McCullough and Ross McKitrick, *Check the Numbers: The Case for Due Diligence in Policy Formation* (Vancouver: Fraser Institute, 2009), which includes a number of case studies of compromised work and its influence on policy.

80.   Judith Curry, "Reasoning About Climate Uncertainty," *Climate Change* 108:4 (2011), 724, 730.

81.   Daniel Sarewitz, "The voice of science: let's agree to disagree," *Nature* 478, October 5, 2011.

include explicit reflection on uncertainties, ambiguities and areas of ignorance, and more openness for dissent. The role of scientists should not be to develop political will to act by hiding or simplifying the uncertainties, explicitly or implicitly, behind a negotiated consensus."[82]

Some governments, cognizant of the increasing complexity of the scientific issues that go into policy making, have determined that they need a "chief" science adviser at the centre of government to advise the president or prime minister and other cabinet officials about science issues writ large. Canada experimented with such a position for a few years but abandoned it after the first incumbent, Arthur Carty, retired (chief science adviser 2004-08). The British government has had a chief science adviser since 1964, now complemented by a network of departmental chief science advisers. The Australian and New Zealand governments also have chief science advisers. US presidents have relied on a White House Office of Science and Technology Policy, with its head popularly known as the president's chief science adviser.

While in many ways an admirable idea, a look at the record suggests that the chief value of the position is to provide a veneer of scientific respectability to a government's policy making. These positions are often filled with political appointees and are chosen because their political preferences align with those of the government. The current chief White House science adviser, John Holdren, may have had a career in science, but he has come to public attention largely for his extreme views on environmental issues, often jointly with Paul Ehrlich; Holdren was trained in aeronautics and plasma physics. As a long-time activist, Holdren is in a poor position to provide dispassionate expert advice. His predecessor, John Marburger, was a Democrat in a Republican administration, a fact for which he gained little credit as he defended the Bush administration's approach to science issues. David King, who served as chief science adviser to both the Blair and Brown governments in the UK, used his role as chief advocate on climate science to the point that few took him seriously on other matters of science. While trained in physical chemistry, King came to public attention as an academic trade unionist and behaved as one. The idea of appointing a scientist to the role of chief science adviser seems to make sense, but experience suggests that the most effective science advisers have been

---

82. Curry, "Consensus distorts the climate picture," *The Australian*, September 21, 2013.

those with considerable experience as both scientists and policy advisers. The range of modern science is such that someone with a highly specialized background in particle physics or microbiology will prove no more knowledgeable on atmospheric physics or nuclear medicine than an experienced policy adviser with a broad science background.

Governments have also long relied on advisory committees to help them address difficult scientific issues and have learned that the advice of an expert committee can generally pave the way for broader political support, but not always. Typically, if the committee comes up with advice that falls within the government's political preferences, the government can then use the advice to justify its position. If the advice is not congenial, governments will often delay, indicating that the commission's report was helpful but that the issue needs more study. In order to tilt the balance in favour of the first outcome rather than the second, members of the commission and supporting staff are carefully chosen to ensure that the chair and most members are already disposed to support the government's position. The idea that such a commission will offer dispassionate, politically neutral expert advice is an important fiction to maintain, but it is still a fiction. There have been commissions that have presented governments with very valuable advice and ideas, but there have been many more whose reports are gathering dust in library basements.

Sheila Jasanoff learned that the most valuable advice comes from committees whose members represent more than the narrow expertise demanded by the issues being considered and who also have the experience and wisdom to accept that their role is one of *policy* advice rather than more limited *technical* advice. Committees premised on the fiction that it is possible to separate the science from the policy generate more conflict than those that adopt a broader perspective. Advice from a committee that includes scientists from a variety of disciplines as well as non-scientists will rely on informal negotiation among committee members to arrive at a reasonable consensus that respects the range of opinion among the scientists. One-sided reports emanating from a narrowly based expert committee rarely command the same respect as those of more broadly based ones. Jasanoff notes that the frequent durability of broadly-based reports may be a puzzle, "for they are founded neither on testable, objective truths about nature, nor on the kind of broadly participatory politics envisaged by liberal democratic theory."[83] Rather, they reflect the experience and wisdom of a well-chosen and briefed committee. These committees, in Jasanoff's opinion, can serve "many of the same functions as judicial review. Indeed, the questions posed to advisory committees by agencies closely parallel those that litigants have traditionally posed to the courts. Is the analysis balanced? Does it take account of the relevant data? Do the conclusions follow rationally from the evidence? Is the analysis presented clearly, coherently, and in a manner that is understandable to non-specialists?"[84]

Whether governments rely on independent commissions, prestigious science advisers, or adopt some other approach, the bottom line is that science on its own is not capable of being the arbiter of complex public policy issues that can neither be resolved on the basis of technocratic advice nor on the basis of broad public participation. Ultimately the issue must be decided on the basis of political values. What politicians *do* deserve, however, is the assurance that the science has been developed on the basis of an open process and is not the result of one-sided or activist science.

Often forgotten in the discussion of controversial science-based policies is that decisions are not a matter of science alone. Governments must also weigh the costs and benefits and the potential broader impact of a policy. A policy that may make a lot of sense to a committee of scientists may, once economists have examined the probable costs, make a lot less sense. The direct costs to government of many regulatory decisions are often relatively modest; the costs to the economy, on the other hand, can be enormous. Clean air, long a goal of most governments in advanced economies, was achieved on the basis of policies whose costs were borne by industry and ultimately by consumers. More can probably be done, for example, to remove even finer particulate matter from the atmosphere but at costs that could prove prohibitive. Legislation to protect endangered species is embedded in the laws of most developed economies. A decision to list an individual bird, lizard, fish, or plant, however, can prove ruinously expensive, often because policy advisers do not take sufficient account of the indirect costs. Good policy advising is more than a matter of science; it is also a matter of considering the wider ramifications of an issue. Scientists, no matter

---

83. Jasanoff, *The Fifth Branch*, 234.
84. Jasanoff, *The Fifth Branch*, 241.

how expert they are in their fields, are rarely the right people to perform this part of the advisory process.

Conclusion

What's a policy analyst to do? Wildavsky's advice in the early 1990s
remains valid today: without an adequate understanding of the underlying scientific issues, a policy analyst is unlikely to be able to
separate valid science from activist or junk science. The analyst's
task is to provide the "bridge from the empirical and analytic to the
prescriptive."[85] It is, therefore, critical to learn enough about the
subject to make an informed assessment and to avoid falling into
the traps set by post-normal, virtual science, and its fellow travelers
– fraud, junk science, and advocacy science. Political conflict re-
volves around competing values and priorities, different perceptions of justice, and rival views on the allocation of resources. Adding scientific disputes to the mix rarely makes an issue easier to resolve. In such circumstances, the scientific issue often stands in as a
proxy for political differences. As Sarewitz points out, "Arguing
about science is a relatively risk-free business; in fact, one can simply mobilize the appropriate expert to do the talking, and hide behind the assertion of objectivity. But talking openly about values is
much more dangerous, because it reveals what is truly at stake: ...
the future economic path of the post-industrial world, population
growth and distribution, patterns of land use, the distribution of
wealth and resources among nations, and the vulnerability of poor
nations to natural and anthropogenic hazards."[86]

Between the scientist and the politician stand the expert policy
advisers and analysts. Within government, much discussion takes
place among competing bureaucratic interests, each acting as proxies for external interests. Officials from departments of agriculture
are prone to represent farm interests, those from departments of industry will argue on behalf of industrial interests, departments of
the environment are aligned with environmental interests, and departments of foreign affairs tend to see the merits of foreign interests. There is nothing sinister in this, just as there is nothing sinister
in competing interests vigorously arguing for their positions before
legislative committees. Within government, policy advice to politi-

---

85. Jasanoff, *The Fifth Branch,* 241, 242.

86. Daniel Sarewitz, "Science and environmental policy: an excess of objectivity," in
     R. Frodeman, ed., *Earth Matters: The Earth Sciences, Philosophy, and the Claims of
     Community* (New York: Prentice Hall, 2000), 91.

cal decision-makers emerges gradually from this brokered market. Politicians rely extensively on this advice but look as well to their own sources of information, including both partisan and non-partisan advisers.

On technically difficult issues in particular, politicians need expert advice and count on their officials to be sufficiently conversant with the nuances of the technical details to provide them with that advice. When officials are prone to offer advocacy rather than non-partisan advice, ministers are not well-served. In extreme cases, ministers and their staffs will pick up the difference. Real problems emerge, however, when officials pretend that they are more comfortable with the technical nuances of an issue than they are. Those working with science-based files are particularly prone to this conceit. Neither ministers nor their senior officials are eager to display the extent of their scientific illiteracy and will thus pretend that they understand more than they do. The result is often a course of action that satisfies no one and leaves the government exposed to charges of incompetence.

Democratic politics is of necessity a messy process, one that often tends to frustrate both scientists and activists, all of them believing that they have offered a solution to a problem and not understanding why it has not been implemented. At the same time, politicians can become frustrated with scientists who cannot supply them with clear, unambiguous facts that line up with their political preferences. Astute policy advisers can help to broker this gap if they are sufficiently acquainted with both the demands of policy making and the subtleties and limits of scientific insights. At a minimum, they need to be able to discern when a scientist, another official, or even a politician is advocating political preferences by calling them "scientific," attempting thereby, as Michael Gough warns, "to place them outside the realm of political discussion, debate, and compromise. But this is an illusion. All policy matters involving human health and the environment are political. The more that political considerations dominate scientific considerations, the greater the potential for policy driven by ideology and less based on strong scientific underpinnings."[87]

The science of climate change provides a case study of policy advice gone off the rails due to the lack of attention of policy advisers overwhelmed by the perceived complexity of the issues and

---

87. Michael Gough, "Science, Risks, and Politics," in Gough, ed., *Politicising Science: The Alchemy of Policymaking* (Stanford, CA: Hoover Institution Press, 2003), 3.

unwilling to make the investment in becoming sufficiently conversant with the basic science to put the demands of activists into a more balanced perspective. David Henderson tells of his own failure, as head of the Economics Section of the OECD, to inform senior finance officials from member governments of the need to pay greater attention to the ongoing work by officials in other departments in preparing for the 1992 Rio Summit. He recalls: "I should have told them that the so-called Rio Earth Summit was an important and worrying event; that it and the developments it gave rise to could well bring serious economic consequences; and that they ought accordingly to inform themselves, to monitor developments, and to take a continuing interest in the issues and processes that were involved. Such a warning would have been justified by events. The Rio Earth Summit of June 1992 marked a major victory for what I call 'global salvationism'."[88] For many of these officials, Rio represented an issue that would engage foreign, environment, and development officials but would be of minor significance in the greater scheme of things that preoccupy Treasury officials. Had they been better versed in science and the economic consequences of the science that lay behind the Rio Summit, they would have recognized the extent to which the wool was being pulled over ministerial eyes by activists in and out of government.

Many of the institutional and ideological problems discussed above are an integral part of climate science: insistence on consensus and the sanctity of peer review, the drive to gain and maintain funding, evidence of many of the characteristics identified by Irving Langmuir as typical of pathological science, i.e., dependence on virtual science based on incomplete, questionable data, and heroic assumptions, reliance on advocacy science and observational studies, claims of post-normal science, and resort to the claims of official science.[89] Fifty years ago sociologist of science Harriet Zuckerman observed that the scientific enterprise relies on two critical norms: rigorous adherence to a methodology that values experimentation,

---

88. David Henderson, "Economics, Climate Change Issues, and Global Salvationism," *Economic Education Bulletin,* XLV:6 (June 2005), 1.

89. In a similar vein, Nina Teicholz provides a thorough dissection of the similar pathway followed in the dietary fat-high cholesterol-heart disease hypothesis that dominated the science and treatment of heart disease from the 1950s until recently. Evidence-based science has now fully discredited every link in this hypothesis, and official science's commitment to the hypothesis is only now eroding. See *The Big Fat Surprise: Why Butter, Meat, and Cheese Belong in a Healthy Diet* (New York: Simon and Schuster, 2015).

close observation, verification, quantification, replicability, and falsifiability, while at the same time maintaining a firm commitment to absolute honesty and scepticism in reporting the results.[90] Many climate scientists, however, seem to have adopted a rather cavalier attitude to the demands of scientific methodology while vilifying scepticism and insisting on full acceptance of a consensus view of the issues. The result is a discipline easily exploited by those with more sinister motives.

❧❧❧❧❧

---

90. See the discussion in Harriet Zuckerman, "Deviant behaviour and social control in science," in Edward Sagarin, ed., *Deviance and Social Change* (Beverley Hills, CA: Sage,1977).

4 | The Science of Anthropogenic Climate Change

One cannot rigorously rule out significant global warming due to increasing greenhouse gases. Indeed, it is logically impossible to prove anything to be absolutely impossible. It nonetheless seems peculiar to base policy on something for which there appears to be no evidence.[1]

MIT Atmospheric Physicist Richard Lindzen

Climate science is a relatively new discipline, combining insights from many well-established branches of human enquiry and knowledge. It has to date made significant advances in the scientific understanding of many climatic phenomena, but it has yet to develop a theory of climate. Hubert Lamb, founder of the Climatic Research Unit (CRU) at what would become the University of East Anglia, remained deeply sceptical of the more extravagant claims advanced by some climate scientists and advised caution in the pursuit of climate treaties and policies. Another pioneer, Reid Bryson, advised that "we can say that the question of anthropogenic [i.e., human] modification of the climate is an important question – too important to ignore. However, it has now become a media free-for-all and a political issue more than a scientific problem."[2]

---

1. Richard Lindzen, "Written evidence submitted by Professor Richard Lindzen," UK House of Commons Energy and Climate Change Committee, December, 2013.

2. Bryson, "Global Warming? Some common sense thoughts," 2004. For Lamb, see the preface to the second edition of his *Climate, History and the Modern World* (London: Routledge, 1994). For a full discussion of the scepticism of both Lamb and Bryson, see Bernie Lewin, *Hubert Lamb and the Transformation of Climate Science* (London: Global Warming Policy Foundation, 2015). Lewin provides a

Despite their calls for caution, climate science has become a matter of controversy largely due to claims by some practitioners that the human impact on the climate is malign but can be rectified. In response, policy makers and the engaged public interested in understanding the public policy implications of climate change will need to gain a reasonable grasp of the claims advanced by these scientists and their activist supporters as well as the objections to those claims made by other scientists. Those in the alarmist community are upholders of the official science of climate change; they represent a much more monolithic group than the diverse range of scientists who make up the sceptical community. The alarmists are closely allied with governments (either directly or through funding), with the United Nations Intergovernmental Panel on Climate Change (IPCC), and with environmental non-governmental organizations (ENGOs), while the critics are largely independent and scattered throughout research groups, think tanks, and universities. They include many retired scientists and other specialists who have found the controversy interesting and who want to lend their expertise and experience to the discussion.[3] To the extent that consensus has any role to play in a scientific controversy, there appears to be a measure of consensus within the alarmist community but not among its critics. The latter encompass a wide range of views, from those critical of one or two aspects of the official science to those prepared to deny any scientific basis for the so-called greenhouse gas warming hypothesis.

The emergence of an international climate science community

Scientific interest in what drives long-term climate patterns goes back some two centuries, and awareness that the climate changes began much earlier. None of this enquiry, however, excited much

---

thoroughly documented account of Lamb's role in the evolution of climate science and his view that historical climatology did not support the alarmism that grew out of the work of modellers focused almost exclusively on the role of anthropogenic factors, particularly that of GHGs.

3. The oft-repeated canard that critics of the AGW hypothesis are funded by fossil-fuel interests is without foundation. Many are retired and not funded at all. Others receive research funding from a variety of sources, including governments. A few foundations, NGOs, and think tanks have received funding from corporate sources but in lesser amounts than corporate contributions to ENGOs. Alarmists, on the other hand, have received billions in funding from governments, foundations, and corporate interests. The issue of funding is discussed further in chapter ten.

public concern, let alone calls for government measures to mitigate climatic changes.[4] Of long standing, however, is media alarm about changing weather patterns, whatever the direction of change. In 1922, for example, the Washington *Post* carried an AP story about unprecedented warming in the Arctic: "The Arctic ocean is warming up, icebergs are growing scarcer and in some places the seals are finding the water too hot ... Reports from fishermen, seal hunters and explorers all point to a radical change in climate conditions and hitherto unheard-of temperatures in the Arctic zone. .... Within a few years it is predicted that due to the ice melt the sea will rise and make most coastal cities uninhabitable."[5] The following year, *Time Magazine* worried that "the discoveries of changes in the sun's heat and southward advance of glaciers in recent years have given rise to the conjectures of the possible advent of a new ice age."[6] Ten years later, however, *The New York Times* reported: "America in Longest Warm Spell Since 1776; Temperature Line Records a 25-year Rise."[7] Neither science nor the media seemed to have had a good grip on the direction of climate change, but there was enough interest for the media to express alarm, something that continues to this day.

Until the 1940s only a few scientists devoted much attention to the issue, but the demands of a global war spurred increased interest in meteorology, an interest that continued after the war had ended. As a result, the discipline of meteorology steadily advanced as scientists gained a better understanding of the composition and dynamics of the atmosphere and of earth systems more generally. For most scientists in the field, this was a period of building basic knowledge in efforts to develop a theoretical understanding of climate dynamics. Weather forecasting, which depends on understanding the evolution of short-term atmospheric patterns, steadily improved. Climatology, the understanding of longer term patterns, remained rather crude. Canadian climatologist Tim Ball explains

---

4. See James Rodger Fleming, *Historical Perspectives on Climate Change* (Oxford: Oxford University Press, 1998); Spencer Weart, *The Discovery of Global Warming*, revised and expanded edition (Cambridge, MA: Harvard University Press, 2008); and Lewin, *Hubert Lamb and the Transformation of Climate Science*.

5. *Washington Post*, November 2, 1922, as quoted in Kirk Myers, "Arctic Ocean warming, icebergs growing scarce, *Washington Post* reports," *Seminole County Environmental Examiner*, March 2, 2010.

6. *Time* Magazine, Sept. 10, 1923, as quoted in Myers, "Arctic Ocean Warming."

7. The New York *Times*, March 27, 1933, as quoted in Myers, "Arctic Ocean Warming."

The Science of Anthropogenic Climate Change   119

that "climate science is the work of specialists working on one small part of climatology. It's a classic example of not seeing the forest for the trees, amplified when computer modellers are involved. They are specialists trying to be generalists but omit major segments, and often don't know interrelationships, interactions, and feedbacks in the general picture."[8] In these early days, dozens of ideas were advanced, some of them involving heroic assumptions and questionable speculation leading to alarming results. Most scientists in the field, however, did not believe that they had a sufficiently solid grasp of the science to warrant some of the speculative conclusions and alarm reached by a few. On the whole, this was a period of reflection and discussion, led initially by American scientists but gradually involving scientists from Canada, Europe, and elsewhere.

An important stimulus to research was the designation by many scientific societies of 1957 as the International Geophysical Year (IGY). This incentive led to greater public financial support for climate science, particularly with the establishment of a number of US programs, including the National Aeronautics and Space Administration (NASA, 1958), the National Center for Atmospheric Research (NCAR, 1960), the Goddard Institute for Space Studies (GISS, 1961), and the National Oceanic and Atmospheric Administration (NOAA, 1970). Each of these institutions provided scientists with funding as well as an institutional home to pursue much more detailed and expensive projects, particularly the gathering and organization of better data. Canada, Britain, Japan, and Europe caught up by establishing similar programs in the 1970s and 1980s. By the early 1980s, scientists had developed a global network of specialists working on climate-related issues, many of them in publicly funded institutions.[9]

Starting in the 1960s, scientists from these institutions began to meet at international gatherings, often sponsored by the United Na-

---

8. Tim Ball, "The Important Difference Between Climatology and Climate Science," *WattsUpWithThat*, November 29, 2013. The article provides a useful overview of the evolution of climate science and the problems created by overspecialization at the expense of maintaining a view of the whole.

9. See David M. Hart and David G. Victor, "Scientific Elites and the Making of US Policy for Climate Change Research, 1957-74," *Social Studies of Science*, 23:4 (November 1993), 643-80. Weart provides a rather Whiggish account of this period in *The Discovery of Global Warming*, identifying many scientists who contributed to the development of the science that became the heart of the IPCC perspective. Lewin, *Hubert Lamb and the Transformation of Climate Science*, provides an effective antidote to Weart's account.

tions or one of its agencies, in order to compare notes. Some of them worried that their findings were pointing to significant problems with human impacts on the composition of the atmosphere that could lead to major, malign changes in future climate patterns. By the 1970s, one concern was that rapid postwar industrialization had polluted the atmosphere with soot, aerosols, and noxious gases that blocked the amount of solar radiation reaching the Earth's surface, leading to cooling and possibly a new ice age. Advances in developing integrated historical temperature series on global climate confirmed this possibility. The warming that had been evident in the 1920s and 1930s had given way to cooling from the 1940s into the 1970s.[10] Journal articles expressing concern became more frequent, and the media picked up on that concern.

Hubert Lamb had devoted his career to studying the extent to which climate changes, often in erratic and chaotic ways, and how these changes could have profound effects on human civilization. His successors in leading the CRU took a different approach. Trevor Davies, Tom Wigley, and Phil Jones used the resources of the CRU to develop the first comprehensive series of global temperatures, collaborating with governments, researchers, and meteorologists around the world to get the raw data and then massaging the numbers into time series that illustrated the evolution of global temperatures dating back to the middle of the 19th century. Working with incomplete and often questionable raw data, they developed global temperature series calibrated to a tenth of a degree Centigrade. It was in many ways a singular accomplishment but, as many critics have since pointed out, it projected a level of certainty and knowledge that was not warranted by the raw data from which the series had been built.[11] Wigley, Jones, and their colleagues also pioneered early modelling, focused on the role of anthropogenic influences, and would become central players in the work of the IPCC.

---

10.   Many of the most enduring maximum temperature records were set in the 1920s and 1930s. US temperatures in the 1930s remain the highest in the US temperature record. The harsh winter of 1939-40 brought this period of warming to an end, with more brutal winters to follow.

11.   Statistician William Briggs has been among the most relentless critics of the claims that these time series represent anything other than a model of global climate patterns, given the extent to which the very poor raw data are aggregated, manipulated, and averaged. The idea that this can be done to a tenth of a degree is, in his view, ridiculous. See his many posts on "Time series & Data Handling," at *WmBriggs.com*.

James Hansen and his team at the GISS, in association with the Earth Institute at Columbia University, embarked on a similar venture and achieved similar results. Like Lamb, Robert Jastrow, the founder of GISS, became highly sceptical of the claims of his successor.[12] Hansen and his team were also engaged in developing computer climate models. Both temperature series remain central to the work of climate scientists and are subject to constant revision based on new algorithms, assumptions, and techniques, making a mockery of the claim that these series can be characterized as data. Both of these global temperature series show that the globe has indeed warmed about 0.7°C since 1880 in a series of steps, including a modest decline from about 1880 to 1910, a steady rise from 1910 to 1945, a modest decline from 1945 to 1977 and then a steady rise again to 1997, followed by a period of stasis to the present. Researchers at both institutions were interested in determining what caused these changes, starting from an assumption opposite that of Lamb: the climate system was largely in equilibrium unless disturbed by an external forcing agent. Like many of their colleagues in the field, they suspected that human activity was responsible for these changes, particularly the postwar cooling and more recent warming. Both research centers became central to the alarmist movement.

Computers were increasingly being used to develop temperature series, and early, but crude, models were being developed to test, for example, the impact of anthropogenic influences such as industrial aerosols and greenhouse gases (GHGs) on future climate patterns. It seemed at times that a theory of climate was developing on the basis of what could be modelled rather than on the basis of what had been observed. For some scientists – particularly the generalists with an overview of climatology in all its complex detail – the answers seemed too simple and suggested that climate scientists were far from understanding cyclical factors, such as variations in the intensity of solar radiation or the role of coupled atmospheric-oceanic cycles as well as those of other terrestrial forces such as clouds and precipitation.[13]

---

12. See Lewin, *Hubert Lamb and the Transformation of Climate Science*, 35.

13. See, for example, the discussion in the original edition of Lamb, *Climate, History, and the Modern World*, chapter 16: "The Causes of Climate's Fluctuations and Changes," which provides a much more complete picture than is found in the monotonic views of many alarmist climate scientists. A few years later, Richard Lindzen, who by this time had established himself as one of the most knowledgeable and published atmospheric physicists, began to express his own

Two separate lines of scientific enquiry were being pursued with their own research programs but with insufficient interaction between them. Atmospheric physicists were primarily interested in the changing composition of the atmosphere and its impact on the radiative balance of the climate system. Oceanographers, on the other hand, were examining the dynamics of ocean currents, their ability to absorb or release heat, and the impact of these forces on climate change. As Hart and Victor point out:

> If the oceans have a high thermal diffusivity then the entire hydrosphere must be heated before atmospheric temperature rises; the oceans would have a large 'thermal inertia' effect. On the other hand, if diffusivity is low, atmospheric temperature will respond more rapidly to increased concentrations of greenhouse gases. But oceanic thermal diffusivity was the territory of oceanographers, and [atmospheric modellers] looked only at equilibrium atmospheric temperature levels, rather than at the specific process and timing by which those levels would be reached. By design, equilibrium calculations ignored the rate of change.[14]

Well into the 1990s these two groups of scientists stayed on separate paths. Once they realized the importance of each others' work, many assumptions had already become deeply embedded, making it difficult to backtrack and to realize that the role of coupled atmospheric-oceanic systems was more important than the modellers had realized, rendering their dire predictions unrealistic.

Two dominant strains of thought had initially developed among the modellers. One group believed that the main concern should be that the globe was cooling, perhaps presaging the end of the Holocene and the beginning of a new ice age sometime in the not too distant future. Others thought that the cooling effect of aerosols would be overwhelmed by the increasing concentration of atmospheric greenhouse gases, leading to global warming. Kenneth Watt, an ecologist at UC Davis in California and a key leader of the early environmental movement, declared on Earth Day 1970, "the world has been chilling sharply for about twenty years. If present

---

doubts about the alarmism of some of his colleagues. In 1990 he wrote: "one may reasonably ask how the issue of global warming has generated such dramatic concern. At least part of the answer lies in the fact that the Greenhouse hypothesis fits conveniently into the agenda of many groups who see that fear of this illusive phenomenon may help generate support for a wide range of activities." "A sceptic speaks out," *EPA Journal* 16 (1990), 47. See also "Some Coolness Concerning Global Warming," *Bulletin of the American Meteorological Society*, 71:3 (March 1990).

14.   Hart and Victor, "Scientific Elites and the Making of US Policy," 655.

trends continue, the world will be about four degrees colder for the global mean temperature in 1990, but eleven degrees colder in the year 2000. This is about twice what it would take to put us into an ice age."[15] A number of popular books captured the alarm inherent in the competing perspectives: Nigel Calder, *The Weather Machine and the Threat of Ice* (1974); Howard A. Wilcox, *Hothouse Earth* (1975); and Lowell Ponte, *The Cooling* (1976). None of these books had the impact of similar alarmist literature a few decades later, but enough copies were sold to make them widely available in the used book market. *Time* sought to capture the issue in "Another Ice Age?"[16] Ponte also made it clear, perhaps unwittingly, that climate change could be an all-purpose bugaboo when he wrote, "It will be more hot, more cold, more wet and more dry, just as it was in the seventeenth century," the depth of the Little Ice Age.[17]

A more serious effort at synthesizing the emerging science and providing an overview of the dire consequences that could flow from mankind's interference with the biosphere was provided by Stephen Schneider, *The Genesis Strategy: Climate and Global Survival*, which concluded: "I have cited many examples of recent climatic variability and repeated the warnings of several well-known climatologists that a cooling effect has set in – perhaps one akin to the Little Ice Age – and that climatic variability, which is the bane of reliable food production, can be expected to increase along with the cooling."[18] Schneider, trained as an engineer and atmospheric physicist, had by this time become fascinated with computer modelling of the climate and had established himself as a major player in the alarmist movement and one of its leading pessimists in the mould of one of his close friends and Stanford colleagues, Paul Ehrlich.

By the early 1980s, however, the slight cooling evident from the 1940s into the 1970s had again given way to warming. The dominant view then became that aerosol-induced global cooling was in-

15. Quoted by Ronald Bailey, "Earth Day, Then and Now," *Reason*, May 1, 2000. Watt was speaking at Swarthmore College, April 19, 1970.

16. "Another Ice Age?" *Time*, June 26, 1974, 86.

17. Lowell Ponte, *The Cooling: Has the next ice age already begun? Can we survive it?* (Englewood Cliffs, NJ: Prentice-Hall, 1976), 40. This was a seriously conceived book, with a foreword by Senator Claiborne Pell, a preface by Reid Bryson, an endorsement by Stephen Schneider, and a 28-page bibliography pointing to the scholarly literature.

18. Stephen Schneider with Lynne Mesirow, *The Genesis Strategy: Climate and Global Survival* (New York: Plenum Press, 1976), 90.

deed a problem but was being overwhelmed by the even more malign effect of rising atmospheric GHGs, also the product of industrialization. GHGs were considered a potentially much more serious
matter than earlier aerosol pollution because of their suspected long
atmospheric residence time and their capacity for absorbing and redistributing heat, thus increasing near-surface temperatures. Scientists had by now also learned that raising the alarm about their findings was a potent way to loosen public purse strings to fund more
research, whereas more dispassionate findings did not have the
same effect.

While scientists were debating whether the planet was warming
or cooling, interested non-scientists could be excused for concluding
that global climate in particular and the earth system in general
must be extremely fragile, balanced on a knife edge between too
warm or too cold. The image projected by various alarmist books
and articles was that climate was generally stable but operated
within a very narrow band that made the planet liveable. Any deviation from that balance spelled imminent disaster requiring immediate efforts to mitigate the danger. This diagnosis came as a surprise to many earth scientists whose fundamental frame of reference
had always been that of a planet in a constant state of flux. Thomas
Moore was prompted to wonder: "given that mankind, over the last
million or so years, has evolved in climates that were both hotter
and colder than today's, how is it that we in the 20th century are so
fortunate as to have been born into the ideal global climate?"[19]

The time was indeed ripe for just such a question. As Moore
continued: "If global climate change is viewed as a threat, environmental organizations can raise more support from the public; politicians can posture as protectors of mankind; newspapers can write
more scary stories, thus increasing circulation; and scientists, even
those most sceptical, can justify research grants to study the issue.
... Apocalyptic forecasts catch people's attention; predictions of
good weather elicit no more than a yawn."[20] By the end of the 1980s
many were prepared to accept that life on Earth was fragile and in
imminent danger of collapse, greatly facilitating the sounding of a
specific alarm that could generate further public concern.

As climate scientists were busy teasing out the contours of an
hypothesis that predicted dangerous global warming, environmen-

---

19.   Moore, *Climate of Fear*, 2.
20.   Moore, *Climate of Fear*, 3-4.

talists had developed a grand narrative that provided context for more specific concerns. That grand narrative had created a widely shared public consciousness about negative human impacts on a fragile planet and had led to the creation of dedicated agencies in most OECD countries charged with considering programs and policies to protect a planet that was now viewed as vulnerable to human disruption.[21] The grand narrative had also led to the first UN conference on the environment in Stockholm in 1972 and the establishment of a UN environmental program (UNEP). Critical to this narrative was the wholly unscientific idea that nature was generally in equilibrium until disturbed by external forces, such as a meteor or super volcano. To many environmentalists and sympathetic scientists, the rapid progress of industrialization and technology had exposed a fragile biosphere to the disruptive forces of civilization, upsetting nature's equilibrium.[22]

Climate change alarmism fit in well with the emerging anxiety about man's malign impact on the environment. Following Rachel Carson's *Silent Spring*, scientists on both sides of the Atlantic had begun to focus on what they believed were the limits to the planet's ability to absorb the abuses heaped on it by mankind's growing presence. In the United States, a group of academic Jeremiahs, mostly biologists, emerged to preach doom and gloom about the pending ecological disaster: Barry Commoner (Washington University in St. Louis), Lamont Cole (Cornell), Paul Ehrlich (Stanford), Garrett Hardin (UC Santa Barbara), Eugene Odum (Georgia), and Kenneth Watt (UC Davis). All pointed to pollution, population growth, and technology as growing threats to Earth's fragile ecosystem.[23] The dangers they identified included the rise in atmospheric carbon dioxide and its probably malign effect on the climate system.

In the summer of 1970 a group of some 70 scientists came together at Williams College in Massachusetts to study critical environmental problems and to draft a joint report, *Man's Impact On The*

---

21. The UK created a Department of the Environment in 1970, the United States established the Environmental Protection Agency in the same year, and Canada followed suit with Environment Canada in 1971. The European Commission set up its own Environment Directorate in 1973.

22. For a thorough discussion of this grand narrative and its many shortcomings, see Jeffrey Foss, *Beyond Environmentalism: A Philosophy of Nature* (Hoboken, NJ: John Wiley & Sons, 2009).

23. See John McCormick, *The Global Environmental Movement* (New York: John Wiley, 1995), 83ff.

Global Environment. A similar exercise in the UK organized by a new journal, *The Ecologist,* produced its own report, *A Blueprint for Survival.* They were followed by the organization of the Club of Rome and preparations for the 1972 UN Conference on the Human Environment in Stockholm. Not one of these reports held out any hope for the future of mankind unless governments took radical steps to end the claimed assault on nature.[24]

The climate change crisis was thus not "discovered" as a problem in the late 1980s but had been carefully nurtured over a period of some twenty years with roots that went back much deeper.[25] Integral to the collective pessimism of the post-Rachel Carson environmental movement was the idea that changes in the planet's climate were part of the human degradation of the global commons. Many scientists approached the study of climate change as part of a broader quest to understand the perceived environmental crisis and provide the science to justify the radical, i.e., collectivist steps required to solve it. Schneider and other early alarmists were not attracted to climate science because they were curious about atmospheric physics; they were looking, rather, for scientific explanations of what they were convinced was an emerging ecological disaster. Theirs was a classic example of motivated reasoning. It may be true that global environmental issues, such as stratospheric ozone depletion, climate change, and loss of biodiversity came to the forefront of the international agenda in the 1980s, but deliberate efforts to bring them to prominence had been in train for more than a generation.

Climate-based alarm thus added a further compelling argument reinforcing the more general alarm: human activity in the form of industrialization was not only leading to unprecedented economic

---

24. Study of Critical Environmental Problems (SCEP), *Man's Impact On The Global Environment: Assessment and Recommendations for Action* (Cambridge, MA: MIT Press, 1970) and The Ecologist, *A Blueprint for Survival* (London: Penguin, 1972). Carroll Wilson, one of the organizers of the Williams College symposium, was also active in promoting the need for a global environmental conference and a follow-up action plan to address global environmental issues. Following the Williams symposium, Wilson brought together 35 atmospheric scientists from 15 countries in Stockholm to produce *Inadvertent Climate Modification: Report of the Study of Man's Impact on Climate.* Wilson was not a scientist but a professor of business at MIT's Sloan School of Management. See "Activist on the World Stage: Carroll Wilson Remembered," *MIT Tech Review,* February/ March 1984.

25. In modern social science jargon, it was "socially constructed." See Mary E. Pettenger, "Introduction: Power, Knowledge, and the Social Construction of Climate Change," in Pettenger, ed., *The Social Construction of Climate Change: Power, Knowledge, Norms, Discourses* (Aldershot, UK: Ashgate, 2007).

and population growth straining the planet's presumed carrying capacity but was also the external forcing agent disrupting long-term climate stability.[26] Years of media reporting about climate instability as typical of nature's fickleness now gave way to the belief that it was not nature that was fickle; rather, it was man who was destabilizing the climate. From now on, all departures from a mythical norm would be used to reinforce the new narrative: human activity was responsible for cooling and warming, storms and heat waves, floods and droughts, and all other climatic extremes. The solution was the same, whether the planet became a hothouse or entered a new ice age: only bigger governments, more central planning, further political regulation, higher taxes, and less individual liberty could save humanity from a looming climate Armageddon. Sceptical climatologist Fred Singer added that "most of these 'compulsive utopians' have a great desire to regulate – on as large a scale as possible. To them global regulation is the 'holy grail'."[27]

Raising alarm turned out to be a simple enough thing to do, as long as one relied on three important all-purpose words: if, could, and might. *If* global temperatures rose 3°C, Arctic temperatures *could* rise by as much as 10°C and *if* Arctic temperatures rise by this much, the Greenland ice sheet *might* become unstable and collapse into the sea, which *could* raise sea levels by up to 10 feet. Even if research scientists were not always at the forefront of these claims, the science advisers employed by ENGOs were less reticent, as were some politicians. Al Gore, an established serial exaggerator even as a member of the US Congress and as Vice President, has made a lucrative post-politics career out of hyping climate science alarm, as have numerous others from Bill McKibben to David Suzuki. If nothing else, climate alarmism has loosened the purse strings of governments and foundations to finance studies that might add some precision and clarity to mostly speculative claims. Over the course of the Reagan administration, the US government was spending $50 million a year on climate-related science. The four years of the first Bush administration saw funding rise from $134 million to $2.8 billion per year.[28] It has continued on that upward path ever since.

---

26. See the discussion in Rupert Darwall, *The Age of Global Warming: A History* (London: Quartet, 2013), 334–50, particularly the discussion of Schneider's view of his role as a scientist.

27. S. Fred Singer, "My adventures in the ozone layer," *National Review* (June 1989).

28. Weart, *The Discovery of Global Warming*, 143ff. and Booker, *The Real Global Warming Disaster*, 314. Weart characterizes the amount of money as pitifully small

Figure 4-1: 2,000 years of climate change

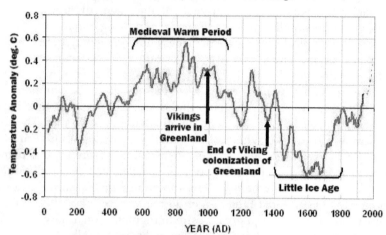

This graph shows the average of 18 non-tree ring proxies of temperature from 12 locations around the Northern Hemisphere, published by Craig Loehle in 2007, and later revised in 2008. It clearly shows that natural climate variability happens, and these proxies coincide with known climate-related events in human history.
Source: Roy Spencer, "2,000 Years of Global Climate Change," at http://www.drroy spencer.com.

The basics of climate change science

The only constant about climate is change. The science to explain that change is concerned with the most complex, coupled, non-linear, multivariate, chaotic natural system known. *Weather* varies from day to day and from season to season. *Climate* also changes over time – from seasonal to annual, from decadal to millennial. Over its four-and-a-half-billion-year history, planet Earth has experienced both long periods as a snowball and other periods as a hothouse. Over the past few million years, there have been short periods of warmth of 10,000 or so years, followed by ice ages of 100,000 plus years. The current interglacial period known as the Holocene is now some 12,000 years long and is among the cooler of recent interglacials. During the Holocene there have been centuries that were either warmer or colder than now. Warmer periods are called *optima* because for human civilization, and much of nature, warm is better than cold. We know from historical ev-

given the immensity of the task, but to governments faced with many other urgent needs, including other areas of scientific research, $50 million is not an insignificant amount and $2.8 billion buys a lot of research.

idence of the Roman and medieval optima that temperatures in Europe may have averaged 2-3 degrees centigrade (C) higher at that time than they are now; there is good evidence from paleoclimatology that the same held true in other parts of the planet. There have also been some colder periods, e.g., the Maunder (ca. 1645-1715) and the Dalton (ca.1790-1830) minima. Our current climate is the result of some two centuries of steady, but not linear, warming from the trough of the last cold phase, known as the Little Ice Age (ca. 1350-1800) (see figure 4-1).

Fossil-based fuels can be found in the Arctic basin because at one time the climate there supported carbon-rich vegetation and animal life. The Romans cultivated grapes in England and even in Scotland during the first century AD. Greenland gained its name because during the Medieval Climate Optimum, Nordic adventurers settled on its south-western coastal plain and were able to sustain colonies there for a number of centuries before they had to abandon their farms due to advancing glaciers and winter cold. The aboriginals who first populated the Americas in the waning years of the last ice age probably migrated from the Eurasian land mass over the land bridge that then joined Alaska and Siberia due to much lower sea levels; a rise of 130 metres (400 feet) in sea level since then has separated the two continents at the Bering Strait. Climate changes, *always*. The idea that at some point there has been or ever could be a stable climate around a long-term norm is a political rather than a scientific assertion.

Weather is what we experience on a daily basis; *climate* is the accumulation of that experience into broad regional patterns that can last from a few decades to centuries. Both involve changes in temperature, wind, precipitation, and other factors. *Weather* changes for a variety of interconnected reasons, from the seasonal tilt of the earth as it orbits around the sun, to fluctuating atmospheric and ocean currents, wind and cloud formation, volcanic eruptions, and more. *Climate* also changes; explanations include natural cycles, e.g., the tilt of the earth's axis and changes in the shape of the earth's orbit (the so-called Milankovitch Hypothesis – see figure 4-2), solar sunspot cycles, the interaction between solar activity and cosmic rays, cloud cover, ocean heat and current circulation cycles (e.g., Pacific Decadal Oscillation (PDO), the El Niño Southern Oscillation (ENSO), and the North Atlantic Oscillation (NAO)).[29]

---

29. For a good discussion of evolving scientific theories about the origins of ice ages and interglacials, see Claire Parkinson, *Coming Climate Crisis: Consider the Past, Beware the Big Fix* (Lanham, MD: Rowman and Littlefield, 2010), 215-24.

Figure 4-2: Milankovitch cycles

Schematic of the Earth's orbital changes that drive the ice age cycles. Changes in the distance between the sun and the earth, and the resultant variation in total solar radiation reaching the earth, are known as the Milankovitch Hypothesis. These include: a) earth's orbital eccentricity (E), i.e., changes in the earth's orbit from almost circular as it is at present to an extreme ellipse approximately 22,000 years ago; b) changes in the axial tilt of the earth from about 21.5° to 24.5° (T); and c) changes in the direction of the axis tilt at a given point of the earth's orbit due to the fact that it is not a perfect sphere (P). The critical factor in inducing an ice age is summer insolation, which determines whether winter snow and ice will survive through the summer. If it does, then the snow pack will gradually accumulate into continental-sized glaciers. 'T' denotes changes in the tilt (or obliquity) of the Earth's axis, 'E' denotes changes in the eccentricity of the orbit and 'P' denotes precession, that is, earth's wiggle as it rotates around its axis. Source: IPCC, AR4, *Technical Summary*, 56.

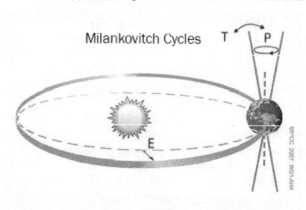

Both weather and climate result from the manner in which the Earth absorbs and distributes energy from the sun. Since the Earth is a sphere, most of the energy received is concentrated in the region between 20° north and 20° south of the equator, much of which is made up of ocean. Solar energy also reaches farther north and south, but at a steeper angle and thus with less intensity. At all latitudes, the Earth radiates heat back to space, cooling the planet, but the tropical zone does not radiate back to space all the energy that it receives. If it did, the extra-tropical regions would be much, much colder than they are. As illustrated in figures 4-3 and 4-4, the planet, through atmospheric and oceanic circulation, maintains a more comfortable temperature by constantly moving heat from the tropics to the poles. Atmospheric circulation controls the shorter-term variations in this pole-ward transport. The oceans can carry much more heat than the atmosphere, and ocean circulation influences long-term variation. Coupled atmospheric-oceanic cycles are critical to understanding variations in weather and climate.

Figure 4-3: Atmospheric circulation patterns

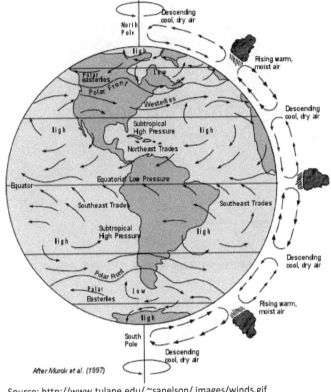

After Murok et al. (1997)

Source: http://www.tulane.edu/ ~sanelson/ images/winds.gif

Of these various coupled atmospheric-oceanic systems, the El Niño Southern Oscillation (ENSO) is the most prominent source of inter-annual variability in global weather and climate patterns. Its two alternating signatures, La Niña and El Niño, represent the negative and positive phases of the oscillation and are defined as sustained sea surface temperature anomalies greater than 0.5°C across the central tropical Pacific Ocean. When ENSO is in its negative phase, the Pacific trade winds cause sun-warmed surface water to pile up against Australia and Indonesia, while cool subsurface water rises in the east. During its positive phase the trade winds falter, and warm water spreads out eastwards across the Pacific Ocean. The Pacific Ocean trade winds set up cloud and rainfall patterns globally with enormous energy amounts transferred between ocean

Figure 4-4: Ocean circulation patterns

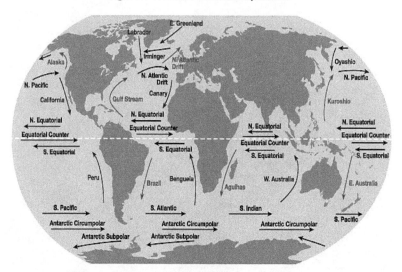

Source: http://fany.savina.net/wpcontent/ uploads/2010/03/Surface_currents.jpg

and atmosphere. ENSO varies between La Niña and El Niño states over 2 to 7 years but also over periods of decades to centuries. Superimposed on the alternation of La Niña and El Niño are longer term variations in the frequency and intensity of El Niño and La Niña, known as the Pacific Decadal Oscillation (PDO). A period of more frequent and intense La Niñas between the mid-forties and 1975 was followed by more frequent and intense El Niños between 1976 and 1998. The pattern can be seen in centuries of proxy series, such as tree and coral rings, sedimentation and rainfall, and flood records. Global surface temperatures indicate a similar trajectory, falling from 1946 to 1975, rising from 1976 to 1998, and declining slightly since.[30] Other, less well understood oscillating patterns include the North Pacific Oscillation and the Atlantic Multidecadal Oscillation (AMO).

Over the past couple of decades, scientists have become better informed about coupled heat-distributing atmospheric-oceanic systems. ENSO and the PDO are the most studied and probably the most important because of the size of the Pacific Ocean, but other

---

30. See Robert Ellison, "ENSO Variations and Global Climate," *American Thinker*, November 28, 2007.

patterns, such as the North Atlantic Oscillation (NAO), are becoming better understood and are adding to the ability of meteorologists to forecast their impact on short-term weather and longer term climate patterns. Much of what we now know of ENSO and its cognates is the work of scientists not affiliated with the IPCC, and IPCC scientists are only now beginning to acknowledge that perhaps they need to take greater account of these short and long-term natural patterns in their modelling and prognostications. Less well understood is what causes the shifts in these patterns from positive to negative phases and back.

Human activity can also influence climate patterns, including changes in land use, urbanization, and industrialization. Many factors can cumulatively drive changes in both the short and the long term. Scientific understanding of these factors is increasing, as are some of the relationships among them, but there remains considerable scope for scientific disagreement and for competing theories. The extent of heterogeneity in the study of climate is a healthy sign that scholars are engaged in vigorous research; the effort by climate alarmists to suppress some of this research by denying access to peer-reviewed journals is indicative of the emergence of an unhealthy cult with an extra-scientific agenda.

Is there a global temperature?

Much of the popular discussion of climate change is focused on what is known as the global temperature anomaly, i.e., the deviation of the global temperature from a previously established periodic average, for example, that from 1961-1990. The focus on temperature enables scientists to develop data that can be used in their models and other studies. As scientists Mark Handel and James Risbey explain, "despite the fact that most discussions of greenhouse change focus on the heat balance, changes in the hydrological cycle will have a much greater effect on humans and the biosphere than changes in temperature. The emphasis on temperature change is mostly due to a lack of confidence in [scientific] knowledge of water issues."[31]

The idea of a global temperature and a global climate is an artefact of climate research. The planet does not have a temperature or a climate. Instead, temperature can be measured at specific points on the planet or in the atmosphere at any given point in time. Near sur-

---

31. Mark David Handel and James S. Risbey, "Reflections on More than a Century of Climate Change Research," *Climatic Change* 21 (1992): 91-96.

face air temperatures vary enormously from place to place, hour to hour, day to day, and season to season, from a high of +58° C recorded in the Libyan desert to a low of -93° C recorded in Antarctica, i.e., a range of over 150°C.[32] Temperature is a local and transient phenomenon; aggregating the characteristics of thousands, even millions, of local temperature readings into a global temperature may serve analytical goals but does not make temperature or climate a global phenomenon.[33] Once analysts move beyond the raw data and construct a time series, do regression analyses, calculate averages or any of the other tasks that take up their time and energy, they are engaged in modelling, and their results are no longer data but models. All that can be said about temperature is that it is always going up or down, spatially and temporally.[34]

The temperature at any spot on the earth's surface at any point in time, ultimately derived from the energy of the sun, is the result of many forces that are constantly in flux as a result of the earth's atmosphere, surface characteristics, and rotation. During daylight hours, solar energy penetrates the atmosphere and strikes the earth's surface, heating it up, usually reaching a maximum in mid-afternoon at any specific location. From then on, the surface at that location cools, reaching a minimum sometime during the early morning hours. Local air temperatures reflect not only the direct impact of the sun but also the impact of surface radiation and albedo, convection, evaporation, condensation, precipitation, wind,

---

32. Even the city of Ottawa's modern temperature record varies from a low of -38.9°C to a high of +37.8°C, a range of nearly 77°C. On any given day, it can vary 15 to 20°C. Twenty thousand years ago, Ottawa was covered by 1,500 metres of ice, and at some point in the future it may be again.

33. Christopher Essex, Ross McKitrick, and Bjarne Andresen write: "There is no physically meaningful global temperature for the Earth in the context of the issue of global warming. While it is always possible to construct statistics for any given set of local temperature data, an infinite range of such statistics is mathematically permissible if physical principles provide no explicit basis for choosing among them. Distinct and equally valid statistical rules can and do show opposite trends when applied to the results of computations from physical models and real data in the atmosphere. A given temperature field can be interpreted as both "warming" and "cooling" simultaneously, making the concept of warming in the context of the issue of global warming physically ill-posed." Abstract, "Does a Global Temperature Exist?" *Journal of Non-Equilibrium Thermodynamics,* 32 (2007).

34. For a full discussion of the extent to which the climate community abuses data and confuses data with models, see the many posts by statistician William Briggs at his weblog, *wmbriggs.com.*

cloud cover, and other dynamic forces that act to reduce the impact of direct energy from the sun and to remove heat from the surface, redistributing it throughout the atmosphere until it eventually escapes to outer space. The redistribution of heat is uneven and often chaotic but operates within boundaries that tend to average out over the planet as a whole. One winter may be mild in parts of North America but brutally harsh in parts of Europe. The previous year, the pattern may be different, and the following year may be different again. As Duke University physicist Robert Brown explains:

> The surface itself is being heated directly by the sun part of the time, and is radiatively cooling directly to space (in at least some frequencies) all of the time. Its temperature varies by degrees K[elvin] on a time scale of minutes to hours as clouds pass between the location and the sun, as the sun sets, as it starts to rain. It doesn't just heat or cool from radiation – it is in tight thermal contact with a complex atmosphere that has a *far greater influence* on the local temperature than even local variations in insolation [the sun's radiation].[35]

Defining an average temperature for any given place for any particular period involves many decisions and calculations. How many observations are needed to determine an average for a day? Is the average the mean between the high and the low? Is it the average of 24 equally spaced hourly observations? The arbitrary nature of defining an average local temperature becomes apparent the moment one begins to ask such questions. If the planet is warming, is it happening because we are observing a rise in nightly lows or in daytime highs, or both? Aggregating local temperatures into a global temperature presents formidable methodological and measurement challenges and is presumed to be more informative than it is. MIT atmospheric physicist Richard Lindzen argues that the global temperature anomaly represents little more than the residue of averaging millions of local observations, a task that presumes the proper siting of well-tuned measuring equipment, an accurate recording of their observations, and a common methodology for their computation. None of these factors can be presumed to be true.[36]

---

35. "Global annualized temperature – 'full of [snip] up to their eyebrows,'" *WattsUpWithThat*, March 4, 2012.

36. Lindzen, "Global Warming: How to Approach the Science," Seminar at the [UK] House of Commons Committee Rooms, Westminster, London, 22nd February 2012.

Figure 4-5: HadCRUT 4 temperature anomaly, 1850-2014, based on a 1961-90 base period.

The graph points to two major increases: from the 1910s to the 1940s and from the late 1970s to the end of the century. Temperatures from the 1940s to the 1970s and since 2000 have been much more stable, a fact suggesting that the relationship between GHG forcing and temperature increases is not direct. The difference between the southern and northern hemispheres reflects the difference in land mass and perhaps urbanization.
Source: University of East Anglia, Climatic Research Unit, October 10, 2014

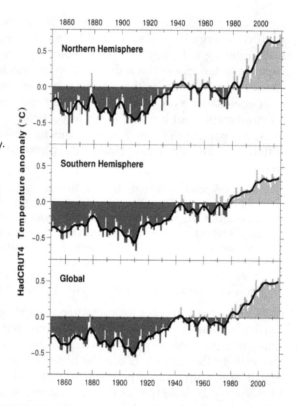

Further complicating public understanding of climate change is that averaging temperature on an annual basis obscures the fact that the planet's temperature varies over the course of the year as a result of the much larger land mass of the northern hemisphere that dominates the build-up of heat during the northern summer. The southern hemisphere, which is dominated by a much larger percentage of ocean, heats up less during the southern summer, reducing the globe's temperature by as much as 3°C during the northern winter. Lindzen et al. point out that "the globally averaged surface temperature shows a strong seasonal cycle. ... The size of this variation (almost 3°C) is, by the standards of climate change, huge." Ramanathan and Inamdar explain that "the extra-tropical and global annual cycle is most likely dominated by the hemispherical asymmetry in the land fraction. During the northern-hemisphere summer (JJA), the large land masses warm rapidly (with about a one-month lag) which dominates the hemispherical and global mean response;

Figure 4-6: The urban heat-island effect

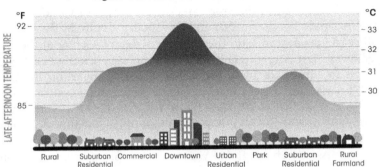

As a result of the increased absorption of heat by concrete, pavement, and other urban surfaces, heat builds up more in cities than in the country side. Even a weather station located on the outskirts of a large city will exhibit this effect, distorting the true surface temperature exhibited at rural monitoring stations. By some estimates, half or more of the global warming recorded by surface stations is due to the urban heat-island effect.
Source: www.c3headlines.com/global-warming-urban-heat-island-bias.

---

however, during the southern-hemisphere summer, the relatively smaller fraction of land prevents a corresponding response. Thus, the globe is warmest during June/July and is coldest during December/January."[37] As illustrated in figure 4-5, the modern increase in global temperature is much more pronounced for the northern than for the southern hemisphere.

Within the climate science community, so-called global temperature anomalies are calculated by aggregating average surface air temperature observations reported by various stations around the world. These stations are not uniformly placed, are on land only, and thus leave out 70 per cent of the earth's surface. Sea surface temperatures can be added to the calculation, based on another set of observations sampling water temperatures near the surface, but until recently these observations were not uniformly distributed or calculated. The quality of the information supplied by surface stations is also far from uniform. Many are contaminated by the urban heat island effect (UHI) (see figure 4-6) and other factors, such as poor siting and substandard equipment: a significant number are not continuous, and a

---

37. Richard S. Lindzen, et al., "Seasonal Surrogate for Climate," *Journal of Climate* (June 1995), 1681. V. Ramanathan and A. Inamdar, "The radiative forcing due to clouds and water vapour," in J.T. Kiehl and V. Ramanathan, eds., *Frontiers of Climate Modelling* (Cambridge: Cambridge University Press, 2006), 141.

large number of observations from rural stations disappeared with
the Soviet Union. The time series reported by NASA's Goddard Institute for Space Studies (GISS) and the Hadley Centre at the UK Met
Office in conjunction with the Climatic Research Centre at the University of East Anglia (HadCRUT) rely on very controversial algorithms to correct for these problems. The IPCC relies on the Had-
CRUT series and argues that the global temperature increased by a
net 0.8°C from 1850 to 2010 (see figure 4-5). Needless to say, the quality of these time series is an important part of the scientific controversy about the extent of anthropogenic global warming.[38]

Discussions about global warming take place in terms of temperature anomalies, i.e., deviations from a norm, which are calibrated in tenths of a degree centigrade, a level of precision not warranted by the quality of the raw data. The extent and sign of the deviation, of course, depend on the base adopted for the norm. Global
warming advocates, for example, long used a surface-based temperature record normalized to a 1951-1980 base period. Many of
those years formed part of a cooling trend, which ended in 1977,
which was followed, using surface data, by a warming trend culminating in the 1997-98 El Niño spike. More recently, temperature series have been normalized to a 1961-1990 base period, as in figure 4-
5. Since 1997, the temperature series have indicated neither cooling
nor warming. Lindzen concludes:

> 'Global Warming' refers to an obscure statistical quantity, globally av-
> eraged temperature anomaly, the small residue of far larger and mostly
> uncorrelated local anomalies. This quantity is highly uncertain but may
> be on the order of 0.7C over the past 150 years. This quantity is always
> varying at this level and there have been periods of both warming and
> cooling on virtually all time scales. On the time scale of from 1 year to

---

38. The full extent of the problems with these records is discussed in Joseph D'Aleo,
     "A Critical Look at Surface Temperature Records," in Don Easterbrook, et al., eds.,
     *Evidence-Based Climate Science: Data Opposing CO2 Emissions as the Primary Source of
     Global Warming* (Amsterdam: Elsevier, 2011). The extent of warming in the last
     quarter of the 20th century is well within the error rate of these time series. See al-
     so Patrick J. Michaels and Robert C. Balling, Jr., *Climate of Extremes: Global Warming
     Science They Don't Want You to Know* (Washington: Cato Institute, 2009). The
     Berkeley Earth Surface Temperature Project (BEST), led by physicist Richard Mul-
     ler, has developed a new record freed of some of the problems plaguing the other
     series but has to date met with limited acceptance due to the formidable problems
     that need to be overcome as well as the reality that the raw data, in Muller's own
     words, are "largely awful". The results to date differ only marginally from the
     other data sets. See Richard A. Muller, "The Case Against Global-Warming Skep-
     ticism," *Wall Street Journal*, October 21, 2011.

100 years there is no need for any externally specified forcing. The climate system is never in equilibrium because, among other things, the ocean transports heat between the surface and the depths. To be sure, however, there are other sources of internal variability as well. Because the quantity we are speaking of is so small, and the error bars are so large, the quantity is easy to abuse in a variety of ways.[39]

Starting in 1978, NASA has used satellites to collect data from around the whole planet based on microwave sounding units. The record from satellites, while still an artificial construct, provides a more reliable base for looking at changing temperature and climate patterns on a global scale in response to increased GHGs, in part because satellites gather data on near surface temperatures that are not affected by the UHI effect and also because they survey the planet as a whole – over both land and sea – on a uniform basis. Any climatic response to increased GHGs takes place in the lower atmosphere and is best measured there. The satellite record, maintained for NASA at the University of Alabama at Huntsville (UAH) by climatologists John Christy and Roy Spencer, shows considerable seasonal, regional, and annual variation, but it also indicates much less long-term change in temperature trends than does the surface record (see figures 4-7 and 4-8).[40]

Year-to-year and seasonal variation is most pronounced at the poles, but longer term trends are marginal. The same data are also collected by Remote Sensing Systems (RSS), a private research firm in California. Its output tracks the UAH results closely, with some minor variations resulting from slightly different assumptions and algorithms. Sea-based buoys that feed information to satellites are another more recent and more reliable innovation that should improve the quality of databases.[41] None of these series are robust enough to support the many claims and conclusions that climate scientists seek to tease out of them. Again, as Duke's Brown indicates: "One of many, many problems with modern climate research is that the researchers seem to take their thermal reconstructions far

39. Richard Lindzen, "Global Warming: How to Approach the Science."

40. See Paul Gattis, "7 Questions with John Christy and Roy Spencer: Climate Change Skeptics for 25 years," *AL.com*, April 1, 2015.

41. For a discussion of the challenges in measuring sea surface temperatures by an experienced oceanographer, see Robert E. Stevenson, "Yes, the Ocean Has Warmed; No, It's Not 'Global Warming'," *21st Century Science and Technology Magazine,* Summer 2000.

Figure 4-7: UAH satellite-based temperature of the global lower troposphere.

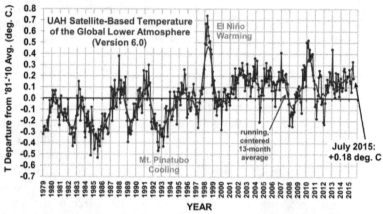

The graph shows changes in the temperature on a monthly basis, not the actual temperature, normalized to the 1981-2010 average. It shows that much of the post-1978 warming took place during the 1997-8 super El Niño.

Source: Roy Spencer at http://www.drroyspencer.com/latest-global-temperatures/

Figure 4-8: Satellite-based temperature anomaly as measured in five bands of latitude: the tropics, the extra-tropics, and the poles.

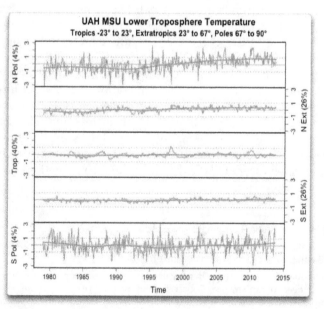

The planet's temperature is most volatile at the poles; the two temperate zones show the least volatility; the El Niño pattern can be seen in the tropical zone.

Source: Willis Eschenbach, "Should We Be Worried?" at *WattsUpWithThat*, January 29, 2014

too seriously and assign completely absurd measures of accuracy and precision. ... The problem becomes greater and greater the further back in time one proceeds, with big jumps (in uncertainty) 250, 200, 100 and 40 odd years ago. ... To claim accuracy greater than 2-3 K is almost certainly sheer piffle, given that we probably don't know *current* 'true' global average temperatures within 1K, and 5K is more likely."[42]

The greenhouse effect

Despite the misleading nature of the metaphor,[43] the so-called greenhouse effect is important for understanding both weather and climate. Temperature on the earth would vary more dramatically between day and night and make the planet largely uninhabitable without this effect, as is the case on the moon, which has no atmosphere. The existence and impact of the greenhouse effect are largely – but not completely – uncontroversial. Energy from the sun strikes the earth's atmosphere; some of that energy is reflected back into space, but enough penetrates the atmosphere, strikes the earth, and heats its surface; this heat ultimately radiates back into the atmosphere, and eventually out into space. Greenhouse gases absorb and scatter infrared radiation, the radiant heat energy that the earth naturally emits from the surface in response to solar heating and that keeps the atmosphere and surface warmer than would otherwise be the case. The net effect of all this is the energy balance upon which life on earth depends. The most important greenhouse gas is water vapour (97 percent by volume; 75 percent or more by impact); others include carbon dioxide and such other trace gases as methane, nitrous oxide, and ozone.

Water, oxygen, and carbon dioxide are critical to life on earth. All of life depends on them. All animals, for example, breathe in oxygen and exhale carbon dioxide, which in turn is used by plants in photo-

---

42. "Global annualized temperature – 'full of [snip] up to their eyebrows'."

43. McKitrick et al. point out that "while use of the term 'greenhouse' is nowadays unavoidable, the term 'greenhouse effect' is an inappropriate metaphor since it suggests a parallel between the mechanism that causes warming in an actual greenhouse and the influence of infrared-active gases, like water vapour and carbon dioxide, on the Earth's climate system. The two mechanisms are quite distinct, and the metaphor is misleading. It leaves out the complexities arising from the nonlinear, dynamic processes of our climate system, namely evaporation, convection, turbulence and other forms of atmospheric fluid dynamics, by which energy is removed from the Earth's surface. Simplistic metaphors are no basis for projecting substantial surface warming due to increases of human-caused carbon dioxide concentration in the atmosphere." *Independent Summary for Policy Makers* (Vancouver: Fraser Institute, 2007), 9.

synthesis, which converts carbon dioxide back into oxygen and carbon. Oxygen and carbon, in various combinations with hydrogen, make up the principal components of plants. All the CO_2 emitted by burning fossil fuels was at one point part of the atmosphere before prehistoric plants and animals converted it into organic matter. As these biota decomposed and were subjected to heat and pressure, they were transformed into the coal, oil, and gas that humans are now recovering and exploiting as stored solar energy.

Table 4-1. Annual Global Carbon Dioxide Emission/Absorption

	Natural	Anthropogenic	Total	Absorption
Million Metric Tons	770,000	23,100	793,100	781,400
Percent of Total	97.1	2.9	100	98.5

Source: IPPC, *Climate Change 2001: The Scientific Basis* (Cambridge, UK: Cambridge University Press, 2001), Figure 3.1, 188.

Carbon in the earth's envelope is in constant motion between four reservoirs – the biosphere, lithosphere, atmosphere, and hydrosphere. Table 4-1 indicates the marginal role of human activity in the IPCC's calculation of the annual global carbon cycle. The increased level of atmospheric CO_2 in the second half of the 20th century was one of the contributors to the "green" revolution, i.e., the revolution in agricultural productivity that made a mockery of the spectre of mass starvation that was central to earlier alarmism. Satellite studies now indicate a significant increase in the earth's biomass over the past quarter century.[44] The biosphere would be severely compromised if CO_2 levels fell below 200 parts per million by volume (ppmv), and experiments have demonstrated that it would thrive with carbon levels of 1000 ppmv or even more, the level often maintained in greenhouses to stimulate growth.

The amount of CO_2 in the atmosphere is very small – about 400 ppmv, or 0.04 percent of the atmosphere by volume at the end of 2013, an increase of perhaps 0.01 per cent since the middle of the 19th century. Relative to the total volume of the atmosphere, the millions of tons of annual CO_2 emissions that environmentalists worry about remain marginal. The extent to which CO_2 is uniformly distributed in the atmosphere is also not clear. Scientists rely on meas-

---

44. Randall J. Donohue, et al., "CO_2 fertilisation has increased maximum foliage cover across the globe's warm, arid environments," *Geophysical Research Letters*, May 31, 2013.

urements at one location: the observatory on Mauna Loa in Hawaii. Its database shows steady but not uniform growth over the past fifty years.[45] Less than half of the CO_2 that humans produce burning fossil fuels stays in the atmosphere, accumulating at the rate of less than one molecule per 100,000 every five years. The atmosphere is vast, and CO_2 is a tiny part of it.[46]

The biosphere's nominal ability to absorb at least half of manmade CO_2 remained constant over the second half of the 20th century despite the increasing amount of CO_2 emitted by human activity. Large forest tracts act as sinks when trees absorb CO_2 through photosynthesis and then release oxygen. Table 4-1 indicates that only a small part of the earth's estimated CO_2 flux – less than three percent – comes from the burning of fossil fuels and other human activities. There is much discussion among scientists as to how much of the possible 0.01 per cent increase in atmospheric CO_2 since 1850 is due to human activity.

Figure 4-9 illustrates the theoretically derived values of the various flows in the earth's energy budget, assuming that the energy budget is largely in equilibrium on most time scales – something on which not all scientists are agreed. Climate scientists such as Lindzen and Spencer maintain that there is enough dynamism within the climate system to account for most of the variability that preoccupies IPCC scientists. The IPCC view is that forcings and feedbacks affect the equilibrium either positively or negatively, leading to warming or cooling. Increases in GHGs lead to a positive forcing

---

45. The Mauna Loa Observatory sits at 4,000 meters above sea level. Its daily calibrations are subject to adjustments that smooth out fluctuations throughout the day. It is difficult to say whether it is typical of the global atmosphere as a whole or whether the level is atypical of the past. See Tom Quirk, "Sources and Sinks of Carbon Dioxide," *Energy & Environment* 20:1-2 (January 2009), 105-121 and H. Thomas, et al., "Changes in the North Atlantic Oscillation influence CO2 uptake in the North Atlantic over the past 2 decades," *Global Biogeochemical Cycles,* 22 (2008), 1-13. Both discuss the extent to which rising levels of atmospheric CO2 can in large measure be explained by natural phenomena.

46. Alarmists will occasionally add methane to the pantheon of gases heating up the atmosphere. Methane (CH4) is a potent greenhouse gas, about 34 times more powerful than CO_2 per molecule, but its share of the atmosphere borders on the insignificant (about .0018 per cent). Additionally, its absorption spectrum overlaps that of water vapour; as a result, much of its impact is overwhelmed by that of water vapour. Whether it increases or decreases – as a result of the flatulence of ruminants such as cattle – will have little or no net greenhouse effect. Worrying about the release of methane by melting tundra sounds scary but will again have little impact on the greenhouse effect.

Figure 4-9: The global mean annual energy budget (March 2000 - May 2004) (W/m²).

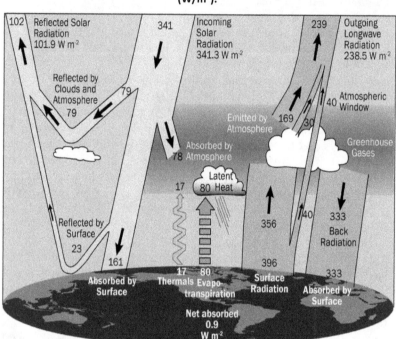

The broad arrows indicate the theoretically derived flows of energy in proportion to their importance. Incoming, short-wave solar radiation is partially reflected by clouds, the atmosphere and the earth's albedo with less than half absorbed by the earth's surface. The surface radiates energy back into the atmosphere; some of that energy is absorbed and scattered by greenhouse gases, which keeps the atmosphere warmer than it would otherwise be.

Source: NCAR at http://www.cgd.ucar.edu/cas/Topics/Fig1_GheatMap.png,

which, in turn, leads to further positive feedbacks, particularly from increases in water vapour, thus leading to increased warming. Figure 4-10 indicates the IPCC's calculations of the size of various forcing factors and their role in the observed warming over the second half of the 20th century.[47]

47. "Radiative forcing is a flux of energy (i.e., an energy flow per unit area), and sensitivity is a ratio of temperature change to this flux. High sensitivity means that a small flux eventually produces a large temperature change, but, because the flux is small, the change will take a long time." Richard Lindzen, "Global warming, models, and language," in Alan Moran, ed., *Climate Change: The Facts* (Melbourne, Australia: Institute of Public Affairs, 2015), kindle edition.

Figure 4-10: Radiative forcing components as understood by the IPCC

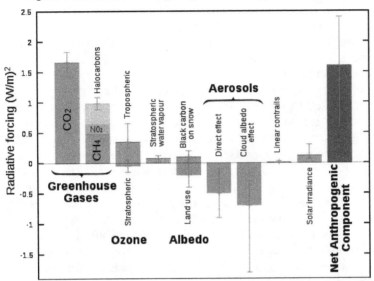

Source: Leland McInnes, Wikemedia Commons, www.gnu.org/copyleft/fdl.html

This hypothesis of the earth's energy budget and its calculation of the impact of various forcing factors are the basis for the computerized general circulation climate models. They are used to run scenarios projecting the evolution of the climate system in response to various forcings and feedbacks over many years into the future, based on differing assumptions about the amount of various forcing agents and feedbacks. The numbers are the result of theoretical calculations rather than observations. Recent observations indicate large errors in the most widely used models due to incorrect calculations of solar radiation and the solar zenith at the top of the atmosphere.[48] More generally, IPCC scientists underestimate the role of variations in solar radiation.[49] Similarly, models rely on an incomplete, and probably misleading, understanding of the role of clouds and albedo as part of the climate system's self-regulating capacity to modulate the impact

48. Linjiong Zhou et al., "On the Incident Solar Radiation in CMIP5 Models," *Geophysical Research Letters*, published online March 5, 2015.

49. Willie Soon points out that of the "38 co-authors and three review editors of the IPCC's solar sub-chapter [ch. 8 in AR5, WG1], only one is an expert on solar physics." "Sun shunned," in Alan Moran, *Climate Change: The Facts* (Melbourne, Australia: Institute of Public Affairs, 2015).

of, for example, changes in atmospheric levels of CO_2.[50] An increasing number of papers in the refereed journals show a growing array of problems with model assumptions.

Only the first 200 ppmv of atmospheric CO_2 have a significant direct effect on the climate system. Each doubling leads to only a ten percent increase in its absorptive capacity. The IPCC theory of global warming depends on the sensitivity of the climate system to feedbacks resulting from the increase in anthropogenic GHGs and industrial aerosols, as well as from other forcings and feedbacks that affect the Earth's radiative balance. If the sensitivity to GHGs is high, the impact of increased CO_2 may be significant; if it is low, its impact disappears among the many other factors. The sign of the feedback is also important: positive feedbacks lead to high sensitivity; negative feedbacks lead to low sensitivity or even to negative impacts. This is the crux of the scientific controversy. All the rest flows from the assumption of high sensitivity but, as Lindzen and others have demonstrated, it is at least as plausible to make the case for low sensitivity as for high. As Lindzen puts it, "the basic agreement frequently described as representing scientific unanimity concerning global warming is entirely consistent with there being virtually no problem at all. Indeed, the observations most simply suggest that the sensitivity of the real climate is much less than found in models whose sensitivity depends on processes which are clearly misrepresented (through both ignorance and computational limitations)."[51] He adds that the sensitivity of the climate system to small changes "is not an indication of the delicacy of the climate system, but rather an indication of the ease with which the system can adjust to changes."[52] As discussed in the next chapter, recent research has steadily reduced estimates of the climate system's sensitivity to increases in CO2.

University of Alabama climate scientist Roy Spencer points out that his "group's government-funded research suggests global warming is mostly natural, and that the climate system is quite insensitive to humanity's greenhouse gas emissions and aerosol pol-

---

50.   See, for example, the review article by Graeme L. Stephens, et al., "The albedo of earth," *Reviews of Geophysics* 53 (March 2015), online.

51.   *Issues in the Current State of Climate Science* (Washington: Center for Science and Public Policy, March 2006), 35.

52.   "Global warming, models, and language," *Climate Change: The Facts*, kindle edition.

lution."[53] He concludes that whatever impact increased GHGs have on the Earth's heat/energy budget is lost in the much larger impact of clouds, precipitation, ocean circulation, and other factors. "Observations suggest that the feedback is close to zero and may even be negative. That is, water vapour and clouds may actually diminish the already small global warming expected from CO_2, not amplify it. The evidence here comes from satellite measurements of infrared radiation escaping from the earth into outer space, from measurements of sunlight reflected from clouds and from measurements of the temperature of the earth's surface or of the troposphere."[54] As discussed in more detail in chapter seven, other scientists maintain that small fluctuations in the output of the sun can have a larger impact as well, either directly, or as a result of the sun's modulation of cosmic rays and their impact on cloud formation. Carleton University earth scientist Tim Patterson concludes: "The geologic record clearly shows us that there really is little correlation between CO_2 levels and temperature. Although CO_2 can have a minor influence on global temperature the effect is minimal and short-lived as this cycle sits on top of the much larger water cycle, which is what truly controls global temperatures. The water cycle is in turn primarily influenced by natural celestial cycles and trends."[55]

Human Influence

Human influence on both climate and weather is multiple and varied, from land use to pollution and urbanization. The urban heat island effect (i.e., increased energy absorption due to urban land and energy use), for example, can have a significant impact on *local* temperatures, particularly in winter and at night but has at best a marginal impact on *global* temperatures and climate.[56] Aerosols re-

---

53. Roy Spencer, "Global Warming: Natural or Manmade?" at *drroyspencer.com*.

54. "Climate Change," Statement before the US Senate Environment and Public Works Committee, Feb. 25, 2009.

55. R. Timothy Patterson, "The Geologic Record and Climate Change," Remarks at Conference "Risk: Regulation and Reality," October 7, 2004.

56. For a full discussion of human impacts, see William R. Cotton and Roger A. Pielke, Sr., *Human Impacts on Weather and Climate*, 2nd edition (New York: Cambridge University Press, 2007). Humans do have a significant effect on global surface temperature series. Since historically most weather stations were located in or near urban centres, the temperature reported may be as indicative of a growing population as of a warmer climate. Many critics indicate that neither GISS nor HadCrut take sufficient account of this phenomenon. See, for example, Ross R. McKitrick, and Patrick J. Michaels, "Quantifying the influence of an-

sulting from industrial activity can block the sun's radiation and reduce temperatures. Changes in forest cover can affect *regional* temperature and precipitation patterns. The gradual decline of the glacier on top of Mt. Kilimanjaro in Africa, for example, is probably not
due to global warming – temperature observations on and around
the mountain do not indicate warming – but rather due to changes
in precipitation patterns resulting, some believe, from the conversion of land from forest to farming in the surrounding area.[57] A
glacier builds at the top from new snow and recedes due to melting
in its tail and pressure from higher up. In the absence of new snow,
the process of sublimation will also shrink a glacier. Glaciers thus
recede unless they are fed by new snow and are influenced more by
changes in precipitation than by temperature patterns. Similarly, the
extinction of the golden toad in Costa Rica was not due to global
warming but may have resulted from land use changes, which
raised the altitude of cloud cover on which its niche mountainside
ecosystem depended. Again, the overall temperature of Costa Rica
has not greatly changed, but local microclimate and precipitation
patterns have.[58]

Critics of the catastrophic anthropogenic global warming theory
do not deny human influence on climate, but they do question the
dominant role of carbon dioxide (CO_2) and other greenhouse gases
in warming the temperature of the planet as a whole to a significant
degree. Critics do not agree that the modest changes seen over the
course of the last quarter of the 20th century will have catastrophic
impacts on either the biosphere or human civilization. There has
been *no statistically meaningful global warming* since the turn of the
century and perhaps earlier. Scientists may well indicate at some
point in the future that the role of GHGs is more significant than has
been demonstrated to date.[59] At this point in time, however, their
role is largely a matter of theoretical conjecture and is not based on

---

thropogenic surface processes and inhomogeneities on gridded surface climate
data," *Journal of Geophysical Research-Atmospheres* 112 (2007), 1-14.

57. See P.W. Mote and G. Kaser, "The shrinking glaciers of Kilimanjaro: Can global
warming be blamed?" *American Scientist* 95:4 (2007), 318-325.

58. R.O. Lawton, et al., "Climatic impact of tropical lowland deforestation on near-
by montane cloud forests," *Science* 294 (2001), 584-587 and Keith, Sherwood, and
Craig Idso, "The Drying of Costa Rican Tropical Montane Cloud Forests," *CO2
Science* 9: 47, 22 November 2006.

59. For a succinct summary of many of the objections to the GHG hypothesis, see
Foss, *Beyond Environmentalism*, 111-65.

any direct observational evidence. Given the extent of the controversy, it is important that serious work testing the greenhouse gas hypothesis continue. Such work would be more credible, however, if it proceeded as a matter of open scientific investigation on a level playing field with other issues that need further investigation, such as the role of the sun, cosmic rays, ocean circulation, atmospheric oscillations, clouds and precipitation, and direct human influences such as aerosols and land use.[60]

The Earth is a dynamic planet; change is the only constant in its geosphere, lithosphere, biosphere, hydrosphere, cryosphere, and atmosphere. No observational evidence exists that modern rates of change – for example, in temperature, precipitation, snow cover, ice volume, cyclones, or sea levels – lie outside historic bounds, let alone geological ones. Humans have an effect on local climate but, despite the expenditure of more than US$100 billion looking for it since 1990, no globally summed human effect has ever been measured on the basis of credible observational evidence. Whatever the size of the human signal – from the cooling effect of aerosols to the warming effect of GHGs – there is good scientific evidence that at the global level it is probably overwhelmed by natural variability in the climate system. For example, over the course of an El Niño year, the oceans can put as much additional CO_2 into the atmosphere as does human industry over the course of that year, while a La Niña year can take as much out again. The two together also illustrate an important causal relationship: heat in the oceans leads to more CO_2 in the atmosphere and colder ocean surface temperatures lead to less, not the other way around.

Climate change models

The case for catastrophic human-caused global warming rests largely on the projections of computer models, i.e., abstract, mathematical representations of a complex reality dependent on the assumptions built into each model. Australian physicist John Reid points out that climate alarmism is grounded in deterministic (linear) models trying to capture a stochastic (unstable) world, an impossible task because it is not physics. The planet's climate operates within clear boundaries, which have varied over geological time, but behaves erratically between those boundaries. In this chaotic real world, as Edward Lorenz demonstrated in his early work with weather models, "slightly differ-

---

60. The work of scientists pursuing these phenomena and the alternative perspectives that they give rise to is discussed in chapter seven.

ing initial states can evolve into considerably different states." Reid further observes, "determinism has long been there, underlying Western Christian thought, but it has recently come to dominate (or perhaps replace) scientific thinking. I believe that this is an unintended consequence of numerical modelling which is now widespread in science. Computers have, in general, been such a boon to science that no one any longer questions the validity of some applications, particularly those numerical models which are based on differential equations. All such models rest on certain assumptions – assumptions which are very rarely questioned or even acknowledged. These include assumptions about the complete absence of discontinuities – cliffs and fronts and shocks – which are, in reality, widespread in nature."[61]

Without computer model projections – the average of which have consistently been in the range of a 1.5°C to 4.5°C global temperature rise over the course of the 21st century from a doubling of carbon dioxide concentration since the 19th century – there would be no basis for alarm.[62] Each computer model represents an hypothesis, i.e., a series of tentative conclusions awaiting testing and observation. Additionally, modellers have yet to integrate the two most important issues identified by climate scientists more than forty years ago as critical to reaching a satisfactory understanding of the global climate system: the role of clouds and of the oceans.[63] Australian earth scientist Robert Carter observes: "computer models predict future climate according to the assumptions that are programmed into them. There is no established Theory of Climate, and therefore the potential output of all realistic computer general circulation models

---

61. Edward Lorenz, "Deterministic Non-periodic Flow," *Journal of the Atmospheric Sciences* 20:2 (March 1963), abstract. John Reid, "A Physicist ponders the Pause," *Quadrant Online,* October 20, 2014.

62. See the detailed and accessible discussion of the shortcomings of models in William Kininmonth, "Illusions of Climate Science," *Quadrant Online,* October 7, 2008. The first major modelling exercise was performed for the US National Academy of Science's 1979 Charney Report. Using the most advanced computer at that time – one that today does not come even close to the computing power of a modest home computer – the report acknowledged the limitations of computer modelling but concluded that the United States would experience a 1.5-4.5°C warming from a doubling of CO2. Little has changed since then. The limitations remain as do the projections despite quantum leaps in computing power because the assumptions have not kept up with what can be learned from an analysis of the observations. See *Carbon Dioxide and Climate: A Scientific Assessment* (Washington: National Academy of Sciences, 1979).

63. See Garth Paltridge, "Uncertainty, sceptics, and the climate system," in Moran, *Climate Change: The Facts.*

encompasses a range of both future warmings and coolings, the outcome depending upon the way in which they are constructed. Different results can be produced at will simply by adjusting such poorly known parameters as the effects of cloud cover."[64]

Feeding dubious data into a computer and then running multiple regressions based on the model to approximate future climate states do not constitute proof. All they show is what might happen given certain assumptions. They also require what have become known as fudge factors, i.e., data and assumptions about industrial aerosols, for example, fed into the model to approximate the recent past before running it to forecast the future. As the then editor of *Science* pointed out in describing a new model developed at NCAR: "Climate modellers have been 'cheating' for so long it's almost become respectable. The problem has been that no computer model could reliably simulate the present climate. Even the best simulations of the behaviour of the atmosphere, ocean, sea ice, and land surface drift off into a climate quite unlike today's as they run for centuries [into the future]. So climate modellers have gotten in the habit of fiddling with fudge factors, so-called 'flux adjustments,' until the model gets it right."[65]

General circulation models have improved over the years, but they remain fixated on the GHG hypothesis and radiative forcings and continue to rely on fudge factors and parameter tuning. All the models relied on by the IPCC have been tuned to produce global warming; none are run to test alternative hypotheses. The fact that each run by a climate model produces a *different* result, and that no two climate models have ever produced the *same* results, indicates the extent to which models do not reflect the complex reality of planet Earth. Instead they reflect the simplified and imperfect mathematical approximations of the climate system and the dodgy data fed into the models. Patrick Michaels suggests the continued reliance on obviously flawed models provides an excellent example of Thomas Kuhn's observation that scientific communities are reluctant to abandon a failing paradigm in which they have invested significant intellectual and professional capital, as well as billions in

---

64. Robert Carter, "The Futile Quest for Climate Control," *Quadrant Online*, November 1, 2008.

65. Richard A. Kerr, "Climate Change: Model Gets It Right – Without Fudge Factors," *Science* 276:5315 (May 16, 1997), 1041.

research funding.[66] Nevertheless, an increasing number of papers in the refereed scientific press are finding that the climate system's sensitivity to increasing GHGs is considerably lower than IPCC estimates, including papers by IPCC contributors.

Taking averages of the results of an "ensemble" of models, as the IPCC does, and then attributing confidence levels to the result is to enter the world of conjuring.[67] Each model is based on different assumptions, algorithms, and data; they do not all test the same parameters. Averaging these estimates of possible climate conditions some 30, 50, or a hundred years into the future is statistically irresponsible, has no physical meaning, and does no more than mislead governments and the public. In the forty or more years that models have been employed by the climate science community, their projections have drifted farther and farther from observed climate patterns (see figure 5-4 in the next chapter), leading some modellers to question the quality of the *observations*.

Models, while useful as an analytical tool, are only as good as the assumptions and the data used to build and run them. Of necessity, models simplify reality. Over time, as more is learned and data improve, they may become more sophisticated and more useful, but the fundamental problem of deterministic models trying to capture a stochastic reality will remain. General circulation climate models are among the most difficult and expensive to build and remain rather primitive at this stage. They are not yet sensitive enough, for example, to take into account the thousands of thunderstorms that take place every day. Earlier models were unable to take account of ocean circulation patterns, possibly the most important drivers of weather and climate. They are still not capable of accounting for cloud formation, evaporation, and precipitation, factors critical to mitigating temperature extremes. Both the hottest and the coldest days, for example, are cloudless days; precipitation is one way in which nature cools the at-

---

66. Patrick Michaels, "Why Climate Models are Failing," in Moran, *Climate Change: The Facts*. Michaels provides an overview of some of this literature.

67. Climate scientist Judith Curry cautions that, given the one-sided nature of IPCC analysis, assigning confidence levels "is not convincing unless it includes parallel evidence-based analyses for competing hypotheses ... . Any evidence-based argument that is more inclined to admit one type of evidence or argument rather than another tends to be biased." Curry and P.J. Webster, "Climate Science and the Uncertainty Monster," *Bulletin of the American Meteorological Society* 92 (December 2011), 1667-82. See also Curry, "Reasoning about climate uncertainty." *Climatic Change* 108 (2011), 723–732.

mosphere and moderates temperatures.[68] One way to test the validity of a model is to use historical data and see if it can hind-cast the results. The record to date of climate models is very disappointing, unless the modellers introduce fudge factors. Their ability to forecast, therefore, needs to be approached with a high level of humility. Princeton physicist Freeman Dyson points out:

> All the fuss about global warming is grossly exaggerated. Here I am opposing the holy brotherhood of climate model experts and the crowd of deluded citizens who believe the numbers predicted by the computer models. ... The models solve the equations of fluid dynamics, and they do a very good job of describing the fluid motions of the atmosphere and the oceans. They do a very poor job of describing the clouds, the dust, the chemistry and the biology of fields and farms and forests. They do not begin to describe the real world that we live in. The real world is muddy and messy and full of things that we do not yet understand. It is much easier for a scientist to sit in an air-conditioned building and run computer models, than to put on winter clothes and measure what is really happening outside in the swamps and the clouds. That is why the climate model experts end up believing their own models.[69]

The most telling point, however, may be the inconvenient fact that global temperatures and other observable climate and related phenomena are failing to match the models' projected values and characteristics. As more than one scientist has pointed out, field studies and observations repeatedly *contradict* model predictions. In these circumstances, the problems are more likely to be found in the models and their underlying hypotheses, assumptions, and data than in real-world observations. Again, it suggests that the burden of proof lies with those convinced that the models have it right. Nevertheless, the media, as with their reporting of epidemiological studies of health phenomena, have become so accustomed to reporting modelling projections as scientific forecasts of the future that non-experts can be forgiven for falling into the trap of accepting

---

68. Climate models should not be confused with models used to forecast weather. The latter have become increasingly sophisticated and rely largely on the movement of atmospheric forces along well-established parameters. Even *their* predictive ability, however, is limited to a few days, with longer predictions relying on historic patterns. Climate models, however, must take account of a much wider range of factors that move or change at different rates, few of which have been well parameterized.

69. Dyson, "Heretical Thoughts about Science and Society," *Edge* 219 (August 9, 2007).

model results as actual evidence of climate change and its many purported impacts.

Climate change theory, much like epidemiology, depends on correlation, i.e., on the search for a coincidence among various observations and then the running of regression analyses to observe their robustness. This is a common form of intellectual enquiry but is no substitute for determinations of cause and effect based on evidence drawn from real-world observations. As was clear in the discussion of epidemiology, correlation may well exist between two observations, but that does not establish cause and effect. All it does is point to an issue ripe for further investigation. The work of paleo-climatologists, for example, indicates that in earlier periods, increases in temperature preceded increases in CO_2 rather than the other way around, with a time lag of many years. This is consistent with basic physics. As oceans heat up, their ability to absorb CO_2 goes down and the flow reverses, with oceans emitting rather than absorbing CO_2 (as anyone can demonstrate with a glass of soda water; as it warms, it releases CO_2 and goes "flat").

Much is made of "statistical significance" in the regression analyses of various data series looking for correlations. *But,* as statisticians never tire of pointing out, this is a much abused concept. Economist Stephen Ziliak points out: "A statistically significant departure from an assumed-to-be-true null hypothesis is by itself no proof of anything. Likewise, failure to achieve statistical significance at the .05 or other stipulated level is not proof that nothing of importance has been discovered. ... the null hypothesis test procedure – another name for statistical significance testing – produces many such errors, with tragic results for real world economies, law, medicine, and even human life."[70]

IPCC climate scientists are also prone to make linear projections from what are cyclical phenomena. As discussed above, there is clear evidence of cyclical patterns in solar insolation, from the 11-year sunspot cycle to the much longer Dansgaard-Oeschger and Gleissberg cycles. Coupled atmospheric-oceanic forces are also cyclical, probably related to solar cycles. In the 1970s, when the Pacific Decadal Oscillation was in its cool mode, some climate scientists, such as Stephen Schneider, were convinced that the cooling was a prelude to human-induced glaciation. In the 1980s, the same scientists, based on a linear

---

70. Stephen T. Ziliak, "Unsignificant Statistics," *Financial Post*, June 10, 2013. Together with Deirdre McCloskey, Ziliak authored *The Cult of Statistical Significance* (Ann Arbor: University of Michigan Press, 2008).

extrapolation of a warming PDO, started worrying about long-term warming.

University of Alaska scientist Syun-Ichi Akasofu concludes on the basis of years of observation of the Arctic climate that, before proceeding to any discussion of the role of human activity in influencing global climate change, it is important to identify the extent of *natural* change.[71] He believes that there is ample evidence of a natural warming of 0.5C° per century since the end of the Little Ice Age. Superposed on this linear trend is a natural oscillation of plus or minus 0.2C° reflecting the impact of such phenomena as the PDO and the NAO. As Akasofu argues, it is within the parameters of these verifiable observations that scientists need to consider what influence, if any, is exerted by human activity, including emissions of CO_2. The continuation of this trend, absent any human influences, suggests a possible temperature increase of less than a degree over the course of the 21st century. Given the chaotic nature of the climate system, there is no reason to assume that the warming trend of the past two centuries will not end at some point in the future and plunge the planet back into another Little Ice Age. Akasofu's point, however, is that the burden of proof lies with those who are convinced that natural factors alone are insufficient to explain changes over the past half century.

Many scientists are particularly annoyed by the claim that "the debate is over," a political statement that is at odds with the fundamentals of the scientific method. As discussed in the previous two chapters, science advances on the basis of scepticism, i.e., by constant testing of a theory by experimentation and observation rather than by a show of hands. It is interesting that support for the GHG hypothesis among climate scientists was probably stronger a decade or more ago than now, even as political support has increased. Many climate scientists then thought the hypothesis offered an elegant explanation but required more evidence. Failure of the evidence to materialize and advances in understanding of other factors have gradually eroded support among climate specialists, even as other, non-climate scientists, invested in projects to study AGW's impact, have become more supportive.

Arguments from authority advanced by those with a stake in the hypothesis have further undermined support. These arguments often involve rather arbitrary assertions as to which scientists are authoritative and which are not. Experts from prestigious institutions, alt-

---

71. Syun-Ichi Akasofu, "On the Recovery from the Little Ice Age," *Natural Science* 2:11 (2010), 1211-24.

hough not necessarily climate specialists, are routinely used to validate claims of AGW or of AGW-induced impacts, but woe to an expert from the same or similar prestigious institution who questions
AGW orthodoxy. That person is immediately vilified as a shill for
some special interest. Some expertise is apparently more credible or
authoritative than others. The real issue, of course, should be whether
or not an hypothesis can be validated or falsified through evidence
and observation rather than on the number of scientists, politicians,
or celebrities who may or may not agree with a particular theory.

Gerrymandering the data and exaggerating the impacts

Given that direct evidence of anthropogenic global warming is
weak, it is not surprising that its enthusiasts have been caught
repeatedly cooking the books and manipulating the data in order
to align them with model forecasts. One climate scientist, for example, lamented in an email that in order to strengthen the case
for action, something had to be done about the medieval warm
period and justified this strategy on the moral imperative of addressing a looming crisis.[72] Michael Mann and his colleagues
then obliged, creating the infamous "hockey stick" graph that
became central to the IPCC 2001 report (see the discussion in the
next chapter) by erasing the Medieval Warm Period–Little Ice
Age–Modern Warm Period climate cycle, a phenomenon based
on a wide range of compelling historical, cultural, archaeological,
and paleoclimatic data points (See figure 4-1 above for a schematic illustration).

There is no compelling evidence that late 20th century temperatures were warmer than those of the medieval period from
the ninth to the thirteenth centuries. Paleoclimatologists are
among the most critical of the IPCC case. It is interesting, as van
Kooten points out, that "paleoclimate reconstructions of past
climate are not necessary to make a scientific case for global
warming. Rather, reconstructions such as the hockey stick graph
are important only from a political standpoint, because, if it is
possible to demonstrate that current temperatures are higher
than those experienced in the past, it will be easier to convince
politicians to fund research and implement policies to address

---

72. David Deming, testimony to US Senate Committee on Environment and Public
     Policy, December 6, 2006. The relevant Climategate emails are discussed in
     A.W. Montford, *The Hockey Stick Illusion: Climategate and the Corruption of Science*
     (London: Stacey International, 2010), 420-4.

climate change."[73] But, as Montford concluded after his exhaustive analysis of the hockey stick controversy, "what the hockey affair suggests is that the case for global warming, far from being settled, is actually weak and unconvincing."[74]

A new paper by Marcott et al. vindicating the Mann hypothesis was published in March 2013, purporting to show that global temperatures over the Holocene period had peaked 9,500 years ago and gently declined until the 20th century when they experienced a sharp uptick.[75] The paper was met with much media fanfare and furious commentary in the blogosphere. By early April it had been thoroughly discredited, with Toronto mathematician Stephen McIntyre playing a prominent role. The method and evidence used did not allow the authors to reach any conclusions about 20th century temperatures, as they subsequently admitted. Extending the conclusions to the 20th century came, in the view of some observers, perilously close to scientific misconduct.[76]

James Hansen, a principal adviser to Al Gore, has often been considered the "father" of climate change alarmism since his dramatic testimony before the US Senate in 1988. He has been shown, however, to have repeatedly adjusted the climate series for which he was responsible at NASA's Goddard Institute for Space Studies. That series – as well as the associated climate model – is usually the outlier. The raw data on which the series are built are frequently adjusted to take account of many factors, a process that provides scope for considerable mischief. Hansen's "corrections," for example, always move in a direction favourable to the AGW hypothesis by lowering early 20th century temperatures and raising later ones.[77]

---

73. Cornelis van Kooten, *Climate Change, Climate Science, and Economics: Prospects for an Alternative Energy Future* (New York: Springer, 2013), 95.

74. Montford, *The Hockey Stick Illusion*, 390.

75. Shaun A. Marcott, et al., "A Reconstruction of Regional and Global Temperature for the Past 11,300 Years," *Science* 339: 6124 (March 2013), 1198-1201.

76. See Roger Pielke, Jr., "Fixing the Marcott Mess in Climate Science," rogerpielkejr.blogspot.co.uk, 31 March 2013. Ross McKitrick provides a description of the issue in "We're Not Screwed?" *Financial Post*, April 1, 2013.

77. Good insight into the extent of the controversy can be gleaned by visiting some of the principal web sites of the two camps. Michael Mann and Gavin Schmidt (one of James Hansen's principal collaborators) blog at *RealClimate.com*, while Steve McIntyre maintains the blog ClimateAudit.org. Hansen's former supervisor at NASA, Dr. John S. Theon, upon retirement, suggested that Hansen had embarrassed the Agency and was never muzzled as Hansen has often claimed.

The publication of the Mann, Marcott, and similar papers is the most egregious example of the lengths to which the alarmist community has been prepared to go to demonstrate the extreme nature of current global climatic conditions. The problem, however, goes much deeper. Scientists in the alarmist community have repeatedly refused to share their data with investigators who do not agree with their perspective. In science, the ability of one scientist to replicate and thus validate the work of another is critical to the credibility of the scientific process. Willingness to share basic data is thus not just a matter of common courtesy but an integral part of the advancement of scientific knowledge. Those who refuse to share data and methods raise legitimate questions about the integrity of their work. Internet bloggers repeatedly expose this problem within the climate science community.[78]

Early in January 2015 the media trumpeted, based on a press release from GISS, that 2014 was the warmest year on record. A little examination shows that this was a misleading if not a fraudulent statement. First, the GISS "record" is at best 135 years long, a mere blip in the history of the planet and its climate. Second, the record was claimed by GISS, the weakest and most controversial of the global temperature series, long headed by Hansen and now by his long-time collaborator, Gavin Schmidt. As we have seen, the raw data come from problematic land stations as well as from even more problematic data from sea surface temperature measurements (responsible for 70 percent of the planet's surface). It studiously ignores the better data from radiosondes based on weather balloons (now dating back 70 years), and satellites (now dating back 35 years). None of the other series identified 2014 as the warmest year. Schmidt belatedly admitted – although this was not reported in the mainstream media – that he only had 38 percent confidence that 2014 was the warmest year. By his calculation, the temperature was two-hundredths of a degree higher than in 2010. The global temperature series are calculated to one hun-

---

See "James Hansen's Former NASA Supervisor Declares Himself a Sceptic," *WattsUpWithThat,* January 27, 2009. On the many problems with Hansen's data, see D'Aleo, *United States & Global Data Integrity Issues.*

78. Chris Horner, *Red Hot Lies* (Washington: Regnery, 2008), provides a good overview of the many instances of data tinkering and outright dishonesty in the alarmist community. Exchanges of emails among IPCC contributors about data sharing are discussed in Montford, *The Hockey Stick Illusion,* as well as in Steven Mosher and Thomas W. Fuller, *Climategate: The CRUtape Letters* (CreateSpace, 2010).

dredth of a degree but are based on raw data that are at best accurate to half a degree. As statisticians never tire of pointing out, homogenized statistics can never be more accurate than the raw data on which they are based. The satellite record, the best we have, shows no net warming since 1998. The GISS press release was little more than an exercise in public relations to advance a particular political agenda. Not coincidentally, it also justified President Obama's claim in his January 20, 2015 state-of-the-union address that 2014 was the hottest year on record.[79]

Needless to say, global warming has become a matter of intense scrutiny and debate. Unlike most scientific controversies, however, this one is playing out not only in learned journals but even more in semi-popular journals, the general media, and the blogosphere. Curiously, the proponents of the controversial and still-to-be proven GHG hypothesis have succeeded in making theirs the default position. All those who fail to accept it in full have been labelled "deniers." Usually, the shoe is on the other foot, but in this case a Kuhnian paradigm shift has taken place without the requisite testing and gathering of evidence to validate the new hypothesis. The reason can in part be found in the astute political manipulation of the issue and success in making it the focus of an international intergovernmental, i.e., political, process early in the development of the hypothesis, as discussed in chapters 11-13.

It is not necessary to question the GHG-induced global warming hypothesis to reject the many alarmist predictions. Piling every environmental threat onto the global warming cause may be an interesting technique for environmental NGOs, but it is not a strategy that does any service to serious scientists committed to understanding the drivers of global climate change, as some of these scientists are beginning to recognize. One of the senior scientists at the UK Met Office's Hadley Centre, for example, complains that "for climate scientists, having to continually rein in extraordinary claims that the latest extreme is all due to climate change is, at best, hugely frustrating and, at worst, enormously distracting. Overplaying natural variations in the weather as climate change is

---

79. Claims that 2014 would be the hottest year had already started in the months leading up to the Lima climate conference in December. It was widely reported in the media while only a few news outlets bothered to check or report the low confidence level and tiny change. Not until page 8 of a Google search are there any stories expressing some doubt. See Anthony Watts, "2014 The Most Dishonest Year on Record," *WattsupWithThat*, January 20, 2015.

just as much a distortion of the science as underplaying them to claim that climate change has stopped or is not happening."[80]

The observation that the planet has warmed marginally – perhaps as much as 0.7 to 0.8°C – over the past century and a half is both trivial and largely uncontroversial. Controversy starts in efforts to explain the drivers of the warming. A number of hypotheses have been advanced, some more difficult to test with observation than others, from increases in greenhouse gases due to human activity to changes in the output of the sun. None at this point in time enjoy what could properly be called a consensus. Science is about probabilities, and the probability that a single factor can explain the changes in a complex, chaotic, coupled, non-linear system is at the low end of possibility. Anderson, Goudie, and Parker, in their standard text on climate change through the ages, conclude:

> No completely acceptable explanation of climate change has ever been presented, and it is also clear that no one process acting alone can explain all scales of climatic changes. Some coincidence or combination of processes in time is probably required, and at different times in Earth's history there have been unique combinations of factors involved. This makes it very difficult to generalize about the causes of climate change. ... Given these considerations, it is clearly impossible at the present state of knowledge to make any safe prognosis of the climatic developments of the future.[81]

❧❧❧❧❧

---

80. Vicky Pope, "Scientists must rein in misleading climate change claims," *The Guardian*, February 11, 2009.

81. David E. Anderson, Andrew S. Goudie, and Adrian G. Parker, *Global Environments through the Quaternary* (New York: Oxford University Press, 2007), 305. This cautious assessment is developed in great detail in Alan Longhurst, *Doubt and Certainty in Climate Science*, a pdf of which is freely available at curryja.files.wordpress.com/2015/09/longhurst-clean.pdf.

5 | The Science and Politics of the IPCC

It is extremely likely that human influence on climate caused more than half of the observed increase in global average surface temperature from 1951–2010. There is high confidence that this has warmed the ocean, melted snow and ice, raised global mean sea level, and changed some climate extremes, in the second half of the 20th century.

UNIPCC, *Summary for Policymakers, Climate Change*

Starting with a handful of scientists in obscure fields sucked up into the save-the-world politics of UNEP, we have now arrived, after only a few short decades, with corruption spread across our great institutions of science. Whether knowingly or not, and whatever their motivations, these scientists opened a gap for huge political forces to overwhelm their principles and processes, and to empower those among them willing to participate in the corruption.

Australian climate sceptic Bernie Lewin[1]

Since 1988 the central player in the unfolding drama of the science and politics of catastrophic anthropogenic climate change has been the United Nations Intergovernmental Panel on Climate Change (IPCC). Once activist scientists in Europe and North America, initially brought together by the World Meteorological Organization (WMO), had satisfied themselves that they had a convincing hypothesis to explain that climate change was man-made, they turned to the United Nations to help them make the political case for remedial action. UN leaders grasped the opportunity with alacrity. Progressives at the UN and their supporters around the world had

---

1. "The scientists and the apocalypse," final paragraph of chapter 18 in Alan Moran, ed., *Climate Change: The Facts* (Melbourne: Institute of Public Affairs, 2015), Kindle edition.

long sought a powerful narrative with which to advance their ambitious agenda of global governance. Harnessing the growing appetite among western environmentalists for a concerted campaign to halt and reverse the perceived rape of the planet could provide such a narrative. The environmental issue, particularly its climate dimension, was ideally suited to becoming the central organizing principle of the UN's campaign to eradicate global injustice and inequality by pursuing "sustainable development." This nebulous phrase was invented in the 1970s by Barbara Ward and adopted in the 1980s by the UN's Brundtland Commission to encapsulate the whole of the UN's justice, economic, and environmental agenda. As political scientist Peter Haas describes it: "Sustainable development urges a simultaneous assault on pollution, economic development, unequal distribution of economic resources, and poverty reduction. It argues that most social ills are non-decomposable, and that environmental degradation cannot be addressed without confronting the human activities that give rise to it. Thus sustainable development dramatically expanded the international agenda by arguing that these issues needed to be simultaneously addressed, and that policies should seek to focus on the interactive effects between them."[2] The solutions to all these problems, from the perspective of the UN and its progressive supporters, lay in central planning, state control, and global governance.

The campaign had its origins in the preparations for the 1972 Stockholm Conference on the Human Environment, inspired and chaired by Canada's Maurice Strong. It reached maturity at the 1992 Earth Summit in Rio de Janeiro, again chaired by Strong. At Rio, government leaders were stampeded into endorsing the full UN sustainable development agenda, particularly two new Conventions devoted to the twin environmental "crises" of the late 20th century concocted by progressive scientists: global warming and species diversity.[3] Few government leaders appreciated the extent to which

---

2.   Peter M. Haas, "When does power listen to truth? A constructivist approach to the policy process," *Journal of European Public Policy* 11:4 (August 2004), 570.

3.   Jeffrey Foss provides an excellent summary of the shortcomings of the species diversity issue in *Beyond Environmentalism: A Philosophy of Nature* (Hoboken, NJ: Wiley and Sons, 2009), 35-7. G. Cornelis van Kooten and Erwin H. Bulte provide a more detailed discussion in *The Economics of Nature: Managing Biological Assets* (Oxford: Blackwell, 2000), 270-318, including a discussion of the limits of our knowledge of species diversity. Climate change and conservation come together in UN efforts through its REDD+ program, which provides incentives to devel-

these two conventions were based on advocacy science as practiced by environmental scientists relying on computer models rather than on verifiable observational studies. [4] While the conventions' proponents and supporters may have been convinced that both issues presented real problems, their more important purpose was to advance the UN's progressive agenda. In order to strengthen the scientific case, activist scientists and their supporters had convinced the UN and member governments to establish a process that would engage them in providing periodic reports setting out in clear and convincing terms the looming threat of anthropogenic climate change. To that end, the UN created the IPCC, jointly supported by two established agencies: the WMO and the UN Environmental Program (UNEP – a product of the 1972 Stockholm conference and initially headed by Maurice Strong) with a mandate "to assess on a comprehensive, objective, open, and transparent basis the scientific, technical and socio-economic information relevant to understanding the scientific basis of risk of *human-induced* climate change, its potential impacts and options for adaptation and mitigation."[5] [emphasis added]

The focus on the anthropogenic dimension of climate change, to the exclusion of natural factors, was based on the assumption that natural factors were stable and well understood. That assumption played well with environmental activists and those committed to the salvationist agenda but became increasingly threadbare as less politically committed scientists continued to explore the world of climate change and found that much could be explained on the basis of natural factors alone, from changes in the sun's output to cycles in coupled atmospheric-oceanic circulation systems.

---

oping countries to preserve forests as a climate sink and as critical to maintaining species diversity.

4.   Foss concludes that "environmental science as actually defined and practiced is not pure science in pursuit of knowledge and truth, but advocacy science applied to the pursuit of environmental objectives." *Beyond Environmentalism: A Philosophy of Nature* (Hoboken, NJ: John Wiley & Sons, 2009), 90.

5.   "Principles Governing IPCC Work," at www.ipcc.ch/pdf/ipcc-principles/ipcc-principles.pdf. The idea that the IPCC is an "independent" panel of experts, as frequently asserted by the UN, governments, the media, and climate activists is, of course, a fiction. The best full examination of the Panel's work and composition can be found in two books by Toronto investigative journalist Donna Laframboise, *The Delinquent Teenager Who Was Mistaken for the World's Top Climate Expert* (Toronto: CreateSpace Independent Publishing Platform, 2011) and *Into the Dustbin: Rajendra Pachauri, the Climate Report & The Nobel Peace Prize* (Toronto: CreateSpace Independent Publishing Platform, 2013).

Box 5-1: A Torrent of Words: IPCC Reports 1990-2014

To date, the IPCC has prepared some 28 volumes of material setting out the science, impacts, and mitigation and adaptation measures it believes are required.

Major Assessment Reports

1990	AR 1 – Full reports from WG1, WG2, and WG3	1,042 pp
1992	AR 1 – Supplements from WG1 and WG2	530 pp
1995	AR 2 – Full reports from WG1, WG2, and WG3	1,914 pp
2001	AR 3 – Full reports from WG1, WG2, and WG3	2,660 pp
2007	AR 4 – Full reports from WG1, WG2, and WG3	2,835 pp
2013-14	AR 5 – Full reports from WG1, WG2, and WG3	ca 3,000 pp

Special Reports

1994	Climate Change 1994: Radiative Forcing of Climate Change and An Evaluation of the IPCC 1992 Emissions Scenarios	339 pp
1997	The Regional Impact of Climate Change: An Assessment of Vulnerability	517 pp
1999	Aviation and the Global Atmosphere	373 pp
2000	Land Use, Land-Use Change, and Forestry	375 pp
2000	Emissions Scenarios	570 pp
2000	Methodological and Technological Issues in Technology Transfer	432 pp
2005	Safeguarding the Ozone Layer and the Global Climate System	478 pp
2005	Carbon Dioxide Capture and Storage	431 pp
2011	Renewable Energy Sources and Climate Change Mitigation	1,075 pp
2012	Managing the Risks of Extreme Events and Disasters to Advance Climate Change Adaptation	582 pp

All reports come with summaries for policy makers as well as synthesis reports and technical summaries. All can be downloaded in whole or in part in pdf format at: http://ipcc.ch/publications_and_data/ publications_ and_ data_reports.shtml. Many are published in book form by Cambridge University Press.

The IPCC's mandate was limited to working with the existing knowledge base and using it to find the anthropogenic influence on climate change. Governments were invited to nominate their best scientists to form an intergovernmental panel that would review the available, peer-reviewed literature and prepare periodic reports summarizing the state of knowledge of the human impact on the climate system. These reports would be reviewed by government officials, presumably scientifically literate ones, who would prepare summaries of the science for policy makers. The IPCC would thus provide the intellectual and evidentiary underpinnings for action by governments, both domestically and internationally. The panel was sold to the world as an independent, objective source of advice to governments, and for the next 25 years the media faithfully echoed that myth as the panel poured out one "authoritative" report after another (see Box 5-1). In reality, the panel's leadership was chosen from the activist scientists who had from the beginning been closely involved in developing the catastrophic climate change story. Many of the scientists who contributed to its reports formed part of a closely knit group of researchers who shared the alarmist perspective. A few scientists who were not part of the "in" group participated in the early days but soon wore out their welcome and concentrated on their own work or became much-maligned critics.[6]

The often-hyped work of the IPCC has been proven to be seriously flawed.[7] Its *Summary for Policy Makers* is less the work of scientists than of officials appointed by their governments to produce such a summary. It partakes of the characteristics of both advocacy science and official science. Already in 2002 Essex and McKitrick pointed out:

> We do not need to guess what is the worldview of the [IPCC] leaders. They do not attempt to hide it. They are committed, heart and soul, to the Doctrine. They believe it and they are advocates on its behalf. They have assembled a body of evidence that they feel supports it and they travel the world promoting it. There would be nothing wrong with this if it were only one half of a larger exercise in adjudication. But governments around the world have made the staggering error of treating the [IPCC] as if it is the only side we should listen to in the adjudication process. What is worse, when on a regular basis other scientists and scholars stand up and publicly disagree with the [IPCC], governments panic because they are afraid the issue will get complicated, and undermine the sense of certainty that justifies their policy choices. So they label alternative views 'marginal' and those who hold them 'dissidents'."[8]

---

6. See John McLean, *Prejudiced Authors, Prejudiced Findings*, An SPPI Original Paper, July 2008 and Laframboise, *The Delinquent Teenager*. A good antidote to the 2007 IPCC Summary is Ross McKitrick et al., *Independent Summary for Policy Makers* (Vancouver: Fraser Institute, 2007).

7. See Laframboise, *The Delinquent Teenager*.

8. Chris Essex and Ross McKitrick, *Taken by Storm* (Toronto: Key Porter, 2002), 305.

Scientists who contributed to some of the underlying technical reports have found their work misrepresented and changed. Reviewers who failed to support the IPCC's perspective have found their contributions ignored. The *Summary for Policymakers* of the 2007 assessment of the physical science (WG1), which appeared six months before the underlying technical report was ready, drove changes in the final version of the underlying science report rather than the other way around. That pattern was repeated for the 2013 *Summary*.[9] Even the idea that the underlying work represents the consensus of 2,500 scientists is false. Australian researcher John McLean did the counting for the 2007 report and concluded that its principal findings are the work of a tightly knit network of climate change modellers, many of whom have worked and published together and have reviewed each other's work. Most of the other participants, who add up to fewer than 2,500 due to extensive double counting, have contributed to the impact analysis but often very narrowly and with the assumption that the principal conclusions of the WG1 climate analysis are credible. Most of them are not climate scientists and have no specialized knowledge related to climate science. Many are social scientists or biologists who contribute to the IPPC's reports as reviewers rather than as principal authors. Others are associates of principal investigators rather than prominent researchers in their own right.[10] In short, the so-called consensus is a manifestation of official science on a grand scale. That being said, the *Summaries* have also proven a tremendous political success, and that is a large part of the problem.

---

9.   Once the *Summary* had passed scrutiny by government officials, the IPCC issued a 10-page document setting out the changes that would be made in the underlying technical chapters in order to bring them into line with the *Summary*. "Changes to the Underlying Scientific-Technical Assessment to ensure consistency with the approved Summary for Policymakers," Thirty-Sixth Session of the IPCC Stockholm, September 26, 2013, IPCC XXXVI/ Doc. 4 (27.IX.2013) Agenda Item: 3.

10. As McLean points out: "The IPCC is a single-interest organisation, whose charter *presumes* a widespread human influence on climate, rather than consideration of whether such influence may be negligible or missing altogether. ... More than two-thirds of all authors of chapter 9 of the IPCC's 2007 climate-science assessment are part of a clique whose members have co-authored papers with each other and, we can surmise, very possibly at times acted as peer-reviewers for each other's work. Of the 44 contributing authors, more than half have co-authored papers with the lead authors or coordinating lead authors of chapter 9." McLean, *Prejudiced authors, Prejudiced findings*.

The work of the IPCC, similar to that of UNCTAD a generation earlier, may be useful, even with its bias and political overlay. It makes sense to summarize what is known about a complex issue. Much of what the assessment reports have summarized is useful, if far too one-sided. Nevertheless, there are serious problems with the IPCC's reports. The first is that the *Summaries for Policy Makers* are not only one-sided but also very political and often go well beyond what the underlying scientific reports attest. It is an exemplar of official science, stripped of the qualifications and uncertainties that characterize normal science. Second, the principal scientific report (WG1), the one focused on the extent of future climate change on which the rest of the scientific reports are premised, is based on scenarios or story lines that are wildly unrealistic regarding population growth, economic growth, energy consumption, and similar phenomena. The reasons are not difficult to discern. Only extreme scenarios fed into computer models will result in a rate of climate change that is sufficiently dramatic to capture the political imagination, a fact that Sir John Houghton, the first chair of WG1, acknowledged: "If we want good environmental policy in future, we'll have to have a disaster. It's like safety on public transport. The only way humans will act is if there's been an accident."[11]

The road leading to the establishment of the IPCC explains how it became a single-purpose, one-sided panel and why governments embraced it.[12] As we saw in the previous chapter, in the development of climate science as a distinct discipline, one group learned quickly that recognition, funding, and government support came more easily if it could be demonstrated that they were investigating phenomena that were both unprecedented and threatening. Rather than relying on national initiatives, those climate scientists who believed that climate change was a problem of planetary proportions requiring global action turned to the United Nations. They needed to harness international organizations and put them at the vanguard of a global campaign. Such a top-down approach might stand a better chance of convincing more reluctant and electorally sensitive national politicians to act.

---

11. Interview of Houghton, "Me and my God," *Sunday Telegraph,* September 10, 1995.

12. See William N. Butos and Thomas J. McQuade, "Causes and Consequences of the Climate Science Boom," forthcoming in *The Independent Review,* for discussion of the problems created for climate science by the emergence of a lavishly funded, single-purpose body such as the IPCC.

From speculative science to UN activism

The first step in engaging governments in a global campaign was to develop regular channels of communication among like-minded scientists. Garth Paltridge, an Australian participant in some of those early meetings, reports that the discussion focused on two issues that needed to be resolved in order to improve long-term forecasting: the role of clouds and the role of the oceans. Clouds influence the balance between the heat generated by incoming solar radiation and the cooling effect of outgoing infrared radiation. Oceans are the planet's main heat reservoir and their internal fluctuations affect both short- and long-term climate variations. These issues *remain unresolved* despite more than 40 years of discussions and billions spent on trying to understand the climate system.[13] More critically, discussion rapidly turned to the anthropogenic dimension, a topic which activist scientists believed might prove more fruitful. To this end scientists looked to the UN, which in a 1961 resolution charged the WMO to work with the International Council for Science (ICSU) to develop a Global Atmospheric Research Program (GARP).

The resolution may have looked innocuous at the time, but it became the vehicle through which activist scientists could launch new cooperative research efforts, apprise governmental officials of their concerns, and seek further funding (figure 5-1).[14] The emerging grand narrative on mankind and nature also held tremendous appeal for the UN. The end of the Cold War had largely marginalized its primary mission of world peace. In response, UN officials had turned to its second mission: economic development. That mission had engaged it in one program and policy initiative after another and had steadily made its leaders and active supporters increasingly attracted to progressive and utopian causes. In the 1960s and 1970s, it had concentrated its energy on what became known as the New International Economic Order (NIEO) through such organs

---

13. See Garth Paltridge, "Uncertainty, sceptics, and the climate system," in Alan Moran, *Climate Change: The Facts* (Melbourne, Australia: Institute of Public Affairs, 2015).

14. A good chronology of the evolution of UN involvement in climate change issues can be found in John W. Zillman, "A History of Climate Activities," *WMO Bulletin* 58:3 (July 2009). Zillman represented Australia at many of these meetings. Over time, he became uncomfortable with the one-sided nature of the discussion. See his address at the launch of William Kininmonth, *Climate Change: A Natural Hazard?* at *Lavoisier.com*, August 2008.

Figure 5-1: The UN and climate change discussions

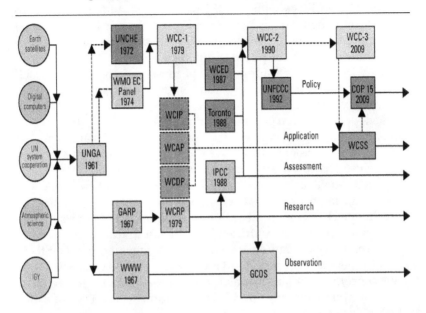

The emergence of climate as an international scientific and policy issue: the five major scientific, technological and geopolitical developments on the left converged to
inspire UN General Assembly (UNGA) Resolution 1721 (XVI) which triggered the establishment of the WMO, World Weather Watch (WWW) and the WMO/ICSU Global Atmospheric Research Program (GARP) and, later and less directly, the convening
of the 1972 United Nations Conference on the Human Environment (UNCHE). The
1974-1977 WMO EC (Executive Committee) Panel of Experts on Climate Change, set
up at the request of the sixth special session of the UNGA, triggered the convening
of the 1979 World Climate Conference (WCC1) and the establishment of the four-
component World Climate Program (WCP), including the WMO/ICSU World Climate
Research Program (WCRP). The 1987 report of the World Commission on Environment and Development (WCED), the 1988 Toronto Conference and the First Assessment Report of the Intergovernmental Panel on Climate Change (IPCC) shaped
the agenda of the 1990 Second World Climate Conference (WCC2), which led to the
establishment of the Global Climate Observing System (GCOS) and the negotiation
of the UN Framework Convention on Climate Change (UNFCCC). The chart also de-
picts the proposed evolution of the service-oriented components of the WCP into a
more integrated World Climate Services System (WCSS), built on GCOS and WCRP,
to produce a new Global Framework for Climate Services.
Source: Zillman, "A History of climate activities," *WMO Bulletin,* 58:3 (July 2009).

as the United Nations Conference on Trade and Development (UNCTAD). In Stockholm in 1972 the UN had added an environmental component and spawned the United Nations Environmental Program UNEP). In the 1980s it sought to combine these two programs into a single, all-purpose one focused on sustainable development. Many of the leading lights of these initiatives were associated with the Club of Rome and similar alarmist groups; they shared a common outlook of pessimism about the future and pushed progressivism as the solution. The fear of anthropogenic climate change wreaking ever greater havoc on the planet spurred them to action and created a close and mutually supportive alliance between activist scientists and progressive utopians

While scientists were still struggling to work out the direction, extent, and impact of climate change, those committed to ensuring that climate change would become an integral part of efforts to save the planet were already sufficiently confident to feature the issue in their broader campaign. In their 1972 book prepared for the Stockholm Conference, Barbara Ward and René Dubos had clearly nailed their concern to the mast of alarm about the climate's future path. Any deviation from a mythical narrow norm would lead to catastrophe. They wrote:

> We encounter another fact about our planetary life: the fragility of the balances through which the natural world that we know survives. In the field of climate, the sun's radiations, the earth's emissions, the universal influence of the oceans, and the impact of the ice are unquestionably vast and beyond any direct influence on the part of man. But the *balance* between incoming and outgoing radiation, the interplay of forces which preserves the average global level of temperature appear to be so even, so precise, that only the slightest shift in the energy balance could disrupt the whole system. It takes only the smallest movement at its fulcrum to swing a seesaw out of the horizontal. It may require only a very small percentage of change in the planet's balance of energy to modify average temperatures by 2°C. Downward, this is another ice age; upward, a return to an ice-free age. In either case, the effects are global and catastrophic.[15]

Scientists had not yet reached this level of certainty or alarm, but less finicky environmental campaigners were ready. At this ear-

---

15.   Barbara Ward and René Dubos, *Only One Earth: The Care and Maintenance of a Small Planet* (New York: Norton, 1972), 192. The 2°C threshold in either direction has no physical meaning, but it would take on increasing importance as the movement maintained that anything beyond this threshold would doom the planet, despite millions of years of geological evidence to the contrary.

ly stage, much of the scientific discussion was focused on competing views of the likely impact of such influences as GHGs and aerosols on future climate patterns. From the start, GARP's focus was on long-term developments and the impact of *human activity*. This work led in 1974 to the UN's commissioning a report on climate change, which was completed in 1977 and confirmed the growing concern – at least, among the climate scientists involved in the discussion – of global warming due to GHGs rather than cooling due to aerosols. Without dismissing the short-term cooling effect of aerosols blocking solar radiation, the long-term threat was considered to be the influence of rising GHGs, now clearly confirmed by the data generated since 1957 at the observatory at Mauna Loa in Hawaii.[16]

In 1979 the WMO, in cooperation with a number of other UN agencies, principally UNEP, organized a World Climate Conference (WCC) in order to give more precision and urgency to what was beginning to be described as the climate crisis. Maurice Strong, the UNEP's first executive director and an inveterate environmental campaigner who never saw an issue that would not benefit from more supranational effort, had ensured from its inception that the UNEP would play an active role in the climate campaign. As Haas points out, "Strong believed that 'the policy is the process': that is by generating an open political process in which states are exposed to consensual science, government officials may be persuaded to adopt more sustainable policies, and individual scientists may gain heightened political profiles at home which may ultimately increase their effectiveness as well."[17] In other words, the UN process of frequent meetings in itself advances the agenda and leads to the development of policy measures congenial to those committed to global governance.

Strong's leadership of the UNEP ensured that climate would remain a central component of the broader UN campaign promoting what would become known as sustainable development. In order to give the movement momentum, Strong welcomed the participation of NGOs in governmental meetings, convinced that their moral fervour and earnest activism would help push governments towards a more favourable stance on his – and the UN's –

---

16. See W. J. Gibbs et al., "Technical Report by the WMO Executive Council Panel of Experts on Climate Change," *WMO Bulletin*, 26:1 (January, 1977) 50-55.

17. Haas, "When does power listen to truth?" 578.

various developmental and environmental schemes.[18] In preparation for Stockholm, Strong collaborated with a Swedish atmospheric physicist and key climate scientist, Bert Bolin, who had devoted his career to understanding the role of greenhouse gases, particularly CO_2, in changing climate patterns. As often happens in science, when a scientist has decided to dedicate his career to discovering the role of a particular factor, he finds it and focuses on it to the exclusion of all others. Bolin *was* such a scientist, but he was also a consummate scientist-diplomat and by the mid-1970s a thoroughly politicized scientist. In 1967 Bolin had been appointed as the first chair of GARP's organizing committee; for the next forty years he played a central role in organizing and directing international efforts and drafting reports to address the problem of human impacts on climate change.[19]

The 1979 World Climate Conference presumptuously called on governments "to *foresee* and *prevent* potential man-made changes in climate that might be adverse to the well-being of humanity"[20] [emphasis added] and to that end established the World Climate Program (WCP) under the WMO's auspices but with support from other UN agencies as well as national and international science organizations. The Program included two research components: Climate Change and Variability Research (led by the WMO) and a Climate Impact Study Program (led by the UNEP). These became additional vehicles for funding alarmist research and conferences and for building a stronger base for international action. At the end of the first week, a smaller group of participants met and hammered out a WCC Declaration summing up their view of the basic objectives of further research and discussion:

> Having regard to the all-pervading influence of climate on human society and on many fields of human activities and endeavour, the Conference finds that it is now urgently necessary for the nations of the world:
>
> > To take full advantage of man's present knowledge of climate;
> > To take steps to improve significantly that knowledge; [and]

---

18. For an appreciation of the extent of Strong's utopian scheming, see Peter Foster, *Why We Bite the Invisible Hand: The Psychology of Anti-Capitalism* (Toronto: Pleasaunce Press, 2014), chapter 15.

19. Weart, *The Discovery of Global Warming*, 105, 108, 129, 145-6, 156. See also John McLean, *Climate Science Corrupted*, SPPI Original Paper, November 20, 2009.

20. UN, "The international response to climate change," *Climate Information Sheet* 17, at UNFCCC.int.

To foresee and prevent potential man-made changes in climate that might be adverse to the well-being of humanity.[21]

On this basis, scientists now enjoyed a strong international mandate – and national funding – to pursue an intensive research program with a particular focus on the role of increasing anthropogenic atmospheric concentrations of greenhouse gases in inducing global warming. It is difficult not to be struck by the stark juxtaposition of the deep pessimism among leaders of the emerging environmental movement about the state of the planet and of humanity's role in its imminent demise and its hubristic optimism about their capacity to solve the planet's problems and restore it to a path of sustainable growth.

Meetings among participating scientists now took place in Villach, a city in southern Austria, and Bellagio, Italy, where planning proceeded for a follow-up to the mandate of the 1979 WCC. At the 1985 Villach conference, chaired by Environment Canada official Jim Bruce, the assembled scientists confidently affirmed their prognosis of impending doom from the impact not only of CO_2 but also of methane, nitrous oxide, fluorocarbons, and other GHGs. They then proceeded to set out what they called a new international consensus: "In the first half of the next century a rise of global mean temperature *could* occur which is greater than any in man's history. ... While some warming of climate now appears inevitable, due to past actions, the rate and degree of future warming *could* be profoundly affected by governmental policies."[22] [emphases added] As one science writer concluded after the Villach Conference, "many experts were frantic to persuade the world of what was about to happen."[23]

The scientists who had been involved in the discussions believed that they had reached the stage at which governments needed to take urgent steps to address the problems they had identified. The extent of those problems called for mitigation measures to slow or even reverse the warming to more "normal" patterns, measures that could be pursued on the basis of existing technologies and at moderate cost. There is no record to suggest that the assembled scientists had given

---

21. Quoted in Zillman, "A history of climate activities."

22. Quoted in Weart, *The Discovery of Global Warming*, 146. While it purported to be a consensus statement, some participants later indicated their dissent from it. See John McLean, "Submission to the UK Parliament Select Committee on Energy and Climate Change," December 17, 2013.

23. Weart, 146, fn 8.

any serious consideration to either the technical or economic challenges that such measures entailed nor that they had invited technical or economic experts to work with them. By involving the UN and its various organs, however, alarmist scientists were pushing on an open door insofar as their solutions involved the same progressive measures already favoured by the UN system as a whole: more government regulation by technocrats based on an expanding network of intergovernmental agreements predicated on wealth redistribution and greater political control of human activity. Not until the establishment of the IPCC were these measures added to the research agenda and then elaborated to ensure a result consistent with the broader UN program: global governance to restrain humanity's self-destructive impulses as manifested in capitalism, industrialization, consumerism, individualism, and all the perceived ills of modern civilization. As the late Alexander Cockburn colourfully put it: "By the late 1980s the UN high brass clearly perceived the 'challenge' of climate change to be the horse to ride to build up the organization's increasingly threadbare moral authority and to claim a role beyond that of being an obvious American errand boy. In 1988 it gave us the IPCC."[24]

The IPCC begins its task

The culmination of the informal efforts to forge a global consensus on climate science in the 1960s and 1970s, many of them sponsored by the UN, had resulted in the establishment of the IPCC, led by the same scientists who had been instrumental in the earlier effort and involving many of the same players. The only addition to this in-group were scientists from developing countries who were needed to provide regional balance. The latter faced two disadvantages: lack of funding for research in their home countries, which meant they had little new to contribute, and a need to catch up with the in-group. Leading scientists from China, India, and elsewhere would soon catch up – and not always by singing from the same hymn-book – but the early phases of the IPCC's work were little more than a continuation of the earlier arrangements, now with the added veneer of being "official." Bolin, by now a seasoned and fully committed veteran of the UN process, became the chair and Britain's John Houghton, another veteran, took on the leadership of Working Group 1, which was studying the physical science basis of global

---

24. Alexander Cockburn, "Who Are the Merchants of Fear?" *The Nation*, May 28, 2007.

warming. Working Group 2, chaired by Russia's Yuri Izrael, would examine the impacts of climate change, while Working Group 3, chaired by Fred Bernthal from the US, would consider possible ways and means of addressing the problems created by adverse impacts. The work of both WG2 and WG3 was critically dependent on what Houghton and his group could establish and would not really come into its own until the third assessment in 2001.

Figure 5-2: Schematic diagrams of global temperature variations since the Pleistocene

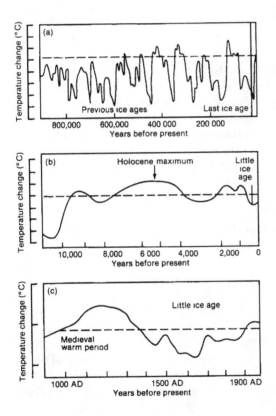

(a) the last million years (b) the last ten thousand years and (c) the last thousand years. The dotted line nominally represents conditions near the beginning of the twentieth century.
Source: IPCC, AR1, WG1, Fig 7.1, 202.

Within two years of its establishment the IPCC produced a report of some weight and consequence (AR1). The most important part, the report of WG1, runs to 414 pages, including a 40-page *Policymakers Summary*. The reports of WG2 and WG3 added a further 296 and 332 pages respectively, including *Policymakers Summaries* of 5 and 40 pages each. By later standards, all three volumes are relatively cautious in their assessments. While remaining consistent with the emerging themes of the climate alarm movement, the scientists and officials who worked on each volume took a credible, if one-sided, approach. At this point, few in the movement were prepared to say that the science was settled. Houghton, for example, frankly admitted that "as in any developing scientific topic, there is a minority of opinions which we have not been able to accommodate, [but] peer review has helped to ensure a high degree of consensus amongst authors and reviewers regarding the results presented. Thus the Assessment is an *authoritative* statement of the views of the international scientific community at this time."[25] [emphasis added] As a result, the first report still included contributions from scientists who would later become uncomfortable with the tone and direction of the Panel's work and decline further participation. The report, for example, relied on widely shared scholarship, advanced no extreme claims, and reproduced the well-known paleoclimatological charts prepared by Hubert Lamb and colleagues (Figure 5-2).

The report also acknowledged that the scientific case was still riddled with uncertainty. In WG1's critical chapter eight, the authors admit: "quantitative detection of the enhanced greenhouse effect using objective means is a vital research area, because it is closely linked to the reduction of uncertainties in the magnitude of the effect and will lead to increased confidence in model projections. The fact that we are unable to reliably detect the predicted signals today does not mean that the greenhouse theory is wrong or that it will not be a serious problem for mankind in the decades ahead."[26]

If anything, the WG1 report showed the huge gaps that remained in scientific knowledge and data. The discussion of available climate data, for example, illustrated the weak foundations on which global temperature anomaly time series were being developed, with most of Asia, Africa, Latin America, Antarctica, the Arctic region and much of the ocean surface not covered at all and dependent on interpolations and algorithms. Similarly, the scientists admitted that their model projections remained at a primitive level. The relatively more confident tone of the *Policymakers Summaries* belied the high degree of uncertainty that the scientists acknowledged in the main body of the report. Any suggestions of danger were based on future developments as projected by models fed "plausible" story lines rather than on any observational evidence. The report's bottom line was that James Hansen's 1988 assertion that the human signal had been detected was

---

25. IPCC, AR1, WG1, "Chairman's Foreword," v.

26. IPCC, AR1, WG1, chapter 8: "Detection of the Greenhouse Effect in the Observations," 243.

premature and had yet to be demonstrated (see chapter 11 for a more detailed discussion of Hansen's premature assertion).

WG1's report, setting out the greenhouse gas theory of climate change science and the anticipated path of future climate change, was clearly the most important part of the report. Until the scientific case for catastrophic anthropomorphic climate change could be clearly established, it would be hard to make a credible case for the malign impacts and the need for cooperative mitigation and adaptation measures that were the subjects of WG2 and WG3. As a result, the case for those was set out in largely speculative terms, enough in the authors' opinion to make the case for a framework convention, a version of which was set out in the report of WG3 and was being negotiated as the assessments were being prepared.[27]

As written and presented, the assessments, including the *Policymakers Summaries,* were not prepared with an eye to the broader public. They were written in a language and style that could only be appreciated by specialists, including fellow scientists and officials focused on the issue. Illustrations, charts, tables, and graphs were technical in nature, in black and white, and not easily interpreted by lay readers. Few of the specialists would have ploughed through all 1,042 pages, and the idea that journalists might sit down and read through all that material is hard to credit. The media relied, instead, on press releases and referred to the reports as "authoritative" without taking the trouble to actually read them. The authors had not yet considered that the IPCC itself could be the main communicator of the science and the dangers of climate change and that its reports could, additionally, serve as the principal vehicle for raising global alarm.

The reports also suffered from a built-in handicap, the tendency to equate climate change with anthropogenic climate change and thus to minimize the important role of the many dimensions of *natural* climate change, dimensions that were not well-known at the time of the IPCC's first assessment report in 1990 and that remained full of uncertainties at the time of its fifth assessment in 2013-14.

Within two years, in time for the Rio Summit in 1992, the IPCC prepared a supplement to its 1990 assessment, aimed at filling a

---

27. Canadian senior environmental official Elizabeth Dowdeswell was vice-chair of WG3, and Bob Rochon from the legal bureau of the Department of External Affairs was the lead official in preparing the draft text. Dowdeswell went on to become the executive director of the UNEP (1993-98) and was appointed lieutenant governor of Ontario in 2014.

number of gaps identified during the negotiation of the UNFCCC. Specifically, the IPCC was asked to prepare:

1. An assessment of national net greenhouse gas emissions, including sources and sinks of greenhouse gases and global warming potentials;
2. Predictions of the regional distributions of climate change and associated impact studies, including model validation studies, updates of regional climate models, and analysis of sensitivity to regional climate change;
3. Energy and industry-related issues;
4. Agriculture and forestry-related issues;
5. Vulnerability to sea-level rise; and
6. Emissions scenarios.[28]

WG1 and WG2 both worked on task 2, while task 6 was discharged by WG1 and tasks 3-5 by WG2. There was no supplementary report from WG3. The 1992 Supplementary Report added a further 350 pages to the 1990 report but did not materially change the analysis or the evidence marshalled by the IPCC in favour of early action by the UN and its members in pursuing a global approach to climate change policy. Much of the Report is repetitive but with some updated information based on more recent technical work. The report of WG2 remained consistent with the views of its chair, Russia's Yuri Izrael, as set out in his preface: "The estimates already available suggest that if continued emission of greenhouse gases persisted through the next century and, in particular, if CO_2 in the atmosphere doubled, there would not be a global catastrophe due to climate change. However, there would be severe impacts in those regions of the world least able to adapt and substantial response measures would need to be taken."[29]

Over the following three years the three working groups continued their efforts to strengthen the case for action and in 1995 released a new assessment (AR2). In the interim, the IPCC had published another special report, *Radiative Forcing of Climate Change,* which examined in depth the mechanisms that govern the relative importance of human and natural factors in radiative forcing, the main "driver" of climate change according to IPCC scientists. The report was largely an exercise in refining the data and models that were used to make

---

28. IPCC, *The IPCC 1990 and 1992 Assessments: Overview and Policymaker Summaries,* Introduction, 5.

29. IPCC, *Climate Change 1992: The Supplementary Report to the IPCC Impacts Assessment,* WG2, Preface by the Chairman, xi.

the case for the malign impact of increasing atmospheric GHGs. The 1995 second assessment, which had at its disposal a further five years of data and specialist literature, largely covered the same ground as the first assessment in 1990 but made a greater effort to identify the human fingerprint, allegedly found by comparing modelled and observed data of near surface temperatures at finer than global scales. At a minimum, the new report had to advance the case beyond that made in 1990 when the panel had concluded that the "the unequivocal detection of the enhanced greenhouse effect from observations is not likely for a decade or more."[30] The 1995 report was largely an exercise in reflecting the progress IPCC-affiliated scientists had made in their modelling of climate change since 1990. On that basis, the report concluded that despite many uncertainties and qualifications, "the balance of evidence suggests that there is a *discernible human influence* on global climate[emphasis added]."[31] For governments and the media, this was the money quote. The rest was detail. It *was* a controversial statement, particularly since the "evidence" came largely from models and could only be discerned by the modellers.

The tension between the science and the political agenda was apparent from the start. For the first assessment report, it was difficult to cobble together 1,042 pages of text involving hundreds of scientists and even more reviewers while avoiding some internal contradictions. In the *Policymakers Summary*, for example, the authors of WG1 insisted:

> We are certain ... [that] there is a natural greenhouse effect which already keeps the Earth warmer than it would otherwise be [and] emissions resulting from human activities are substantially increasing the atmospheric concentrations of the greenhouse gases carbon dioxide, methane, chlorofluorocarbons (CFCs) and nitrous oxide. These increases will enhance the greenhouse effect, resulting on average in an additional warming of the Earth's surface. The main greenhouse gas, water vapour, will increase in response to global warming and further enhance it.[32]

On the following page, however, the same authors indicated that it was their judgment that:

> Global mean surface air temperature has increased by 0.3°C to 0.6°C over the last 100 years, with the five global-average warmest years being in the 1980s. ... The size of this warming is broadly consistent with predic-

---

30. IPCC, AR1, WG1, "Executive Summary," xii.

31. IPCC, AR2, WG1, *Summary for Policymakers*, 5.

32. IPCC, AR1, WG1, "Executive Summary," xi.

tions of climate models, but it is also of the same magnitude as natural climate variability. Thus the observed increase could be largely due to this natural variability; alternatively this variability and other human factors could have offset a still larger human-induced greenhouse warming. The unequivocal detection of the enhanced greenhouse effect from observations is not likely for a decade or more.[33]

The first quotation is a conclusion drawn from theoretical considerations and explored in models; the second is based on observations, a fact that may have been noticed by scientists but was lost on politicians and policy makers. Drawing conclusions on theoretical grounds that are *not* confirmed by observations means that the IPCC knew they were treading on tenuous scientific ground. Bert Bolin subsequently argued that their mission was to be both true to the science and relevant to the Panel's policy purposes.[34] That task would increasingly drive them to compromise the science in order to remain policy relevant.

In the second assessment report, the authors were comfortable with indicating this ambiguity in a single paragraph:

> Our ability to quantify the human influence on global climate is currently limited because the expected signal is still emerging from the noise of natural variability, and because there are uncertainties in key factors. These include the magnitude and patterns of long-term natural variability and the time-evolving pattern of forcing by, and response to, changes in concentrations of greenhouse gases and aerosols, and land surface changes. Nevertheless, *the balance of evidence suggests that there is a discernible human influence on global climate.* [emphasis added.][35]

To its critics, the IPCC had gone too far. Up to this time, criticism of the IPCC's elaboration of the anthropogenic greenhouse warming theory had been muted while criticism of its methods and procedures had been rare, but in 1996, US physicist Frederick Seitz, former president of the US National Academy of Sciences, charged that Ben Santer, the convening lead author of chapter 8 for the 1995 report of WG1, had made changes in the text of the chapter after it had been approved by all the contributing authors and well after the deadline for comments. Santer did not deny that changes were made but insisted that they had been made at the request of governments in order to clarify the text, remove redundancies and am-

---

33.   IPCC, AR1, WG1, "Executive Summary," xii.

34.   Bert Bolin, "Trust the Science," *Our Planet* 7:2 (1994), 23-4 and "Science and Policy Making," *Ambio* 23:1 (1994), 25-9.

35.   IPCC, AR2, WG1, "Summary for Policymakers," 22.

biguities, and bring it into line with the money quote in the *Summary for Policymakers*. Seitz and his colleague, Fred Singer, would have none of this and insisted that the integrity of science was at stake.[36] The final sentence of the paragraph made a claim that was not supported by the science in the main body of the text as approved by the authors in Madrid in August 1995. In the end, Seitz's and Singer's complaint enjoyed only a limited shelf life. Veteran critics of the AGW hypothesis, however, remembered this episode when the embarrassing emails among some of the key climate scientists involved in writing the reports of WG1 for all five assessments came to light in 2009. The chicanery that was alleged in 1996 was now bared for all to see and for the scientists involved to explain.[37]

While Seitz and Singer did not succeed in derailing the IPCC's mission, they did succeed in alerting fellow scientists that something was amiss and that the IPCC's pronouncements were taking a clearly political turn. Until that time, the main criticism had come from the energy industry, at times working through the Saudi and other oil-rich governments. Their reservations were clearly based on economic considerations. Seitz and Singer, however, criticized the *science* and, more importantly, expressed concern about the *integrity* of the science. The conclusions in WG1's second assessment report that there was a discernible human influence had been based on an as-yet unpublished work by Santer, which pointed to a human fingerprint emerging only in the second half of the 20th century. When Santer's article appeared later in 1996, it immediately attracted criti-

---

36. The most detailed analysis of this episode and its implications can be found in a series of blog posts by Australian Bernie Lewin at *Enthusiasm, Skepticism and Science*, starting with "Madrid 1995: Was this the Tipping Point in the Corruption of Climate Science?" April 21, 2012. Seitz's original complaint can be found in "A Major Deception on Global Warming," *The Wall Street Journal*, June 12, 1996, and Santer's response – signed by 40 colleagues – in a letter to the *Journal*, June 25, 2012, accompanied by a letter from Bert Bolin and John Houghton. Seitz and Singer responded in the July 11, 2012 edition of the *Journal*. An earlier exchange took place in the pages of *Energy Daily*, starting with a May 22, 2012 article, "Doctoring the Documents," by Dennis Wamsted, its editor, and followed with a letter from Santer and colleagues on June 3, 2012. This is the first documented instance of the IPCC juicing up its reports in order to get the political message across, but, as Donna Laframboise demonstrates in *The Delinquent Teenager*, it would not be the last. Steven Schneider and Paul Edwards defend Santer and the IPCC in "The 1995 IPCC Report: Broad Consensus or 'Scientific Cleansing'?" *Ecofable/Ecoscience* 1:1 (1997), 3-9.

37. See A.W. Montford, *The Hockey Stick Illusion: Climategate and the Corruption of Science* (London: Stacey International Publishers, 2010).

cism from Patrick Michaels and Chip Knappenberger, who argued that the data were of dubious quality and had been carefully selected to get the desired results.[38] In an earlier paper Santer and his colleagues had set out just how difficult it is to find the human fingerprint, concluding: "If the *paleo* data are reasonably correct and representative of large regions of the planet, then the *current* model estimates of natural variability cannot be used in rigorous tests aimed at detecting anthropogenic signals in the real world."[39] For scientists in the know, finding the human fingerprint remained an elusive quest, but for purposes of building the necessary political momentum to "save the planet," Bolin, Houghton, and the other IPCC principals had shown that they were prepared to shape the science to meet broader political goals.[40] That willingness became more apparent in the next report but, for the moment, the inclusion of the phrase *discernible human influence* was enough to carry the day at the meeting of the Conference of the Parties to the UNFCCC at Kyoto that agreed to the Kyoto Protocol. Left for another day was stronger language that would pave the way for a climate treaty with teeth.

It was now clear that the IPCC, rather than being a purely scientific body assessing the science, was the key institution to which the UN, national governments, and environmental groups looked to

---

38.   Santer et al., "A search for human influences on the thermal structure of the atmosphere," *Nature* 382 (July 4, 1996), 39-46, and Patrick Michaels and Chip Knappenberger, "Human effect on climate?" *Nature* 384 (December 12, 1995), 522-3.

39.   T.P. Barnett et al., "Estimates of low-frequency natural variability in near-surface air temperature," *The Holocene* 6 (1996), 255-63. CRU's Phil Jones and Keith Briffa were among the co-authors, and all five authors were involved in writing chapter 8 of IPCC AR2 WG1, and thus fully aware that Santer's new evidence was not very robust. Three years later the same group reviewed the literature on the detection and attribution of the human fingerprint and conceded that they were far from isolating it. T.P. Barnett et al., "Detection and Attribution of Recent Climate Change: A Status Report," *Bulletin of the American Meteorological Society* 80:12 (December, 1999), 2631-59.

40.   As Lewin clearly shows, there was also a gap between the carefully nuanced and qualified discussion in the specialist literature and what had been included in the IPCC *Summary for Policymakers*. Lewin's account draws heavily on internal Australian delegation reports provided to him by John Zillman, the chief Australian delegate, and one of the few participants in the IPCC meetings who took exception to some of the heavy arm-twisting required to get the claim of a discernible human fingerprint into the report. Lewin's account also illustrates why consensus or settled science is an oxymoron as the scientists and officials wrestle to find politically acceptable formulations of issues that reflect profound differences about the science.

strengthen the case for more meaningful mitigation measures. Even so, the bottom line that emerged from the more than 3,500 pages of text encompassing the first two reports plus the supplement was that the case for human-induced global warming remained weak, Santer's efforts to push the envelope notwithstanding. Lewin characterizes the Santer intervention as "the tipping point ... when political exigencies – the enemies of science – broke through the lines and went on to overrun all its institutions."[41] Perhaps this is so, but it only became apparent well after the fact. Until then, the IPCC's first two reports had done little more than ensure that governments were now sufficiently engaged to conclude the UN Framework Convention on Climate Change (UNFCCC) and its Kyoto Protocol. IPCC leaders realized that something more was required of them, particularly in light of UNFCCC article 4.2(d), which committed the parties to consider more onerous obligations based on the best available science. If more were to be done, the IPCC's next report would have to make the case.

Raino Malnes argues that it should not be surprising that occasionally "scientists tailor their opinion to their interest with a view to procuring funds or securing positions," but that it should "come as a big surprise if it turns out that scientists regularly compromise intellectual concerns," as appeared to be the case in the second IPCC assessment report. He adds, "on becoming members of [the] IPCC, scientists undertook to contribute to policy making. ... It is likely that they felt an urge to provide determinate results."[42] David Hart and David Victor had already argued that this was precisely the case with the global warming movement from the outset:

> The conversion of the greenhouse effect into an environmental problem deemed worthy of sustained public research support in the early 1970s illustrates [the] interactions between [scientific and political objectives]. Elite oceanographers and atmospheric scientists helped to define anthropogenic climate change as an environmental issue; in the process, they broadened the scope of environmentalism, and secured public resources for research on the problem. They also engineered a retooling of non-elite scientists, so that, by 1974, a coherent new thrust had emerged around such problems as detection of climate change.[43]

---

41. Lewin, "Madrid 1995."

42. Raino Malnes, "Imperfect Science," *Global Environmental Politics* 6:3 (August 2006), 61 and 69.

43. Hart and Victor, "Scientific Elites and the Making of US Policy for Climate Change Research, 1957-74," *Social Studies of Science*, 23:4 (November 1993), 661.

In preparation for the next assessment report, due out in 2001, there was a changing of the guard at the IPCC. Bert Bolin stepped down as chair and was replaced by Robert Watson, a British scientist with experience on both sides of the Atlantic and an outspoken advocate of environmental causes. He had worked at NASA, in the Clinton White House, and at the World Bank and had been a contributor to the first two reports of WG1. He assumed his new duties in 1997 and added a harder edge to the Panel's work. John Houghton stayed on as chair of WG1, but Yuri Izrael was replaced by Harvard's James McCarthy while the task of chairing WG3 fell to the Netherlands' Bert Metz.

The key task lay with WG1 and the authors of what would become chapters two and twelve in the next report: "Observed Climate Variability and Change," and "Detection of Climate Change and Attribution of Causes." Together, these two chapters would provide the key findings that demonstrated the human dimension responsible for accelerating climate change. Santer was again among the contributors but this time serving as a contributing author, together with a new star, Michael E. Mann, a recently minted Yale PhD on the faculty of the University of Massachusetts. His claim to fame was a pair of articles written together with two more experienced paleoclimatologists, R.S. Bradley and M.K. Hughes.[44] Those two articles purported to demonstrate that the paleo record for the northern hemisphere, made up largely from tree rings and some ice and sediment cores, showed that for the previous thousand years, temperatures had varied between 0.2 and 0.5 C° below the 1961-1990 average and then, starting at the turn of the 20th century, had begun to shoot up to reach 0.7° above that average: the hockey stick graph (See Figure 5-3).

It did not take long for the IPCC's leaders to recognize a gem. The hockey stick graph had more explanatory power for the public than dozens of articles in learned journals. Properly drawn, it would convince all but the most hardened sceptics that the planet was warming at an alarming rate. When combined with the projections

---

44. Michael Mann, R.S. Bradley, and M.K. Hughes, "Global-scale temperature patterns and climate forcing over the past six centuries," *Nature* 392: 6678 (1998), 779–787, and Mann, Bradley, and Hughes, "Northern hemisphere temperatures during the past millennium: inferences, uncertainties, and limitations," *Geophysical Research Letters* 26:6 (1999), 759–762. Bradley and Hughes were also members of the IPCC WG1 writing team. The most devastating critique of Mann and his work can be found in Mark Steyn, *A Disgrace to the Profession* (Stockade Books, 2015).

Figure 5-3: The IPCC hockey-stick graph

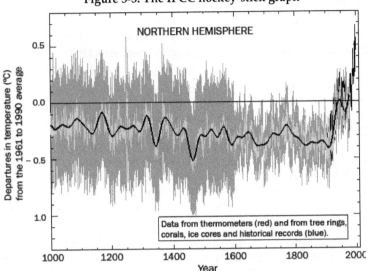

The shaded areas indicate the error bars. To the experienced observer of such graphs, the error bars make the black line look much weaker.

Source: IPCC, AR3, WGI, *Summary for Policy Makers* (2001), 3.

of the models, it was dynamite, particularly when compared with the Lamb graph used in AR1. (See Figure 5-2) Gone were the Medieval Climate Optimum and the Little Ice Age. After 1000 years of stable temperatures, rapidly growing greenhouse gases could push temperatures as high as 5C° above those in which modern civilization had thrived. Who would deny that this was a frightening prospect? The original graph was featured five times in the third assessment report, and an enhanced version featured in the less widely read synthesis report. IPCC scientists now had a sure-fire money quote and the lead for the opening pages of WG1's third assessment report:

> New analyses of proxy data for the Northern Hemisphere indicate that the increase in temperature in the 20th century is likely to have been the largest of any century during the past 1,000 years. It is also likely that, in the Northern Hemisphere, the 1990s was the warmest decade and 1998 the warmest year. Because less data are available, less is known about annual averages prior to 1,000 years before present and

for conditions prevailing in most of the Southern Hemisphere prior to 1861.[45]

Combining the findings from paleoclimatology (chapter 2) with those of the scientists working on detection and attribution (chapter 12), the *Summary for Policymakers* was able to conclude that:

> There is a longer and more closely scrutinised temperature record and new model estimates of variability. The warming over the past 100 years is very unlikely to be due to internal variability alone, as estimated by current models. Reconstructions of climate data for the past 1,000 years also indicate that this warming was unusual and is unlikely to be entirely natural in origin. ... There are new estimates of the climate response to natural and anthropogenic forcing, and new detection techniques have been applied. Detection and attribution studies consistently find evidence for an anthropogenic signal in the climate record of the last 35 to 50 years.[46]

For the previous 20 years the approach to detection had largely been a matter of eliminating natural causes, thus leaving human influence as the only plausible explanation for the increase in temperature over the second half of the century. The paleo record now made that process even more convincing: natural forces had kept the average global temperature within a narrow band of less than half a degree for at least a millennium; starting at mid-century it had escaped that stable band, and models indicated that there was much more to come. The only satisfactory explanation, according to the IPCC, was the enhanced forcing from rising GHGs – the result of human activity – driving temperatures well beyond historical experience.

It was a powerful argument, but it was also bogus, and it did not take long for suspicious scientists, statisticians, and mathematicians to discover why. Two Canadian investigators, mathematician Stephen McIntyre and economist Ross McKitrick, doggedly pursued Mann in order to determine both the quality of his data and the methodology he had used to arrive at his amazing results.[47] The

---

45. IPCC, AR3, WG1, *Summary for Policymakers*, 2.

46. IPCC, AR3, WG1, *Summary for Policymakers*, 10.

47. McIntyre founded a blog in 2003, *Climateaudit.com*, originally devoted almost exclusively to debunking Mann's work but extended to other dubious claims made by climate scientists as the years have gone by. McIntyre takes no position on whether the GHG hypothesis is valid or not, rather devoting his considerable forensic skills to determining whether the data and methods used by climate scientists actually demonstrate what is claimed. In too many instances he has demonstrated that either the data, the methodology, or both, do not hold up.

data were not originally Mann's but were the work of various other investigators. Some of the data were wrongly interpreted by Mann (e.g., turning one series of numbers from lake sediments upside down), others had been carefully cherry-picked, and data for the period before 1400 depended on a single miraculous tree. Building chronologies from tree-ring data – dendrochronology – was also a matter of controversy, with many specialists indicating that without proper validation of the data from instrumental records, tree-ring chronologies were of little value. Determining whether tree-ring widths are due to temperature, precipitation, fertilization, or other influences is also a matter of controversy, suggesting that tree-ring data need to be interpreted with great care. Finally, suggesting that the graph represented the average temperature of the Northern Hemisphere was a stretch, given the limited provenance of the tree-ring data used in building the chronology.

Much more damning, however, was the discovery of two elements that bordered on fraud. The first came to light once McIntyre had determined Mann's methodology and discovered that it depended on novel statistical techniques that no experienced statistician would ever use. As McIntyre demonstrated, it did not matter what numbers were fed into Mann's methodology, the result would have come out in the shape of a hockey stick. Even more damning was the fact that Mann had relied on a subterfuge that became known as "Mike's trick." It turned out that the tree-ring data did not confirm the rapid increase in temperature in the second-half of the 20th century. This was known among climate scientists as the divergence problem: the tree-ring record diverged from the instrumental record. Mann solved this problem by truncating his paleo data and replacing it with instrumental data to accentuate the blade. If he had used only paleo data, the hockey stick would have lacked the prominent blade. The fact that even with his questionable methodology he could not produce a convincing hockey stick suggests that the paleo data on which he had relied did not provide a very reliable indication of temperature variations. Taken together, these discoveries completely destroyed the credibility of the star witness to rapid anthropogenic warming. Mann and his colleagues have repeatedly tried to rehabilitate their method and data with predictable

---

Together with McKitrick, he has also contributed to the peer-reviewed literature pursuing the same theme. McIntyre provides a retrospective on his saga in "IPCC and the 'Trick'," *Climateaudit.com*, December 10, 2009.

results: McIntyre and others have consistently demolished those results.[48]

Even more damning is the fact that it has become increasingly clear, particularly after the release of the Climategate e-mails, that many of Mann's fellow IPCC authors knew full well that including his graph was an exercise in deception. They also knew that Mann had only recently completed his PhD, was extremely ambitious, and was very defensive of his work when challenged. He was far from the seasoned climate expert portrayed later by the IPCC's defenders. Nevertheless, the allure of the potential impact of his chart was such that they chose to go with it and ignore all the contradictory data at their disposal. John Christy, one of Mann's lead co-authors of chapter two, but not one of the in-group, provided a devastating account to the US Congress in 2011. He made clear that, as time went by, the process became more and more political, with the science taking a back seat to the politics of climate change. The need for some deception in order to get the story across became routine, and the continuing uncertainty was swept from the IPCC's pages. In his testimony, Christy provides disturbing details of how the chapter two team gradually cobbled together what they all knew was a highly questionable graph. He told the congressional committee:

> To many [among the chapter two team], this appeared to be a "smoking gun" of temperature change proving that the 20th century warming was unprecedented and therefore likely to be the result of human emissions of greenhouse gases. ... The Hockey Stick was prominently featured during IPCC meetings from 1999 onward. I can assure the committee that those not familiar with issues regarding reconstructions of this type (and even many who should have been) were truly enamoured by its depiction of temperature and sincerely wanted to believe it was truth. Scepticism was virtually non-existent. Indeed it was described as a "clear favourite" for the overall *Policy Makers Summary*.[49]

---

48. In order to salvage his career, Mann has devoted much of his energy to suing his critics and trying to block access to his e-mails and data prepared with public funds. The most notable court case is one involving columnist Mark Steyn, which is wending its way slowly through the courts of the US District of Columbia and which is chronicled in Steyn's columns available at his website, *Steynonline.com*.

49. John Christy, "Testimony to the US House Committee Examining the Process concerning Climate Change Assessments," March 31, 2011. The testimony as a whole provides critical insight into the operations of the tightly knit group of scientists who are at the centre of the IPCC's work, the same people who are the stars of the Climategate e-mails. Appended to Christy's testimony are some of his earlier, critical remarks on problems with the IPCC as well as testimony to

Whatever their private misgivings at the time, the chart was a smash success and became an icon around the world that proved that global warming was happening and was threatening to become much worse unless steps were taken to curb greenhouse gas emissions. Unfortunately, the IPCC had also set a precedent that would come back to haunt it as time went on. The 1996 effort to change an ambiguity into a certainty could perhaps be forgiven, but the 2001 use of the hockey stick was clearly conceived to deceive and a telling example of the science becoming the handmaiden of the IPCC's political agenda. By the time of the fourth assessment report, the number of deceptive practices rose to a flood, and the critics had a field day finding them and pointing them out to all who would listen, a number that steadily increased with time.

AR3 depended heavily on models, simulations, and scenarios to make its case. By using general circulation climate models, IPCC scientists claimed that they could simulate both natural and anthropogenic forcings and that only by combining both types of forcings could the models track the observed global climate over the previous 150 years. The argument, of course, is entirely dependent on the assumptions and data fed into the models. These limitations were not featured in the *Summary for Policymakers,* and only the most dedicated investigators would find them. Consequently, the SPM had become something not originally intended: a vehicle for summarizing and concentrating the core findings without all the nuances and caveats, thus strengthening the message with governments, the media, and the wider public.

For the third assessment report the IPCC hired editors and graphics specialists to enhance its accessibility to the lay reader. The graphics are especially well done and more likely to maximize alarm. They lend themselves well to repackaging in slick brochures and booklets, as Canadian officials did, prominently featuring the hockey stick graph and the conclusion that the 1990 decade was the hottest one in a millennium with 1998 the hottest year. The combination of frightening temperature data and attractive graphics brought global warming alarm home to a much wider public and

---

that effect from Ross McKitrick. Christy also observes that "the L.A. [lead author] of this particular section [Mann] had been awarded a PhD only a few months before his selection by the IPCC. Such a process can lead to a biased assessment of any science. But, problems are made more likely in climate science, because, as noted, ours is a murky field of research – we still can't explain much of what happens in weather and climate."

provided the movement with the momentum it desperately needed. The media, ENGOs, and government websites still feature both, ignoring the storm of valid criticism that they unleashed.

In light of the dramatic impact of AR3's WG1 report, the impact of the equally lengthy reports from WG2 and WG3 proved to be anticlimactic. WG2 discusses the sensitivity, adaptive capacity, and vulnerability of natural and human systems to climate change as well as the potential consequences of climate change, issues on which experts can have many opinions but which are essentially unknowable. Computer models can make the speculation more systematic and impart an air of scholarship, but in the final analysis they remain speculations. The evolution of complex natural and socio-economic systems and their interactions are difficult to model, and previous experience with computer-based prognostications does not inspire confidence. In its 2001 report the IPCC extended its speculation on these themes to 890 pages. The most important conclusion is that much more money is needed to fund the research required to fill the gaps in human knowledge. The IPCC appears to have discovered one of the holy grails of the academy: a process capable of generating research that provides compelling reasons to fund more research.

WG3's report worked out to a modest 437 pages that assess the scientific, technical, environmental, economic, and social aspects of the mitigation of climate change. The report is basically an exercise in demonstrating how much better off the world would be if it fully implemented the UN/IPCC progressive agenda. In the report's own words: "The effectiveness of climate change mitigation can be enhanced when climate policies are integrated with the non-climate objectives of national and sectorial policy development and turned into broad transition strategies to achieve the long-term social and technological changes required by both sustainable development and climate change mitigation."[50] The report envisions a world of extensive central planning in order to ensure high levels of local, national, and international coordination and reminds governments once again that the climate change problem "involves complex interactions between climatic, environmental, economic, political, institutional, social and technological processes. This may have significant international and intergenerational implications in the context of broader societal goals such as equity and sustainable develop-

---

50.   IPCC AR3 WG3, *Summary for Policymakers*, 12.

ment."[51] Computer modelling played a more critical role in this third iteration of the mitigation challenge, indicating the impact of various scenarios on economic development, energy usage, emissions, atmospheric stabilization and abatement of temperature increases to an "acceptable" level. Here we find some of the most egregious internal contradictions: prognostications, for example, of economic progress in developing countries that will result in national incomes well above those of many industrialized countries today but at the same time the need of these same developing countries for extensive aid to overcome the malign impacts of climate change. Only people fully immersed in UN/IPCC-think could write such a seeming non sequitur. Similar to WG2's report, the WG3 *Summary for Policymakers* concludes with a "high priority" agenda for more research and an implied appeal for funding.

The IPCC ups the ante

The third assessment report had given the global warming cause a huge initial boost but one that dissipated as the controversy surrounding the hockey stick graph took its toll. The next report would have to undo that damage and show that scientists were more confident than ever and that they were on the right track. By the time the report came out – in three tranches in February, April, and November 2007 – the movement was on a roll: the Kyoto Protocol had come into force, and governments were in the midst of preparing for an ambitious conference in Bali that would advance the prospects of a much more aggressive agreement. It would replace Kyoto and have the intended effect of stabilizing the world's climate around an average global temperature no higher than 2°C above the pre-industrial norm, an arbitrary number first advanced by Barbara Ward and René Dubos in their book for the Stockholm Conference that had since been adopted as the political goal of the movement.

The fourth assessment report, AR4, did not disappoint. The IPCC, now under the leadership of Rajendra Pachauri, an Indian railroad engineer, pulled out all the stops to ensure that the report gained worldwide publicity. At 2,835 pages, AR4 set a new record for verbosity and for repeating itself. This time, not only did each of the three working groups provide a report including a *Summary for Policy Makers* (SPM) as well as a *Technical Summary* (TS), but the IPCC's leaders also produced a *Synthesis Report* (SR) in November. The SR is

---

51.   IPCC AR3 WG3, *Summary for Policymakers*, 3.

no more than a Reader's Digest version of the reports of the three working groups, with emphasis on the juicy, alarming bits. The American Enterprise Institute's Hayward, Green, and Schwartz found that "the levels of confidence and alarm cited in the new [*Synthesis Report*] exceed not only that of the underlying primary climate research, but even that of the three working group reports upon which the SR is ostensibly based. ... In general, the three working group reports do an admirable job of reviewing and evaluating an enormous body of scientific work and are well worth careful reading. A careful reading, however, will disabuse any fair-minded reader that many important aspects of climate science are 'settled' and beyond argument. ... the terms 'uncertain' and 'uncertainties' appear more than 1,300 times in the 987-page full report of WG1 [alone]."[52] But then, few people would ever read all of the report. The purpose of the SPMs and the *Synthesis Report* was to remove the nuances and caveats and to craft as alarming a picture as possible. They are political documents created by the IPCC's leadership, some of whom are actually scientists. The leadership sees its primary mission as the delivery of a succinct and convincing political message that will confront policy makers with the need for urgent action.

Given the length of AR4's primary reports and their SPMs, the leadership decided that the shorter version set out in the *Synthesis Report* integrating the findings of all three reports might appeal more to ENGOs, journalists, and, most of all, officials. Politicians, of course, would get their own even shorter versions from the briefing notes prepared by their officials, shielding them from learning about the continuing levels of uncertainty and the hundreds of qualifications and nuances actually expressed throughout the three reports by the scientists themselves.

All three working group reports were more confident than ever, although one would look in vain for any new evidence to support that confidence, i.e., evidence based on physical observations rather than on computer models. One would also look in vain for the hockey-stick graph. It had been banished, although the IPCC offered no explanation for this lacuna and continued to assert that "paleoclimatic reconstructions show that the second half of the 20th century was likely the warmest 50-year period in the Northern Hemisphere in the

---

52. Stephen F. Hayward, Kenneth P. Green, and Joel Schwartz, "Politics Posing as Science: A Preliminary Assessment of the IPCC's Latest Climate Change Report," American Enterprise Institute, December 4, 2007, 1, 3.

last 1300 years,"[53] accompanied by new graphs that purported to show the range of climate for the past 1300 years on the basis of multi-proxy data. It is less misleading than the original but is still based on data that have been cherry-picked to get the right results and that make use of "Mike's trick," i.e., using proxy data and grafting instrumental data on the end, thus comparing apples and oranges.[54]

Instead of focusing on the paleoclimate record, the authors of both WG1 and WG2 concentrated on so-called secondary evidence of the impacts of global warming, much of that evidence again the work of computer models which, in the world of the IPCC, is often confused with reality. To the extent that any of this is reliable, physical evidence might point to some warming at the regional level, but there has been no net warming globally since the 1997/98 El Niño. Even then, the physical evidence could not differentiate between *natural* and *anthropogenic* warming.[55] If anything, the evidence illustrated nature's dynamism rather than any malign impact on the part of humans. Physical evidence linking human activities with climate change remained elusive. Subsequent investigation indicated that much of the information about secondary impacts had been sourced from so-called grey literature – largely reports from ENGOs – rather than from peer-reviewed literature and had no basis in any scientific work based on observations.[56] For example:

- claims that the Himalayan glaciers would melt by 2035 came from a non-peer reviewed article in *New Scientist* and was based on a guess expressed in a phone interview by an Indian glaciologist; the IPCC author responsible for this tidbit subsequently admitted he put it in to scare people; none of his co-authors or re-

---

53. IPCC, AR4, WG1, 702.

54. The paleoclimate data for the last 2000 years are discussed in WG1, chapter 6, 466-83. Mann is no longer listed as a contributor, but some other stars of the Climategate emails continued to be on the writing team, including Keith Briffa and Jonathan Overpeck.

55. The problem with secondary evidence is discussed in the next chapter.

56. See Donna Laframboise, *Findings of the Citizen Audit of the 2007 IPCC Report*, at *NoFrakkingConsensus*, April 14, 2010. The audit found that 5,587 of the 18,531 references cited in the 2007 IPCC report were not from peer-reviewed literature. Use of grey literature was more prevalent in the reports of WG2 and WG3 than in that of WG1, but no chapter was exempt. In 13 chapters, more than half the references were to grey literature. Only 8 chapters, all in WG1, relied on grey literature for less than 10 per cent of their references.

viewers were sufficiently familiar with the science to catch this exaggeration.[57]

- a claim that up to 40 percent of the Amazon rainforest was vulnerable to fire due to lack of precipitation came from an unsubstantiated WWF report;
- claims that crop yields in Africa would be halved by 2020 came from a Canadian ENGO, with no basis in science; and
- claims linking rising property loss due to climate extremes were not based on any scientific literature and were contradicted by the underlying reports.[58]

These and other gaffes are indicative of a process determined to provide as much alarm as possible and careless enough to let such obvious nonsense slip through the review process.[59] Once again the in-group responsible for the reports decided to ignore all science that was not predicated on the IPCC's GHG hypothesis and its modelled impacts. The comments of reviewers whose opinions did not fit the template were ignored. Instead, the IPCC preferred to rely on the work of scientists and other "experts" with close ties to ENGOs whose financial health is entirely dependent on maintaining the climate scare.[60] Based on a determination to concentrate on human forcings, little or no attention was paid to increasingly interesting work on the role of the sun, cosmic rays, clouds, aerosols, coupled atmospheric-oceanic forces, and other natural factors, as discussed in chapter seven. Scant effort was devoted to such contro-

---

57. Chris Essex, "Deceived and manipulated," *Quadrant Online*, March 21, 2010. In a later article, Tony Thomas dissects the full extent of the Himalayan gaffe. He notes: "The IPCC authors also got the area of Himalayan glaciers wrong (33,000 square kilometres, not 500,000!), and the number of glaciers wrong (9000 to 12,000, not 15,000). The Pindari glacier shown as having annual shrinkage of 135 metres is actually shrinking only by 25 metres a year." He also draws attention to how long it would take for other glaciers to melt at current rates. "The Fictive World of Rajendra Pachauri," *Quadrant Online*, March 1, 2012.

58. See, for example, Jeffrey Ball and Keith Johnson, "Climate Group Admits Mistakes," *Wall Street Journal*, February 10, 2010.

59. See the catalogue compiled by John McLean of articles published in 2007 pointing out the many failings. "The IPCC under the Microscope," at mclean.ch/climate/IPCC.htm.

60. Donna Laframboise has meticulously documented the extent to which the IPCC has been "colonized" by environmentalists at *NoFrakkingConsensus*. Many IPCC staffers and volunteers are active members of such ENGOs as Greenpeace and the World Wildlife Fund. See also her testimony to the UK House of Commons Energy and Climate Change Committee, "The Lipstick on the Pig: Science and the Intergovernmental Panel on Climate Change," December 10, 2013.

versial issues as the quality of the data on which the authors relied or the role of the urban heat island effect.

The key chapter this time was chapter 9 of WG1: "Understanding and Attributing Climate Change," prepared under the leadership of Gabrielle Hegerl (Germany and US) and Francis Zwiers (Canada). Over the course of 71 small-print pages, including charts, graphs, and illustrations, the authors review the voluminous literature on detection and attribution of climate change. Fascinating as this literature may be, none of it offers direct observational evidence to provide the basis for the confident claim set out in the *Summary for Policymakers*:

> Warming of the climate system is unequivocal, as is now evident from observations of increases in global average air and ocean temperatures, widespread melting of snow and ice, and rising global average sea level. … Most of the observed increase in global average temperatures since the mid-20th century is *very likely* due to the observed increase in anthropogenic greenhouse gas concentrations. This is an advance since [AR3's] conclusion that 'most of the observed warming over the last 50 years is *likely* to have been due to the increase in greenhouse gas concentrations.' Discernible human influences now extend to other aspects of climate, including ocean warming, continental-average temperatures, temperature extremes and wind patterns. [emphasis added].[61]

The tenuous nature of the detection and attribution of the human signal was clearly set out in AR3 but buried this time in an avalanche of words. A more honest assessment would have explained why, despite the continuing efforts of hundreds of scientists, observational evidence of the human fingerprint remained elusive. Upgrading the assessment from *likely* to *very likely* was a purely political decision.

The SPMs made a mockery of some of the careful nuances in the main reports. The *Synthesis Report*, for example, claimed that "there is high agreement and much evidence that all stabilization levels assessed can be achieved by deployment of a portfolio of technologies that are either currently available or expected to be commercialized in coming decades, assuming appropriate and effective incentives are in place for their development, acquisition, deployment and diffusion and addressing related barriers."[62] As discussed later (chapter eight), there was no basis for such a claim in 2007, nor is there today. The authors of the *Synthesis Report* knew this because the re-

---

61. IPCC, AR4, WG1, *Summary for Policymakers*, 5 and 10.

62. IPCC, AR4, *Synthesis Report*, 68.

port of WG3 had clearly stated that "many studies have indicated that the technology required to reduce GHG emissions and eventually stabilize their atmospheric concentrations is not currently available."[63]

The IPCC's extensive reliance on grey literature substantially undermines the claim that it is based on the peer-reviewed work of thousands of scientists around the world. Since the IPCC has no mandate to conduct any research of its own, it relies exclusively on existing literature. Propaganda from ENGOs and similar sources hardly qualifies as the latest scientific research. The fact that all the dubious claims helped to make the reports more alarming also suggests that they were included deliberately and approved at senior levels. It is also curious that such mistakes managed to make it into a report that had been checked and rechecked by a team of authors and then sent out for review and comment by other scientists. As Donna Laframboise learned when she did some more digging, many contributors to the reports believe "the use of grey literature is *essential, necessary,* and *unavoidable* in the preparation of IPCC reports. According to [IPCC insiders], the IPCC has relied on grey literature *extensively* for some time."[64] She also learned that "a significant number of IPCC insiders believe many of their colleagues possess inferior scientific credentials. They believe these people's participation in the IPCC is a result of concerns that have nothing to do with science."[65]

The claim that AR4 represented the work of hundreds of scientists, who in turn had mined the work of thousands more, is also misleading. In fact, each chapter is often the work of fewer than half a dozen people and was reviewed by perhaps another half dozen. In many cases, they were relying on their own work and, when necessary, prepared papers that were rushed into publication on the basis of peer review by fellow authors in order to fill in gaps. Most chapters represented the views of small in-groups who were committed to advancing their own work and freezing out competing

---

63.   IPCC, AR4, WG3, 485.

64.   Laframboise, "Grey Literature: IPCC Insiders Speak Candidly," *NofrakkingConsensus,* January 21, 2011.

65.   Laframboise, "IPCC Nobel Laureates Lack Scientific Credibility," January 20, 2011, at *NoFrakkingConsensus.* She found this and other damning information about the IPCC's working procedures in the background information to the October 2010 Report of the InterAcademy Council, *Climate change assessments: Review of the processes and procedures of the IPCC.*

views. As Andrew Lord Turnbull points out: "While the IPCC presents itself as a synthesis of the work of over 2,000 scientists it appears that in practice it is a process in which a much smaller number of scientists, whose work and careers are intertwined, dominate the assessment and seek to repel those who are situated elsewhere in the spectrum of scientific opinion. There is no transparent process for selection of participants in the assessments. Its handling of uncertainty is flawed and outcomes that are highly speculative are presented with unwarranted certainty. Use is made of non-peer-reviewed material without identifying it as such."[66]

The most damning assessment of AR4 came from an IPCC insider, economist Richard Tol, who concluded: "I have read most of AR4, and by and large it is able but uninspiring. Climate policy may be one of [the] greatest challenges of our time – it has been 20 years since the IPCC was formed, and 10 years since the Kyoto Protocol was signed, but climate policy has achieved close to nothing – and one would hope that AR4 would teem with intellectual energy in an attempt to solve the many questions that are still open. Instead, it is a rather dull read, with little news even in those areas that I do not follow on a daily basis."[67] Precisely! At a cost of millions of dollars, the IPCC had by now produced over 7,000 pages of text in its four assessment reports, plus thousands more in its special reports, much of it repetitive, uninspiring, and misleading, but it had failed to find any convincing evidence based on observations that the modest step-change in global temperature in the late 1970s, analogous to a similar step-change earlier in the century, had been the result of human agency. Instead, it continued to use models to hammer away at an unproven assumption and to apply the results to model impacts and solutions that were even more difficult to take seriously than the results of the climate models.

Despite the many readily identified problems with AR4, both the scientific and popular media enthusiastically welcomed each report as strongly confirming the need for governments to act

---

66. Lord Turnbull, "Foreword," to Andrew Montford, *The Climategate Inquiries* (London: Global Warming Policy Foundation, 2010), 4. Turnbull was a senior British official who served as head of the British civil service and secretary to the cabinet as well as permanent secretary of the Treasury and of the Environment Department. See also John McLean, *Prejudiced Authors, Prejudiced Findings*, SPPI Original Paper, July 2008, and McLean, *The IPCC Can't Count its 'Expert Scientists:' Author and Reviewer Numbers Are Wrong*, SPPI Reprint, July 2009.

67. Richard S.J. Tol, "Biased Policy Advice from the Intergovernmental Panel on Climate Change," *Energy & Environment* 18:7/8 (2007), 929.

quickly and to ward off decisively the looming catastrophe of global warming. *Nature*, for example, devoted ten articles in one issue to the latest assessment of WG1. It editorialized that "the IPCC report, released in Paris, has served a useful purpose in removing the last ground from under the climate-change sceptics' feet, leaving them looking marooned and ridiculous."[68] A few pages later, reporter Jim Giles enthused that "the disturbing predictions about global warming in the latest report from the Intergovernmental Panel on Climate Change (IPCC) mark a turning point. That's not because of the figures themselves, which are largely in line with previous IPCC forecasts, but because the science behind them is now certain enough to make a serious response from policy makers almost inevitable. The debate is no longer about whether we can believe the numbers, but what we should do about them."[69] Giles reached this conclusion strictly on the basis of WG1's *Summary for Policy Makers,* since the main report was still being massaged to make it consistent with the SPM. The New York *Times,* "a trumpet that never sounds retreat in today's war against warming,"[70] editorialized that "the world's scientists have done their job. Now it's time for world leaders ... to do theirs. That is the urgent message at the core of the latest – and the most powerful – report from the Intergovernmental Panel on Climate Change, a group of 2,500 scientists who collectively constitute the world's most authoritative voice on global warming."[71] The *Times* also thought it helpful to point out that many of the contributing authors believed the situation was even worse than summarized in AR4.[72] Action clearly could no longer be delayed.

There were a few dissenting voices, but they were generally drowned out by the overwhelming commitment of the mainstream media to the established climate change mantra. In an effort to provide some balance, dissenting climate scientist Fred Singer had organized the Nongovernmental International Panel on Climate Change (NIPCC), and in February 2008 it issued its first dissenting

---

68.   Editorial, "Light at the end of the tunnel," *Nature* 445 (February 8, 2007).

69.   Jim Giles, "From words to action," *Nature* 445 (February 8, 2007).

70.   George Will, "Just the Facts about Global Warming," *Washington Post,* February 27, 2009.

71.   Editorial, "The Scientists Speak," New York *Times,* November 20, 2007.

72.   "Even though the synthesis report is more alarming than its predecessors, some researchers believe that it still understates the trajectory of global warming and its impact." Elizabeth Rosenthal, "UN Report Describes Risks of Inaction on Climate Change," New York *Times,* November 17, 2007.

report indicating the extent to which the work of the IPCC did not provide a convincing basis for the policy changes it recommended. As Frederick Seitz noted in his preface: "It is foolish to [act] when the problem is largely hypothetical and not substantiated by observations. As NIPCC shows by offering an independent, nongovernmental 'second opinion' on the 'global warming' issue, we do not currently have any convincing evidence or observations of significant climate change from other than natural causes."[73] The authors carefully reviewed the literature that had been ignored by the contributors to the IPCC and crafted a report demonstrating that the evidence for anthropogenic global warming was weak while that for the null hypothesis – both warming and cooling due to natural factors – was overwhelming.[74] As Lord Turnbull put it: "the public has been fed a particular variant of the climate change story with many of the caveats stripped out. There is, however, a much richer but more complex story to be told which recognizes the complexities and uncertainties and also recognizes that there are strong natural variations upon which manmade emissions are superimposed."[75] That richer story can be found in the work of the NIPCC and, if taken into account, would have a profound impact on the trajectory of climate policy.

Despite vigorous denials by the IPCC community that any of the criticisms were valid, the impact of the NIPCC and other critics gradually eroded public support to the point that even the *New York Times* acknowledged that there might be a problem. One of its sci-

---

73. Fred Singer et al., *Nature, Not Human Activity, Rules the Climate: Summary for Policymakers of the Report of the Nongovernmental International Panel on Climate Change* (Chicago: The Heartland Institute, 2008), iii. This first NIPCC report closely tracks McKitrick, et al., *Independent Summary for Policy Makers*, prepared a year earlier and involving many of the same players. The full report was released in May 2009, *Climate Change Reconsidered* (Chicago: The Heartland Institute), providing an assessment of over 708 pages of evidence of peer-reviewed literature that contradict the official science of the IPCC. The NIPCC has since provided a series of updates, all of which can be found at its website, *Climate Change Reconsidered*.

74. "While AR4 is an impressive document, it is far from being a reliable reference work on some of the most important aspects of climate change science and policy. It is marred by errors and misstatements, ignores scientific data that were available but were inconsistent with the authors' pre-conceived conclusions, and has already been contradicted in important parts by research published since May 2006, the IPCC's cut-off date." *Nature, Not Human Activity, Rules the Climate*, 1.

75. Turnbull, "Foreword" to Montford, *The Climategate Inquiries*, 4.

ence reporters, John Broder, wrote in 2010 that "for months, climate scientists have taken a vicious beating in the media and on the Internet, accused of hiding data, covering up errors and suppressing alternate views. Their response until now has been largely to assert the legitimacy of the vast body of climate science and to mock their critics as cranks and know-nothings. But the volume of criticism and the depth of doubt have only grown, and many scientists now realize they are facing a crisis of public confidence and have to fight back."[76]

A large part of that doubt was kindled by the release in November 2009 of emails among the core scientists involved in writing the report of WG1 on the physical science.[77] The emails provided a chilling picture of the extent to which core scientists were prepared to go in order to keep dissenting opinion not only out of IPCC reports but out of the peer-reviewed literature altogether. The emails indicated an appalling lack of regard for access to information laws as well as for scientific ethics. As far as these scientists were concerned, dissent was wholly beyond the pale and needed to be squashed and delegitimized. Climatologist Hans von Storch expressed the disappointment of many of his colleagues: "what we can now see is a concerted effort to emphasize scientific results that are useful to a political agenda by blocking papers in the purportedly independent review process and skewing the assessments of the UN's IPCC. The effort has not been so successful, but trying was bad enough."[78] Andrew Montford similarly concludes in his detailed study of the events culminating in the hockey stick graph what the e-mails clearly revealed:

> Senior climatologists have sought to undermine the peer review process and bully journals into suppressing dissenting views. This means that the scientific literature is no longer a representation of the state of

76. John Broder, "Scientists Taking Steps to Defend Work on Climate," New York *Times,* May 2, 2010.

77. The first tranche of more than a thousand emails from a central server at the Climatic Research Unit (CRU) at the University of East Anglia was released in November 2009. The first release also contained documents indicating that much of the CRU temperature data base was a mess and less than reliable. A second tranche of more emails and documents was released two years later in November 2011. No one has yet claimed responsibility, but most IT specialists believe someone at the University of East Anglia, with access to its main servers, released the emails.

78. Hans von Storch, "Good Science, Bad Politics," Wall Street Journal, December 22, 2009.

human knowledge about the climate. It is a representation of what a small cabal of scientists feel is worthy of discussion. ... The IPCC reports represent the outcome of a process in which a relatively small group of scientists produce a biased review of literature they themselves have colluded to distort through gatekeeping and intimidation. The emails establish a pattern of behaviour that is completely at odds with what the public has been told regarding the integrity of climate science and the rigour of the IPCC report-writing process.[79]

Both the University of East Anglia and the University of Pennsylvania felt compelled to launch inquiries to determine whether the leaked emails pointed to any actionable misconduct on the part of the scientists involved in the scandal. Not surprisingly, as is common in such official enquiries and brilliantly satirized in the British comedy series, *Yes Minister* and *Yes Prime Minister,* all three investigations carefully avoided the elephant in the room: the extent to which the emails demonstrated that global warming science had been subverted by their authors. In each case, the chair knew what was required of him. All avoided calling witnesses who might discuss the impact of the scientists' misconduct on the work of the IPCC or on national policy. Instead, the focus was on such narrow issues as whether any of the emails pointed to conduct inconsistent with the terms of funding contracts.[80]

The first enquiry, launched by the University of East Anglia (UEA) and chaired by Sir Muir Russell, included members with clear conflicts of interest, including the chair, a former civil servant who is a fellow of the Royal Society of Edinburgh and a longstanding proponent of AGW alarm. The enquiry's findings of wrongdoing were confined to a slap on the wrist of the CRU scientists for failures to display a proper degree of openness, to properly archive data, and to respond more favourably to information requests. The panel never looked into breaches of the access to information rules, into efforts to suppress publication of dissenting scientific articles, or into the deletion of compromising emails. At best, the report ad-

---

79. Montford, *The Hockey Stick Illusion,* 449.

80. See the dissection of all four inquiries in Montford, *The Climategate Inquiries.* He concludes: "there can be little doubt that none of them have performed their work in a way that is likely to restore confidence in the work of CRU. None has managed to be objective and comprehensive. None has shown a serious concern for the truth. The best of them – the House of Commons enquiry – was cursory and appeared to exonerate the scientists with little evidence to justify such a conclusion. The Oxburgh and Russell inquiries were worse." 6.

mitted that CRU's work had not always followed conventional scientific methodology and ethics.

Given the unfavourable reception accorded the first enquiry, the UEA launched a second, independent Science Appraisal Panel chaired by Lord Oxburgh. It proved equally pliable to the university's wishes to sweep the scandal under the rug. Oxburgh is honorary president of the Carbon Capture and Storage Association, chairman of a company that builds wind turbines, a member of the Global Legislators Organization for a Balanced Environment (GLOBE),[81] involved in a number of other businesses promoting renewables, and deeply committed to the AGW alarm. As with the Russell panel, the Oxburgh panel decided it would conduct its work in closed hearings and largely confine itself to obtaining information from the CRU scientists themselves. The Oxburgh panel displayed the same deference to the scientists as the earlier enquiry with a similar hostility to critics of the science, many of whom sought an opportunity to testify but were denied the courtesy. Not surprisingly, the panel's report glossed over all the interesting questions and exonerated the UEA scientists. It could hardly do more, having avoided learning anything that might be compromising and confining itself to a report of five pages. It did criticize the IPCC for failure to take account of the caveats and nuances in papers prepared by CRU scientists, apparently unaware that the authors of the IPCC chapters in question and of the CRU papers were one and the same.

The Science and Technology Committee of the UK House of Commons conducted its own brief enquiry, focusing as much on the failure of the Muir Russell investigation as on the misconduct of scientists. Its findings "confirm that the Climategate inquiries had serious flaws, lacked balance and transparency and failed to achieve their objective to restore trust and confidence in British climate science."[82] The committee's work was limited to half a day with the dissolution of Parliament at the call of the 2010 election. While the committee issued a brief report, Labour government members ensured that little more would be done.

Similar to the UEA inquiries, the Penn State panel avoided learning anything that would make it difficult for them to exonerate a star scientist. Its focus was on the role played by Michael Mann,

---

81. See the discussion of GLOBE in chapter 10.

82. Press Release, "Flawed Climategate Inquiries failed to restore confidence in UK science," Global Warming Policy Foundation, December 24, 2011.

one of the central players in the Climategate emails. From the perspective of the university, Mann's prowess in raising research funding was his most important attribute and one that the panel should not undermine. In its report, the panel dismissed all the allegations against Mann and cited his fund-raising skills as evidence of his prominence in the climate science community. As journalist Clive Crook saw it: "The report ... says, in effect, that Mann is a distinguished scholar, a successful raiser of research funding, a man admired by his peers – so any allegation of academic impropriety must be false." In Crook's view, "the Climategate emails revealed ... an ethos of suffocating groupthink and intellectual corruption."[83] Each of the panels looking into this ethos chose to hold its collective noses and pretend that all was well in the world of climate science.

In March 2010, after three years of sustained criticism and public scrutiny of the IPCC's flawed fourth assessment report (AR4), the UN tasked the InterAcademy Council (IAC), a multinational body made up of many of the world's science councils, with reviewing the procedures and conduct of the IPCC. Over the course of six months the review committee received submissions, held hearings, and reviewed the working procedures of the IPCC and its subsidiary bodies. The panel discharged its task quickly and thoroughly, meeting with experts around the world, inviting briefs from the interested public, interviewing a range of scholars and officials, and inviting the public to fill in a questionnaire posted on its website. The response was overwhelming, particularly from experts who had participated in the work of the IPCC over the previous 20 years and were critical of its procedures. The panel distilled all this information into a reasonably fair assessment of the IPCC's shortcomings and recommended a number of sensible steps it could take to make its reports more credible to a wider audience. Within the limits of its mandate, the review panel discharged its task much more thoroughly, openly, and credibly than had the four inquiries into the CRU email scandal.

Chaired by the president emeritus of Princeton University, Harold Shapiro, the IAC panel issued its report at the end of August, 2010, finding serious shortcomings in the IPCC's conduct and procedures and recommending major changes to make the process more transparent and its reports more balanced and reliable. In a

---

83. Clive Crook, "Climategate and the Big Green Lie," *The Atlantic,* July 14, 2010.

foreword, IAC co-chairs Robert Dijkgraaf and Lu Yongxiang summed up its recommendations as follows:

> The committee urges that the IPCC management structure be fortified and that the IPCC communications strategy emphasizes [sic] transparency, including a plan for rapid but thoughtful responses to crises. It also stresses that because intense scrutiny from policymakers and the public is likely to continue, IPCC needs to be as transparent as possible in detailing its processes, particularly its criteria for selecting participants and the type of scientific and technical information to be assessed. More consistency is called for in how IPCC Working Groups characterize uncertainty. The committee emphasizes that in the end the quality of the assessment process and results depends on the quality of the leadership at all levels.[84]

Well-intentioned and researched, the IAC report nevertheless fails to come to grips with the fundamental flaws that result inevitably from the IPCC's structure and mandate. While billed as an independent scientific body, the IPCC is in fact an arm of the UN and its members and is far from independent. The Panel is made up of government appointees, and its reports are closely scrutinized and edited by government representatives before they are finalized. Its *Summaries for Policymakers* are negotiated texts which, once completed, are used to "correct" the underlying scientific reports. Scientists who are asked to contribute to its work are carefully scrutinized to ensure that their views will not derail the objective of finding the human influence on climate change and of identifying the malign impacts of that influence. The scientific virtue of scepticism is *not* an asset at the IPCC. In short, its mandate is to produce official science. Ross McKitrick, who closely followed both the IAC process and the fate of its recommendations, concludes:

> Overall, the IAC recommendations, which were understated to begin with, were translated into even more superficial terms by the IPCC Task Groups, after which they received almost no critical scrutiny by the member countries on whom the responsibility of oversight rests, prior to being rubber-stamped at Abu Dhabi. While a superficial impression of reform may have been created, few countries appear to have studied the proposals to an extent sufficient to yield meaningful

---

84. Robert Dijkgraaf and Lu Yongxiang, "Foreword," to InterAcademy Council, "Climate Change Assessment: Review of the Procedures and Processes of the IPCC," at reviewipcc.interacademycouncil.net/report.html.

comments, and even fewer seem to recognize any serious need for reform at all. The whole reform process was unserious and ineffectual.[85]

It probably could not have been otherwise. The fundamental problem faced by the IAC panel and one that faces virtually every official enquiry into the operation of a public institution is the lack of authority to delve into such critical questions as the institution's mandate and how well it has discharged that mandate. Inquiries into procedures are rarely anything but superficial and easily ignored. The IPCC was established on the urging of a small in-group of climate researchers led by Bert Bolin, who had been involved in the UN-sponsored meetings that examined the human impact on climate as far back as 1961. From the beginning it was led by scientists on a mission who ensured that it would never deviate from that mission: to convince governments that they needed to take steps to mitigate the threat posed by the emissions of greenhouse gases, a mission that fit well within the UN's broader socio-economic agenda and its push for global governance on progressive principles. Reforming the IPCC's procedures would not in any way change the built-in tunnel vision of its founders and other prime players. The IPCC was never intended to provide a balanced assessment of the science; rather, its mission was to provide an "expert" justification for a predetermined policy path.

As is often the case, the most interesting aspects of this exercise in damage control can be found not in the final report but in the briefs, questionnaires, and other raw material on which the report is based. No committee could ever capture the full range of criticisms levelled against the organization, but this raw material has provided excellent input for critical assessments by others. Donna Laframboise, for example, has mined this material to demonstrate the extent to which the IPCC is a single-purpose panel with a mandate and contributors dedicated to a single perspective. Those who do not share that point of view do not get a hearing.

Regardless of one's viewpoint, the IAC report provided the IPCC with sage advice, much of it between the lines, which, if taken seriously, would ensure more transparency and even-handedness in future assessment reports. The fifth assessment report would reveal whether that advice was taken into account or whether critics such as McKitrick and Curry were right that the whole idea of an inter-

---

85. Ross McKitrick, *What is Wrong with the IPCC: Proposals for Radical Reform* (London: Global Warming Policy Foundation, 2010), 36.

governmental panel on such a complex issue needed to be re-thought.[86] The very idea that such a panel could provide advice on the complex interface between science and public policy on a global scale was probably unrealistic.

The perils of the IPCC's transition into a self-perpetuating bu-reaucracy have been apparent for some time. David Keith, a Har-vard University professor who recently resigned as an IPCC author, observes: "The IPCC is showing typical signs of middle age, includ-ing weight gain, a growing rigidity of viewpoint, and overconfi-dence in its methods. It did a great job in the early days, but it's be-come ritualized and bureaucratic, issuing big bulk reports that do little to answer the hard questions facing policymakers."[87] Perhaps so, but with so much invested in the movement, governments, UN leaders, scientists, ENGOs, and other hangers-on stubbornly main-tain that the perils of climate change are growing ever more menac-ing, that the time for effective remedial action is shrinking rapidly, and that the IPCC is needed more than ever.

The IPCC's last report?

Over the course of its first four reports, the IPCC had consistently grown bolder in its assessment of the impact of humans on the cli-mate system. It had little choice. The reports of WG2 and WG3 had also grown more confident in projecting the catastrophic impact of the human influence and in assuring governments that solutions were at hand if they were prepared to follow the Panel's advice. Without an increasing confidence in the physical basis of anthropo-genic global warming or climate change, the dire warnings and con-fidence in the proposed solutions were not credible. Inevitably, there-fore, each assessment upped the ante. The critical problem was that the planet was not cooperating. The impact of the 1997-98 El Niño spike, which had made 1998 the "warmest year in more than a mil-lennium," had been neutralized by the equally large La Niña that fol-

---

86. Judith Curry concludes that "we need to put down the IPCC as soon as possible – not to protect the patient who seems to be thriving in its own little cocoon, but for the sake of the rest of us whom it is trying to infect with its disease. Fortu-nately much of the population seems to be immune, but some governments seem highly susceptible to the disease. However, the precautionary principle demands that we not take any risks here, and hence the IPCC should be put down." "IPCC diagnosis – permanent paradigm paralysis," September 28, 2013, at *Climate Etc.*

87. David Keith, as quoted by Fred Pearce in "Has the UN Climate Panel Now Out-lived its Usefulness?" *Yale 360,* September 30, 2013.

lowed. Based on the IPCC's own data, there had been no net warming since 1997. It was then increasingly difficult to make the case that radiative forcing by human emissions of greenhouse gases was the principal driver of climate change. As Patrick Michaels sees it, "Over the years, the IPCC has behaved like a treed cat. Instead of closing its eyes and scurrying to the ground, it climbs onto even higher and thinner branches, while yowling ever louder. How does it back down from a quarter-century of predicting a quarter of a degree (Celsius) of warming every decade, when there's been none for 17 years now?"[88]

Additionally, the planet had refused to provide any direct physical evidence of the human impact. At the outset, Bert Bolin, John Houghton, and their collaborators were confident that their hypothesis was right; since then, they had looked diligently, but the evidence on which the IPCC and its coterie of committed scientists continued to rely could only be found in the climate models based on the data and assumptions supplied by the same scientists. Sceptical scientists, on the other hand, had been steadily building a large body of evidence pointing to the role of solar activity, cosmic rays, cloud formation, and coupled atmospheric-oceanic circulation systems in influencing ever-changing global climate patterns, contradicting the official science of the IPCC. Of particular interest was the extent to which scientists had made progress in understanding the complex issue of the climate system's sensitivity to increases in GHGs. That was the challenge that the leaders of the IPCC faced as they laboured in producing their fifth assessment, due for release in 2013.

Furthermore, widespread criticism of the IPCC's last two reports had eroded public confidence in the credibility of its work, and the IAC's recommendations for reform threatened to make its working procedures much more difficult. Finally, these problems had become steadily more urgent as the politics of the movement had gradually begun to disintegrate after the debacle of the 2009 Copenhagen conference. The United States remained outside the fold, and by 2012 Canada, Japan, and Russia had declined further participation in the Kyoto process. Australia, which had only joined in 2007, left in 2013 with a further change in government, while developing countries, by now responsible for more than half of the world's GHG emissions, remained adamant that they would not entertain any limits until their economies had made sufficient progress

---

88. Patrick Michaels, "The IPCC Political-Suicide Pill," *National Review Online*, September 26, 2013.

to afford to take regulatory action to reduce their emissions. Based on the IPCC's logic, the world might well pass the point of no return before that moment arrived. Rajendra Pachauri and his collaborators had their work cut out for them.

In the face of increasing evidence from recent scientific research that the IPCC's models were failing to capture what was happening in the climate system, IPCC scientists aggressively asserted that they were now more confident than they had been seven years earlier: any remaining uncertainty had declined from 10 per cent to 5 per cent. As the latest SPM put it:

> Warming of the climate system is unequivocal, and since the 1950s, many of the observed changes are unprecedented over decades to millennia. The atmosphere and ocean have warmed, the amounts of snow and ice have diminished, sea level has risen, and the concentrations of greenhouse gases have increased. ... Human influence has been detected in warming of the atmosphere and the ocean, in changes in the global water cycle, in reductions in snow and ice, in global mean sea level rise, and in changes in some climate extremes .... This evidence for human influence has grown since AR4. It is *extremely likely* [95 per cent confidence level] that human influence has been the dominant cause of the observed warming since the mid-20th century. (AR5 WG1 SPM, 4 and 17)

Like all bureaucrats, the idea that they may have been wrong is too horrible to contemplate, so they keep on trucking, churning out the same nonsense that has been their stock in trade for a quarter century. As French physicist Pierre Darriulat puts it: "When writing the SPM, the authors are facing a dilemma: either they speak as scientists and must therefore recognize that there are too many unknowns to make reliable predictions, both in the mechanisms at play and in the available data; or they try to convey what they 'consensually' think is the right message but at the price of giving up scientific rigour. They deliberately chose the latter option. The result is they have distorted the scientific message into an alarmist message asking for urgent reaction, which is quite contrary to what the scientific message conveys."[89]

It did not take long for the critics to pounce. They concentrated their efforts on at least seven serious problems with IPCC science, each of which the latest report from WG1 either ignores, fails to explain, or contradicts in the fine print:

---

89. Pierre Darriulat, "Written Evidence Submitted by Professor Pierre Darriulat," Select Committee into the Intergovernmental Panel on Climate Change, UK House of Commons, December 20, 2013.

- IPCC estimates of the climate system's *sensitivity* to increases in atmospheric greenhouse gases is off by at least a factor of two. Reducing that sensitivity in their models and reports, however, makes the climate crisis disappear and invalidates not only the reports of WG1, but even more those of WG2 and WG3.[90]

- The IPCC has failed to explain the *lack of warming* since 1998, a period now adding up to 18 years over which the pace of human CO_2 emissions has accelerated as countries such as China and India have industrialized. The IPCC admits that warming has slowed to a rate of 0.05°C, with an error factor of plus or minus 0.1°C; i.e., the error factor is larger than the estimated rate of warming. The many – more than 60 at time of writing – explanations advanced to explain the pause range from fanciful to farcical; none are scientific, i.e., falsifiable and based on evidence. Any hypothesis that requires that many *ad hoc* explanations is clearly not scientific but political.

- Only a very small percentage of IPCC models, in the order of 2 percent, probably more a matter of chance than of perspicacity, have tracked recent global temperatures, an indication of failure that should have led to serious questions about their reliability to capture the fundamental contours of the climate system. As renowned Danish physicist Niels Bohr once cautioned: "prediction is very difficult, especially about the future." The lame explanation that these are not "predictions" but "plausible story lines" has not penetrated as far as politicians, the media, and IPCC spokespersons.

- The IPCC has failed to explain why the analogous warming over the course of the first half of the 20th century was largely natural while that since mid-century has been dominated by anthropogenic factors.

- The IPCC staunchly maintains that the warming of the past sixty plus years is unprecedented despite multiple lines of instrumental evidence that global temperatures were similar in the 1930s

---

90. As discussed further below, the climate system's sensitivity to increased atmospheric CO_2 is the key issue for IPCC scientists. In their view, the direct temperature effect of increased atmospheric CO_2 is amplified by a factor of three or more, since a warmer world would be more humid leading to a positive feedback in radiative forcing resulting from the more powerful greenhouse effect of increased water vapour. In their calculations, a doubling in the amount of atmospheric CO_2 would change the equilibrium temperature by 1.5° to 4.5°C, and most likely about 3°C.

and even more evidence from paleoclimatology that they were higher at least eight times over the course of the Holocene. Indeed, it is a mainstay of earth science that the long-term trajectory over the course of the Holocene was one of cooling, not warming.

• Some of the so-called secondary evidence on which the IPCC relies has proven contradictory and inconsistent with IPCC predictions. For example, over the period of satellite records (1979-2015), Arctic ice has shrunk but Antarctic ice appears to have grown more quickly, so that the total amount of ice at the two poles has grown. Buried in the detail is the IPCC's qualification that Arctic temperature anomalies in the 1930s were apparently as large as those in the 1990s and 2000s and that there is still considerable discussion of the ultimate causes of the warm temperature anomalies that occurred in the Arctic in the 1920s and 1930s. Similarly, the rate of sea level increase over the period 1920-1950 appears to have been the same as that from 1993-2012, with the first alleged to be the result of natural warming and the second to be the result of anthropogenic factors. The IPCC's own data do not support its conclusion of a substantial contribution from anthropogenic forcings to the global mean sea level rise since the 1970s.

• Alleged increases in extreme weather events (e.g., storms, droughts, and floods) are not reflected in the actual record and are fewer than during the last period of warming (1920s and 1930s). The rush to attribute all recent, short-term climate phenomena to anthropogenic forcing is undermined by the extent to which IPCC scientists are prepared to admit in the underlying scientific reports that such phenomena are not unprecedented, thus confirming the null hypothesis of natural climate change.

Judith Curry reaches the obvious conclusion that on the basis of these problems alone, WG1's latest report weakens rather than strengthens the IPCC's case for anthropogenic global warming. Curry states: "If you read the fine print (not just the SPM) and compare the AR5 with statements made in the AR4, the IPCC AR5 WG1 Report makes a weaker case for AGW than did the AR4."[91] Of course, none of these problems are reflected in the *Summary for Policymakers* prepared by the AR5 WG1 team for consumption by politicians and the media. As a result, the popular media faithfully report that the case for action is stronger than ever.

---

91.   Judith, "IPCC AR5 weakens the case for AGW," *Climate Etc.*, January 6, 2014.

The failure of the IPCC to reflect recent scientific findings on equilibrium climate sensitivity (ECS) is probably the most serious problem.[92] AR5 concludes that "equilibrium climate sensitivity is likely in the range 1.5°C to 4.5°C (high confidence), extremely unlikely less than 1°C (high confidence), and very unlikely greater than 6°C (medium confidence)." (IPCC, AR5, WG1, SPM, 16) At the same time, Figure 1 of Box 12.2 in the AR5 WG1 report shows that half of the observational-based studies of ECS cite values below 1.5°C in their ranges of ECS probability distribution. In effect, AR5 reflects greater uncertainty and a tendency towards lower values of ECS than AR4, which concluded: "The equilibrium climate sensitivity. . . is likely to be in the range 2°C to 4.5°C with a best estimate of about 3°C and is very unlikely to be less than 1.5°C. Values higher than 4.5°C cannot be excluded." (IPCC, AR4, WG1, SPM, 12). Mathematician Nicholas Lewis told the UK Parliament Select Committee looking into the report of AR5:

> There are two principal issues with the IPCC's handling of the climate sensitivity area. Firstly, the inclusion of sensitivity estimates from flawed observational studies that used unsuitable data, were poorly designed and/or employed inappropriate statistical methodology.[sic] That obscured what should have been a key message from AR5 – that the best observational evidence now points to the climate system being substantially less sensitive to greenhouse gases than previously thought. Secondly, the elevation of computer models over observational evidence. [sic] Virtually all the projections of future climate change in AR5 are based on simulations by GCMs [global climate models] despite these being out of line with the best observational evidence."[93]

---

92. See, for example, Nicholas Lewis and Marcel Crok, *A Sensitive Matter: How the IPCC buried evidence showing good news about global warming* (London: Global Warming Policy Foundation, 2014). As Lewis and Crok demonstrate, IPCC authors, rather than working with observational evidence, continue to insist that model results are more reliable. Patrick Michaels adds: "Since the beginning of 2011, at least 16 separate experiments published by nearly 50 researchers show that the "sensitivity" of temperature to carbon dioxide that is characteristic of the IPCC's climate models is simply too high." "The IPCC Political-Suicide Pill."

93. Nicholas Lewis, "Written Evidence Submitted by Nicholas Lewis," Select Committee into the Intergovernmental Panel on Climate Change, UK House of Commons, December 20, 2013.

Figure 5-4: Comparison of observed global mean temperature anomaly and IPCC models

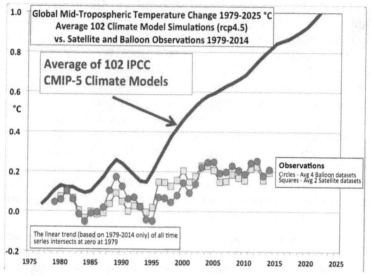

Source: J.R. Christy, University of Alabama at Huntsville, at *AL.com,* March 18, 2015.

The credibility of all five of the IPCC's voluminous reports should be assessed in terms of the extent to which the core projections of the models on which the IPCC relies reflect observations over the period 1951-2013, the period over which the IPCC maintains that human influence became dominant and that global temperatures increased. Patrick Michaels and Chip Knappenberger found that only six of the 114 models used by the IPCC projected the same or a lower decadal rate of temperature increase than was observed over this 62-year period.[94] Even more telling, the models projected a wide range of decadal increases, ranging from 0.068°C per decade, to 0.223°C per decade, suggesting that they reflect the assumptions and data of their creators rather than of nature's behaviour. The same point is made, but more obscurely, in one of the IPCC's own graphics buried in the *Technical Summary* to the 2013 Report of WG1. (See figure 5-4 for a simpler presentation showing the discrepancy between model outputs and observed temperatures).

---

94. "A Clear Example of IPCC Ideology Trumping Fact," *WattsUpWithThat,* April 16, 2014.

Six years earlier, in AR4, the IPCC confidently asserted that the earth would experience a 0.2°C temperature increase *per decade* over the 21st century. To date, there has been none. It has now revised its projection to a range of 0.1°C to 0.23°C per decade. WG1 now concludes that "the hiatus is attributable, in roughly equal measure, to a decline in the rate of increase in ERF [effective radiative forcing] and a cooling contribution from internal variability (expert judgment, *medium confidence*). The decline in the rate of increase in ERF is attributed primarily to natural (solar and volcanic) forcing but there is *low confidence* in quantifying the role of forcing trend in causing the hiatus, because of uncertainty in the magnitude of the volcanic forcing trend and *low confidence* in the aerosol forcing trend." (IPCC, AR5, WG1, 1010, emphasis in original). Again this explanation was tucked away in the scientific report and fails to resolve anything, perhaps indicating why the authors suggest that their expert judgment only merits "medium confidence" (as likely as not). There has been no volcanic activity with significant global impacts since 1992, and the aerosol fudge factor has long been a favourite used by IPCC modellers but, conveniently, cannot be measured with any confidence.[95]

In addition, the IPCC has never dealt satisfactorily with the fact that temperatures also increased during the first half of the 20th century, i.e., before the major rise in population, industrialization, and the quantum leap in the use of fossil-fuelled transportation, all of which supposedly greatly accelerated the release of GHGs and increased their atmospheric concentration. Based on the temperature series used by the IPCC, over the sixty-year period from 1894-1953, temperatures rose 0.48°C while CO_2 only increased by 18 ppm. Over the next 60 years (1954-2013), CO_2 rose 82 ppm while temperatures rose only 0.39°C. (See figure 5-5) Other population and industry-sensitive GHG gases – methane, nitrous oxide, chlorofluorocarbons – also rose much more quickly in the second period than the first, and yet the climate response was much more muted. The IPCC models can, apparently, explain the temperature increase over the first period on the basis of natural forcings, but not over the second, leaving human influence as

---

95. In a recent paper, climatologist Bjorn Steven argues that the models assume too negative a level for aerosols, which in turn helps to justify higher positive sensitivity for CO2. In order to track the observed temperature record, if the cooling effect of aerosols is small, the warming effect of carbon dioxide must also be small. His research suggests both are lower than assumed in IPCC models. "Rethinking the lower bound on aerosol radiative forcing," *Journal of Climate* (March 2015 e-view).

Figure 5-5: Global surface warming, 1894-2013: natural vs. anthropogenic

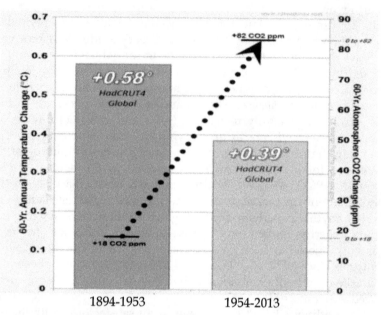

Two 60 year-periods of warming: according to the IPCC, the warming from 1894-1953 was largely the result of natural forcings while that from 1954-2013 was dominated by anthropogenic forcings. During the first period, CO_2 rose only 18 ppmv, while during the second period, it increased by 82 ppmv. Temperatures based on HadCrut 4.

Source: www.c3headlines.com/2014/02/climate-science-consensus-60-years-hadcrut-global-warming-those-stubborn-facts.html.

the only possible explanation. As sceptical scientists, none of whom deny some impact from rising GHGs, point out, the IPCC needs to try harder to understand the natural forces that drive climate change and then determine the extent of any human influence. In the IPCC's description of the influence of natural forcings on the earth's radiative balance, IPCC scientists admit that their understanding of some of these factors ranges from low to medium. Furthermore, understanding of the role of coupled atmospheric-oceanic circulation systems, convection, the water cycle, insolation, cosmic rays, and other forces remains incomplete, making the IPCC's assertion of growing confidence in its model-derived conclusions border on the risible and confirming the extent to which the IPCC is an exercise in policy-based evidence-making, rather than science.

Warming in the late 20th century was not unprecedented. Indeed, the pace of warming in the first half of the 20th century was higher. In the last 12,000 years, within the current interglacial, the climate has had as many as eight episodes of warming and cooling, in recurrent cycles, all in general comparable to our modern warming period, but clearly lacking any influence from anthropogenic CO_2. The only feature that distinguishes the modern warming from previous cycles is that it is cooler. In order to characterize modern warming as unprecedented, the IPCC simply ignores the earlier Holocene warming and cooling cycles of which our modern warming is the last and coolest.

AR5, similar to its predecessors, is marred by another hallmark of advocacy science: a penchant for relying on the argument that one or more lines of evidence or of theoretical modelling are "consistent with" one of its more dubious claims. To the lay reader this may appear to be an argument supporting the claim. To a trained scientist, it means that the claim is weak and that the best AR5's authors could do was point to evidence that does not contradict it.[96] If one ignores IPCC press releases and speeches from senior UN officials, and if one puts aside the *Summaries for Policymakers* and the *Synthesis Reports* issued by the IPCC bureau, one comes away with the astonishing conclusion that the evidence to justify the dire conclusions and warnings is simply not there. It exists only in the models and in the speculative conclusions that IPCC officials and alarmist scientists draw out of these scientific reports.

The IPCC justifies the preparation of these speculative press releases and summary reports by arguing that, having discovered that the climate is changing – a geological and historical commonplace – and having asserted that human activity is an important driver of this change, it then follows that there is a need to do something about it. But, as Richard Lindzen points out, the fact of climate change in and of itself does not lead to the need for remedial action. As he notes, "what is missing is a scientific [i.e., rigorous, objective, evidence-based] assessment of the potential threat posed by climate change balanced against its potential benefits. Without a science-based benefit/risk analysis, a simple scientific finding on its own doesn't merit specific action no matter how scientifically groundbreaking it might be. ... The reason there is no threat assessment accompanying the science academies' statement is that there is no sci-

---

96. See the interesting discussion in Rupert Darwall, "An Unsettled Climate," *City Journal*, Summer 2014.

entific consensus on what level of threat climate change poses."[97] The IPCC, however, was established on the assumption that there is a grave threat, and its task is to determine its extent and what to do about it. It has glossed over risk/benefit analyses and moved directly to describing presumed threats and consequences and to using these as part of a public relations campaign calling for drastic remedial action.

Consistent with much of environmental science, IPCC scientists have put to the side the careful qualifications and caveats that have long been the hallmarks of the physical sciences and have pursued post-normal, advocacy science, letting their political preferences guide them. As Claire Parkinson observes: "I have listened with a mixture of horror and awe as scientists I know skilfully and successfully use carefully chosen superlatives to generate media attention, even when the superlatives are not warranted and the scientists know that they are not valid but feel that some exaggeration is acceptable. ... This is sometimes for the admirable reason that the individual's deep concern about the future of the planet or of civilization overrides considerations such as balance and objectivity. However, ... giving a demonstrably biased story is troublesome and can greatly confuse the issues".[98] Harvard astrophysicist Willie Soon is even more direct. "The central lesson to be learned from this episode in scientific history is that to create an organisation financially and ideologically dependent upon coming to a single, aprioristic viewpoint, regardless of the objective truth, is to create a monster that ignores the truth."[99]

*****

---

97.   Richard Lindzen, *Issues in the Current State of Climate Science*, SPPI, March 2006, at www.scienceandpublicpolicy.org, 27.

98.   Claire L. Parkinson, *Coming Climate Crisis? Consider the Past, Beware the Big Fix* (Lanham, MD: Rowman & Littlefield, 2010), 265.

99.   Willie Soon, "Sun shunned," in Alan Moran, *Climate Change: The Facts* (Melbourne, Australia: Institute of Public Affairs, 2015), kindle edition.

6 | Secondary Evidence and Impacts

I see confirmation bias everywhere in the climate debate. Hurricane Katrina, Mount Kilimanjaro, the extinction of golden toads – all cited wrongly as evidence of climate change.[1]

Matt Ridley, 2011

Detection, attribution, and inflated impacts

For the IPCC, the holy grail of climate science is detection and attribution of the human impact on climate change, but in the face of weak evidence, it has increasingly relied on the alleged *impact* of climate change – so-called secondary evidence – as proof of anthropogenic global warming. Most of that secondary evidence provides no such proof but has become part of the staple of alarmist stories in the media. Their credibility has become increasingly threadbare as the alleged global warming has failed to materialize. Evidence of postwar global warming is now limited to the period from 1977 to 1998. It will not be long before the standstill or pause is longer than the warming. It is also important to keep in mind at the outset that the mere existence of *warming*, locally, regionally, or globally, is not necessarily evidence of *anthropogenic* global warming.

---

1. Matt Ridley, "Scientific Heresy," Angus Millar Lecture of the Royal Society of the Arts, Edinburgh, October 31, 2011.

All of these phenomena, whatever their explanation, are irrelevant to understanding the causes and future course of climate change. Glaciers will advance or retreat, regardless of whether the warming or cooling is anthropogenic or natural. Earth's climate is in a constant state of flux, either warming or cooling, on all spatial and temporal scales. The null hypothesis for this constant state of change is that natural forces are largely responsible. That being said, there is credible evidence for some human impacts such as land use changes and the urban heat island effect. Most of these effects, however, are principally evident at the local and regional levels. Belief in global-scale human influence as a result of increases in atmospheric concentrations of greenhouse gases (GHGs) remains largely hypothetical, a matter of conjecture and computer modelling rather than a matter of observation.

Threatened polar bears

The polar bear is often featured as the canary in the mineshaft. The Worldwide Fund for Nature (WWF), for example, uses it in its logo and continues to propagate the myth of disappearing polar bears. This is pure fantasy; a few of the 19 sub-populations of polar bears around the Arctic Circle today may exhibit signs of stress, largely due to hunting. Globally, populations have grown substantially since wide-scale hunting was curbed in the 1950s and 1960s. Observation-based estimates of the total population have quintupled over the past 70 years, now estimated to be between 20 and 25 thousand, larger than many other large carnivores. Its listing as a "vulnerable species" by the World Conservation Union and as "endangered" under the US Marine Mammal Act is a triumph of activist lobbying and is based on computer projections of habitat loss due to global warming. In 2009-10 scientists sponsored by the government of Nunavut – a Canadian territory in the far north – conducted a comprehensive aerial survey of polar bears in the Foxe Basin, one of the areas in which polar bears are considered to be at risk because it is seasonally ice free. The survey found three times the number of polar bears than previous and less complete surveys had suggested, numbers consistent with surveys twenty years earlier.[2]

Like so much of the virtual world of climate alarm, there is no observational evidence to substantiate the claim of dying polar bears.

---

2.   See S. Stapleton, et al., "Foxe Basin Polar Bear Aerial Survey, 2009 and 2010." *Final Report*, 2012. Department of Environment File Report, Government of Nunavut, Igloolik, Nunavut, Canada.

Figure 6-1: Extent of global sea ice, 1979-2015

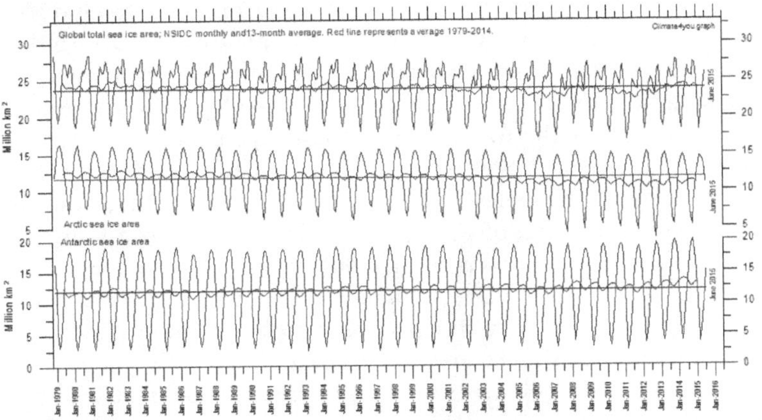

The top panel indicates the extent of annual variation in global sea ice area in millions of square kilometres; the middle panel indicates variation in Arctic sea ice, while the bottom panel indicates variation in the sea ice surrounding the Antarctic continent.
Source: http://www.climate4you.com/images.

The polar bear has survived earlier glacials and interglacials, and there is no basis for the fear that it will not adapt to whatever changes may currently be at play.[3] Idso et al. conclude that "forecasts of dwindling polar bear populations assume trends in sea ice and temperature that are counterfactual, rely on computer climate models that are known to be unreliable, and violate most of the principles of forecasting. ... We find there is no basis for concern that climate change will ever cause the extinction of polar bears."[4]

Arctic and Antarctic melting

There are two issues here: melting sea ice and melting glaciers in Greenland, Alaska, and in Antarctica. Sea ice varies over the course of the year as it builds up in the fall and winter and declines in the summer months in both the Arctic and Antarctic (See Figure 6-1). Arctic temperatures are only slightly above freezing in the short Arctic summer and well below freezing the rest of the year. Recent

---

3.   See Susan J. Crockford, "Healthy Polar Bears, Less Than Healthy Science," Global Warming Policy Foundation, Note 10, 2014. Crockford maintains a blog related to polar bear issues at *PolarBearScience*.

4.   Craig Idso and Fred Singer, eds., *Climate Change Reconsidered* (Chicago: Heartland Institute, 2009), 661.

temperature patterns indicate significant variation during the cold months but stable temperatures averaging just over freezing during the brief summer. Whatever warming is taking place in the Arctic, it is largely a matter of more variation during the winter months that will have little impact on Arctic eco-systems. On the colder Antarctic continent, only the Antarctic peninsula may experience temperatures above freezing. Specialists in polar weather patterns believe that melting at the poles is less driven by temperatures than by wind and by the effect of warmer water from farther south or north entering the polar regions as a result of oceanic circulation.[5] As Richard Lindzen told a UK parliamentary committee: "summer ice depends mostly on how much is blown out of the arctic basin – something that used to be textbook information."[6]

As indicated in figure 6-1, global sea ice waxes and wanes; in Antarctica, sea ice is increasing, but the extent of summer ice in the Arctic decreased over the 1990s and into 2007; the next few winters saw steady recovery of Arctic ice so that by January 2014 global sea ice levels were back to the levels observed at the beginning of the satellite era in January 1980. September sea ice extent in 2013 was 60 percent higher than in the previous year, similar to levels seen in the 1980s.[7] In the 1940s, as indicated by the voyages of Henry Larsen in the RCMP patrol vessel St. Roch during those years, open water in summer in the Arctic reached a low similar to that of the first decade of the 21st century. The natural Arctic cycle was at a similar stage at the beginning of the 20th century when Norwegian explorer Roald Amundsen sailed through the Arctic in 1903. Sea ice levels have no impact on global sea levels, as was demonstrated 2,500 years ago by Archimedes. More generally, Idso et al. conclude that "sea ice, precipitation patterns, and sea levels all fluctuate largely in response to processes that are unrelated to greenhouses gases, and therefore cannot be taken either as signs of anthropogenic global warming or of climate disasters that may be yet to come."[8]

---

5. See Chip Knappenberger, "Arctic Sea Ice Losses," SPPI, 31 October 2008, and "A million square miles of open water," *World Climate Report*, October 22, 2007.

6. Lindzen, "Reconsidering the Climate Change Act – Global Warming: How to approach the science," Presentation to seminar at the [UK] House of Commons Committee Rooms, Westminster, London, February 22, 2012.

7. NASA's National Snow and Ice Data Center (NSIDC) maintains the satellite data on the extent of Arctic and Antarctic sea ice. *WattsUpWithThat's* sea ice reference page provides daily updates of data from various sources on the extent and status of sea ice in both the Arctic and Antarctic.

8. Idso et al., *Climate Change Reconsidered* (2009), 135.

Melting of Greenland and Antarctic glaciers

Arctic and Antarctic glaciers, like all glaciers, lose ice at their edges due to melting in its tail, sublimation, and pressure from higher up – a glacier is in effect a frozen river – and grow as a result of new snow. Ice does not melt rapidly when average high temperatures barely exceed 0°C for only three months of the year, and low temperatures routinely reach –50°C. The annual Greenland temperature average is –11°C. Antarctic average temperatures are even lower. Since the Second World War the ice cover on Greenland has grown substantially. For example, one of a group of planes that made a forced landing on the ice in 1942 was found under 268 feet of new ice when it was recovered fifty years later in 1992 – an *increase* averaging more than five feet per year.[9] Measurements of the net ice mass of both Greenland and Antarctica have shown increases since sophisticated satellite-based surveys began 30 years ago. Scientific study of these ice masses continues, with conflicting reports appearing periodically in the scientific literature, none of which support evidence of rapid, catastrophic melting. Many reports are computer studies based on questionable assumptions. For the record, the Greenland ice mass averages 2,000 meters thick, and some of it surpasses 3,000 metres, and thus most of it rarely experiences temperatures above freezing. The same holds true for the Antarctic. Melting of these two massive stores of water would have a major impact on sea levels, but it would take thousands of years at current melting rates, assuming no replenishment through snow.[10] Claims by Al Gore, James Hansen, Stephen Schneider, and others of massive sea level increases over the 21st century due to the melting of the Greenland and Antarctic ice masses are without scientific foundation, as even the IPCC cautions.

Sea levels and glacier melting

In recent decades, there has been a very small annual rise (in mms) in sea level as the result of a combination of three principal factors:

9. See "Recovery of Glacier Girl," at p38assn.org/glacier-girl-recovery.htm.

10. See "Greenland Climate: Now vs. Then, Part I. Temperatures," *World Climate Report*, October 16, 2007. It is worth recalling the sheer mass of these two ice-bound islands: an average of 3,000 metres of ice covering 14 million square kilometres (Antarctica) and 2,000 metres over 2.2 million square kilometres (Greenland). The territories of Antarctica and Greenland together add up to nearly the equivalent of the lower 48 United States and Canada combined. Given their size and isolation, only satellite-derived data can provide any indication of their changing climatic conditions.

melting glaciers, sinking land, and thermal expansion. Contemporary mountain glacier melting has a small impact given the size of the oceans and the relatively modest amounts of ice remaining in lower altitude glaciers other than those in Greenland and Antarctica.[11] High mountain glaciers wax and wane as a result of natural precipitation cycles. Additionally, post-Little-Ice-Age natural warming dating back to the middle of the 19th century is the principal reason for the gradual shrinkage of many lower-altitude glaciers. In recent years, more sophisticated satellite telemetry of sea levels as well as better tide gauges suggest that there has been no perceptible increase in global sea levels in the 21st century. *World Climate Report*, after surveying recent literature, concluded: "the rate of sea level rise continues to slow. The rate during the most recent 10-yr. period is 2.32 mm/yr. (or about 9 inches per century). This is not much above the 20th century average rate of 1.8mm/yr. (7 inches per century), and *far* below the average rate of 10 mm/yr. required to raise global average sea level by 1 meter (3.25 feet) by 2100 – the new in-vogue value for what activists believe the IPCC should have projected (rather than the ~ 33 centimeters (15 inches) that they did project)."[12]

Since the beginning of the Holocene, sea levels have risen steadily in response to the natural global warming that followed the end of the last Ice Age at a mean rate of 1.3 metres per century and a total of about 130 metres. By the 20th century sea levels were rising worldwide by less than 0.20 metres per century, and Mörner, the world's leading expert on sea levels, says that there is little reason to argue that sea levels will rise significantly faster in the 21st century than they did in the 20th, because virtually all of the land-based ice that once covered North America and northern Eurasia melted long ago, and most of the remaining mountain glaciers are at high altitudes and high latitudes where there is little danger that they will melt any time soon.

Coral bleaching

After the large 1998 El Niño, the media carried stories of widespread coral bleaching and worried that this was indicative of the future under global warming. Less easily impressed scientists

11. See Nils-Axel Mörner, "Setting the Frames of Expected Future Sea-Level Changes by Exploring Past Geological Sea Level Records," in Don Easterbrook, et al., eds., *Evidence-Based Climate Science: Data Opposing CO2 Emissions as the Primary Source of Global Warming* (Amsterdam: Elsevier, 2011), 185-209.

12. "Sea Level Rise: Still Slowing Down," April 17, 2011, *World Climate Report*.

pointed out that living, sub-fossil, and fossil corals all exhibit temporary bleaching associated with stress, including rapid changes in temperatures, whether higher or lower. They indicated as well that this phenomenon is a common occurrence in reef corals and that the 1998 event was well within historical experience. Not surprisingly, all affected corals recovered rapidly, but coral bleaching has since become another favourite alarmist indicator of catastrophic global warming. There is no evidence to indicate that either the frequency or severity of such events has increased. Coral reefs exist in a significant range of temperatures and have survived climatic changes over the course of millions of years. Australian marine scientist Walter Starck points out:

> As for coral bleaching, the central fact never mentioned is that the high surface water temperatures associated with bleaching events are not the result of exceptionally high air temperatures. They result from extended periods of calm weather during which mixing from wave action ceases and [the] surface layer becomes exceptionally warm. Such warming is especially marked in very shallow water such as on reef flats. At the same time the absence of waves also eliminates the wave driven currents that normally flush the reef top. Bleaching conditions require at least a week or more of calm weather to develop and this may happen every few years, only once in a century, or never, depending on geographic location. On oceanic reefs it is less common due to ocean swell and currents even in calm weather. In coastal areas it is more common due to the absence of swell and reduced currents.[13]

Severe weather: hurricanes and tornadoes

There is no evidence that extreme weather events such as tropical storms, hurricanes, or tornadoes are more severe and frequent today than they have been in the past (see figure 6-2); in any event, none of these are consistent with the theory of global warming, as acknowledged by the IPCC. Storms are the result of differences in temperature and pressure between cold and warm fronts. Presumably, as the planet warms, particularly in higher latitudes as claimed by IPCC scientists, that difference will become smaller, reducing the basis for increases in violent storms.[14] Taking into account significant year-to-

---

13. Walter Starck, "The Great Barrier Reef and the Prophets of Doom," *OnLine Opinion,* May 8, 2008. See also Craig Idso, "CO2, Global Warming and Coral Reefs: Prospects for the Future," SPPI, January 12, 2009, Idso et al., *Climate Change Reconsidered* (2009), 596-639, and "Corals and Climate Change," *World Climate Report,* July 7.

14. Some scientists believe that the extra energy resulting from global warming leads to more intense storms, offsetting the reduced difference between warm

Figure 6-2: Global tropical cyclone frequency, 1971-2014

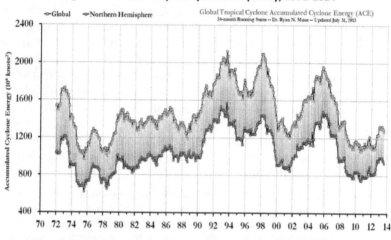

Source: http://policlimate.com/tropical/frequency_12months.png.

year and decade-to-decade variation, there has not been an increase in the number or intensity of land-falling Atlantic hurricanes for well over a century. And the number of *severe* typhoons or tropical cyclones has actually fallen over the past 30 years. Increased damage claims caused by severe weather are a result of greater wealth and of large numbers of people who choose to live in affected areas, for example, along the Florida Coast.[15]

Some recent major storms, such as Katrina in New Orleans, Sandy in New Jersey and New York, and Haiyan in the Philippines, led to dozens of articles claiming that these killer storms all pointed to global warming; some conceded that the storms themselves may not have been indicative but that their intensity *was*. This is a good example of a proposition that cannot be falsified and is thus questionable science. Katrina was a category 3 (out of 5) Atlantic hurricane but hit a vulnerable, unprepared city head on, causing billions of dollars of damage. Sandy was a late storm, creating the unusual confluence of an Atlantic hurricane and an Atlantic nor'easter hitting a very vulnerable area at high tide. By the time it reached landfall, it was no longer a hurricane. Haiyan was a very intense ty-

---

and cold fronts. Despite much controversy, this is a puzzle that will prove difficult to resolve.

15. See "Natural or Anthropogenic Effects on Atlantic Hurricanes, Past, Present, and Future," *World Climate Report*, November 3, 2008.

phoon but not unusual for the north-west Pacific; such typhoons, however, rarely make landfall.

Nevertheless, as Richard Lindzen cautions, "the fact that some models suggest changes in alarming phenomena will accompany global warming does not logically imply that changes in these phenomena imply global warming. This is not to say that disasters will not occur; they always have occurred, and this will not change in the future."[16] More recent efforts have focused on attributing the perceived additional severity of individual storms to the impact of global warming. Thus, Kevin Trenberth and Ben Santer, of the National Center for Atmospheric Research (NCAR), are hard at work tuning computers to be able to discern which percentage of a storm's severity is due to human influence and which percentage is due to natural factors.[17] The desire to attribute extreme weather events to AGW is, of course, consistent with the belief that climate is generally in a stable state unless upset by some "forcing," and such attribution assumes that the data and the computer programs are sufficiently robust to perform such intricate calculations.[18]

Severe weather: droughts and flooding

There is also no evidence that severe droughts and flooding are increasing. Patterns of flood and drought have fluctuated throughout history, and there has been no discernible or significant alteration in these patterns in recent decades. Again, the IPCC scientific reports specifically warn against attributing individual droughts or floods to anthropogenic global warming. Nevertheless, the IPCC and global warming alarmists generally insist that droughts and floods are becoming more frequent, despite evidence in the literature to the contrary. In its review of the literature, CO_2 *Science* reports:

---

16. Richard S. Lindzen, "Earth is never in equilibrium," *Boston Globe*, April 8, 2010.

17. See the discussion of the attribution problem in the context of severe weather on Judith Curry's blog, *Climate Etc.*, January 15, 2011.

18. Gilbert Compo, a lead scientist involved in NOAA's Twentieth Century Reanalysis Project, told the *Wall Street Journal*: "In the climate models, the extremes get more extreme as we move into a doubled CO_2 world in 100 years. So we were surprised that none of the three major indices of climate variability that we used show a trend of increased circulation going back to 1871," Anne Jolis, "The Weather Isn't Getting Weirder: The latest research belies the idea that storms are getting more extreme," February 10, 2011. The project's first report can be found in Compo et al., "The Twentieth Century Reanalysis Project," *Quarterly Journal of the Royal Meteorological Society*, 137 (January 2011), 1-28, Part A.

Over the totality of earth's *land* area, there appears to have been a slight intensification of the hydrologic cycle throughout the 20th century, which may *or may not* have been caused by the concomitant warming of the globe; but it also appears there was *no* intensification of deleterious weather phenomena such as tropical storms, floods and droughts. In addition, [one study] demonstrates that over the period 1979 to 2004, when climate alarmists claim the planet experienced a warming that was *'unprecedented over the past two millennia,'* there was no net change in *global* precipitation (over both land *and water*). Consequently, several of the most basic 'theoretical expectations' of the climate modelling enterprise appear to have no real-world support in 20th-century hydrologic data.[19]

Severe weather: heat waves

The claim that anthropogenic global warming will lead to more heat waves and record temperatures is also not borne out by the relevant data. More critically, that claim represents a misunderstanding of what causes heat waves and extreme high temperatures. Heat waves are caused by stationary, blocking highs, which push the jet stream northward and block cooler air masses from moving southward. Differential layers of air pressure also block warm air from rising, blocking cooler air from replacing the heated surface air. These heat waves can be intensified by prolonged dry conditions leading to less soil moisture, reducing surface capacity to absorb heat, as well as to less atmospheric moisture, which in turn leads to clear skies and increased solar radiation.[20] The confluence of such conditions can lead to prolonged heat waves, such as in France in 2003, Russia in 2010, and the United States in 2012. There is no evidence, however, that the frequency and intensity of such events have increased or that recent heat waves are outside historical boundaries.

---

19. "Predicted Effects of Global Warming on the Global Water Cycle," *CO2 Science* 26 (April 2006). That report is brought up to date in Craig and Sherwood Idso, *Carbon Dioxide and the Earth's Future*, 16-31, which provides an extensive review of more recent scientific literature examining criticisms of the modelling of droughts and floods in general circulation models as well as studies of the frequency and severity of droughts and floods in the real world. See also Andrew Montford, *Precipitation, Deluge and Flood: Observational evidence and computer modelling*, GWPF Briefing 10 (London: Global Warming Policy Foundation, 2014).

20. See Jim Steele, *Landscapes & Cycles: An Environmentalist's Journey to Skepticism* (CreateSpace Independent Publishing Platform, 2013) and the 13 references to heat waves at *World Climate Report*.

Sea surface temperatures

The upper layers of ocean temperatures have now been measured systematically and consistently for a decade on the basis of the Argo network of almost four thousand satellite-linked diving robots. To the consternation of the global warming industry, this much more sophisticated data set is failing to demonstrate the modelled increase in ocean temperatures. Similarly, NASA's Aqua satellite has, since 2002, gathered data on the composition of the atmosphere on a much more rigorous basis than had been possible previously.[21] Interpretation of the resulting data by Roy Spencer suggests a much lower temperature sensitivity than that fed into the general circulation climate models and points to a need to recalibrate the models to take this and other new evidence into account.[22] Efforts by climate scientists such as Kevin Trenberth to explain the "missing heat" as hiding in the deep ocean at times borders on farce.[23]

Ocean acidification

One of the more creative claims advanced by alarmists is that the oceans are turning acidic. Most people were introduced to the concept of alkalinity (or basic) and acidity in introductory chemistry in high school. The dividing line between the two – neutral – has been assigned the value of pH7 and the degree of acidity or alkalinity is determined on the basis of a logarithmic scale. For example, pH4 is ten times more acidic than pH5 and 100 times more acidic than pH6

---

21. Aqua is a NASA Earth Science satellite mission named for the large amount of information that the mission will be collecting about the Earth's water cycle, including evaporation from the oceans, water vapour in the atmosphere, clouds, precipitation, soil moisture, sea ice, land ice, and snow cover on the land and ice.

22. Spencer and his colleagues have written four articles reporting the results of their work on clouds and temperature sensitivity: Roy W. Spencer, William D. Braswell, John R. Christy, and Justin Hnilo, "Cloud and Radiation Budget Changes Associated with Tropical Intraseasonal Oscillations," August 9, 2007; Roy W. Spencer, "Satellite and Model Evidence Against Substantial Manmade Climate Change," December 27, 2008; Roy W. Spencer, "Global Warming as a Natural Response to Cloud Changes Associated with the Pacific Decadal Oscillation (PDO)," October 20, 2008 (updated December 29, 2008); and Roy W. Spencer & William D. Braswell, "Potential Biases in Feedback Diagnosis from Observational Data: A Simple Model Demonstration," November 1, 2008. All are available in pdf. format at http://www.drroyspencer.com/research-articles.

23. See Magdalena A. Balmaseda, Kevin E. Trenberth, and Erland Källén, "Distinctive climate signals in reanalysis of global ocean heat content," *Geophysical Research Letters* 40:9 (May 2013), 1754-9.

Oceans exhibit pH levels in the range of 7.8 to 8.4 – that is, they are alkaline. Oceans are not uniformly the same pH level due to run-off from land and rivers, circulation, rainstorms, temperature, and other factors. Daily, seasonal, and multi-year pH fluctuations at any given location can be on the order of ±0.3 pH units or more. Aquatic life is well-adapted to the range of alkalinity exhibited at various locations. Similar to land-based life, marine plant and animal life depend on the carbon in the ocean for growth and take up as much or more carbon from the ocean as terrestrial biota absorb from the atmosphere.

The oceans are also the largest "sink" for carbon dioxide, containing some 50 times more than the atmosphere. Carbon dioxide is in constant motion between these two reservoirs. As the oceans loses heat, they absorb CO_2, turning it into carbonic acid and thus buffering ocean alkalinity; as they absorb heat, they release CO_2. In general, tropical waters *release* CO_2 to the atmosphere, while cooler oceans farther north and south *absorb* CO_2 from the atmosphere. This is an important part of the carbon cycle. Alarmists are concerned that the combined increase in atmospheric CO_2 and global warming will lead to a gradual decrease in alkalinity; calling this process acidification, however, is nothing more than a scare tactic. Although theoretically possible, the chemistry of the oceans, like most earth systems, is in a constant state of flux and operates within larger boundaries than alarmists are prepared to admit. The oceans are also vast and deep. The IPCC calculates the decline in alkalinity *at the surface* in the order of pH 0.1 as a result of human emissions of CO_2 over the past two and a half centuries. Without baseline observations of historical alkalinity, it is hard to take this number seriously. Not surprisingly, model calculations point to further "acidification."

Part of the challenge in this area of earth science is that we have insufficient data to come to informed decisions about the future. Sherwood, Keith and Craig Idso, after reviewing the literature on the history of ocean pH values, conclude: "In light of these several diverse and independent assessments of the two major aspects of the ocean acidification hypothesis – a CO_2-induced decline in oceanic pH that leads to a concomitant decrease in coral growth rate – it would appear that the catastrophe conjured up by the world's climate alarmists is but *a wonderful work of fiction.*"[24]

---

24. "The Ocean Acidification Fiction," *CO2 Science,* June 3, 2009.

Figure 6-3: Reconstructed global temperature over the past 420,000 years

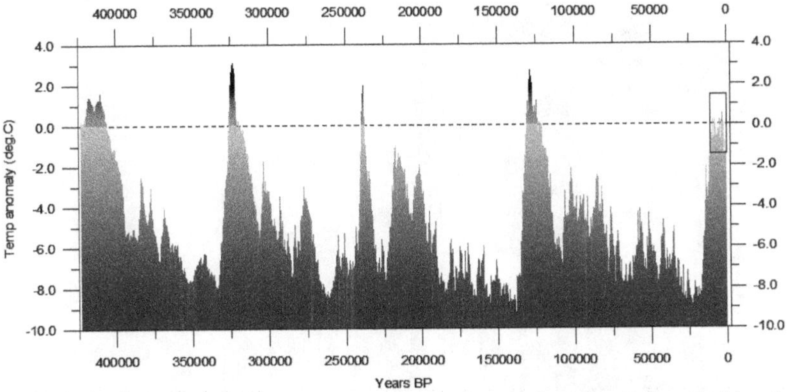

Data derived from the Vostok ice core in Antarctica. The record spans over four glacial periods and five interglacials, including the present. The horizontal line indicates the modern average global temperature. Source: www.climate4you.com/GlobalTemperatures.htm.

Paleoclimate evidence

The hockey stick graph produced by Michael Mann and associates for the IPCC's third assessment report in 2001 is an outlier in the paleoclimatic literature. As indicated in figures 6-3 and 6-4, Arctic and Antarctic ice cores and other proxies for Earth's earlier temperatures all indicate a wide range of climate change over the eons. Proxy records also indicate the extent to which atmospheric concentrations of carbon dioxide have varied over time and the extent to which current levels are low in geologic terms. Significant evidence from ice cores, rock weathering, and geologic features also points to the broadly accepted hypothesis that in geologic time, warming preceded increases in carbon dioxide rather than the other way around, and that, more broadly, there is little correlation between concentrations of atmospheric carbon dioxide and temperature (see figure 6-4).

Australian earth scientist Ian Plimer concludes that "for many environmentalists … it is ideologically impossible to acknowledge that the planet is dynamic and that past natural changes are far greater than anything measured in modern times. The weather dominates daily life. For the last few decades, global warming has replaced the weather. When we did not fry, climate change replaced global warming. And when the climate stubbornly did not [change],

Figure 6-4:
Atmospheric CO₂ and temperature fluctuations over geological timescale

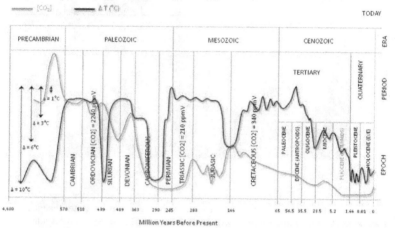

Standard graph showing lack of correlation in geologic time between atmospheric concentrations of carbon dioxide and temperature.

Source: http://www.biocab.org/Carbon_Dioxide_Geological_Timescale.html .

the language has been downgraded to carbon pollution, carbon footprint and carbon-free economy. Meanwhile, the planet has been doing what it always does – change."[25] In the same vein, his countryman, paleoclimatologist Bob Carter, adds: "Earth systems are constantly changing, and its lithosphere, biosphere, atmosphere, and oceans incorporate many complex, homeostatic buffering mechanisms. Changes occur in all aspects of local climate, all the time and all over the world. Geological records show that climate also changes continually through deep time. Change is what climate does, and the ecologies of the natural world change concomitantly in response."[26] Science writer Matt Ridley characterizes the current conceit that climate is in equilibrium unless disturbed by human agency as a special kind of narcissism that afflicts modern environmentalists.[27]

25. Ian Plimer, *Heaven and Earth: Global Warming, the Missing Science* (Lanham, MD: Taylor Trade, 2009), 440.

26. Robert M. Carter, *Climate Change: The Counter Consensus* (London: Stacey International, 2010), 130-1.

27. Matt Ridley, *The Rational Optimist: How Prosperity Evolves* (New York: HarperCollins, 2010), 329.

The IPPC's assessment of the impact of climate change

The idea of using secondary evidence to bolster the basis for an hypothesis is in itself not a problem. In such cases, however, it is important that the secondary evidence actually demonstrate the alleged relationship. As the above examples suggest, such is not the case. In the few instances in which there may be a correlation, there are better explanations for the phenomenon than anthropogenic climate change. Climate change impacts, adaptation, and vulnerability are the responsibility of IPCC Working Group 2. It has now prepared five assessment reports in addition to a special report on extreme events. All follow the working procedures of the IPCC, with co-chairs, lead authors, contributing authors, and review editors tasked with providing a detailed assessment of the state of the science in their respective areas based on a review of all relevant, peer-reviewed literature.[28] These procedures seem to have been more a matter of public relations than of actual practice. As Donna Laframboise and her collaborators have demonstrated for AR4, a large amount of information is sourced from so-called grey literature. Furthermore, the evaluation of what is relevant seems to be dictated by the intended message. Consistent with the alarmist mantra that the science is settled and that virtually all scientists agree, peer-reviewed scientific articles that do not subscribe to the alarmist view are either ignored or given short shrift. Fortunately, this lacuna is addressed in the reports of the NIPCC,[29] whose authors search *all* the literature and present a more balanced perspective.

For example, research by paleoclimatologists indicates that periods of cooling, including over the course of the current Holocene, correlate well with prolonged droughts, stormier weather, and greater volatility while periods of warming are associated with much less volatile weather. The models on which the IPCC relies,

---

28.   A total of some 70 natural scientists, under the leadership of Stanford's Chris Field, were involved in the preparation of WG2's fifth assessment report. Field, a biologist with a strong record of alarmism, has frequently told the media that climate change is "worse than we thought." See, for example, Bjorn Carey, "Climate Change on pace to occur 10 times faster than any change recorded in past 65 million years, Stanford scientists say," *Stanford News*, August 1, 2013, and "Chris Field discusses runaway climate change," *Beyond Zero Emissions*, no date.

29.   All the NIPCC reports are available at its website, *ClimateChangeReconsidered*. Its latest report, on biological impacts, was released March 31, 2014, concurrent with the release of WG2's fifth assessment report.

however, based on "physical principles," project that warming will have the effect of increasing storminess, droughts, and floods. While physical principles provide an important foundation for scientific research, the interaction among these physical principles within the highly complex climate system remains far from fully understood; simplifying these interactions in the models often leads to questionable results, largely because deterministic models cannot capture the random nature of chaotic climatic interactions. Observations from history and geology can help to explain how some of these forces interact. Similarly, historical evidence clearly indicates that during periods of warming civilization advances and people are healthier and more prosperous, but in the IPCC world of modelling, warming leads to poverty, decline, and inferior health.

The IPCC is in the business of threat inflation and of attributing all malign effects, real or imagined, to human agency. The reports of WG2 are no exception. Rupert Darwall pointed out after the release of WG2's fifth assessment report in 2014, "the job of the IPCC is to ratchet up the alarm. This it did in its report ... on the impacts of climate change. It scored a bull's-eye in the *Financial Times*: 'Climate change harms food crops, says IPCC,' the headline ran. 'Climate Signals Growing Louder,' the New York *Times* opined, though the reality is that the volume is being turned up by the IPCC, not the climate itself. For the IPCC, this is mission accomplished – at considerable cost to the body's residual credibility and integrity."[30] Richard Tol, one of the two economists who contributed to the report, asked that his name not be included among the authors of AR5 W2. Among issues he had with the final report was its relentlessly pessimistic tone: "It is pretty damn obvious there are positive impacts of climate change, even though we are not always allowed to talk about them."[31]

AR5 WG1's report on the physical science of climate change increased confidence in the assertion that the human influence on climate was responsible for more than half the post-1951 global warming despite the fact that all the warming over this period was concentrated in the 1979-1998 period. Over the same period, the concentration of atmospheric GHGs increased steadily and relatively uniformly, suggesting little correlation between the rise in GHGs

30.   Rupert Darwall, "Why the IPCC Report Neglects the Benefits of Global Warming," *National Review Online,* April 1, 2010.

31.   Cheryl K. Chumley, "UN climate author withdraws because the report has become 'too alarmist,'" Washington *Times,* March 2014.

Box 6-1: Key risks and vulnerabilities due to climate change, IPCC AR5

The key risks that follow, all of which are identified with high confidence, span sectors and regions. Each of these key risks contributes to one or more reasons for concern.

1. Risk of death, injury, ill-health, or disrupted livelihoods in low-lying coastal zones and small island developing states and other small islands, due to storm surges, coastal flooding, and sea-level rise.
2. Risk of severe ill-health and disrupted livelihoods for large urban populations due to inland flooding in some regions.
3. Systemic risks due to extreme weather events leading to breakdown of infrastructure networks and critical services such as electricity, water supply, and health and emergency services.
4. Risk of mortality and morbidity during periods of extreme heat, particularly for vulnerable urban populations and those working outdoors in urban or rural areas.
5. Risk of food insecurity and the breakdown of food systems linked to warming, drought, flooding, and precipitation variability and extremes, particularly for poorer populations in urban and rural settings.
6. Risk of loss of rural livelihoods and income due to insufficient access to drinking and irrigation water and reduced agricultural productivity, particularly for farmers and pastoralists with minimal capital in semi-arid regions.
7. Risk of loss of marine and coastal ecosystems, biodiversity, and the ecosystem goods, functions, and services they provide for coastal livelihoods, especially for fishing communities in the tropics and the Arctic.
8. Risk of loss of terrestrial and inland water ecosystems, biodiversity, and the ecosystem goods, functions, and services they provide for livelihoods.

Many key risks constitute particular challenges for the least developed countries and vulnerable communities, given their limited ability to cope.

Source: IPCC, WG2, AR 5, *Summary for Policy Makers,* March 31, 2014, 15.

---

and temperature. The lack of net warming since 1997 alone should have produced some uncertainty in WG1's conclusions.[32]

WG2's report, however, was much more cautious in tone and, without stating so directly, suggested that the authors were aware that the little warming that had taken place since 1951 indicated that there might be less warming over the course of the 21st century than

---

32. The large 1997-98 El Niño was offset by the equally large La Niña that followed, and thus the trend line from 1997-2014 is flat.

they had thought earlier and thus fewer dire impacts. WG2's 2007 (fourth) assessment report warned of five major areas of concern: availability of fresh water, ecosystem stress, adequate supplies of food, threatened coastal communities, and threats to health. WG2's latest report identified eight key risks and vulnerabilities due to climate change (see Box 6-1). Unlike the third and fourth reports, however, the fifth report, while not reducing the alarm, suggested that adaptation and resilience might be the best way to prepare for some of the key risks and vulnerabilities it identified. The third and fourth reports had stuck closely to the UN claim that climate change requires mitigation rather than adaptation.

The new report also reached less alarmist conclusions about the economic impacts of climate change than its predecessor and than the infamous 2006 study by the British government under the direction of Nicholas Stern. IPCC authors calculated, with medium confidence, that the overall cost of climate change to the global economy would be between 0.2 and 2.0 percent of GDP for a 2.0 °C temperature increase by the end of this century.[33] By way of contrast, Stern and his colleagues claimed climate change would cost from 5 to 20 per cent of world GDP from now to the end of time. Even so, the WG2 scientists were handicapped in reaching a realistic conclusion about impacts because they had to work with an assumption of substantially more warming by the end of the 21st century as had been claimed by the scientists who had prepared WG1's assessment. More recent estimates based on current observations rather than models suggested a much lower sensitivity for GHG forcings; if taken into account, they should also lead to much lower estimates of impacts, risks, and vulnerabilities.[34] The authors of WG1's 2014 report knew this but, as Ridley notes, waited a few months to "admit that 'estimates derived from observed climate change tend to best fit the observed surface and ocean warming for [sensitivity] values in the lower part of the likely range.' Translation: The data suggest we probably face less warming than the models indicate, but we would rather not say so." Ridley concludes: "Almost every global environmental scare of the past half century proved exaggerated including the population 'bomb,' pesticides, acid rain, the ozone hole, falling sperm counts, genetically engineered crops and killer bees.

33. IPCC AR5 WG2, *Summary for Policy Makers*, 19.

34. See Nicholas Lewis and Marcel Crok, "A Sensitive Matter: How the IPCC Buried Evidence Showing Good News About Global Warming," GWPF Report 13 (London: Global Warming Policy Foundation, 2014).

In every case, institutional scientists gained a lot of funding from the scare and then quietly converged on the view that the problem was much more moderate than the extreme voices had argued. Global warming is no different."[35]

New Scientist, in its coverage of WG2's latest report, noted that "in essence, the predictions are intentionally more vague. Much of the firmer language from the 2007 report about exactly what kind of weather to expect, and how changes will affect people, has been replaced with more cautious statements. The scale and timing of many regional impacts, and even the form of some, now appear uncertain."[36] In a similar vein, Fred Pearce, a veteran science journalist, reports that "careful readers will note a new tone to its discussion of these issues that is markedly different from past efforts. It is more humble about what scientists can predict in advance, and far more interested in how societies can make themselves resilient. It also places climate risks much more firmly than before among a host of other problems faced by society, especially by the poor. That tone will annoy some for taking the edge off past warnings, but gratify others for providing a healthy dose of realism."[37] WG2 co-chair Field, however, while presiding over the preparation of a much more realistic assessment, still insists that:

> The report itself is scientifically bold. It frames managing climate change as a challenge in managing risks, using this characterization as a starting point for two of the report's core themes. The first is the importance of considering the full range of possible outcomes, including not only high-probability outcomes. It also considers outcomes with much lower probabilities but much, much larger consequences. Second, characterizing climate change as a challenge in managing risks opens doors to a wide range of options for solutions.
>
> One of the things I like most about the report is that it combines cold, analytical realism, with a careful look at a broad range of possible solutions. This mapping of not only the serious and admittedly sometimes depressing 'problem space' but also the exciting and potentially uplifting 'solution space' allows the report to assess not only the impacts and challenges but also the opportunities and synergies. *Truly, much of the*

---

35. Matt Ridley, "Climate Forecast: Muting the Alarm," *Wall Street Journal,* March 27, 2014.

36. Michael Slezak, "World must adapt to unknown climate future, says IPCC," *New Scientist,* March 31, 2014.

37. Fred Pearce, "UN Climate Report is Cautious on Making Specific Predictions," *Yale Environment 360,* March 31, 2014.

material in the WG2 report is as much about building a better world as it is
about understanding serious problems.[38] [emphasis added]

The last sentence clearly places this report squarely within the context of the UN's broader purpose in raising and maintaining the alarm: building a better world, a theme discussed in detail in chapter 11. The "message" is now more than a matter of documenting climate change's many dire impacts but also of finding solutions that integrate with the UN's broader goal of social engineering to create a more "sustainable and just" world governed by UN technocrats.

Questionable Future Impacts

If, for the sake of argument, we accept the IPCC's assessment that there will be warming-derived stress for each of the eight areas it identifies, it is striking that the impacts are generally quite modest, develop over time, and are easily addressed by largely local adaptation strategies. Anticipated sea rises of a few millimeters a year, for example, hardly constitute an existential threat, even to low-lying areas. More than a quarter of the Netherlands, for example, lies below sea level, with its lowest point seven meters below the surrounding sea, a problem that was initially solved using 16th century technology. As in so many other instances, the alarmist community lacks both historical and geological perspective, assuming that all disturbing contemporary phenomena are unusual and unprecedented and require immediate governmental remediation.

Access to fresh water is not one of absolute availability but one of distribution. The thriving farm economy of the Central Valley in California, for example, depends wholly on extensive irrigation systems that bring water south from the Sacramento River and the Sierra watershed hundreds of miles to the north. California's growing population has increased demand for that water, creating problems, particularly during drier winters when the Sierra snowpack is not enough. Desalination technologies now make it possible to bring potable water at reasonable cost to areas previously denied access to water, as has been demonstrated by the fresh fruit and vegetable farms now thriving in Israel. Cleaning up fresh water supplies in other areas can similarly be done at lower costs than the costs of proposed climate mitigation measures. Similarly, the idea that all

---

38. Chris Field, Press conference in Yokohama, March 25, 2014. Field is the leading US candidate to replace disgraced IPCC chair Rajendra Pachauri.

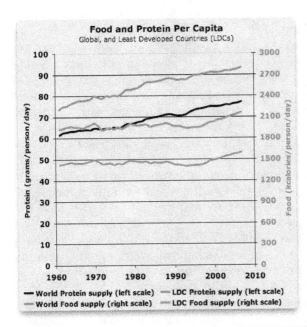

Figure 6-5: Global food and protein per capita, 1960-2010

This graphic, based on data from the FAO, shows that total food and protein per capita have risen both globally and for the world's poorest countries (LDCs)
Source: Willis Eschenbach, "Farmers versus Famine," *WattsUpWith That,* March 26, 2011.

technological and other improvements in agricultural production will cease with global warming and that farmers will not know how to adapt to changing circumstances makes a mockery of the last century and a half of developments in food production. As indicated in Figure 6-5, the availability of food has steadily increased since the 1960s when Paul Ehrlich and his fellow eco-pessimists predicted widespread famine by the 1980s. There is, indeed, still starvation in the world today, but it is due to political interference with distribution systems rather than with a global lack of available foodstuffs.[39]

Part of the IPCC's assessment of reduced global food supplies is derived from computer simulations showing increased flooding and more droughts as a result of the regional redistribution of precipitation patterns. These simulations are even less credible than the computer programs that project rapid increases in global temperatures. They ignore that variability is the nature of weather and climate, requiring farmers to adapt on a regular basis. Australians, for example, who have extensive historical experience with both flooding and droughts, were told confidently during the droughts of the

---

39. See Giovanni Federico, *Feeding the World: An Economic History of Agriculture, 1800-2000* (Princeton: Princeton University Press, 2005).

first decade of the 21st century that this was the new normal as a result of global warming. The next few years, however, brought record rains and concerns about flooding. Since there has been no warming since 1997-98, neither the dry nor wet periods can be the result of global warming but are part of the cyclical nature of Australian precipitation patterns resulting from the influence of cyclical variations in such coupled atmospheric-oceanic circulations systems as the PDO and ENSO.[40] The same holds true for other parts of the globe prone to droughts and flooding. As *World Climate Report* observes, "In nearly every presentation on global warming, we hear that floods and droughts will be more severe as the temperature rises. Believe it or not, and who would not believe it given thousands of websites on the issue, there are many scientists who believe the opposite."[41]

Throughout its assessment reports, the IPCC places considerable emphasis on the extent to which global warming will harm the poor, particularly in developing countries. Consistent with much of the UN's work, the IPCC seeks to strengthen its case by hitching remediation of the climate system to the more general UN campaign for sustainable development. Much of the analysis, however, is circular and unconvincing. To achieve the high emission numbers that lead to catastrophic warming, the IPCC relies on scenarios estimating growth in global population, production, and energy use. These numbers are then fed into the climate models. They are also used to run models on the socio-economic impacts of warming, which point to anticipated harm to the world's poor. Lost in this shell game is that the scenarios which give the desired high emission numbers indicate that per capita incomes in developing countries will be significantly higher than those in industrialized countries today. To reach high levels of warming, the models need high levels of economic development and energy use, but the modellers then seem to forget that in this much richer world, the poor will also be much better off. At these levels of income, it is difficult to understand why some of that wealth could not be spent on public health, improvements in food production, and strengthening of infrastructure, thereby reducing the scope of these imagined catastrophes. In

---

40. Ron Pike, "From food bowl to dust bowl," *Quadrant Online,* August 25, 2011, and Debbie Buller, "Despair on the Land," *Quadrant Online,* September 13, 2011.

41. "Floods, Droughts and Global Cooling," April 24, 2008. See also "Update in Global Drought Patterns (IPCC Take Note)" *World Climate Report,* February 24, 2010.

June 2013, World Bank President Jin Yong Kim joined the chorus of doomsters and committed the Bank to a wholesale assault on the scourge of global warming. "Moving ahead," he said, "we at the Bank will be looking at everything we do through a climate lens."[42] Given the projections upon which the IPCC relies, there will be no need for the Bank in another generation or two, as all countries will boast incomes well above those of today. Matt Ridley suggests that "if there is a 99 percent chance that the world's poor can grow much richer for a century while still emitting carbon dioxide, then who am I to deny them that chance? After all, the richer they get the less weather dependent their economies will be and the more affordable they will find their adaptation to climate change."[43]

Among the more absurd dimensions of the global warming scene is the prominent role of so-called scientific impact studies. As Lindzen notes: "Here, scientists who generally have no knowledge of climate physics at all, are supported [funded] to assume the worst projections of global warming and imaginatively suggest the implications of such warming for whatever field they happen to be working in. This has led to the bizarre claims that global warming will contribute to kidney stones, obesity, cockroaches, noxious weeds, sexual imbalance in fish, etc. The scientists who participate in such exercises quite naturally are supportive of the catastrophic global warming hypothesis despite their ignorance of the underlying science."[44] Many scientists have discovered that research funds flow to those who will advance the global warming story, no matter how tenuous the connection, as long as they toe the party line and forecast dire consequences.[45] As a result, there are studies that as-

---

42. Kim, "US takes key climate change steps, but the world must do more," *Washington Post*, June 27, 2013. See also Cornelis van Kooten, *Climate Change, Climate Science, and Economics: Prospects for an Alternative Energy Future* (New York: Springer, 2013), 276-78.

43. Ridley, *The Rational Optimist*, 333.

44. Richard Lindzen, "Climate Science: Is It Currently Designed to Answer Questions?" at *Global Research*, September 22, 2014. Van Kooten adds: "The IPCC's Working Group I scientists only know what historical temperatures have done and what might happen in a world of higher temperatures, but have no comparative advantage in predicting the extent of damages and social unrest/upheaval that global warming or attempts to mitigate it might cause. Here we are in a fuzzy arena where theology, philosophy and social science trump climate science." *Climate Change, Climate Science, and Economics*, 62.

45. For examples, see Paul Driessen, with Willie Soon and David R. Legates, "Is Global Warming Really Cause for Alarm?" May 22, 2010, at *Townhall.com*.

sume worst-case scenarios, model the results in as alarming a way as possible in fields unrelated to the core issue, and then garner maximum media attention in order to attract more funding. Tropical disease specialist Paul Reiter explains what is involved: "The advent of low-cost computers has propelled mathematical modelling into a major role in the description of complex systems, including ecology, epidemiology and public health. It is now relatively simple to run stochastic models, built of a selected set of variables, with interactions driven by sets of differential equations. Complex systems imply the need for a large number of variables and operators, but as these numbers increase, so does the variance – and the uncertainty – of the models. In a sense, therefore, they remain an extension of the intuitive approach because the selection of variables, the assumptions of the frequency distributions that are involved, the mathematical descriptors of the operators, and the constraints on both are made by the modeller."[46]

Given the apocalyptic, existential tones in which the threat of global warming is couched by alarmists, it is, therefore, interesting to summarize the dire predictions of the impact of global warming made by the IPCC. Trapped by the assumptions built into its scenarios which project global warming of between 1.8° and 4.5° C by the end of the 21st century,[47] the IPCC indicated in its 2007 report that such changes would lead to a possible loss of 5 percent of global GDP by the end of the century. After doing the numbers for the IPCC's gloomiest scenarios, former British Chancellor of the Exchequer, Nigel Lawson, concluded that "the disaster facing the climate is that our great-grandchildren in the developed world would, in a hundred years time, be only 2.6 times as well off as we are today, instead of 2.7 times, and that their contemporaries in the developing world would be 'only' 8.5 times as well off as people in the developing world are today, instead of 9.5 times as well off."[48] Few people appreciate that such is the economic "calamity" that awaits the world's population in the absence of remedial action. Indur Goklany points out that "climate change is a moral and ethical issue.

---

46. Paul Reiter, "Global Warming and Malaria: Knowing the Horse Before Hitching the Cart," SPPI Reprint, December 11, 2009.

47. This is the middle range that the IPCC finds most credible. Its scenarios project warming of both much less and much more, depending on the various assumptions built into the scenarios.

48. Nigel Lawson, *An Appeal to Reason: A Cool Look at Global Warming* (London: Overlook Press, 2008), 36.

But it is a strange ethical calculus that justifies reducing existing gains in human wellbeing, increasing the cost of humanity's basic necessities, increasing poverty, and reducing the terrestrial biosphere's future productivity and ability to support biomass."[49] The 2014 fifth assessment report scales down the apocalyptic tone considerably, indicating that an approximate rise of 2.0°C could lead to an annual reduction in global GDP of between 0.2 and 2.0 percent. To date, this less pessimistic assessment has not found its way into ENGO rhetoric and public policy discussions.

Part of the problem in the economic impact assessments advanced by the IPCC is the conflation of population increases and global warming. IPCC scenarios assume that population growth will continue at rates similar to those that prevailed in the 20th century, ignoring the impact that prosperity has historically had on fertility rates. Natural population growth in OECD countries has for more than a generation been below the replacement rate; a similar slowdown is now apparent in many developing countries, indicating that global population levels will likely peak by mid-century and then begin a slow decline. Even these increases will admittedly create pressures on habitats for flora and fauna or increase property damage from violent storms. Rising prosperity in developing countries, although perhaps not at the pace modelled by the IPCC, will add to these pressures. Nevertheless, these issues are unrelated to global warming; efforts to mitigate increased GHG-induced global temperatures will have no impact on them, except to exacerbate problems by wasting resources that could have been used for other purposes on climate-related adjustment programs and policies.

Despite the relative modesty of the claims made in the IPCC's own reports, some climate scientists, ENGOs, and other alarmists hype them into much more frightening scenarios, ones based on the flimsiest of studies. Gore's over-the-top film and subsequent book, *An Inconvenient Truth*, are built on a catalogue of horror stories bearing no resemblance to anything found in the IPCC's reports. It is

---

49.   Indur Goklany, "The human triumph from fossil fuels," *Financial Post*, July 8, 2015. He has independently modelled projections of future impacts and has concluded that caution and adaptation are the best approaches. He warns that "an aggressive strategy would retard increases in global wealth, which would lead to greater hunger, poorer health, and higher mortality, as well as retard progress toward environmental improvements such as safer water, better sanitation, reduced habitat loss, and lowered threats to biodiversity." *The Precautionary Principle: A Critical Appraisal of Environmental Risk Assessment* (Washington: Cato Institute, 2001), 85.

striking that more sober elements in the global warming camp have not seen fit to distance themselves from this farrago of lies and exaggerations. Instead, they organize conferences, write popular books, and maintain blogs for which the only discernible purpose is to raise the alarm in the hope of spurring governmental action. The media, in turn, publicize every press release describing another computer study that projects the dire consequences of climate change. All of this handwringing would be of little moment except that the preferred public policies sought by the alarmists would require wholesale changes in lifestyles in the developed world and would ensure greater poverty in the developing world. Governments, to their shame, are the main funders of these studies, and public servants use them to justify expensive but ineffective programs, policies, and regulations.

As already noted, the IPCC scenarios themselves are wildly alarmist, not only on the basic science but also on the underlying economic assumptions, which in turn drive the alarmist impacts. The result cannot withstand critical analysis. Economists Ian Castles and David Henderson, for example, show the extent to which the analysis is driven by the desire to reach predetermined outcomes.[50] Other economists have similarly wondered what purpose was served by pursuing such unrealistic scenarios. It is hard to credit the defense put forward by Mike Hulme, one of the creators of the scenarios, that the IPCC is not engaged in forecasting the future but in creating "plausible" story lines of what might happen under various scenarios.[51]

Each scare scenario is based on linear projections without any reference to technological developments or adaptation. If, on a similar linear basis, our Victorian ancestors in the UK, worried about rapid urbanization and population growth in London, had made similar projections, they would have pointed to the looming crisis arising from reliance on horse-drawn carriages and omnibuses; they would have concluded that by the middle of the 20th century, Lon-

---

50. Ian Castles and David Henderson, "The IPCC emission scenarios: An economic-statistical critique," *Energy & Environment* 14: 2-3 (2003), and "Economics, emissions scenarios and the work of the IPCC," *Energy & Environment* 14: 4 (2003). See also David Henderson, "Climate Change Issues and Global Salvationism," *Economic Education Bulletin* XLV (6), 1-6.

51. See L.O. Mearns and M. Hulme, lead authors, "Climate Scenario Development," IPCC, Third Assessment Report (2001), chapter 13, and Mark New and Mike Hulme, "Representing uncertainty in climate change scenarios: a Monte-Carlo approach," *Integrated Assessment* I (2000), 203-213.

don would be knee-deep in horse manure, and all of the southern counties would be required to grow the oats and hay to feed and bed the required number of horses. Technology progressed and London adapted. Why should the rest of humanity not be able to do likewise in the face of a trivial rise in temperature over the course of more than a century?

The work on physical impacts is equally over the top. All the scenarios assume only negative impacts, ignore the reality of adaptation, and attribute any and all things bad to global warming. Assuming the GHG theory to be correct means that its impact would be most evident at night and during the winter in reducing atmospheric heat loss to outer space.[52] It would have greater impact in increasing minimum temperatures than in increasing maximum temperatures. Secondary studies, however, generally ignore this facet of the hypothesis.

The IPCC believes that a warmer world will harm human health due, for example, to increased disease, malnutrition, heat-waves, floods, storms, and cardiovascular incidents. As already noted there is no basis for the claim about severe-weather-related threats or malnutrition. The claim about heat-related deaths gained a boost during the summer of 2003 because of the tragedy of some 15,000 alleged heat-related deaths in France as elderly people stayed behind in city apartments without air conditioning while their children enjoyed the heat at the sea shore during the August vacation. Epidemiological studies of so-called "excess" deaths resulting from heat waves are abused to get the desired results. Similar studies of the impact of cold spells show that they are far more lethal than heat waves and that it is much easier to adapt to heat than to cold.[53] More fundamentally, this, like most of the alarmist literature, ig-

---

52. Freeman Dyson points out that "the warming effect of carbon dioxide is strongest where air is cold and dry, mainly in the Arctic rather than in the tropics, mainly in mountainous regions rather than in lowlands, mainly in winter rather than in summer, and mainly at night rather than in daytime. The warming is real, but it is mostly making cold places warmer rather than making hot places hotter. To represent this local warming by a global average is misleading." Dyson, "Heretical Thoughts about Science and Society."

53. To the extent that extreme weather events contribute to increased mortality, the case for cold weather impacts is much stronger than for heat-related ones. See, for example, Indur Goklany, "Death and Death Rates Due to Extreme Weather Events: Global and U.S. Trends, 1900–2006," in Paul Reiter, et al., *Civil Society Report on Climate Change* (London: International Policy Press, 2007), and "Overstating Health Impacts of Global Warming," *World Climate Report*, November 22, 2005.

nores the basics of the AGW hypothesis: the world will *not* see an exponential increase in summer, daytime heat (and thus more heat waves), but a decrease in night-time and winter cooling, particularly at higher latitudes and altitudes. Based on the AGW hypothesis, Canada, China, Korea, Northern Europe, Australia, New Zealand, South Africa, Chile, and Argentina will see warmer winters and warmer nights. There are clear benefits to such a development, even if there may also be problems, but the AGW industry tends to ignore the positive aspects of their alarmist scenarios.

The feared spread of malaria, a much repeated claim, is largely unrelated to climate. Malaria's worst recorded outbreak was in Siberia long before there was any discussion of AGW. Similarly, the building of the Rideau Canal in Ottawa in the 1820s was severely hampered by outbreaks of malaria due to the proximity of mosquito-infested wetlands in the area. Malaria remains widespread in tropical countries today in part because of the UN's lengthy embargo on the use of DDT, the legacy of an earlier alarmist disaster. Temperature is but one factor, and a minor one at that, in the multiple factors that affect the rise or decline in the presence of disease-spreading mosquitoes. Wealthier western countries have pursued public health strategies that have reduced the incidence of the disease in their countries. Entomologist Paul Reiter, widely recognized as the leading specialist on malaria vectors and a contributor to some of the early work of the IPCC, was aghast to learn how his careful and systematic analysis of the potential impacts had been twisted in ways that he could not endorse. In a recent paper, he concludes: "Simplistic reasoning on the future prevalence of malaria is ill-founded; malaria is not limited by climate in most temperate regions, nor in the tropics, and in nearly all cases, 'new' malaria at high altitudes is well below the maximum altitudinal limits for transmission. Future changes in climate may alter the prevalence and incidence of the disease, but obsessive emphasis on 'global warming' as a dominant parameter is indefensible; the principal determinants are linked to ecological and societal change, politics and economics."[54]

---

54. Paul Reiter, "Global Warming and Malaria: Knowing the Horse Before Hitching the Cart," SPPI Reprint, December 11, 2009, 12. More generally on DDT and malaria, see Paul Reiter, "Global Warming Won't Spread Malaria," *Science & Environment*, April 6, 2007, 52-6; Christopher Booker and Richard North, *Scared to Death: From BSE to Global Warming: Why Scares Are Costing Us the Earth* (London: Continuum Books, 2007), 167-70; and Bjørn Lomborg, *Cool It*, 92-102. The WHO

Catastrophic species loss similarly has little foundation in past experience. [55] Even if the GHG hypothesis were to be correct, its impact would be slow, providing significant scope and opportunity for adaptation, including by flora and fauna. One of the more irresponsible claims was made by a group of UK modellers who fed wildly improbable scenarios and data into their computers and produced the much-touted claim of massive species loss by the end of the 21s century. There are literally thousands of websites devoted to spreading alarm about species loss and biodiversity.

Global warming is but one of many claimed human threats to the planet's biodiversity. The claims, fortunately, are largely hype, based on computer models and the estimate by Harvard naturalist Edward O. Wilson that 27,000 to 100,000 species are lost annually – a figure he advanced purely hypothetically but which has become one of the most persistent of environmental urban myths. [56] The fact is that scientists have no idea of the extent of the world's flora and fauna, with estimates ranging from five million to 100 million species, and that there are no reliable data about the rate of loss. By some estimates, 95 per cent of the species that ever existed have been lost over the eons, most before humans became major players in altering their environment. A much more credible estimate of recent species loss comes from a surprising source, the UN Environmental Program. It reports that known species loss is slowing, reaching its lowest level in 500 years in the last three decades of the

---

finally relented in 2006 and lifted its embargo on DDT, conceding that politics should give way to science.

55. This much-hyped, global warming-related scare was generated by British bio-modeller Chris Thomas and colleagues, "Extinction risk from climate change," *Nature* 427, January 8, 2004, 145-148. He estimated that by the end of the 21st century, as many as 90 per cent of species would be lost. It is also worth recalling that catastrophic climate change and species diversity were the two featured anxieties at the 1992 Rio Earth Summit, with governments pledging not only to tackle AGW through the UN Framework Agreement on Climate Change but also to fight species loss through the UN Convention on Biological Diversity. Both conventions were formally opened for signature at the Summit, and both were celebrated with large doses of hype and renewed anxiety at Rio+20, June 20-21, 2012.

56. The numbers were first advanced in E.O. Wilson, *The Diversity of Life* (Cambridge: Belknap Press, 1992). They built on estimates first made by Norman Meyers in *The Sinking Ark: A New Look at the Problem of Disappearing Species* (New York: Pergamon, 1979), in which he maintained that by the end of the 20th century, the earth would be losing 40,000 species a year. See also E.O. Wilson, ed., *Biodiversity* (Washington: National Academies Press, 1988) for a full rehearsal of the biodiversity issue and its principal proponents.

20th century, with some 20 reported extinctions despite increasing pressure on the biosphere from growing human population and industrialization.[57] The alarmist community has also introduced the scientifically unknown concept of "locally extinct," often meaning little more than that a species of plant or animal has responded to adverse conditions by moving to more hospitable circumstances, e.g., birds or butterflies becoming more numerous north of their range and disappearing at its extreme southern extent. Idso et al. conclude: "Many species have shown the ability to adapt rapidly to changes in climate. Claims that global warming threatens large numbers of species with extinction typically rest on a false definition of extinction (the loss of a particular population rather than entire species) and speculation rather than real-world evidence. The world's species have proven very resilient, having survived past natural climate cycles that involved much greater warming and higher CO_2 concentrations than exist today or are likely to exist in the coming centuries."[58]

♦♦♦♦♦

---

57. Brian Groombridge and Martin D. Jenkins, *World Atlas of Biodiversity* (Berkeley, CA: UNEP and University of California Press, 2002).

58. Idso et al., *Climate Change Reconsidered* (2009), 588. See also C. Loehle and W. Eschenbach, "Historical bird and terrestrial mammal extinction rates and causes," *Diversity and Distributions* (2011), 1-8 and G. Cornelis van Kooten and Erwin H. Bulte, *The Economics of Nature: Managing Biological Assets* (Oxford: Blackwell, 2000), 270-318.

7 The Science is *Not* Settled

The Intergovernmental Panel on Climate Change has spoken: 'Warming of the climate system is unequivocal' and it is 'very likely' due to human activities.[1]

IPCC Lead Author Kevin Trenberth (2009)

The great tragedy of science – the slaying of a beautiful hypothesis by an ugly fact.

Thomas Henry Huxley (1825-95)

IPCC-affiliated scientists and government officials insist that the science explaining anthropogenic climate change is settled and that the few scientists who may differ are cranks or shills for oil and other nefarious interests. They also like to point out that there is no consensus among those scientists who disagree. The first point is false and demeaning; the second is true but unremarkable. The claim of consensus, of course, serves the political goal of marginalizing scientists who disagree with the official science and of strengthening the case for a public policy response to climate change. A significant number of research scientists, however, with few or no ties to any economic or political interests have principled problems with the so-called consensus and the science behind it.

---

1. Kevin Trenberth, in debate with hurricane specialist William Gray, February 19, 2008, at Fort Collins Forum in Colorado.

As we have seen, the consensus or "official" perspective was developed by a relatively small group of climate scientists associated with the IPCC and a number of journals and blogs. Many of these scientists work in government agencies and laboratories or are dependent on government funding for their work. The sceptical perspective involves thousands of scientists, each independently advancing arguments and objections that question the IPCC perspective. They are drawn from a wide range of scientific disciplines, can be found in universities, think tanks, industry, and even government agencies, and include many retired specialists who find the climate change issue an intriguing and challenging way to maintain their critical skills. Their views range from the mildly sceptical about one or two aspects of the official view to those who dismiss the whole idea as a cruel hoax. There are also those who work on highly specialized aspects of climate change and who express little interest in the larger questions but keep plugging away on the issues that count in their own specialized areas of research.

Virtually all sceptics accept that temperatures have increased over the course of the 20th century and that human activity may have contributed to this increase. Most reject, however, the claim that the trivial, cumulative increase in the global temperature – insofar as it can be measured – over the past century and a half, perhaps as much as 0.8°C, presages future catastrophic change. Many further insist that natural factors have been given inadequate attention by those wedded to the official version of climate change. A significant number also believe that the data are not sufficiently robust to underpin the conclusions advanced by the IPCC and its contributors.

In a nutshell, most sceptical scientists question the certainty with which IPCC scientists attribute late 20th century warming to increases in atmospheric greenhouse gas levels. They also question the extent to which IPCC-affiliated scientists dismiss or minimize the role of such natural factors as changes in total solar irradiance, fluctuations in the hydrological cycle, fluxes in cosmic rays, and oscillations in coupled atmospheric-oceanic circulation patterns. IPCC scientists have high confidence that the direct warming effect of CO_2 is amplified by a factor of at least three because a warmer world will lead to higher levels of water vapour, a more powerful greenhouse gas than CO_2. As discussed earlier, much of their confidence is derived from modelling exercises. Sceptics have little confidence in modelling experiments and base much of their scepticism on the extent to which observational evidence contradicts model outcomes. Both sides rely on a wide range of data sources but reach different conclusions about the quality and interpretation of the data.

These differences in perspective have profound implications for the way one views the political and economic consequences of climate change. If IPCC scientists are right and increased greenhouse gases *are* the principal cause of late 20th century global warming and are indicative of more warming to come, then there may be policy steps that can be taken to mitigate the impact of climate change at the global level. If, on the other hand, greenhouse gases are a minor factor and climate change is largely a matter of natural forces, then the need for policy steps can be limited to local adaptation, as required.

Despite the repeated claims that only a small but persistent group of scientists dispute the conclusions of the IPCC, criticism of the IPCC's work and assertions was evident from the beginning and is still widespread. Fred Singer, for example, an atmospheric physicist with extensive experience both as a university and government researcher – he turned 90 in 2014 – responded to the IPCC's first Assessment Report (1990) by organizing the Science and Environmental Policy Project (SEPP) with the specific objective of presenting alternative views. He recruited the late Frederick Seitz, a physicist and former president of the National Academy of Sciences, to chair SEPP's board and organized a series of conferences and petitions to demonstrate that not all scientists shared the IPCC perspective.[2] Since retiring as professor of environmental sciences at the University of Virginia, Singer has devoted most of his energies to promoting wider discussion. He points out that the world of climate science can now be divided into three broad groups: warmistas, sceptics, and deniers. He concludes that from his perspective "we can accomplish very little with convinced warmistas and probably even less with true deniers. So we just make our measurements, perfect our theories, publish our work, and hope that in time the truth will out."[3]

---

2. Both Singer and Seitz have been at the receiving end of campaigns to discredit their views by tying them to support from oil, tobacco, and other presumed nefarious interests. See, for example, Naomi Oreskes and Erik Conway, *Merchants of Doubt: How a Handful of Scientists Obscured the Truth on Issues from Tobacco Smoke to Global Warming* (New York: Bloomsbury Press, 2010). Both men enjoyed distinguished careers as scientists and public servants and deserve better. These *ad hominem* attacks, however, have become all too common from climate alarmists and the ENGO community and raise serious questions about the motives and objectives of the climate alarm movement.

3. S. Fred Singer, "Climate Deniers Are Giving Us Skeptics a Bad Name," *American Thinker*, February 29, 2012.

Singer contributed to the early work of the IPCC as an expert reviewer but found that the opinion of reviewers is taken into account only to the extent that it confirms the objectives of the organization: to verify a malign human influence on climate change. For Singer, the IPCC "depends very much on detailed and somewhat arbitrary choices of model inputs – for example, the properties and effects of atmospheric aerosols, and their temporal and geographic distribution. It also makes arbitrary assumptions about clouds and water vapour, which produce the most important greenhouse forcings. One might therefore say that the IPCC's evidence is nothing more than an exercise in curve-fitting."[4] In an earlier article, he summarized the problem with the data upon which the IPCC relies: "The commonly reported and accepted warming between 1978 and 2000 is based only on thermometers from land surface stations and is not supported by any other evidence that I could find. Specifically, ocean data (from 71 percent of the earth's surface) and global atmospheric data (as recorded by satellites and independent balloon-borne radiosondes) do not show such a warming at all. In addition, most proxy data, from non-thermometer sources such as tree rings, ocean sediments, ice cores, stalagmites, etc., show no warming during this same crucial period."[5]

Together with Dennis Avery, Singer has offered his own view on the principal drivers of long-term climate change, i.e., cyclical patterns in the sun's output. For Avery and Singer, there are multiple lines of evidence from ice cores and other paleoclimatological sources pointing to some 600 climate cycles over the past million or so years, each roughly 1500 years in length, a cyclical pattern first discovered by Danish scientist Willi Dansgaard and his Swiss and French collaborators, Hans Oeschger and Claude Lorius. This cycle brings alternating periods of warming and cooling, largely due to changes in the intensity of the sun. Within this 1500-year pattern, shorter cycles can be observed as a result of the internal dynamics of coupled atmospheric-oceanic circulation patterns, such as the El Niño-Southern Oscillation (ENSO) and the Pacific Decadal Oscillation (PDO). There is also the much longer glacial-interglacial cycle of the past three million years consisting of alternating periods of cold

---

4.   Singer, "Climate Deniers Are Giving Us Skeptics a Bad Name."

5.   S. Fred Singer, "Fake! Fake! Fake! Fake!" *American Thinker*, January 2, 2012. As a word of caution, he adds: "One has to be careful in this analysis since the year 1998 shows a major warming spike caused by a Super-El Niño. But by 1999 and 2000, temperatures had returned to pre-1998 values."

lasting about 100,000 years and brief interglacial periods of some 10,000 years.[6] Singer finds that these periodic, overlapping oscilla-tions offer a fully satisfactory explanation of recent changes and that minor impacts flowing from changes in atmospheric greenhouse gases are overwhelmed by these natural factors.

Similarly, Richard Lindzen, long the Alfred P. Sloan Professor of atmospheric physics at MIT and now retired, has been a persis-tent critic of IPCC science despite the fact that he was nominated by the US government to participate in the IPCC's second and third Assessment Reports. He has characterized much of that work as "an admirable description of research activities in climate science" but finds that the *Summary for Policy Makers* "has a strong tendency to disguise uncertainty, and conjures up some scary scenarios for which there is no evidence."[7] He has subsequently become an an-noying thorn in the side of IPCC scientists, taking every opportuni-ty to point out the limitations of the GHG theory and indicating that natural factors can explain most of the variations in climate over the past half-century. In company with many other physicists, Lindzen views the climate system as more complex and chaotic than IPCC supporters claim and doubts that scientific understanding of its in-tricate internal dynamics is sufficient to make scary predictions about future climate. Both his scientific and policy writings have gained a wide audience within the sceptical community.

Paul Reiter was another early critic, less of the work of WG1 ex-amining the physical science of climate change and more of the findings of WG2 focused on impacts, adaptation, and vulnerability. As one of the world's leading experts on insect-spread diseases, he was nominated by the US government to contribute to the 2001 third Assessment Report but resigned when it became clear that the project was more about politics than about science. Reiter explained, "it [IPCC] is a panel among governments. Any scientist who partic-ipates in this process expecting the strictures of science to reign must beware, lest he be stung."[8] Rejecting the IPCC view that cli-mate change will lead to an increase in tropical and infectious dis-

---

6.   S. Fred Singer and Dennis T. Avery, *Unstoppable Global Warming: Every 1,500 Years*, Updated and Expanded Edition (Lanham, MD: Rowman & Littlefield, 2008).

7.   Richard Lindzen, "Canadian Reactions To Sir David King," *The Hill Times*, Ot-tawa, Feb. 23 - March 1, 2004.

8.   Quoted in Lawrence Solomon, "Bitten by the IPCC," *National Post*, March 23, 2007.

eases, he concludes that the story is much more complex: "The ecology and natural history of disease transmission, particularly transmission by arthropods, involve the interplay of a multitude of interacting factors that defy simplistic analysis. The rapid increase in the incidence of many diseases worldwide is a major cause for concern, but the principal determinants are politics, economics, human ecology, and human behaviour. A creative and organized application of resources to reverse this increase is urgently required, irrespective of any changes of climate."[9]

Singer, Lindzen, and Reiter are not alone. Hundreds of other scientists pursuing research in various disciplines pertinent to the science of climate change harbour reservations of one kind or another about the IPCC hypothesis and its purported impacts. The extent of that disagreement becomes evident when one begins to look beyond the literature that is contained in the journals specializing in climate science. It is also evident in the increasingly specialized quality blogosphere. This chapter surveys the perspectives of some of the principal critics of AGW science, largely organized in terms of their core disciplines.

Climate scientists

Systematic study of climate is relatively new, encompassing no more than two generations of researchers. Its practitioners are drawn from a number of scientific disciplines, and very few are masters of the whole field. Some are trained as meteorologists, others as physicists, chemists, mathematicians, computer modellers, geologists, cosmologists, or astronomers. Those who have focused most of their research efforts on climate issues have gradually coalesced into practitioners of what can now be considered climate science. There is no consensus as to how many "climate scientists" that may include. In a 2011 paper, Anderegg et al. determined that "97-98 percent of the climate researchers most actively publishing in the field support the tenets of [anthropogenic climate change] outlined by the [IPCC]."[10] They achieved this remarkable result by excluding the work of all those with whom they disagreed or who had not published a minimum number of papers in the peer-reviewed journals controlled by the alarmist community. Their paper reflected the

9.   Paul Reiter, "Human Ecology and Human Behaviour," in *Civil Society Report on Climate Change* (London: Civil Society Coalition on Climate Change, 2007), 23.

10.   William R.L. Anderegg et al., "Expert credibility in climate change," *Proceedings of the National Academy of Science* 107:27 (July 2010), 12107–09.

extent to which politics has become central to the work of "consensus" climate scientists.

Because the science is far from mature, different groups of scientists, depending on their core disciplines, are working on various theories and explanations of why and how climate changes and what the impact of those changes might be. This is as it should be: researchers testing various hypotheses to determine which stand up to rigorous examination and which are falsified by new data and observations. Many of these researchers have no view on whether or not the IPCC position accords with their own. As Legates et al. have documented, of 11,944 scientific papers on climate-related topics published over the 21 years from 1991-2011, only 41 explicitly endorsed the IPPC perspective. The vast majority expressed no opinion at all.[11] Their papers focussed, rather, on the authors' specific research interests and let the more politically engaged debate the political issues in the blogs and in the media. For non-experts in the field, it is thus difficult to determine what, in fact, constitutes mainstream science and to assess the extent to which various claims are widely shared. As previously noted, the IPCC scientific reports also focus on the anthropogenic dimension of climate change and the GHG theory and do not adequately reflect work on other hypotheses and issues.

Sceptical climate scientists have focused much of their work on understanding the natural processes involved in the dynamics of constantly changing climate and have made considerable progress in gaining a better understanding of the role of the atmosphere, oceans, and clouds in distributing solar heat around the planet and eventually emitting it back to outer space. The late Reid Bryson, who trained many current climate scientists and meteorologists at the University of Wisconsin, concluded shortly before his death in 2008 that natural factors were paramount in understanding climate change. He told one interviewer: "Climate's always been changing and it's been changing rapidly at various times, and so something was making it change in the past. Before there were enough people to make any difference at all, two million years ago, nobody was changing the climate, yet the climate was changing, okay? All this argument is the temperature going up or not, it's absurd. Of course it's going up. It has gone up since the early 1800s, before the Indus-

---

11.   David Legates et al., "Climate Consensus and 'Misinformation': A Rejoinder to Agnotology, 'Scientific Consensus and the Teaching and Learning of Climate Change,'" *Science & Education*, August 2013. See also Box 7-1 below.

trial Revolution, because we're coming out of the Little Ice Age, not because we're putting more carbon dioxide into the air."[12]

The late French climate scientist Marcel Leroux was the author of one of the standard texts in climatology, in which he summarizes his view, one that was widely shared among older climatologists such as the University of East Anglia's Hubert Lamb and Wisconsin's Bryson:

> The possible causes, then, of climate change are: well-established orbital parameters on the paleoclimatic scale, with climatic consequences slowed by the inertial effect of glacial accumulations; solar activity, thought by some to be responsible for half of the 0.6°C rise in temperature, and by others to be responsible for all of it, which situation certainly calls for further analysis; volcanism and its associated aerosols (and especially sulphates), whose (short-term) effects are indubitable; and far at the rear, the greenhouse effect, and in particular that caused by water vapour, the extent of its influence being unknown. These factors are working together all the time, and it seems difficult to unravel the relative importance of their respective influences upon climatic evolution. Equally, it is tendentious to highlight the anthropic factor, which is, clearly, the least credible among all those previously mentioned.[13]

One of Bryson's students, Roy Spencer, now a principal researcher at the University of Alabama at Huntsville and one of the pioneers with Fred Singer and John Christy in developing satellite-based data, concentrates his research efforts on the role of clouds and, together with researchers at other universities, is well-published in the specialist literature. He has also written two popular books and writes a weblog. Together with Christy, he maintains the University of Alabama satellite temperature record based on data generated by NASA satellites.

Spencer does not dispute the basic tenets of the GHG theory but believes that it provides at best a partial explanation of recent climate dynamics. He points out that water vapour is the principal greenhouse gas and that the hydrological cycle – evaporation, condensation, cloud formation, and precipitation – is key to understanding climate variations. He also points to the impact of coupled atmospheric-oceanic circulation patterns – ENSO, PDO, NAO, and others – for understanding changes in decadal climate patterns. He

---

12. Dave Hoopman, "The Faithful Heretic: A Wisconsin Icon Pursues Tough Questions," *Wisconsin Energy Cooperative News,* May 2007.

13. Marcel Leroux, *Global Warming - Myth Or Reality? The Erring Ways of Climatology* (Berlin: Springer, 2005), 510.

believes that the trivial amount of direct warming flowing from a doubling of CO_2 is overwhelmed by the dynamics within the climate system. Spencer's research also points to the role of socio-economic factors, such as the urban heat island effect, in explaining much of the late 20th century surface warming that shows up in some of the land-based global temperature series.[14]

Christy, who also serves as Alabama's state climatologist, has been both a lead and a contributing author for the IPCC, but he was somewhat embarrassed to share in the 2007 Nobel Peace Prize. Instead, he saw himself as part of a group of people "who remain so humbled by the task of measuring and understanding the extraordinarily complex climate system that we are sceptical of our ability to know what it is doing and why. As we build climate data sets from scratch and look into the guts of the climate system, however, we don't find the alarmist theory matching observations. ... Others of us scratch our heads and try to understand the real causes behind what we see. We discount the possibility that *everything* is caused by human actions, because everything we've seen the climate do has happened before."[15]

In addition to his collaborative work with Spencer, Christy's research is focused on building detailed climate records, particularly at the local and regional levels, in order to determine what, if any, global warming signals exist in those records that exceed natural variability. To date his work shows no such signals. Indeed, the records show increases in night-time temperatures only, a classic example of local responses to socio-economic factors such as urbanization and land-use changes rather than to increases in daytime warming. In Christy's view, "daytime temperature is much more representative of the deep atmospheric temperature where the warming due to the enhanced greenhouse effect should be evident." In testimony before a US House Committee, he was quite scornful of the efforts by various scientists to read anthropogenic global warming signals into the limited data that scientists possess and has found that much of this work suffers from what he labels the non-

---

14.   Much of Spencer's research is discussed at his website, drroyspencer.com, including references to some of his published research. See also Roy W. Spencer, *Climate Confusion: How Global Warming Leads to Bad Science, Pandering Politicians and Misguided Policies that Hurt the Poor* (New York: Encounter Books, 2008) and *The Great Global Warming Blunder: How Mother Nature Fooled the World's Top Climate Scientists* (New York: Encounter Books, 2010).

15.   John R. Christy, "My Nobel Moment," *Wall Street Journal,* November 1, 2007.

Figure 7-1: The missing fingerprint in the lower tropical troposphere

Atmospheric Warming 1979 - 1999

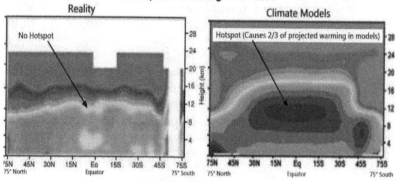

The right panel shows the typical atmospheric hotspot predicted by IPCC climate models between 30° North and 30° South at an altitude of 6-15 Km. The left panel, based on radiosonde and satellite observations, shows no hotspot.

Adapted from: S. Fred Singer, "Lack of Consistency between modelled and Observed Temperature Trends," *Energy & Environment* 22:4 (2011), 378.

falsifiable hypothesis, i.e., everything can be explained by the hypothesis and nothing contradicts it. "These assertions cannot be considered science or in any way informative since the hypothesis' fundamental prediction is 'anything may happen.'... If winters become milder or they become snowier, the hypothesis stands. This is not science."[16]

Christy is also one of the climate scientists who has pointed to the discrepancy between the observational evidence from satellites and radiosondes (probes sent up with weather balloons) of the tropical lower troposphere and the data derived from models. (See figure 7-1) According to all the models, the temperature in the lower troposphere over the tropics should show a typical fingerprint of greenhouse warming: a hotspot. Christy and others indicate that this fingerprint is absent in the observational data. Christy further points out that the biggest response to greenhouse gases is "in the atmosphere, not on the surface. So if you want to measure the response and say that's the greenhouse gas response, you would look in the atmosphere. That's precisely where satellites measure it. ...

16. Written Statement of John R. Christy, Subcommittee on Energy and Power, Committee on Energy and Commerce, US House of Representatives, March 8, 2011.

The response of the climate system is stronger in the atmosphere than on the surface. Carbon dioxide is a greenhouse gas. When you put more of it in the atmosphere, the radiation budget will respond appropriately. … It's just that what we found with the real data is that the way the earth responds is to shed a lot of that heat, not keep it in, which climate models do. So I'd rather base policy on observations than on climate models."[17] A 2010 paper in *Remote Sensing*, co-authored with eight other scientists, including Spencer and Roger Pielke, Sr., concluded that "the majority of [the IPCC fourth Assessment Report] simulations tend to portray significantly greater warming in the troposphere relative to the surface than is found in observations."[18] This issue is hotly debated among climate scientists, with IPCC scientists insisting that the dissenting scientists fail to read the raw data correctly.

That debate heated up further when Richard Lindzen and Yong-Sang Choi published two papers demonstrating another anomaly in the models: data from the ERBE (Earth Radiation Budget Experiment) satellite indicate that "all current [climate] models seem to exaggerate climate sensitivity (some greatly)."[19] Both the Christy et al. and the Lindzen and Choi papers go to the heart of the IPCC hypothesis by demonstrating that the increasingly detailed observational data derived from satellite sensors are at odds with

---

17. John Christy in an interview with Paul Gattis, "7 Questions with John Christy and Roy Spencer: Climate Change Skeptics for 25 years," *AL.com,* April 1, 2015.

18. Christy et al., "What Do Observational Datasets Say about modelled Tropospheric Temperature Trends since 1979?" *Remote Sensing* 2 (2010), 2148-2169. David Douglass, professor of physics at the University of Rochester, an expert on ocean heat content and cloud formation and frequently a collaborator with Christy, Spencer, and Roger Pielke, Sr., examines the extent to which climate models are able to replicate observations and concludes that they do so poorly. See David H. Douglass, "A comparison of tropical temperature trends with model predictions," *International Journal of Climatology* 28:13 (November 2008), 1693-1701.

19. Lindzen and Choi, "On the Observational Determination of Climate Sensitivity and Its Implications," *Asia-Pacific J. Atmos. Sci.,* 47:4 (2011), 377-390. Lindzen remains cautious about the results. In a later communication, he notes: "If one reads [our new] paper, one sees that it is hardly likely to represent the last word on the matter. One is working with data that [are] far from what one might wish for. Moreover, the complexity of the situation tends to defeat simple analyses. Nonetheless, certain things are clear: models are at great variance with observations, the simple regressions between outgoing radiation and surface temperature will severely misrepresent climate sensitivity, and the observations suggest negative rather than positive feedbacks." Communication to Chip Knappenberger at *MasterResource,* June 9, 2011.

the assumptions built into the IPCC climate models and indicate that the climate system is much less sensitive to increased atmospheric greenhouse gas concentrations than the IPCC maintains.[20] To be sure, opposing scientists are vigorously contesting the conclusions of these and other papers published by sceptical scientists as well as making extensive efforts to block their publication.[21] For our purposes what matters is that these and similar controversies within the climate science community point to a lack of consensus and to less certainty than the public policy discussions of the past decade or two would suggest.

Patrick Michaels, who was for many years at the University of Virginia and served as Virginia's state climatologist, is now at the Cato Institute and George Mason University. He is another prominent climatologist who has made no secret of his dissent from the IPCC view. Like Spencer, Christy, and others, Michaels has no difficulty with the basic GHG hypothesis but believes the IPCC exaggerates the sensitivity of the climate system to increased atmospheric greenhouse gases and does not agree that the data support catastrophic increases in future temperatures. He also believes that the IPCC fails to point to the many benefits from modest warming and increased atmospheric GHGs. In addition to his contributions to the scientific literature, Michaels has written or edited five books on global warming for lay audiences. In all of them, he presents what he characterizes as the moderate view of climate change.[22]

The University of Colorado's Roger Pielke, Sr., explicitly rejects the label of sceptic or denier, finding the labels both demeaning and

---

20. Ross McKitrick and Timothy Vogelsang revisit this issue in a 2014 paper showing that the models not only predict far too much warming for the tropical troposphere but also potentially get the nature of the change wrong. The models portray a relatively smooth upward trend over the whole span (1958-2012), while the data exhibit a single jump in the late 1970s, with no statistically significant trend on either side. "HAC robust trend comparisons among climate series with possible level shifts," *Environmetrics* 25 (November 2014), 528-47.

21. Lindzen explains the extraordinary treatment of his paper at *MasterResource,* "Lindzen-Choi 'Special Treatment': Is Peer Review Biased Against Nonalarmist Climate Science?" June 9, 2011.

22. Patrick Michaels, *The Satanic Gases: Clearing the Air about Global Warming* (with Robert Balling, Jr.: Cato Institute, 2000); *Meltdown: The Predictable Distortion of Global Warming by Scientists, Politicians, and the Media* (Cato Institute, 2004); *Shattered Consensus: The True State of Global Warming* (Rowman and Littlefield, 2005); *Climate of Extremes: Global Warming Science They Don't Want You To Know* (with Balling: Cato Institute, 2009) and *Climate Coup: Global Warming's Invasion of Our Government and Our Lives* (Cato Institute, 2011).

confusing. He is firmly convinced that there is a human influence on climate change but that this influence is much more complex than the single factor of increases in greenhouse gases. Rather, he points to factors such as urbanization, land-use changes, and aerosols as being more important than GHG emissions. On his former weblog, he explains: "Humans are significantly altering the global climate, but in a variety of diverse ways beyond the radiative effect of carbon dioxide. The IPCC assessments have been too conservative in recognizing the importance of these human climate forcings as they alter regional and global climate. ... Attempts to significantly influence regional and local-scale climate based on controlling CO_2 emissions alone is an inadequate policy for this purpose."[23] Pielke, Sr., is one of the most published scientists on this aspect of climate research. Together with W.R. Cotton he has written one of the standard texts about human impacts on weather and climate.[24] He criticizes the IPCC for its overreliance on models and for selectively choosing the data and studies supporting its preconceived conclusions. His principal criticism of IPCC scientists is their herd mentality and their unwillingness to admit that there is more than one perspective on the many technical issues making up climate research. On his weblog – now no longer active – he reported on a wide range of climate-related research and maintained discussions with scientists from various perspectives.

Garth Paltridge, an atmospheric physicist and former chief scientist at Australia's Commonwealth Scientific and Industrial Research Organisation (CSIRO), is one of a group of Australian sceptics, many with experience as government scientists, who were among the earliest critics of the IPCC view of climate science. He was involved as an Australian official in the 1979 World Climate Conference that led to the establishment of the IPCC. His scientific criticism is grounded in the IPCC models' inadequate treatment of

---

23. "Roger Pielke, Sr.'s Perspective on the Role of Humans in Climate Change," at PielkeClimateSci, March 3, 2008. Pielke's son, Roger A. Pielke, Jr., trained as a political scientist, has focused his research on the interface between science and policy, particularly climate policy. He shares his father's misgivings about the herd mentality of the IPCC and has written two influential books outlining climate policy issues: *The Honest Broker: Making Sense of Science in Policy and Politics* (Cambridge: Cambridge University Press, 2007) and *The Climate Fix: What Scientists and Politicians Won't Tell You About Global Warming* (New York: Basic Books, 2010).

24. William R. Cotton and Roger A. Pielke, Sr., *Human Impacts on Weather and Climate,* 2nd edition (Cambridge: Cambridge University Press, 2007).

clouds and their critical role in determining the degree of surface warming. His more general concerns were expressed well in his 2009 book, *The Climate Caper*: "the science behind the issue [anthropogenic global warming], and particularly the uncertainty of the science behind the issue, was irrelevant even before the so-called 'IPCC process' got off the ground."[25] In a 2012 column, Paltridge succinctly summed up the problem with both the science and the public policy response: "The broad theory of man-made global warming is acceptable in the purely qualitative sense. If humans continue to fill the atmosphere with carbon dioxide, there can be little doubt that the average temperature of the world will increase above what it would have been otherwise. The argument about the science is, and always has been, whether the increase would be big enough to be noticed among all the other natural variations of climate. The economic and social argument is whether the increase, even if it were noticeable, would change the overall welfare of mankind for the worse."[26]

Meteorologists

Closely allied to climatology is the much larger universe of meteorologists. Similar to sceptical climatologists, sceptical meteorologists regard forecasting years, decades, and even centuries into the future to be little more than witchcraft. Two former TV meteorologists, Joe D'Aleo and Anthony Watts, have been at the forefront of those debunking catastrophic anthropogenic climate change for lay audiences. Both maintain popular weblogs devoted to sharing news about climate developments, publicizing new research, and providing informed comment.[27] *Watts Up With That* has earned the distinction of being the most visited and respected science site in the blogosphere. Both of them specialize in debunking alarmist claims by presenting data – typically from official sources – that show, if anything, the opposite of what is claimed. Both are highly sceptical of the temperature series used by IPCC scientists and have devoted

---

25. Garth W. Paltridge, *The Climate Caper: Facts and Fallacies of Global Warming* (Ballan, Australia: Connor Court, 2009), 11.

26. Paltridge, "Science Held Hostage in Climate Debate," *Australian Financial Review*, June 22, 2012.

27. D'Aleo manages *Icecap* and Watts maintains *WattsUpWithThat*. The latter, in addition to a daily weblog of news and developments, maintains useful reference and resource pages for the lay climate enthusiast. Both have earned the respect of the sceptic community.

their considerable resources to demonstrating the inadequacies of the US surface station network in particular and the global temperature series more generally. Watts has set up a weblog specifically devoted to a survey of all official US weather stations. Hundreds of volunteers have contributed, with the result that Watts and his collaborators have been able to demonstrate how few of the stations meet the required criteria.[28]

Both have also contributed to more specialized literature. D'Aleo has done considerable work on major ocean cycles, such as the ENSO, PDO, and NAO oscillations, and the teleconnections between them, demonstrating the high level of correlation existing between broad global climate patterns and the phases of these oscillations. Watts has teamed up with other researchers to present his findings on the US climate network and to identify the problems with the US and global climate anomaly temperature series.[29]

William Kininmonth, a former head of the National Climate Centre at Australia's Bureau of Meteorology, participated in the 1990 World Climate Conference and the subsequent negotiation of the UN Framework Convention on Climate Change. He has characterized both efforts as a rush to judgment based on flawed science. Kininmonth finds that the IPCC's "radiative forcing hypothesis is simple, but inadequate. It portrays radiation as the central and only important process of the climate system. Those who subscribe to it have been seduced into forgetting elementary school geography; earth is a globe with seasonal patterns of solar heating that generate temperature differences between the tropics and the poles. The one-dimensional energy budget model is a prescription for flat earth physics whose application leads to erroneous conclusions. ... To focus on the chimera of anthropological greenhouse warming while ignoring the real threat posed by natural vulnerability is self-delusion on a grand scale."[30] He also points out: "computer models can be parameterized to project a future based on this simplistic

---

28.　Anthony Watts, "Is the US Surface Temperature reliable?" accessed at *Surface-Stations.org.*

29.　See, for example, Joseph D'Aleo and Don Easterbrook, "Multidecadal Tendencies in ENSO and Global Temperatures Related to Multidecadal Oscillations," *Energy & Environment,* 21:5 (September 2010), 437-60, and Souleymane Fall, et al., "Analysis of the impacts of station exposure on the U.S. Historical Climatology Network temperatures and temperature trends," *Journal of Geophysical Research,* 116, D14120.

30.　William Kininmonth, *Climate Change: A Natural Hazard* (Brentwood, Essex: Multiscience Publishing, 2004), 7 and 10.

view of the climate system, but this is a future unrelated to the real world climate system, which is much more complex and chaotic. As a result, governments have been misled into adopting policies to mitigate climate changes that bear no relationship to reality."[31]

Applied mathematicians and statisticians

Ever since scientists learned to express natural phenomena and relationships in numerical terms, there has developed a symbiotic link between science and mathematics. Statistics has become one of the indispensable tools of scientific investigation. It is virtually impossible today to pursue any branch of science without a working knowledge of calculus and statistics. Climate science is no exception, particularly given the important role of calculus and statistics in gathering and manipulating data and developing models. The increasing sophistication and power of modern computers, which can immensely simplify and accelerate the collection and handling of data, have also become critical to most of the developments in climate science over the past forty years.

It is not surprising, therefore, that some of the most trenchant criticism of climate science has come from mathematicians and statisticians, many of whom, upon looking into the way some climate scientists gather and manipulate data, have come away appalled at what passes for science. Climate science has also become particularly prone to one of Einstein's most famous warnings about the relationship between statistics and science: "not everything that is important can be measured and not everything that can be measured is important." The data upon which much of climate science rests are of dubious quality, in part because much of what is important in climate science is either very difficult to measure or has been estimated to a much larger extent than the literature admits. The alarmist community relies extensively on the average global temperature anomaly as the prime indicator of anthropogenic global warming because, flawed as it is, it is readily understood by policy makers. The more important metric of global ocean heat uptake is much more difficult to measure. Compounding this weakness is a penchant among some climate scientists to express the results of their research with a measure of precision that is unwarranted by the data.

The most critical metric is the hotly debated equilibrium climate sensitivity (ECS) of the global climate to a doubling of atmospheric

---

31. William Kininmonth, "Cold facts dispel theories on warming," *The Australian*, April 29, 2009.

greenhouse gases. Dating back as far as the work of Svante Arrhenius at the beginning of the 20th century, it has been estimated to be between 1.5° and 5° C based on theoretical considerations, with a best estimate usually taken to be about 3.0° C. The complexity of the climate system's many feedbacks and forcings makes it difficult to be more precise. There are also some estimates derived from observational evidence of approximately 1.7°C. UK mathematician Nic Lewis provides a devastating critique of the mathematical contortions used by the IPCC to confirm the higher number – which is used in virtually all models – concluding, "in the area of climate sensitivity then, the IPCC includes many studies that are severely flawed – as regards statistical methodology and/or their design or data used – and therefore provide scientifically unsound estimates."[32]

Temperature, global or otherwise, is not the only metric that is given more precision than the data warrant. Charts and schematic diagrams, for example, showing the earth's radiation budget with numeric precision (see figures 4-9 to 4-10) give the average reader the impression that these are measured quantities, whereas they are no more than educated guesses based on theoretical calculations rather than observations. Similarly, climate scientists assert numeric precision in the earth's carbon cycle, again with a precision suggesting exact observation. Both sets of numbers fall under Einstein's characterization of matters that are important but that cannot be measured with any precision.

While many climate metrics have improved over the years, particularly since the beginning of the satellite era, many climate series still do not have reliable historic base lines that can be used as a foundation for comparing more recent patterns. Thirty-five (satellite) or 150 (instrumental) years of data hardly qualify as indicators of Earth's historic or geologic climate. Proxy data, the basis of many metrics derived from various indices prior to the widespread introduction of instrumentation, can do little more than suggest changes in direction of climate metrics rather than in actually measuring them. Tree ring data (dendrochronology), for example, which are often used as a proxy for climate, may indicate no more than changes in optimal growing conditions related not only to temperature but also to precipitation, fertilization, and competition. Using tree-ring data to provide a temperature chart for hundreds of years into the past may be more than a

32. Nicholas Lewis, Submission to the UK Parliament Select Committee on Energy and Climate Change, December 17, 2013.

stretch. The best climatological data often include error bars, a fine point in graphmanship rarely appreciated by the lay reader.

Bjørn Lomborg, the Danish statistician who made his name by questioning much of the data used by the environmental movement, was also among the earliest critics of the catastrophic global warming hypothesis. His criticism is largely grounded in his discovery that much of the hype that passes for environmentalism is contradicted by such data as exist. In the case of global warming, he is prepared to accept that increases in greenhouse gases would lead to a warmer world but concludes that the solutions on offer, in addition to costing too much, would have little effect.[33]

Ross McKitrick, together with Chris Essex, an applied mathematician at the University of Western Ontario, compiled one of the first major critical assessments of climate science in *Taken by Storm: The Troubled Science, Policy and Politics of Global Warming*.[34] McKitrick, an environmental economist at the University of Guelph who is intimately familiar with modelling and applied statistics as important aspects of contemporary economics, has teamed up with various other researchers to become one of the most articulate critics of the way some climate scientists abuse modelling and statistics by drawing detailed inferences from rudimentary data.

McKitrick, either alone or with various collaborators, has published a variety of articles in peer-reviewed journals, principally in non-climate journals, demonstrating the many statistical weaknesses in the principal global surface temperature series: the Global Historical Climatology Network, the Hadley-CRU temperature series, and the Goddard Institute of Space Studies series.[35] These are not inde-

---

33. See Lomborg, *The Skeptical Environmentalist: Measuring the Real State of the World* (Cambridge: Cambridge University Press, 2001).

34. Toronto: Key Porter Books, 2002. I first read *Taken by Storm* in 2002 because the book was shortlisted, along with my own *A Trading Nation*, for the 2002 Donner Prize. *Taken by Storm* was runner up to the winner, John Helliwell's *Globalization and Well-Being*.

35 McKitrick and Patrick J. Michaels, "A Test of Corrections for Extraneous Signals in Gridded Surface Temperature Data," *Climate Research* 26 (2004), 159-173; McKitrick and Michaels, "Quantifying the influence of anthropogenic surface processes and inhomogeneities on gridded global climate data," *Journal of Geophysical Research-Atmospheres*, 112 (2007); McKitrick, "A Critical Review of Global Surface Temperature Data Products," SSRN Working Paper 1653928, August 2010; McKitrick, "Atmospheric Oscillations do not Explain the Temperature-Industrialization Correlation," *Statistics, Politics and Policy*, I:1 (July 2010); McKitrick and Nicolas Nierenberg, "Socioeconomic Patterns in Climate Data," *Journal of Economic and So-*

pendent series but are closely related to each other and share many of the biases and problems that McKitrick has identified. By going back to much of the raw data used in developing these series, he has also been able to demonstrate that the scientists responsible for them take inadequate account of the socio-economic impacts on temperature, e.g., urbanization, and that much of late 20th century global warming in the data correlates better with socio-economic factors than with GHG emissions.

Due to widespread criticism of the quality of the temperature series relied on by climate scientists, Berkeley physicist Richard Muller spearheaded an effort to develop a new, transparent temperature series by going back to the raw data. The first results of the Berkeley Earth Surface Temperature (BEST) project were released in October 2011 and were met with a chorus of criticism, particularly from statisticians, who pointed out that once again inappropriate statistical techniques were being used in order to develop a product that met the requirements of official climate science. In putting the new series together, Muller admits his group faced a serious problem: "The temperature-station quality is largely awful. ... Using data from all these poor stations, the UN's Intergovernmental Panel on Climate Change estimates an average global 0.64°C temperature rise in the past 50 years, 'most' of which the IPCC says is due to humans. Yet the margin of error for the stations is at least three times larger than the estimated warming."[36]

Statisticians Doug Keenan and William Briggs conclude that because the raw material is so unreliable, the final result cannot be better, no matter how much statistical manipulation is used to massage the data. They also find that many of the techniques used to manipulate and interpret the data do not pass the statistical smell test. They are particularly critical of three aspects of climate science statistical manipulation: the averaging of averages, the use of arbitrary data start-and-end points to compute averages for the purpose of establishing anomalies, and the inappropriate use of smoothed time series. As Briggs writes: "If, in a moment of insanity, you *do* smooth time series data and you *do* use it as input to other analyses, you dramatically increase the probability of fooling yourself! This is because smooth-

---

cial *Measurement*, 35:3,4 (2010), 149-175); and McKitrick, "Encompassing tests of socioeconomic signals in surface climate data," *Climatic Change,* June 2013.

36. Richard A. Muller, "The Case Against Global-Warming Scepticism: There were Good Reasons for Doubt, Until Now," *Wall Street Journal,* October 21, 2011.

ing induces spurious signals – signals that look real to other analytical methods."[37]

Physicists

The basis for much of climate science is the physics of the atmosphere, and many climatologists are trained as atmospheric physicists. It is not surprising, therefore, that physicists generally have taken more interest in the anthropogenic global warming hypothesis than any other major group of scientists. Many have found the hypothesis less than convincing, from major icons of the discipline such as Freeman Dyson, Ivar Giaever, Robert Jastrow, and Frederick Seitz, to lesser known physicists such as Duke's Robert Brown, Rochester's David Douglass, California's Hal Lewis, and Princeton's Will Happer.

Objections range from wholesale rejection to milder criticisms that the AGW hypothesis is less than sufficient to explain the intricacies of the most complex, dynamic, coupled, non-linear natural system known. As Brown points out: the climate system is, "quite literally, the most difficult problem in mathematical physics we have ever attempted to solve or understand! Global Climate Models are *children's toys* in comparison to the actual underlying complexity, especially when (as noted) the *major drivers setting the baseline behaviour are not well understood or quantitatively available*" (emphasis in original).[38] Famed Princeton physicist Freeman Dyson similarly emphasizes how little scientists know and how much they have to learn about earth systems before prescribing solutions. In a 2011 interview, he said: "When I listen to the public debates about climate change, I am impressed by the enormous gaps in our knowledge, the sparseness of our observations and the superficiality of our theories. Many of the basic processes of planetary ecology are poorly understood. They must be better understood before we can reach an accurate diagnosis of the present condition of our planet. ... We need to observe and measure what is going on in the biosphere, rather than relying on computer models."[39]

---

37. William M Briggs, "Do not smooth times series, you hockey puck!" *WmBriggs. com*, September 6, 2008. Keenan similarly assesses statistical issues at his weblog, *Informath*. An hour spent at wmbriggs.com will prove both rewarding and entertaining, as Briggs demolishes one climate change assertion after another based on impeccable logic.

38. Robert G. Brown, "A response to Dr. Paul Bain's use of 'denier' in the scientific literature," *WattsUpWithThat*, June 22, 2012.

39. In conversation with Russ Roberts at *EconTalk*, March 2011.

Nobel laureate Ivar Giaever resigned from the American Physical Society (APS) over its claim that the science of global warming was "incontrovertible," a word he found wholly inappropriate in the context of science and more suitable for a religious discussion. In his letter of resignation, he points out that "the claim … is that the temperature has changed from ~288.0 to ~288.8 degrees Kelvin in about 150 years, which (if true) means to me that the temperature has been amazingly stable, and both human health and happiness have definitely improved in this 'warming' period."[40]

The University of California's Hal Lewis similarly resigned from the APS in 2010 a few months before his death after paying his dues for 67 years, citing that the Society had changed its mission from promoting scientific knowledge to suppressing science in order to protect funding, particularly from government sources. A specialist in high energy physics, including cosmic rays and elementary particles, Lewis had a keen sense of the trade-offs between policy and science. In his letter of resignation, he pointed to the "global warming scam, with the (literally) trillions of dollars driving it, that has corrupted so many scientists, and has carried APS before it like a rogue wave. It is the greatest and most successful pseudoscientific fraud I have seen in my long life as a physicist."[41]

Antonino Zichichi, a member of the Pontifical Institute of Sciences, is confident that natural factors such as the sun are the main players in climate change. An expert on the physics and mathematics of complexity, he dismisses climate modelling as being both mathematically and scientifically naïve. He told a seminar at the Vatican in 2007 that models used by the IPCC are incoherent and invalid from a scientific point of view. On the basis of actual scientific fact "it is not possible to exclude the idea that climate changes can be due to natural causes," and that it is plausible that "man is not to blame."[42]

Will Happer, who at one time was the senior official in the first Bush administration supervising all climate research in the US Energy Department, has long criticized the alarmist community for

40. "Nobel Prize Winning Physicist Resigns Over Global Warming," *FoxNews*, September 14, 2011.

41. John Ingham, "Global warming is 'the greatest fraud in 60 years'," *Daily Express*, October 12, 2010.

42. "Some Restraint in Rome," *National Post*, May 23, 2007. Zichichi was obviously not one of the advisers who helped Pope Francis prepare his climate-change encyclical, *Laudato Si'* (discussed in chapter 10), which placed the Vatican squarely, and controversially, in the alarmist camp.

exaggerating what it knows and hyping the possible impacts of climate change. He recently wrote:

> Let me summarize how the key issues appear to me, a working scientist with a better background than most in the physics of climate. CO_2 really is a greenhouse gas and other things being equal, adding the gas to the atmosphere by burning coal, oil, and natural gas will modestly increase the surface temperature of the earth. Other things being equal, doubling the CO_2 concentration, from our current 390 ppm to 780 ppm will directly cause about 1 degree Celsius in warming. At the current rate of CO_2 increase in the atmosphere—about 2 ppm per year—it would take about 195 years to achieve this doubling. The combination of a slightly warmer earth and more CO_2 will greatly increase the production of food, wood, fiber, and other products by green plants, so the increase will be good for the planet, and will easily outweigh any negative effects. Supposed calamities like the accelerated rise of sea level, ocean acidification, more extreme climate, tropical diseases near the poles, and so on are greatly exaggerated.[43]

Physicists who specialize in the science of the solar system have been particularly dubious about attributing late 20th century warming to greenhouse gases. Duke University's Nicola Scafetta and the US Army's Bruce West, for example, find that:

> The average global temperature record presents secular patterns of 22- and 11-year cycles and a short timescale fluctuation signature ... both of which appear to be induced by solar dynamics. ... The non-equilibrium thermodynamic models we used suggest that the Sun is influencing climate significantly more than the IPCC report claims. If climate is as sensitive to solar changes as the above phenomenological findings suggest, the current anthropogenic contribution to global warming is significantly overestimated. We estimate that the Sun could account for as much as 69 percent of the increase in Earth's average temperature, depending on the TSI [total solar insolation] reconstruction used."[44]

---

43. William Happer, "The Truth About Greenhouse Gases," *First Things,* June/July 2011.

44. Scafetta and West, "Is climate sensitive to solar variability?" *Physics Today,* March 2008, 50-1. The article summarizes an extensive research effort looking at solar influence on climate on all time scales and published in various other journals. See, for example, Scafetta, "Empirical evidence for a celestial origin of the climate oscillations and its implications," *Journal of Atmospheric and Solar-Terrestrial Physics* (2010).

Figure 7-2: Sunspot numbers since 1610, showing the 11-year cycle

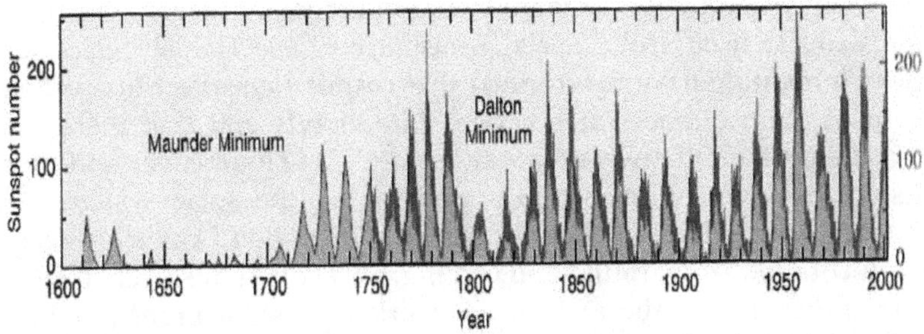

Source: C. de Jager, "Solar activity and its influence on climate," *Netherlands Journal of Geosciences,* 87:3 (2008), 208.

Similar findings can be found in the work of other solar scientists, such as Khabibullo Abdusamatov, David Archibald, Sallie Baliunas, Rhodes Fairbridge, Cornelis de Jager, George Kukla, Gerald Marsh, Alexander Shapiro, and Willie Soon. For these scientists a solar grand maximum – a period of high insolation – covering much of the 20th century was one of the principal drivers of recent global warming. The sun appears now to be going into a quieter phase, presaging a potential cooling period, perhaps similar to earlier quiet phases such as the Dalton minimum in the early 18th century or the even more extended Maunder minimum in the 16th and throughout the 17th century (see figure 7-2). Abdusamatov, for example, points out that parallel warming on Earth and Mars can only be explained in terms of the sun and certainly not in terms of changes in the Earth's atmosphere. He shares the view of many other scientists that various cyclical patterns, such as the Dansgaard-Oeschger, Bond, Heinrich, Gleissberg, and Schwabe cycles, are critical to understanding climate patterns not only on earth but also throughout the solar system.[45]

Their work has since been given an interesting boost by the findings of a group of Danish physicists who have discovered that the ebb and flow of total solar insolation in and of itself may not be sufficient to explain climate fluctuations but that it is amplified by the influence of related fluxes in the solar wind and its impact on

---

45. See Solomon, *The Deniers,* 161ff. Astrophysicist Willie Soon provides a catalogue of recent literature indicating the extent to which IPCC scientists misrepresent the role of the sun in recent climate change. "Sun shunned," in Alan Moran, *Climate Change: The Facts* (Melbourne, Australia: Institute of Public Affairs, 2015).

the penetration of galactic cosmic rays into the solar system, an idea first suggested by the University of Minnesota's Edward Ney. Eigil Friis-Christensen and Henrik Svensmark of the Danish Space Research Institute have determined that cosmic rays affect the nucleation of atmospheric water vapour into clouds and that there is a good correlation between variations in global cloud cover and fluxes in the level of cosmic rays penetrating the solar system and Earth's atmosphere. In turn, fluxes in global cloud cover correlate well with global warming and cooling. Subsequent research, including at the CERN Laboratory in Switzerland, have demonstrated the role of cosmic rays in cloud formation, adding further credibility to this additional explanation for solar-induced climate change. As Svensmark writes: "During the last 100 years cosmic rays became scarcer because unusually vigorous action by the Sun batted away many of them. Fewer cosmic rays meant fewer clouds – and a warmer world."[46]

Israeli astrophysicist Nir Shaviv and the University of Ottawa's Jan Veizer have extended the Svensmark hypothesis by looking at the fluxes in cosmic rays that are related to the solar system's passage through the arms of the Milky Way. Shaviv explains:

> Various climate indices appear to correlate with solar activity proxies on time scales ranging from years to many millennia. ... Moreover, it now appears that there is a climatic variable sensitive to the amount of tropospheric ionization – clouds. If this is true, then one should expect climatic variations while we roam the galaxy. This is because the density of cosmic ray sources in the galaxy is not uniform. In fact, it is concentrated in the galactic spiral arms. ... Thus, each time we cross a galactic arm, we should expect a colder climate. ... The main result of this research is that the variations of the flux, as predicted from the galactic model and as observed from the iron meteorites, are in sync with the occurrence of ice-age epochs on Earth.[47]

Veizer expanded this line of research by looking at reconstructed paleo records for seawater temperatures. By comparing them to variations in cosmic ray fluxes as the solar system moves through the Milky Way, Veizer found statistically significant correlations.

---

46. Henrik Svensmark and Nigel Calder, *The Chilling Stars: A New Theory of Climate Change* (Cambridge, UK: Icon Books, 2007), cover blurb.

47. Nir J. Shaviv, "The Milky Way Galaxy's Spiral Arms and Ice-Age Epochs and the Cosmic Ray Connection," *Science Bits*, March 30, 2006, provides an explanation of the theory in plain language and includes references to the peer-reviewed literature.

He also found that "cosmic rays, when hitting the atmosphere, generate a cascade of cosmogenic nuclides that then rain down to the Earth's surface and can be measured in ice, trees, rocks and minerals. Such records over the past 10,000 years correlate well with the highly variable climate."[48]

One of the more interesting aspects of solar climate theories is that IPCC-associated scientists dismiss them as the work of amateurs unfamiliar with the physics of the atmosphere. They feel fully confident in dismissing the findings of highly credentialed solar scientists as being largely irrelevant to understanding climate change, insisting that the science is settled. As Georgia Tech's Judith Curry points out, after reviewing some of the solar literature at her website, "there has been an implicit assumption by the IPCC that natural forcings are of minor importance. In my opinion this has been to the great detriment of our understanding of the climate system."[49]

Another set of physical scientists has focused on changes over time in atmospheric concentrations of CO_2. Their conclusions similarly point to the sun as the principal driver of global climate change. The late Polish scientist Zbigniew Jaworowski, a specialist in examining the climate record revealed by ice cores, insisted that many of the studies of ice cores demonstrating the make-up of the pre-20th century atmosphere are ill-founded. For Jaworowski, "the basis of most of the IPCC conclusions on anthropogenic causes and on projections of climatic change is the assumption of low levels of CO_2 in the pre-industrial atmosphere. This assumption, based on glaciological studies, is false." He based this criticism on the allegation that scientists have adopted an 83-year fudge factor in order to align ice core records with modern observations. He charged that "improper manipulation of data, and arbitrary rejection of readings that do not fit the pre-conceived idea on man-made global warming are common in many glaciological studies of greenhouse gases."[50]

Related criticisms were raised by the late German scientist Ernst-Georg Beck, who rejected the proposition that the curve of increasing CO_2 based on spectroscopic readings at the observatory on Mauna Loa since 1958 – the Keeling curve – could be extrapolated

48. Jan Veizer, "Climate Change Isn't Settled," *The Australian*, April 24, 2009. See also Veizer, "Celestial climate driver: a perspective from four billion years of the carbon cycle," *Geoscience Canada*, 32 (2005), 13-28.

49. Curry, "21st century solar cooling," *Climate Etc.*, March 10, 2012.

50. Jaworowski, "Climate Change: Incorrect information on pre-industrial CO2," at *warwickhughes.com/icecore*.

in linear fashion back to the 19th century. He collected thousands of pre-1958 chemical measurements of the atmosphere which showed much more variation than the 280 ppm for pre-industrial CO_2 assumed by IPCC scientists. In Beck's view, the post-1958 increase in the atmosphere's CO_2 cannot be attributed simply to the burning of rising quantities of coal and oil in the post-war period. His data showed a good correlation between atmospheric CO_2 levels and sea-surface temperature with a one-year lag. Not surprisingly, he also disputed the straight-line connection between CO_2 levels and global temperature.[51]

Two Australian scientists, Tom Quirk and Murray Salby, have recently revisited the issue of anthropogenic CO_2 in the atmosphere. In a 2009 paper Quirk concluded that "CO_2 emissions from fossil fuels do not make it to the 'well-mixed' atmosphere but are probably fixed locally. The increase in CO_2 is driven by other processes related to the natural variability of the climate. Some of this CO_2 variability is correlated with and may be related to ENSO events."[52] Salby, in a revision of his text on the physics of the atmosphere, comes to the conclusion that scientists do not have sufficient evidence to determine the extent to which increases in atmospheric CO_2 are of anthropogenic or natural origin. By revisiting the issue of the ratio of the common ^{12}C isotope of carbon and the much rarer ^{13}C isotope, which is presumed to be more common in CO_2 of anthropogenic origin, Salby concludes that the difference has been exaggerated. The addition of anthropogenic CO_2 to the atmosphere can be calculated to a reasonable degree by measuring industrial and transportation activity. Natural fluxes in CO_2, however, are not measurable, and thus the Keeling curve, which measures the total increase in atmospheric CO_2, cannot determine whether the increase is anthropogenic or natural in origin or a combination of the two. He points out that IPCC scientists assume that virtually all of the post-1958 increase is due to the burning of fossil fuels and the production of cement.[53]

---

51. See Beck, "180 years of Atmospheric CO2 Gas Analysis by Chemical Methods," *Energy & Environment*, 18:2 (2007).

52. Quirk, "Sources and Sinks of Carbon Dioxide," *Energy & Environment*, 20:1-2 (2009), 105-121.

53. Murray Salby, *Physics of the Atmosphere and Climate*, 2nd edition (Cambridge: Cambridge University Press, 2012). See the profile of Salby and his problems with the climate science community in Rupert Darwall, "An Unsettling Climate," *City Journal*, Summer 2014. Roy Spencer has similarly questioned whether natural fluxes in the carbon cycle are sufficiently understood to conclude that

One line of criticism advanced by some theoretical physicists has not gained much traction, i.e., the problems that the greenhouse theory poses for the first and second laws of thermodynamics.[54] The problem with this particular line of criticism is that it attacks a straw man, the ill-posed and poorly described greenhouse effect. Unfortunately, IPCC scientists have opened themselves up to this kind of criticism because they persist in using the greenhouse metaphor and make simplified presentations with schematic pictures that give the impression that the atmosphere warms the earth, despite the fact that the atmosphere is cooler than the earth's surface. Greenhouse gases in the atmosphere keep the earth from cooling to the same extent that it would in the absence of these gases. Increasing greenhouse gases to the atmosphere will *add* to this effect to the extent that their effect has not yet reached the saturation point. Climatologists such as Lindzen, Singer, and Spencer all cringe when they hear criticism of this fundamental characteristic of earth's climate system, and they have succeeded to some extent in expressing in plain English what climatologists mean by the greenhouse effect and its consistency with well-established physical principles, including the laws of thermodynamics. In large measure, this issue is more a matter of semantics than of physical principles and detracts from the more fundamental criticisms of IPCC climate science by many physicists.[55]

Earth scientists

As a discipline, earth scientists have been among the most consistent critics of the catastrophic anthropogenic global warming hypothesis. As students of the planet's geological history dating back as far as four and a half billion years, they find it difficult to accept

---

the late 20th century increase in atmospheric CO2 is largely anthropogenic in origin. See "Roy Spencer on how Oceans are Driving CO2," a*WattsUpWithThat*, January 25 and 28, 2008.

54. A full exploration of this problem was set out in Gerhard Gerlich and Ralf D. Tscheuschner, "Falsification Of The Atmospheric CO2 Greenhouse Effects Within The Frame Of Physics." *Int.J.Mod.Phys.* B23:275-364, 2009. The issue was subsequently taken up by a group of authors in Tim Ball et al., *Slaying the Skydragon: Death of the Greenhouse Gas Theory* (Stairway Press, 2010). Despite their preoccupation with the second law of thermodynamics, other aspects of their criticism of the greenhouse theory are apt and well taken and fall well within those of many other critics.

55. See, for example, Lindzen, "Climate v. Climate Alarm," presentation to the American Chemical Society, August 28, 2011.

that there is anything extraordinary or threatening about current or
projected climatic conditions. From their perspective, earth's climate
is always changing on all spatial and temporal scales. If anything,
the Holocene, the current interglacial, has been among the most stable climatic periods in earth's long history. Within that period, current climatic conditions are neither the warmest nor the coldest. In
fact, the earth is recovering from what was probably the coldest period of the last nine thousand years, the Little Ice Age. From the
longer perspective of the last four or five million years, today's climate is far from the warmest, and the overall pattern is of steady
cooling as a result of long glacial periods interrupted by short interglacials, as illustrated by figures 6-3 and 6-4 in the previous chapter.
Nothing going on today is remarkable from a geological perspective.

One of the most comprehensive criticisms of IPCC science
comes from a 2009 book by a leading Australian geologist, Ian
Plimer.[56] He believes that the IPCC climate science community is
largely ignorant of the planet's geological history and thus pays little attention to a body of knowledge that has long concluded that
climate change on all time scales is one of Earth's defining characteristics rather than a recent aberration flowing from human emissions of fossil fuels. As Plimer points out, "underpinning the global
warming and climate change mantra is the imputation that humans
live on a non-dynamic planet. On all scales of observation and
measurement, sea level and climate are not constant. Change is
normal and is driven by a large number of natural forces. Change
can be slow or very fast."[57] The IPCC's preoccupation with the GHG
hypothesis and human influence disposes it to pay insufficient attention to the lessons of the Earth's geological history. Natural factors have long driven climate change, including both terrestrial and
solar forces. Plimer writes: "Calculations on super-computers, as
powerful as they may be, are a far cry from the complexity of the
planet Earth, where the atmosphere is influenced by processes that

---

56. Ian Plimer, *Heaven and Earth: Global Warming – The Missing Science* (Lanham,
    MD: Rowman & Littlefield, 2009). The book hit home and led to a furious coun-
    terattack by warmists, alleging many errors and misrepresentations. *Heaven and
    Earth* may well contain some errors and bloopers in its nearly 500 pages of close
    print, but many of the errors identified by bloggers are little more than well-
    worn articles of alarmist faith that have long been discredited by earth scien-
    tists. Reading the blogosphere leaves the impression that Plimer is a raving lu-
    natic rather than a well-respected and credentialed geologist.

57. Plimer, "The Past is the Key to the Present: Greenhouse and Icehouse over
    Time," SPPI Reprint Series, 2009, 8.

occur deep within the Earth, in the oceans, in the atmosphere, in the Sun and in the cosmos. To reduce modern climate change to one variable (CO_2) or, more correctly, a small proportion of one variable – human-produced CO_2 – is not science, especially as it requires abandoning all we know about planet Earth, the Sun and the cosmos. Such models fail."[58]

Plimer's perspective is widely shared among earth scientists. Its most articulate expression can be found in the work of Bob Carter, until recently adjunct research professor in earth sciences at Australia's James Cook University. An active member of the NIPCC, together with Singer, McKitrick, and others, Carter has also written one of the most accessible overviews of the science from a sceptical perspective.[59] His research is focused on paleoclimatology based on data collected from various sediment cores drilled in the seas around Australia. As a result, he has gained a strong appreciation for the overwhelming importance of natural causes in explaining the variable pattern of climate change. Carter points out:

> Climate change is self-evidently a natural process. Warmings, coolings, cyclones, floods, droughts and bushfires have been coming and going since long before human industrial processes started adding carbon dioxide to the atmosphere; and, indeed, since before there were humans at all. The appropriate question is therefore not whether climate change is 'real', but the more specific one of whether human-related greenhouse emissions are causing dangerous global warming.

> Instead, tens of thousands of scientific papers published in reputable journals delineate changes in climate and the environment, and ecological responses, that are entirely consistent with the null hypothesis of natural causation. In contrast, not a single paper exists that demonstrates an evidential cause-effect link between change in an environmental variable (be that more or less [sic] storms, floods, droughts, cyclones, honeyeaters or even polar bears) and warming caused by human-related carbon dioxide emissions.[60]

Carleton University paleoclimatologist Tim Patterson is another active researcher piecing together high resolution climate records from sediment cores drilled off Canada's west coast, in high Arctic lakes and bogs, off the coast of Ireland, and in various Ontario bogs and lakes. His work points to a close connection between climate

58. Plimer, "The Natural History of Climate Change," *IPA Review* (August 2009), 1.

59. Robert Carter, *Climate: The Counter Consensus* (London: Stacey International, 2010).

60. Carter, "Climate Review I," *Quadrant Online*, February 6, 2012.

and solar cycles, consistent with the views of solar scientists. In a 2007 opinion article, he wrote:

> Climate stability has never been a feature of planet Earth. The only constant about climate is change; it changes continually and, at times, quite rapidly. Many times in the past, temperatures were far higher than today, and occasionally, temperatures were colder. As recently as 6,000 years ago, it was about 3°C warmer than now. Ten thousand years ago, while the world was coming out of the thousand-year-long 'Younger Dryas' cold episode, temperatures rose as much as 6°C in a decade – 100 times faster than the past century's 0.6°C warming that has so upset environmentalists. Climate-change research is now literally exploding with new findings. Since the 1997 Kyoto Protocol, the field has had more research than in all previous years combined and the discoveries are completely shattering the myths.[61]

Don Easterbrook, emeritus professor of geology at Western Washington University, is another hands-on researcher. The focus of his research is the geology of the US Northwest and the clues it provides for such phenomena as the PDO. A recent book, written with seven other scientists, *Evidence-Based Climate Science,* gathers and analyzes scientific data concerning patterns of past climate change, influences of changes in ocean temperatures, the effect of solar variation on global climate, and the effect of CO_2 on global climate. The book clearly presents evidence ignored by IPCC-affiliated scientists that points to natural patterns of climate change. In the opening chapter Easterbrook sums up the view of the eight co-authors:

> Recent global warming (1978-1998) has pushed climate changes into the forefront of scientific inquiry with a great deal at stake for human populations. With no unequivocal, "smoking gun", cause-and-effect evidence that increasing CO_2 caused the 1978-1998 global warming, and despite the media blitz over the 2007 IPCC report, no tangible physical evidence exists that CO_2 is *causing* global warming. Computer climate models *assume* that CO_2 is the cause and computer model simulations are all based on that assumption. ... Abundant physical evidence from the geologic past provides a record of former periods of recurrent global warming and cooling that were far more intense than recent warming and cooling. These geologic records provide a clear evidence of global warming and cooling that could not have been caused by increased CO_2.[62] [emphases in original]

---

61. Tim Patterson, "Read the Sunspots," *National Post,* June 20, 2007.

62. Don Easterbrook, "Geologic Perspectives," chapter 1, in Easterbrook, et al., *Evidence-Based Climate Science: Data Opposing CO2 Emissions as the Primary Source of Global Warming* (Amsterdam: Elsevier, 2011), 4.

This book and an abundance of similar literature from earth scientists whose research is focused on past climate patterns indicate that the small group of paleoclimatologists associated with the IPCC are in a distinct minority. If there is a consensus about climate in geological time, it is the opposite of that espoused by the IPCC.

Plimer, Carter, Patterson, and Easterbrook have all spent their careers doing what Freeman Dyson points out is so critical to advancing science's knowledge of earth systems: getting their hands dirty and their feet frozen collecting observational evidence and then subjecting that evidence to close scrutiny in their labs. At the same time, they make full use of the power of computers and other modern equipment to tease out evidence about the distant past in an effort to understand current climate patterns. Patterson's work, for example, has been funded by the government of Canada in order to gain better insight into the effect of cyclical climate patterns on migratory fish species on the West Coast and on the long-term viability of the Tibbitt to Contwoyto winter ice road in the Northwest Territories. This kind of work may not endear them to the alarmist community, but it is vital to fishermen on the West Coast and mining companies in Canada's far north. None of the evidence that these earth scientists have collected suggests that recent climate patterns fall outside the boundaries of previous experience during the Holocene, let alone much deeper into geological time.

Other scientists

Sherwood, Craig, and Keith Idso are three research scientists with extensive knowledge of the role of CO_2 in the atmosphere, biosphere, and lithosphere, and together they maintain the Center for the Study of Carbon Dioxide in Tempe, Arizona, and its website, *CO₂Science*. The Center was "created to disseminate factual reports and sound commentary on new developments in the world-wide scientific quest to determine the climatic and biological consequences of the ongoing rise in the air's CO_2 content. ... In this endeavour, the Center attempts to separate reality from rhetoric in the emotionally-charged debate that swirls around the subject of carbon dioxide and global change."[63] All three Idsos are active researchers and widely published in agronomy and in the biology of CO_2. Their most important contribution to the anthropogenic climate change debate has been the maintenance of a detailed and easily accessible

---

63. Mission Statement at *CO2Science*. See also the Idsos' statement "Carbon Dioxide and Global Warming: Where We Stand on the Issue," at *CO2Science*.

database of articles on various aspects of the impact of climate change on the atmosphere and biosphere. The Idsos have also contributed extensively to the work of the NIPCC.

One of the more important issues that the Idsos have thoroughly debunked is the assertion by many proponents of the GHG hypothesis that the Medieval Climate Optimum was a local phenomenon limited to western Europe. The Idsos have diligently catalogued hundreds of studies by thousands of researchers providing evidence of a Medieval Climate Optimum on all continents and in the oceans; each entry includes a short description of the study and references.[64] The site also provides access to all the primary temperature databases, records on ocean alkalinity, and an archive on published papers examining the response of the biosphere to enhanced atmospheric CO_2. Finally, the site provides access to a number of major reports prepared by the Idsos and their collaborators on such issues as ocean acidification, estimates of global food production in the 21st century, and the impact of enhanced levels of atmospheric CO_2 on the biosphere.

Nils-Axel Mörner, a Swedish geophysicist, was long recognized as one of the world's leading experts on sea level, a subject in which he has been immersed for more than forty years. His reputation, however, has taken hard hits from the IPCC community because he insists that IPCC assessments of sea level rise do not accord with his own measurements and those of other field researchers. In his view, the move towards relying on satellite telemetry rather than tidal gauges is potentially a large step forward, but only if the raw data are widely shared. Currently, too many corrections are necessary to bring the data into line with computer modelling. Based on his own data and taking into account such factors as subsiding and rising land, Mörner does not believe that there has been much of a rise in sea level over the past thirty or forty years and does not expect that there will be much of an increase over the *next* thirty or forty years. He points out that concern over sinking islands in the Pacific and Indian oceans is completely misplaced because they are coral-based islands with no recent record to indicate that their shorelines are being inundated.[65] Mörner's charges and concern reflect the broader

---

64. "Medieval Warm Period Project," at *CO2Science*.

65. See Mörner, "Setting the Frames of Expected Future Sea Level Changes by Exploring Past Geological Sea Level Records," and "The Maldives: A Measure of Sea Level Changes and Sea Level Ethics," in Easterbrook et al., *Evidence-Based Climate Science*, 185-209.

concern of other sceptics that the data are not good enough to support many of the conclusions reached by IPCC scientists and require extensive manipulation to obtain the required results.

Marine scientist Walter Starck is one of the world's leading experts on coral reefs. He is a prolific author of both popular and scientific articles as well as a producer of documentaries on corals and oceans. He has spent years investigating coral reefs on his research vessel *El Torito*. Originally from Florida, Starck settled in Australia and has become an active and bluff critic of the excesses of environmentalism in general and of the global warming mantra in particular. He has contributed some blistering essays on both topics to the Australian opinion journal, *Quadrant Online*; the following excerpt is a typical example of his views:

> Environmentalism has seriously damaged scientific research as well as the credibility of science among the public. The diversion of research effort away from seeking fundamental new understanding about the world and towards the production of evidence to support political agendas has seriously affected the development of new knowledge. It has also fostered an atmosphere wherein evidence is selected, distorted, suppressed and occasionally fabricated to accord with what is perceived to be an environmentally correct perspective. Worse yet, many scientists have come to accept such dishonesty and even view it as righteous if it is in support of what is deemed to be a higher good.

> Science has been badly corrupted and scientists are no longer to be trusted just because they are scientists. The only immediate solution is to go back to the basics of the scientific method. Claims of authority and expert consensus, personal degradation of dissenting opinion, and pissing contests over credentials aren't good enough. Show us the evidence. If it can't be produced or does not support the claims being made, funding should be cut off.[66]

He has been particularly critical of the IPCC's exaggerated impact assessments, such as the threatened demise of coral reefs through bleaching and ocean acidification. Based on Starck's experience diving on coral reefs and studying marine life, he sees no evidence of either phenomenon other than as part of the normal ebb and flow of marine life. He points out:

> Most modern reef coral genera have fossil histories going back from 5-10 million years to over 100 million years. During this time they have survived both ice ages and periods when climate was warmer than even the most extreme predictions for warming from CO_2 emissions.

---

66. Starck, "Fishy Science on the Great Barrier Reef," *Quadrant Online*, July 8, 2012.

Geological evidence indicates they thrived when CO_2 was at 5-10 times current levels. This is far higher than we might reach before running out of fossil fuels. In some areas modern day reefs with healthy corals flourish where the pH is as low as 7.8 and disaster for the GBR [Great Barrier Reef] at this level is more the perverse hope of alarmists than it is a probability.[67]

Dissecting the "consensus"

The vast majority of scientists active in the field of climate science are apolitical. They pursue research in their narrow specialty, appreciate the extent to which the political debate has made it easier to get funding for their research, join with others to write highly technical papers published in the specialist literature, and keep their heads down when journalists show up. This mindset has been borne out by in-depth surveys. Political scientist Dennis Bray and climate scientist Hans von Storch, for example, prepared a 76-question survey in 2008 and sent it to 2,681 climate scientists. Three hundred seventy-three scientists from 36 countries responded, an 18.2 percent response rate, suggesting that more than 80 percent of those surveyed were not sufficiently engaged on the politics of the issue to reply. Their analysis of the 373 responses indicated that polarization is not as prevalent as media stories suggest. The vast majority had no affiliation with the IPCC but had published in peer-reviewed journals. A little more than half were affiliated with academic institutions, and three-quarters were involved in the physics and modelling of climate. For most questions, generally scored from strongly agree to strongly disagree, the answers were crowded in the middle. Most agreed, however, that climate change is a problem requiring a public policy response, without specifying what that response should be.[68]

Communications specialist Robert Lichter at George Mason University did a similar survey in 2007 of 489 self-identified climate scientists and found: "Overall, only 5 percent describe the study of global climate change as a 'fully mature' science, but 51 percent describe it as 'fairly mature,' while 40 percent see it as still an 'emerging' science. However, over two out of three (69 percent) believe there is at least a 50-50 chance that the debate over the role of hu-

---

67.   Starck, "Observations on Growth of Reef Corals and Sea Grass Around Shallow Water Geothermal Vents in Papua New Guinea," March 4, 2010 at *WattsUp-WithThat*.

68.   See Dennis Bray and Hans von Storch, "CliSci 2008: A Survey of the Perspectives of Climate Scientists Concerning Climate Science and Climate Change," at *Academia.edu*.

man activity in global warming will be settled in the next 10 to 20 years. Only 29 percent express a 'great deal of confidence' that scientists understand the size and extent of anthropogenic [human] sources of greenhouse gases, and only 32 percent are confident about our understanding of the archaeological climate evidence."[69]

Both these surveys contradict a number of surveys done to "prove" that there is a consensus around the findings of the IPCC. A 2004 literature survey done by historian Naomi Oreskes, for example, demonstrating the extent of the "consensus," has been thoroughly debunked, although AGW enthusiasts continue to cite it.[70] As already noted, the literature survey by Anderegg et al. achieved its consensus by dismissing as climate scientists all those who had not published in the peer-reviewed journals controlled by "mainstream" climate scientists. Another survey done by Doran and Zimmerman in 2009 found that 97 percent of climate scientists agreed with the IPCC position. This was achieved with a very small sample size (79) and two general questions: has the mean global temperature increased and has human activity contributed significantly to this warming.[71] Patrick Michaels observes that whatever is the truth about who agrees with the IPCC, 100 percent of scientists now know that there has been no statistically significant warming since the 1997-98 El Niño.[72]

Ironically, the core of climate change alarmists privately indicate that they are uncertain and even sceptical about some of the issues which, officially, they insist have long been settled. One of the most damaging revelations that resulted from the release of the so-

---

69. S. Robert Lichter, "Climate Scientists Agree on Warming, Disagree on Dangers, and Don't Trust the Media's Coverage of Climate Change," April 29, 2008, at stats.org/stories/2008/global_warming_survey_apr23_08.html.

70. See "Beyond the Ivory Tower: The Scientific Consensus on Climate Change," *Science* 306, no. 5702 (3 December 2004), 1686. It was originally shown to be nonsense by British political scientist Benny Peiser and subsequently by others. See Benny Peiser, "The Dangers of Consensus Science," *National Post*, 17 May 2005. Oreskes, together with Erik Conway, defends her findings and expands on them in *Merchants of Doubt.*

71. P.T. Doran and M. Kendall Zimmerman, "Examining the Scientific Consensus on Climate Change," *Eos Trans. AGU*, 90:3 (2009). The original survey was much larger, but the authors decided to eliminate most of the respondents as being insufficiently credible as climate scientists, limiting their analysis to the 79 who met their narrow criteria. The research question was disingenuous since few scientists deny human influence. (See Box 7-1)

72. Patrick Michaels, "If 97% of Scientists Say Global Warming is Real, 100% Say It Has Nearly Stopped," *WattsUpWithThat*, November 18, 2014.

called Climategate emails among this group of climate scientists is the extent of their own doubts and their criticisms of each other's work. Their public insistence that the science is settled is a strategy in a political campaign aimed at enforcing official science. As John Brignell points out: "It is to some extent forgivable when people adopt extreme positions out of misapprehension or delusion. It is quite another matter if they mislead others by deliberate falsehood. ... In science, up to recent times, there is no circumstance in which a deliberate falsehood is justifiable. It requires at a minimum being drummed out of one's learned society. ... The global warming religion changed everything. ... As for the accompanying slogan 'The science is settled,' if it is settled it is not science and if it is science it is not settled."[73] Judith Curry, in a draft paper posted for comment on her website, captured the issue well in her choice of title: "Climate Change: No Consensus on Consensus." She introduces the paper with an apt quote from the late Israeli diplomat, Abba Eban: "Consensus means that everyone agrees to say collectively what no one believes individually." She points out that in science the issue of consensus only arises at times of controversy. Areas of genuine consensus do not require such assertions. The very fact that consensus is an issue can be taken as indicative of a lack of consensus. Curry notes that consensus can play a constructive role in legitimizing policy based upon scientific research, but, at the same time, "it under-exposes scientific uncertainties and dissent, making the chosen policy vulnerable to scientific errors; and it limits the political playing field in which players can present different policy perspectives."[74]

The extent and depth of dissent from the basic hypothesis put forward by the IPCC and affiliated scientists illustrate well that climate science is full of controversies. To be sure, alarmist scientists will insist that each of these criticisms is ill-founded and, over time, this may well prove to be the case. The work of so-called mainstream scientists may equally prove to be misguided over time. This is what one would expect in an emerging science trying to come to grips with an immensely complex set of problems on the basis of data that are far from ideal. Normally, such controversies are a matter for the scientific literature and scientific societies to pursue and perhaps, over time, work out.

---

73. John Brignell, "How we know they know they are lying," at *NumberWatch.com*.
74. Judith Curry, "No consensus on consensus: Part II," *Climate Etc.*, July 13, 2012.

For most scientists, the real issue is not whether humans influence climate, but by how much. Very few now accept the IPCC conclusion that "it is extremely likely that human influence on climate caused *more than half* of the observed increase in global average surface temperature from 1951–2010." [emphasis added] Developments over the past few decades have made this an increasingly tenuous position as the evidence has failed to materialize. Most scientists point to some human influence but now believe that, among the many climatic forces, the human signal is proving very difficult to isolate. A small group of IPCC climate scientists, however, continue to insist not only that anthropogenic warming now dominates the climate system but also that the issue goes far beyond science and threatens the continued existence of human civilization and the planet. They have the support of hundreds of ENGOs and their members, of international organizations, and of national and local governments. In addition to raising the alarm, many also insist that they have solutions that are technically feasible, economically affordable, and politically tenable.

*****

Box 7-1: The "97 percent agree" mantra

Economist Paul Krugman took to the pages of *The New York Times* on Sunday in order to regurgitate Sierra Club talking points regarding global warming and to castigate the Republican Party for being "anti-science." ... Krugman's central thesis is that the theory that mankind is causing catastrophic climate change has to be true, because "97 to 98 per cent of scientists" agree that it's true. You'll see the "97 to 98 per cent" number appearing quite often now. It's become a key talking point of the alarmist crowd, as they struggle to regain relevance in a world that has a harder and harder time taking them seriously. But where does that amazing number come from? It arises from a 2009 survey that two University of Illinois researchers conducted [Doran and Zimmerman]. 10,257 Earth scientists responded and, much to the U of I professors' chagrin, the results were far from satisfying to the alarmist crowd.

Many of the respondents indicated that they believe that natural forces are much more important than mankind's paltry contributions to climate trends. Some questioned the validity of the models that have been used to predict massive forcing attributable to carbon dioxide and other greenhouse gases. All in all, it wasn't the kind of response that the researchers were looking for when they were trying to prove consensus. So, the professors decided that 10,180 of the scientists who responded weren't qualified to comment on the issue because they were merely solar scientists, space scientists, cosmologists, physicists, meteorologists, astronomers and the like. Of the remaining 77 scientists whose votes were counted, 75 agreed with the proposition that mankind was causing catastrophic changes in the climate. And, since 75 is 97.4% of 77, "overwhelming consensus" was demonstrated once again.

This attempt to silence dissent across scientific disciplines is a sad and troubling feature of the global warming alarmist movement. As a scientist and a sceptic, I often hear alarmists tell me that I'm not qualified to opine on global warming because I'm merely a chemist. I'm not a climatologist, so my vote should not count. Now, having specialized in air quality work for the past thirty years, having run many dispersion models (related to, but not the same as, climate models) and knowing a fair bit about thermodynamics, I'll flatter myself to think that I know a whole lot more about the issue than 99% of the people writing about it in the mainstream media. And yet, people like Krugman feel no shame when they speak authoritatively about an issue they don't understand in the slightest. I'll make Mr. Krugman a deal: I won't write about exchange rate instability if he will take a pass on atmospheric science.

Excerpted from Rich Trzupek, "Krugman Fails Climate Science 101," *FrontPage Magazine,* August 30, 2011.

8 | The Limits of Mitigation Strategies

The whole problem with the world is that fools and fanatics are always so certain of themselves but wiser people so full of doubts.

Philosopher Bertrand Russell

Energy, civilization, and prosperity

At its most fundamental level, the demonization of greenhouse gases amounts to a rejection of civilization and of material progress. Much of modern civilization and human welfare relies on the ability to control combustion and use the energy released to do the work otherwise done by humans and beasts of burden. As early humans discovered, controlled combustion is a powerful source of energy. Over time, it became the basis for political, economic, social, and cultural developments, that is, for *civilization*. Controlled combustion provided humans with a reliable source of light and heat to cook their food – an important contributor to the enlargement of their brains – to extend the day, to make metal tools, to burn clay for ceramic pots, to clear forests, and generally to expand their range and activities.

Most productive combustion involves the rapid oxidation of stored carbon, producing energy in the form of heat and light and releasing carbon dioxide. The carbon comes largely from trees and other plant matter, ranging from recently deposited carbon used in a wood or dung fire to carbon laid down eons ago and now available as coal, oil, and gas. The carbon dioxide released is critical to the cycle of life, because plants rely on it to build the cellulose, sugar, and other carbon-rich compounds that give plants structure and to lay down carbon for future useful purposes; nearly half of plant matter is carbon. Millions of years ago, when the planet's atmosphere was warm, moist, and rich in carbon dioxide, plant matter thrived and, as it died and decomposed, this carbon eventually became the basis for modern industrial-scale combustion.

Combustion is also the basis of human and animal life. Just as plants require carbon dioxide, humans and other animals need carbon derived from plants and animals as the basis of their own energy. In their bloodstreams oxygen from the air combines with carbon-based molecules to provide energy. As with other forms of combustion, carbon dioxide is the principal by-product of this process. Every day, the average adult exhales 400-500 litres of carbon dioxide. The tortured logic of the environmental movement now classifies carbon dioxide as a pollutant and as the enemy of continued life on earth.[1] As Bishop Peter Forster and Lord Bernard Donoughue point out: "The discovery of new ways to release the energy stored in fossil fuels was integral to the Industrial Revolution upon which modern western society is based. Let us not forget that fossil fuels are nature's primary, and very efficient, means of storing the energy of the sun. Burning them has everywhere diverted human beings from burning wood, killing whales and seals, and damming streams: there were therefore genuine environmental benefits to be gained from the switch to fossil fuels. Nature is in most trouble in societies that have not yet made the switch."[2]

From the beginning of the industrial revolution, as illustrated in figure 8-1, it was human ability to exploit fossil fuel sources of energy – a matter of both science and economics embedded in a sociopolitical framework – that provided the basis for rapid material progress. In the words of economic historian Sir Tony Wrigley, fossil fuels allowed man to create "a world that no longer follows the

---

1.   Paul Driessen provides a spirited overview of carbon dioxide's many benefits in *Miracle Molecule: Carbon Dioxide Gas of Life* (Washington: CFACT, 2014).

2.   Peter Forster and Bernard Donoughue, "The Papal Encyclical: A Critical Christian Response," Global Warming Policy Foundation, Briefing 20, 2015. Donoughue is a British Labour peer, and was formerly a senior government adviser. Forster is Bishop of Chester.

Figure 8-1: Fossil fuels and human progress

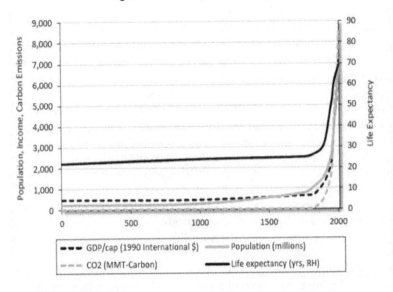

Global progress over past 2000 years, as indicated by trends in world population, gross domestic product per capita, life expectancy, and carbon dioxide emissions from fossil fuels. Source: John Ray at antigreen.blogspot.ca, August 5, 2013

rhythm of the sun and the seasons, a world in which the fortunes of man depend largely upon how he himself regulates the economy and not upon the vagaries of weather and harvest, a world in which poverty has become an optional state rather than a reflection of the necessary limitations of human productive powers."[3] Water, wind, and animal power were quickly replaced by more reliable hydrocarbons, starting with coal and eventually with oil and gas. Industrialization's steady spread from its origins in Britain led to a massive increase in human prosperity and well-being, first in Western Europe and North America, then in Oceania and Japan, and more recently in the rest of the world. In every instance, its spread has depended on the ability to exploit carbon-based energy.

The massive increase in prosperity flowing from industrialization and modernization has led to people living longer and healthier lives, exercising more choices about careers, where and how to live, how to spend their leisure time, and much more. As Peter

---

3.   E. A. Wrigley, *Continuity, Chance and Change: The Character of the Industrial Revolution in England* (Cambridge: Cambridge University Press, 1988), 9.

Glover and Michael Economides point out, "the relative wealth and poverty of nations is [sic] entirely definable by its per capita energy consumption."[4] Alan Pasternak demonstrated more than a decade ago that there is a high level of correlation between annual per capita electricity use and economic development. Pasternak concluded: "The estimates of electricity use associated with high levels of human development presented in this analysis argue for substantially increased energy and electricity supplies in the developing countries and the formulation of supply scenarios that can deliver the needed energy within resource, capital, and environmental constraints. Neither the Human Development Index nor the Gross Domestic Product of developing countries will increase without an increase in electricity use."[5]

Most people give little thought to the many products that make their lives easier, healthier, more productive, and more fulfilling, from the many appliances that simplify life at home to the communications, entertainment, and computing devices they use on a daily basis. All require energy, as do the cars, trucks, trains, planes, and automobiles that underpin commerce and bring us closer together and the pharmaceuticals and technologies that give us longer, healthier lives. Climatologist John Christy told the US Congress: "Oil and other carbon-based energies are simply the affordable means by which we satisfy our true addictions – long life, good health, plentiful food, internet services, freedom of mobility, comfortable homes with heating, cooling, lighting and even colossal entertainment systems, and so on. Carbon energy has made these possible."[6] Despite quantum leaps in the efficiency with which we use energy, Canadian per capita reliance on energy has grown more than tenfold since Confederation. Similar patterns hold true in other advanced economies. Given a choice between living now or living in an earlier age, most people would choose now. They readily embrace the cornucopia of machines and gadgets that allow them to do more things faster – all made possible by access to low-cost power derived largely from the energy stored in hydrocarbons.

---

4. Glover and Economides, *Energy and Climate Wars* (New York: Continuum, 2010), 5.

5. "Global Energy Futures and Human Development: A Framework for Analysis," US Department of Energy, Lawrence Livermore National Laboratory, October 2000, 17.

6. John Christy, "Testimony," US House Energy and Power Subcommittee, 20 September 2012, 33.

For most people living before the full flowering of industrialization, life fit Thomas Hobbes' famous description: nasty, brutish, and short. The idea of primitive man living in harmony with nature that is touted by some environmentalists exists only in their imaginations.[7] It is no accident that many leaders in the alarmist camp are deeply critical of modern technology and industry and their contributions to human welfare. From their perspective, humans and all their modern material accomplishments are a matter of regret rather than of celebration. The prospect of a sharp reduction in material well-being is taken as a necessary but not regrettable outcome of solving the climate "crisis." In effect, for alarmists, climate mitigation policy is as much a means of achieving their larger goals as it is a matter of addressing a possibly serious issue.

Rachel Carson, in the book that launched modern environmentalism, believed that the "road [man] has long been travelling is deceptively easy, a smooth superhighway on which we progress with great speed, but at its end lies disaster. The other fork in the road – the one less travelled by – offers our last, our only chance to reach a destination that assures the preservation of our earth."[8] According to many of Carson's admirers, humans should quit exploiting nature for *their* benefit and, instead, should adapt to nature and its vicissitudes, foregoing the benefits of industrialization and civilization. Entomologist E.O. Wilson goes farther, arguing that man has evolved far too rapidly and thus beyond his evolutionary niche, with dire consequences for the planet and the rest of life.[9] The Club of Rome declared in 1972 that "the earth has cancer, and the cancer is man."[10] Twenty years later, two of its members explained that "in searching for a common enemy against whom we can unite, we came up with the idea that pollution, the threat of global warming, water shortages, famine and the like, would fit the bill. In their totality and their interactions these phenomena do constitute a com-

---

7. See Paul Seabright, *The Company of Strangers: A Natural History of Economic Life* (Princeton: Princeton University Press, 2004) for a discussion of life before markets and civilization.

8. *Silent Spring* (Boston: Mariner, 2002), 277.

9. Edward O. Wilson, *The Social Conquest of Earth* (New York: Liveright, 2012).

10. Mihaljo Mesarovic and Eduard Pestel, *Mankind at the Turning Point: The Second Report of the Club of Rome* (Boston: Dutton, 1974), 1. The idea was first developed by Alan Gregg in "A Medical Aspect of the Population Problem," *Science* 121 (1955), 681-2. Much of climate alarmism has its origins in similar Malthusian analyses of demographic impacts.

mon threat that must be confronted by everyone together. ... All these dangers are caused by human intervention in natural processes, and it is only through changed attitudes and behaviour that they can be overcome. The *real enemy then is humanity itself.*"[11] [emphasis added]

There is no objective basis for this dark image of modern man and nature. At the beginning of the industrial revolution, as economic historians such as David Landes, Angus Maddison, and Gregory Clark have shown, 90 percent or more of the world's population lived at or near a subsistence level;[12] today, based on World Bank studies, less than 15 percent remain at that level. In absolute terms, that 15 per cent still adds up to nearly a billion people, about the same number of people as lived at the beginning of the industrial era. Prosperity and technological breakthroughs have fuelled rapid demographic increases all over the world, including in developing countries and, as economist Surjit Bhalla has shown, have led to prosperity even in those countries.[13] By the beginning of the 21st century, even the absolute number of people living in abject poverty had finally begun to decline.

The ability to harness the energy stored in hydrocarbons has been critical in underwriting this miracle.[14] The alarmist community believes that this miracle, with its disastrous consequences for the planet, was all a mistake and hence proposes radical changes to wean the world off fossil fuels. Environmentalists like to clothe their

---

11. Alexander King and Bertrand Schneider, *The First Global Revolution: A Report by the Council of the Club of Rome* (New York, 1993), 75. Robert Zubrin provides a full discussion of the anti-human dimensions of environmentalism in all its forms in *Merchants of Despair: Radical Environmentalists, Criminal Pseudo-Scientists and the Fatal Cult of Antihumanism* (New York: Encounter, 2012). Rupert Darwall, in *The Age of Global Warming: A History* (London: Quartet Books, 2013), discusses anti-humanism specifically in the context of the climate change debate.

12. David S. Landes, *The Wealth and Poverty of Nations: Why Some Are So Rich and Some Are So Poor* (New York: Norton, 1998); Angus Maddison, *The World Economy: A Millennial Perspective* (Paris: OECD, 2001; and Gregory Clark, *A Farewell to Alms: A Brief Economic History of the World* (Princeton: Princeton University Press, 2007).

13. Surjit Bhalla, *Imagine There's No Country: Poverty, Inequality, and Growth in the Era of Globalization* (Washington: Institute for International Economics, 2002).

14. See Alex Epstein, *The Moral Case for Fossil Fuels* (New York: Portfolio/Penguin, 2014) for a through discussion of the many benefits of fossil fuels and their role in underwriting modern prosperity and well-being. See also Robert Bryce, *Power Hungry: The Myths of "Green" Energy and the Real Fuels of the Future* (New York: Public Affairs, 2010), particularly 13-44.

arguments in the mantle of morality. The moral dimension of climate alarmism is stark indeed, but it does not support the alarmist side of the debate. Paul Driessen and Willie Soon point out:

> Sub-Saharan Africa remains one of Earth's most impoverished regions. Over 90 percent of its people still lack electricity, running water, proper sanitation and decent housing. Malaria, malnutrition, tuberculosis, HIV/AIDS, and intestinal diseases kill millions every year. Life expectancy is appalling, and falling. And yet UN officials, European politicians, environmentalist groups and even African authorities insist that global warming is the gravest threat facing the continent. … Warming alarmists use the 'specter of climate change' to justify inhumane policies and shift the blame for problems that could be solved with the very technologies they oppose. Past colonialism sought to develop mining, forestry and agriculture, and bring better government and healthcare practices to Africa. Eco-colonialism keeps Africans 'traditional' and 'indigenous,' by insisting that modern technologies are harmful and not 'sustainable' in Africa. … So this is where radical climate change alarmism has taken us. When the health of Planet Earth is at stake, human life means little – even if the 'disasters' are nothing more than worst-case scenarios conjured up by computer models, headline writers, Hollywood, and professional doomsayers like Gore, Hansen and NOAA alarmist-in-chief Susan Solomon.[15]

For most people, the moral implications of this perspective are repugnant. Nigel Lawson points out that "the ethical issue is not just about how much we care about distant future generations; it is also about how much we care about the present generation, not least in the developing world, and its children. Certainly, for the governments of those countries, the question of how great a sacrifice the present generation and their children should make, in terms of unnecessary poverty, malnutrition, disease and premature death, in the hope of benefiting substantially better off generations a hundred or two hundred years hence, is not a difficult one, either in ethical or indeed in political terms."[16] It is a mystery, therefore, why a call for action grounded in such a dark and loathsome view of mankind has gained such traction in governments, universities, and even in some Christian churches. It is an even greater mystery why the current pope has taken the advice of Malthusian and other

---

15. Paul Driessen and Willie Soon, *Eco-Colonialism Degrades Africa*, SPPI Commentary and Essay Series, February 14, 2009. Driessen develops this theme in much more detail in *Eco-Imperialism: Green Power, Black Death* (Bellevue, WA: Free Enterprise Press, 2003-04).

16. Nigel Lawson, *An Appeal to Reason: A Cool Look at Global Warming* (London: Overlook Press, 2008), 95.

alarmists and crafted an encyclical that aligns the Church with those who have long advocated abortion and birth control and who believe that there is a moral imperative to deny the poor of the Third World the benefits of reliable energy.[17]

Whatever one thinks of climate change and regardless of whether the planet warms or not, it is important to keep in mind that the rest of the world wants in and, in the short to medium term, it can only get in by burning hydrocarbons. From a developing world perspective, the world does not use enough energy. For much of the world outside of North America, Europe, Japan, and Oceania, per capita energy usage is at levels that would take Canadians and Americans back to the late 19th century or even earlier. The single most important brake on economic development and increased welfare in Africa, Latin America, and still much of Asia is lack of access to reliable, low-cost energy. As University of Victoria economist Cornelis van Kooten concludes: "It does not matter what rich countries do to reduce their emissions of carbon dioxide. Their efforts will have no impact on climate change, but they will have an adverse impact on their own citizens. Whether the climate change story is real or not, whether the climate model projections are accurate or not, fossil fuels will continue to be the major driver of economic growth and wealth into the foreseeable future."[18]

More than a billion people in poor countries still rely largely on biomass – dung, wood, and other recent plant matter – to cook their food and heat their homes. The health implications of this limitation alone require urgent attention. The idea that further modernization in the poorer parts of the world poses a threat to the future of the planet must be one of the most bizarre ideas ever to gain widespread currency. It is but the latest version of what Robert Zubrin calls the ideology of anti-humanism. In his view, many alarmists see human sacrifice – that is, denial of low-cost, widely available energy to developing countries – as critical to solving the global climate "crisis".[19]

✦✦✦✦✦

---

17. Pope Francis, *Laudato Si' (Praise be to You)*, June 18, 2015.

18. Cornelis Van Kooten, *Climate Change, Climate Science and Economics: Prospects for an Alternative Energy Future* (New York: Springer, 2013), 322.

19. Zubrin, *Merchants of Despair*, 233.

Growing energy demand and "finite" supplies

Fortunately, despite all the hand wringing and international happy talk over the past few decades, world energy consumption continues to increase by about 2-3 percent a year. Much of the recent growth in consumption is the result of the rapid modernization and urbanization of China, India, Brazil, Indonesia, and other recently industrializing economies. Fossil fuels account for some 85 percent of total world consumption, nuclear and hydro nearly 10 percent, and renewables the final five percent, three-quarters of which still come from traditional sources such as wood, dung, and other biomass. So-called modern renewables – wind, solar, geothermal, and biofuels – account for a little over one percent. Since the 1973 oil crisis, consumption has doubled for all types of primary energy, as have CO_2 emissions, despite efforts to reduce fossil fuel use through both national and international efforts. Canada, blessed with abundant rivers, relies less on hydrocarbons to generate power than most other advanced economies. In 2012, Canada sourced 28.5 per cent of its total primary energy from renewables, of which hydro, nuclear, and wood waste constituted 98 per cent. Fossil fuels provided the remaining 61.5 per cent. Wind provided 0.6 per cent, biofuels another 0.6 per cent, municipal waste 0.2 per cent, and solar less than 0.1 per cent. Total installed wind capacity at the end of 2013 was 7,803 MW, and total solar capacity was 1,210 MW, but delivery from both sources was less than 2,000 MW.[20] In the US, 82 per cent of primary energy was derived from fossil fuels in 2012.

Even as governments earnestly discuss the need to reduce GHG emissions at international conferences and implement various strategies to achieve lofty renewable energy and emissions reduction goals at national and sub-national levels, the same governments plan or approve building new hydrocarbon-fuelled electrical generating facilities, only a few of which are mandated to include capture and sequestration technologies. If nothing else, such action demonstrates that here, as in other difficult areas of public policy, consistency is sacrificed to expediency, and rhetoric takes precedence over action.[21]

---

20. Natural Resources Canada, *Energy Markets Fact Book 2014-15* (Ottawa: Natural Resources Canada, 2014.

21. The World Resources Institute claims that "1,231 new coal plants with a total installed capacity of more than 1.4 million MW," are at various planning stages world wide, including in the United States, China, Vietnam, South Africa, India, and Turkey. See Lisa Friedman, "India Has Big Plans for Burning Coal," *Scien-*

Concerns about the eventual depletion of finite, carbon-based fuels – or other resources, for that matter – may have some basis in reality but remain premature at this point in time. Alarmists like John Holdren, Stephen Schneider, Paul Ehrlich, and Lester Brown have been hyping this concern for years and predicting the end of industrial civilization. Experience has demonstrated, however, that markets and human ingenuity are more resilient and inventive than alarmists have been – and are – prepared to admit. The cost of extracting finite resources is a critical dimension. Prices and markets will stimulate more discovery and technological developments and keep this concern at bay.[22]

World reserves of cheap coal can fuel civilization for generations to come. The World Coal Association estimates that, at current rates of extraction, proven coal reserves, i.e. reserves that can be economically recovered based on current technology, will last well into the 22nd century. Proven oil and gas reserves are estimated to be capable of supplying energy for about 46 and 59 years respectively of current production, but both technological and economic developments are likely to expand these horizons. Light, easily extracted crude oil, for example, may be less abundant, but oil embedded in shale and oilsands, or in less accessible parts of the planet, remains substantial. Cambridge Energy Research Associates, the leading global energy analysts, calculated in 2010 that historical oil production had reached 1.1 trillion barrels and estimated that 1.9 trillion barrels of conventional oil and 2.4 trillion barrels of non-conventional oil remained for post-2010 extraction. The shale gas revolution has greatly enlarged the scope for gas and reduced its price significantly in countries that welcome fracking.

ExxonMobil's 2040 Outlook for global energy consumption is projecting steady growth in hydrocarbon use and modest growth for renewables, hydro, and nuclear over the next two decades, with much of the total growth taking place in non-OECD countries and

---

tific American, September 17, 2012. At the same time, UBS AG reported: "European utilities are poised to add more coal-fired power capacity than natural gas in the next four years, boosting emissions just as the era of free carbon permits ends. Power producers from EON AG to RWE AG will open six times more coal-burning plants than gas-fed units by 2015." Matthew Carr, "Coal Era Beckons as Carbon Giveaway Finishes," *Bloomburg Businessweek,* September 21, 2012.

22. Economist Julian Simon successfully taught this lesson to the alarmists in his famous wager with Paul Ehrlich. Simon bet Ehrlich that the world price of an agreed list of five raw materials – copper, chromium, nickel, tin, and tungsten – would be lower in 1990 than they were in 1980. Simon handily won the bet.

Figure 8-2: Global consumption of energy by primary source, 1800-2014

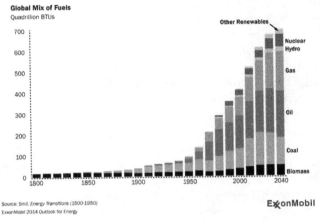

Source: ExxonMobil, *2014 Outlook for Energy*

growth of renewables largely confined to the OECD countries. Even in the OECD, however, enthusiasm for renewables is limited. (See Figure 8-2). Longer-term projections by the IEA, the US EIA, and the US GAO indicate that recoverable reserves of all three fossil fuels remain very large and that steady improvements in both the technology and economics of extraction suggest that estimates of recoverable reserves will continue to be subject to upward revisions for many more years.[23]

The issue for current and medium-term policy makers, therefore, is not whether they need to prepare for an era when fossil fuels will have become sufficiently scarce to warrant serious concerns about replacements but whether the impact of continued reliance on fossil fuels will have catastrophic impacts on the composition of the atmosphere and on long-term climate change. Concerns about peak oil and gas are diversions from that issue. If there is any rationale for urgently looking at alternative sources of energy, it has to be found in the claims of climate science.

Maintaining, let alone spreading, modern civilization on a basis other than carbon would be a monumental undertaking and can only be justified on the strongest of evidence that it is both necessary

---

23. See the discussion in Bryce, *Power Hungry*, 53-79. See also Vaclav Smil, *Energy Myths and Realities* (Washington: AEI Press, 2010) and Glover and Economides, *Energy and Climate Wars*.

and feasible. At a minimum, it would take many years, and attempts to reduce those years would be very costly. Using government funds to support fundamental research on alternative forms of energy may be a prudent strategy; using government funds to subsidize premature or unproven forms of energy that cannot compete on their merits is *not*. The idea that there is a cornucopia of green jobs and profits available from a move to non-carbon energy is the latest variant in the world of do-it-yourself economics. To date, as Donald Hertzmark indicates: "The arithmetic for green jobs is ineluctable and grim. For each utility worker who moves from conventional electricity generation to renewable generation, two jobs at a similar rate of pay must be foregone elsewhere in the economy; otherwise the funds to pay for the excess costs of renewable generation cannot be provided. Moreover, by raising costs throughout the economy, high-cost green energy will reduce the competitiveness of US exporters, thereby destroying (presumably well-paying) jobs in such industries."[24]

Using government regulations to penalize use of low-cost, readily accessible, efficient energy sources borders on the irresponsible, particularly for developing countries. The case for CO_2-driven global warming is based on controversial science, and the predicted dire impacts rest on even weaker foundations. The solution – driving up the cost of fossil fuels and replacing them with other forms of energy – is neither technologically realistic nor economically responsible within a foreseeable time frame. Additionally, as the authors of the *Civil Society Report on Climate Change* point out, "Technocratic plans for the climate (whether driven by global agencies or national governments) are predicated on the same fatal conceit that led to the failure of socialism: that government is better able to identify and act upon information that is ultimately only available to individual economic actors within society."[25]

Strategies to reduce atmospheric CO_2

The alarmist movement offers five immediate solutions that will reduce atmospheric CO_2 and stabilize global temperatures: carbon taxes, carbon cap-and-trade schemes, carbon capture and sequestration, carbon offsets, and terrestrial sinks. Each one is fraught with problems. Some alarmists are also convinced that alternative, non-

24. Hertzmark, "Green Jobs: Making Society Poorer," *MasterResource*, April 6, 2009.
25. Reiter, et al., *Civil Society Report on Climate Change*, 14.

carbon-based energy sources are coming on stream, ensuring that living standards and economic development will not be adversely affected by efforts to reduce carbon intensity. Unfortunately, none of the claimed alternatives is close to being ready to perform this function. Additionally, each alternative is opposed by environmentalists on various grounds that should raise serious questions about their long-term goals and objectives.

For many environmentalists, switching to new forms of energy is still a problem because it delays de-industrializing advanced economies. James Speth, a leading American environmentalist, maintains that: "The prioritization of economic growth is among the roots of our problems. ... Economic growth may be the world's secular religion, but for much of the world it is a god that is failing – underperforming for most of the world's people and, for those in affluent societies, now creating more problems than it is solving."[26] The media's current favourite Malthusian, Bill McKibben, adds in a familiar lament: "growth may be the one big habit we finally must break."[27] For anti-growth environmentalists, industrialization, capitalism, and population growth are the satanic trinity that must be exorcised in order for the planet to survive. Even sustainable development places too great a burden on the planet, and the solutions on offer would involve massive increases in government regulatory activity and control leading to significant costs for economies that are already stretched from fiscal overreach due to the continuing growth of government programs.[28]

Carbon Taxes

There is already substantial experience with carbon taxes. European governments have long imposed high gasoline taxes to reduce consumption, promote public transportation, spur the production of more fuel-efficient automobiles, and fund infrastructure. Calibrating the tax to have the desired effect, however, can be tricky. European fuel taxes have certainly had an impact on European preferences for smaller, more fuel-efficient cars and perhaps for public

---

26. James Speth, "America is the Best Country in the World at Being Last," *Orion Magazine*, January 27, 2015.

27. Bill McKibben, *Eaarth: Making a Life on a Tough New Planet* (New York: St. Martin's, 2011), 48.

28. Peter Foster provides a compelling overview of the warped economic ideas of the environmental movement in *Why We Bite the Invisible Hand: The Psychology of Anti-Capitalism* (Toronto: Pleasaunce Press, 2014).

transport. Other factors, however, complicate the capacity to reach clear conclusions on the benefits and impact of fuel taxes. Consider one major difference: Europeans live in more compact societies and denser cities that make public transit a much more viable option than for most cities in North America. Even so, Europeans have not significantly reduced car ownership to levels below those of North America. How much higher would carbon taxes need to go in order to have an impact on Europe's future carbon "footprint"? British Columbia's modest carbon tax was just enough to give the Campbell government green credentials but not high enough to irritate most BC motorists or to have a significant impact on consumption.[29] Sales and excise taxes on fuel, both federal and provincial, already make up a third or more of the price in most provinces and in many US States. A much higher tax will be needed to alter fundamental consumption patterns, and political resistance to much higher fuel taxes, at least in North America, is deeply ingrained. Additionally, Pigovian taxes raise costs throughout the economy, often with undesirable effects requiring further government interference.[30]

On July 1, 2012, after five years of political wrangling, Australia introduced a national carbon tax set at A\$23 per tonne of emitted CO_2. The price was scheduled to be increased by 2.5 percent per year before being replaced by a more ambitious emissions trading scheme in 2015. The tax proved to be highly controversial, even though it was limited to the 500 largest emitters, mainly mining companies, heavy industry, and generating facilities, and the goal

---

29. The starting rate of \$10 per tonne of carbon emissions in July 2008 gradually rose by \$5 a year to \$30 per tonne by 2012. For diesel and home heating oil, the rate started at about 2.7 cents per litre, rising to 8.2 cents by 2012, and for gasoline at an extra 2.4 cents on a litre at the pumps, increasing to 7.24 cents per litre. Rural British Columbians were among those who were reported to be irritated. As has been demonstrated during various short-term price spikes, it takes significant and sustained price increases to begin to affect consumer behaviour for essential products like gasoline and heating oil. See Philip Cross, "The carbon tax illogic," *Financial Post,* January 13, 2015, for a review of some of the literature claiming the efficacy of the tax.

30. Economist Arthur Pigou was the first to propose that governments could address negative "externalities," for example, environmental effects, by imposing a tax that would raise a product's price and thus reduce its consumption. Additionally, such taxes could be used to fund mitigation strategies and reduce the impact of the undesirable externalities. For a more complete discussion of economic instruments to reduce carbon emissions, see van Kooten, *Climate Change, Climate Science, and Economics,* 285-324. See also Ross McKitrick, "An Evidence-based Approach to Pricing CO_2 Emissions," Global Warming Policy Foundation Report, July, 2013.

was to induce investment in cleaner technologies. The defeat of the Labour government in 2013, in part because of the carbon tax, led to the repeal of the scheme. In addition to the EU's emissions trading scheme discussed below, a number of EU members have introduced various carbon or CO_2 taxes, mostly focused on major emitters. India introduced a carbon tax on coal set at 50 rupees per tonne, a modest amount that will raise some revenue but will have little impact on the use of coal in generating electricity.

Determining carbon intensities for individual products and calculating the level of tax required are formidable challenges to imposing a carbon tax. In recognition of these difficulties, most proponents suggest that a carbon tax be levelled on producers and be limited to major carbon emitters. Such a tax would be easier to administer but would create serious problems for domestic firms competing with imports or for firms active in export markets, as has been demonstrated in Australia. To get around this problem and level the international playing field, some proponents have suggested that in addition to taxing domestic emitters, governments should introduce border tax adjustments by imposing a carbon tariff on competing products imported from countries that fail to impose a carbon tax of their own and then by remitting taxes on products exported to those countries. Border tax adjustments would thus prevent free riders from gaining an unfair advantage. Unfortunately, this beguilingly simple solution would contravene the international trade rules embedded in the World Trade Organization (WTO) and various regional trade agreements. Characterizing the new tariff as a carbon tax does not change the fact that it is a tax imposed on imports, i.e., a tariff, while taxes that are remitted upon exports constitute prohibited export subsidies.

Some argue that countries not applying carbon taxes would, in effect, be subsidizing their exports.[31] Accordingly, countervailing import duties should be applied to level the playing field. Again, the international trade rules governing the application of countervailing duties are designed to prevent this kind of abuse. Of course, governments could ignore the trade rules and simply apply countervailing duties to all imports from countries not applying carbon taxes at some arbitrarily determined level and then face the consequences of lost export markets as affected countries retaliated. As

---

31. Jeff Rubin and Benjamin Tal argue that "in effect, a carbon tariff is a countervail against unfair energy subsidies." "The Carbon Tariff," *StrategEcon,* CIBC World Markets, March 27, 2008.

public policy, this approach has little to recommend it. Governments could certainly seek changes to the WTO and other agreements to provide for carbon import tariffs, but it is difficult to envisage what would induce other countries, particularly those that have decided not to adopt measures to reduce carbon emissions, to agree to changes that would see their exports reduced.[32]

The idea that carbon taxes can be imposed on a revenue-neutral basis also needs to be assessed cautiously. Governments do not have a very good record of curbing their appetite for revenue. Once money arrives at national treasuries, the temptation to do things with that money is very large, even if it is only to redistribute it in socially – read politically – desirable ways. Raising the cost of carbon will have significant multiplier effects throughout the economy that will quickly lead to special interests looking for relief based on the revenue generated by the carbon taxes. A carbon tax, therefore, may be a market-based approach to reducing carbon intensity, but it will inevitably generate extensive government programming and intervention to deal with its effects, leading to multiplying deadweight losses. And there will be nothing "neutral" about these programs. Some of the money raised from Peter's consumption of carbon may be returned to Peter, but more of the money is likely to find its way to Paul, Mary, and assorted other more politically worthy citizens.

Finally, it should be noted that imposing carbon tax regimes in industrialized countries but not in developing countries, as envisaged in the Kyoto Protocol and advocated by the IPCC and other UN organs, can have perverse consequences. The UN is right that the world's poor need more energy, and fossil fuels are the right choice for them until such time as non-carbon sources become competitive. Making carbon-rich energy more expensive only in industrialized countries may reduce consumption and spur transition to non-carbon sources, but, unless the policy is pursued on a global basis, the effect will be undesirable market distortions, increasing the incentives to shift carbon-based production from developed to

---

32. For a full discussion of carbon taxes, cap and trade, subsidies, and the international trade rules, see Gary Clyde Hufbauer, Steve Charnovitz, and Jisun Kim, *Global Warming and the World Trading System* (Washington: Peterson Institute for International Economics, 2009). More generally, see Van Kooten, *Climate Change, Climate Science and Economics*, 293-309, and Cameron Hepburn, "Carbon Taxes, Emissions Trading, and Hybrid Schemes," in Dieter Helm, and Cameron Hepburn, eds., *The Economics and Politics of Climate Change* (New York: Oxford, 2009), 365-84.

developing countries. The net result may well be a global increase in CO_2 emissions rather than a decrease, which is surely not what climate alarmists have in mind.

Cap and trade

Cap-and-trade schemes are among the solutions favoured by environmentalists. Given their innate suspicion of markets and price-based measures, they are more comfortable with command-and-control measures than with taxes. Nevertheless, they like to claim that cap and trade is a market-based solution. Former Czech president Vaclav Klaus, trained as an economist, is blunt in dismissing this sleight of hand. He notes: "We should not deceive ourselves. A cap-and trade-scheme is a government intervention par excellence, not a 'market solution.' How much 'to cap' is the decision of the government. ... The size of the cap defines the price of carbon and this price is nothing else than a tax imposed upon citizens of the country."[33] Most large firms also prefer cap and trade because this strategy makes it easier to plan and pass costs on to their customers. They part company with environmentalists on how best to allocate emissions permits. Environmentalists insist that "polluters" must pay, while firms believe costs should be allocated on the basis of historical performance, letting markets determine them. Either way, permits will accrue value which, in turn, will be reflected in costs and prices throughout the economy.

Under a cap-and-trade scheme, governments would set limits on allowable GHG emissions and provide individual firms and industries with tradable permits adding up to the permissible level for the country as a whole. The cap would need to be set initially at a reasonable level and then steadily reduced in order to have the desired effect. Proponents of cap and trade claim that it would encourage industries to devote resources to finding alternative energy sources and energy efficiencies. Firms that could not adapt to their permitted levels would need to purchase permits from those that *have* learned to adapt to lower levels of carbon intensity. The superficial resemblance of cap-and-trade schemes to the market has convinced some governments that they may be politically more attractive than direct taxes. Most environmentalists do not appreciate that the price effect of cap-and-trade schemes is similar to that of a tax compounded by higher dead-weight losses. In other words, selling

---

33.   Vaclav Klaus, "Advancing the Global Debate Over Climate Change," Speech at the Washington *Times'* Briefing, Washington, November 4, 2009.

permits to major emitters would amount to a major indirect tax on consumption.

Experience to date with cap-and-trade schemes is not encouraging. Setting the necessary limits for individual firms and industries is a major undertaking. The incentive for firms to move their energy-intensive operations to non-cap-and-trade jurisdictions is significant. Additionally, it is difficult to manage cap-and-trade schemes without the risk of significant fraud and corruption. The prospect of establishing a global cap-and-trade regime is mind-boggling.

The European Union launched its Emissions Trading Scheme (ETS) in 2005, aiming to decrease emissions gradually by reducing permitted emissions of major emitters such as oil refineries, manufacturing facilities, and electrical generating plants. The ETS extends to all 28 EU member states plus Iceland, Norway, and Liechtenstein. The 11,000 facilities covered by the ETS in 2012 were collectively responsible for half of EU emissions of CO_2. Covered facilities were granted tradable emissions credits for each of three trading periods (2005-2007; 2008-2012; and 2013-2020) with a view to reducing emissions by 21 percent by 2020. For the initial two trading periods, each member state was responsible for allocating allowances based on national limits set by the EU Commission. For phase three, allowances were centrally allocated. Installations exceeding their limits could purchase credits from those having leftover credits throughout the ETS trading area.[34] Prices were initially set at €30/tCO_2 but rarely reached that level; prices hovered around €22/tCO_2 in 2008, dropped to €13/tCO_2 in 2009 and continued to decline, hitting €10/tCO_2 in 2012, too low to have any effect on emissions levels. Other problems, such as theft, fraud, over-allocation, and windfall profits, have plagued the system. EU officials believe that the first two periods, while fraught with difficulties, provided important lessons needed to improve the system, lessons which will allow the third phase to meet its objectives. Nevertheless, as *The Economist* points out, the combination of the global recession, the Euro crisis, and over-generous allocations has made the ETS largely irrelevant over the course of its first two trading periods.[35] By the beginning of 2014, European sources were indicating that both the Commission

---

34. See Europa: Climate Action – Emissions Trading System at ec.europa. eu/ clima/policies/ets/ index_en.htm.

35. "Complete Disaster in the Making," *The Economist*, September 15, 2012.

and member states were in full retreat, reducing existing commitments and unable to agree on future commitments.[36]

Experience with the UN's Clean Development Mechanism has been worse. A panel of experts set up by the UN to review the CDM's initial operations found that prices had fallen precipitously and that the CDM itself was in danger of collapsing unless governments committed themselves to much higher mitigation targets, adapted the CDM to new market and political circumstances, reformed its operating procedures, and strengthened and restructured its governance. The Panel, set up by the UN and composed of people highly sympathetic to the CDM's objectives, issued a scathing report on its operations in September 2012, advocating far-reaching reforms. The prospect of governments adopting serious climate mitigation programs, including cap and trade schemes, remains unlikely, suggesting that the CDM may well fade away, another victim of idealistic UN efforts to remake the world. On balance, while it may not have had much impact on the global climate or on reducing carbon emissions, the CDM did provide a vehicle for financing some worthwhile projects that might not otherwise have found funding. By 2014, the Secretariat was running out of projects to manage because governments in eligible countries were losing interest. With carbon prices now trading in pennies rather than dollars, there is little prospect of renewed investor interest. The Bonn-based staff of about 150 are staying on, drawing down remaining capital by preparing studies in the hope that a future global emissions program will provide the CDM with renewed life.[37]

The now defunct Chicago Climate Exchange was a private sector effort to develop a market for carbon credits. At its height, some 400 firms were members and traded carbon credits and offsets. Car-

---

36. See, for example, "EU in full retreat on climate policy," *Global Warming Policy Foundation,* January 14, 2014, "Backsliding on the Climate," New York *Times,* January 21, 2014, and Dominic Lawson, "Admit it, greenies: the game's up for renewable energy," *Sunday Times,* January 26, 2014.

37. See "Climate Change, Carbon Markets and the CDM: A Call to Action," Report of the High-Level Panel on the CDM Policy Dialogue; and Susan Twidale and Ben Garside, "UN agency brokering carbon credits spends millions on staff despite massive drop in projects," *National Post,* April 28, 2014. In a study commissioned by the UN, Vivid Economics indicated that without fresh demand, 4.7 billion credits would need to be taken out of the system by 2020, potentially costing €11.7 billion (US$15 billion) in order to lift prices just to €2.50. In the first five months of 2013, just 72 schemes had been registered, down 94 per cent from more than 1,100 in the first five months of 2012. See Susanna Twidale, "UN study reveals $15-bln price tag to save the CDM," *PointCarbon,* June 11, 2013.

bon prices never attained the value its promoters envisaged; the exchange closed in 2010 when the price fell to 5-10 US cents per tonne of CO_2, from a high of $7.50 per tonne in 2008. Extensive trading in 2008 was predicated on the expectation that the US Congress would pass a cap-and-trade bill. Its demise was inevitable once it was clear that there would be no federal regulatory regime to back it up.[38]

Unless and until governments agree, at a global level, to establish emissions caps with a stable regime for trading emissions credits, a prospect that looks increasingly slim, cap and trade will remain a "promising" solution without prospects.

Carbon capture and sequestration

The third solution involves clean-fuel technologies, including the capture and sequestration of carbon (CCS) from fossil-based energy generation, particularly coal. On the surface, this is an attractive solution. Coal remains the most abundant and cheapest fossil fuel. Unfortunately, clean coal remains an unrealistic solution due to both cost and technological shortcomings. The necessary technologies to make this a feasible and cost-effective solution exist only in the imaginations of environmentalists and emerging "green" entrepreneurs. Much progress has been made in addressing the pollutants which are the undesirable by-products of burning coal to generate electricity: sulphur dioxide, nitrogen oxides, mercury, and particulates. Conventional air pollution from coal-burning power plants is now largely a bad memory in industrialized countries.[39] Concern with acid rain in the 1980s contributed importantly to this development. Classifying CO_2 as a pollutant, however, has again stigmatized coal as a dirty source of energy and the target of aggressive environmental regulation.

Massive amounts of money have already been spent exploring solutions but, alas, technologies that can be applied on the scale nec-

---

38. Ed Barnes, "Collapse of Chicago Climate Exchange Means a Strategy Shift on Global Warming Curbs," *Fox News*, November 9, 2010.

39. I am not suggesting that coal-burning plants currently produce no conventional pollutants: rather, that the amount produced by new plants is significantly less than that of older plants, and that even older plants have been retrofitted to remove most pollutants from their exhausts. Environmentalists demand that more be done, but even to them the real issue is now carbon dioxide, a natural component of the air we breathe. For an environmentalist perspective on EPA regulations aimed at putting additional pressure on coal plants, see Susan Tierney, "Electric Reliability under New EPA Power Plant Regulations: A Field Guide," World Resources Institute, January 18, 2011.

essary to have the desired impact have yet to be realized. Costs remain astronomical, and no politically acceptable answers have been found to the problem of storing the massive amounts of CO_2 that would need to be captured and sequestered.[40] Whether retrofitting or building a new plant, significant additional investment is needed to add carbon capture to a coal-based electrical generating facility. The amount of energy that would need to be diverted to power the capture technology would make coal uncompetitive as a source of energy for base-electrical generation. Adding as much as 25 percent more generating capacity for the sole purpose of capturing CO_2 seems a foolhardy venture. While clean-coal technologies may eventually prove feasible and cost-effective, they remain unrealistic. The best that can be said is that such technologies are sufficiently promising to justify modest government expenditures on research and demonstration projects.[41] Even then, Vaclav Smil concludes, "carbon sequestration on a scale sufficient to affect the earth's climate ... would be a task of an unprecedented magnitude."[42]

US government websites provide an optimistic spin on ongoing laboratory work on these issues but admit that there remain significant hurdles to deploying CCS technology on a major scale. The Obama Administration has proposed spending billions on clean-coal technologies, and other governments have done the same, but the results to date have been minimal. EPA regulatory efforts have put increasing pressure on conventional US coal-based electricity generators to reduce their carbon emissions. For most operators, however, conversion to gas is the least costly option, particularly following the significant reduction in North American gas prices after the result of widespread adoption of fracking technologies.[43]

---

40. Smil calculates that capturing and storing just 15 per cent of 2008 global CO_2 emissions from fossil fuel combustion would require "a gathering, transportation, and storage industry whose annual throughput would be (depending on the stored gas density) 1.2-2.2 times that of the annual volume throughput of the world's crude oil industry with its immense networks of wells, pipelines, compressor stations, tankers, and above- and underground storages. ... at a[n operating] cost of close to $300 billion a year." *Energy Myths and Realities*, 91-3.

41. For a more detailed discussion, see Howard Herzog, "Carbon Dioxide Capture and Storage," in Helm and Hepburn, eds., *The Economics and Politics of Climate Change*, 263-83.

42. *Energy Myths and Realities*, 96.

43. See, for example, "Technologies: Carbon Sequestration," at DOE.gov. On the EPA effort, see Matt Cover, "EPA Regulations Will Close Coal Plants, Raise Electricity Prices, GAO Says," CNS News, August 22, 2012. Robert Bryce sug-

Nevertheless, US utilities continue to build new high-efficiency coal plants to generate electricity;[44] in 2013, 39 percent of US electricity was generated with coal.[45] Germany, which was in the vanguard of efforts to move away from coal, is opening new, mega, coal-fired generating plants, having decided that nuclear is too dangerous, wind and solar too unreliable, and gas from Russia too expensive.[46] China is building coal plants at the rate of one or two a month, using the best of conventional technologies and relying almost exclusively on coal to generate its electricity; it is now the world's number-one consumer of all energy and of coal-generated, electrical energy in particular.[47] Thus, at the same time that governments speak earnestly of reducing CO_2 emissions, new coal-fired, electrical-generating capacity continues to be added. In short, coal, while much vilified, remains the principal energy source for generating electricity in much of the world, and new investments indicate that it will continue to be so for many years to come.[48]

Carbon offsets

Environmentalists have convinced themselves that offsets are an important part of the solution. To date, this approach has amounted to little more than a massive scam reminiscent of the indulgences sold by rogue priests for the expiation of sins during the years leading up to the Protestant Reformation. As an example, while some conscience-stricken consumers may feel better purchasing offsets when buying an airline ticket, they had better not inquire too close-

---

gests: "when it comes to carbon capture and sequestration, Americans are hearing lots of dreams and precious few facts." *Power Hungry*, 160.

44.   Modern, UltraSuperCritical, coal-powered, electrical-generating plants are achieving reductions of 15-17 per cent in CO_2 emissions over the previous generation of coal-fired plants, while producing the same amount of electricity. See Anton Lang, "Ultra Super Critical Coal Fired Power gives a 15% CO2 Emissions Reduction," March 28, 2013, *Joannenova.com*.

45.   "Energy Explained," US Department of Energy, EIA.

46.   Julia Mengewein, "Steag Starts Coal-Fired Power Plant in Germany," Bloomberg, November 15, 2013; "Merkel's Green Shift forces Germany to burn more coal," *Financial Post*, August 20, 2012; and Roland Nelles, "Germany Plans Boom in Coal-Fired Power Plants – Despite High Emissions," *Spiegel Online*, March 21, 2007.

47.   Keith Bradsher, "China out-paces US in cleaner coal-fired plants," New York *Times*, May 10, 2009; "Lights and action," *Economist*, April 29, 2010; and CIA World Fact Book: China.

48.   As noted above, globally 1,231 new coal plants are at various stages in the planning cycle.

ly as to what in fact happens to the money.[49] Some trees may be planted but far fewer than the number needed to offset the carbon that their trip will produce. More likely the money will end up in the pockets of savvy ENGOs and will pay for their lobbying efforts for more draconian measures. Al Gore's example is not very reassuring. He told Congress that he had purchased offsets to mitigate the gargantuan carbon footprint of his Nashville mansion, calculated to be 20 times larger than that of the average American home, as well as his two other homes and houseboat, his SUV and luxury cars, and his frequent airline trips. The company that had so obligingly sold him the offsets turned out to be owned by Al Gore. Its business plan was shrouded in mystery.[50]

A creative variant on offsets is offered by a Canadian firm, Bullfrog Power. It urges electricity consumers to sign up with it for a small daily premium – typically 50¢ a day for private customers – and in return they can reduce their "energy pollution," support "local green energy projects," and "create a clean, healthy energy future." Bullfrog Power "... ensures the energy going onto the systems on your behalf is from clean, renewable sources, displacing energy from polluting sources."[51] In most Canadian provinces, power is distributed by a provincial utility, either directly to consumers or to local distribution companies. Ontario, for example, has a complex, highly inefficient, and costly system distributing power that it generates in its own facilities or that it buys from contractors in and outside the province. It currently generates and buys more than required and sells the excess to US state and other provincial grids at distress prices. It pays high premiums to wind and solar contractors. Ontario consumers pay among the highest rates in North America. Bullfrog cannot guarantee what is delivered to specific customers; on typical days, the share of green energy fed into the grid amounts to less than one percent because the wind fails to

---

49. For a full discussion of the problems, see Joanne Nova, *Carbon Credits: Another Corrupt Currency?* SPPI, February 2, 2009.

50. Gore buys offsets from Generation Investment Management, of which he is the principal shareholder. It, in turn, buys offsets from Carbon Neutral Company, which admits that its activities do not reduce GHG emissions. Rather, these activities "(1) demonstrate commitment to taking action on climate change; (2) add an economic component to climate change; (3) help engage and educate the public; and (4) may provide local social and environmental benefits that help to encourage the use of low-carbon technologies." In other words, it is a scam. See Steven Milloy, "Al Gore's Inconvenient Electric Bill," *Fox News*, March 12, 2007.

51. Bullfrogpower, "About us," *Bullfrogpower.com*.

blow or the sun to shine. Rather, Bullfrog provides additional premiums to wind and other contractors to encourage their production,
with the company taking an undisclosed share for administrative
costs and profit. It amounts to a feel-good scheme with little or no
impact on power generation or emissions reductions.[52]

Terrestrial sinks

Finally, environmentalists believe more can be done to increase natural sinks capable of absorbing atmospheric CO_2.[53] Much more carbon is contained in the oceans and embedded in the Earth's surface
than in the atmosphere. Some scientists believe that additional progress can be effected through more aggressive conservation management of forests, fields, lakes, water reservoirs, and soil, and that
more CO_2 can be drawn down from the atmosphere and stored naturally in biomatter and soils. The amount of carbon that can be
stored in the soil through terrestrial sequestration depends on vegetation types and other factors. It is of interest that over the past two
or three decades, as the atmosphere's CO_2 content has increased, nature has drawn down more of it on its own, leading to a greener
planet. Reducing deforestation, allowing more land to lie fallow,
growing plants with higher levels of carbon uptake and fixation in
the soil (e.g., trees), and seeding oceans to encourage more algae
blooms can all increase natural carbon sequestration. More sequestration of carbon in the oceans has some serious disadvantages for
the alarmist community because it purportedly leads to one of their
other nightmares, "acidification" of the oceans.

The carbon cycle is critical to understanding how the evolution
of the Earth system has, over the years, made the planet suitable for
life. Different parts of the carbon cycle move at different speeds.
Within the lithosphere, the cycle moves at the pace of millions of
years. Within the biosphere, it does so on an annual basis. Scientists'
understanding of the carbon cycle and its short- and long-term
budgets has increased significantly over the past few decades. Nevertheless, given the complexity of the Earth system, the metrics of

---

52. See "Where is my electricity coming from?" at media.cns-snc.ca/ontario elec-
    tricity/ontarioelectricity.html. The site regularly updates the capacity utilization
    of the various sources of energy. See the many articles in the *Financial Post* by re-
    tired banker Parker Gallant on the failings of Ontario's electricity system.

53. See Krister P. Anderson, Andrew J. Plantinga, and Kenneth R. Richards, "The
    National Inventory Approach for International Forest-carbon Sequestration
    Management," in Helm and Hepburn, eds., *The Economics and Politics of Climate
    Change*, 302-24.

that cycle remain largely a matter of educated guesswork. As Falkowski et al. conclude: "Our knowledge of the carbon cycle within the oceans, terrestrial ecosystems, and the atmosphere is sufficiently extensive to permit us to conclude that although natural processes can potentially slow the rate of increase in atmospheric CO_2, there is no natural 'saviour' waiting to assimilate all the anthropogenically produced CO_2 in the coming century. Our knowledge is insufficient to describe the interactions between the components of the Earth system and the relationship between the carbon cycle and other biogeochemical and climatological processes."[54] In the geological past, atmospheric CO_2 reached levels 20 to 30 times today's level, and the planet thrived. Under current climatic conditions, scientists can only guess at what point natural sinks may reach a saturation point.

Under the Kyoto Protocol, members were allowed to count practices as described above as part of their contribution. Reforestation and re-introducing forests on grass and crop lands were both eligible to be counted as certified emissions reductions for which countries could take credit. The UN's Clean Development Mechanism allowed developing countries to cash in on this technique.

"Green" energy sources

All five of the measures advocated by alarmists to reduce GHG emissions are predicated on reducing reliance on fossil fuels and replacing them with alternative sources of energy that are both clean and renewable, i.e., sources that do not emit greenhouse gases and do not deplete finite resources; wind, solar, biomass, hydro, and geothermal are the most popular of these resources. Nuclear power, while not renewable, operates largely free of GHG emissions but raises a host of other issues that trouble environmentalists. The story for each of the proposed alternative sources of energy is as uninspiring as the likelihood of successfully implementing GHG reduction strategies. As Glover and Economides conclude: "The stark reality is that current technology offers no realistic hope of seeing the current generation of alternative energy sources replacing hydrocarbons for decades to come, *if ever*" [emphasis in original].[55] Not that this would not be desirable. Burning hydrocarbons does have some drawbacks, but doing so remains the only realistic basis for

---

54. P. Falkowski, et al., "The Global Carbon Cycle: A Test of Our Knowledge of Earth as a System," *Science* 290 (October 13, 2000), 291-6.

55. Glover and Economides, *Energy and Climate Wars*, 11.

civilization and prosperity. Until such time as science and engineering can develop alternatives that are both cleaner and as cost-effective and efficient as hydrocarbons and that are capable of providing energy in the many applications relying on hydrocarbons, fossil fuels will remain critical to civilization.[56]

The UN IPCC, in a special report on the contribution of renewable energy to meeting desired mitigation goals, asserts that "historically, economic development has been strongly correlated with increasing energy use and growth of GHG emissions, and renewable energy can help decouple that correlation, contributing to sustainable development. ... Renewable energy offers the opportunity to contribute to social and economic development, energy access, secure energy supply, climate change mitigation, and the reduction of negative environmental and health impacts."[57] In its *Summary for Policy Makers,* the report offers a relentlessly optimistic assessment of the status of the technology and the economics of renewable energy. As with its work on the science of climate change, the IPCC relies on models and scenarios. For this report, it reviewed 164 scenarios from 16 large-scale integrated models to reach its assessment. It concludes: 1) that renewable energy has a large potential to mitigate GHG emissions; 2) that growth in renewable energy will be widespread around the world; 3) that there is no obvious single dominant technology at a global level; and 4) that the global overall technical potentials do not constrain the future contribution of renewable energy.[58] There are reasons to question this assessment.

Smil, for example, takes his readers through a number of earlier energy visionaries, from Amory Lovins to professors at the Harvard Business School, all of whom had confidently asserted that the United States could satisfy anywhere from a third to all of its energy needs by the year 2000 on the basis of decentralized, electrical generation by using wind, solar, and hydro. The year 2000 has come and gone and all their visions have proven as trustworthy as instant weight-loss programs. Even a decade later, the United States still relied on hydrocarbons for 82 percent of its primary energy consump-

56. For more detailed discussions of wind, solar, and biomass, see Van Kooten, *Climate Change, Climate Science and Economics,* 391-400, and Richard Green, "Climate Change Mitigation from Renewable Energy: Its Contribution and Cost," in Helm and Hepburn, eds., *The Economics and Politics of Climate Change,* 263-83.

57. UN IPCC, *Renewable Energy Sources and Climate Change Mitigation,* Special Report, 2012, 18.

58. UN IPCC, *Renewable Energy Sources and Climate Change Mitigation,* 20-26.

Figure 8-3: US primary energy consumption by source, 1949-2014
(Quadrillion BTUs)

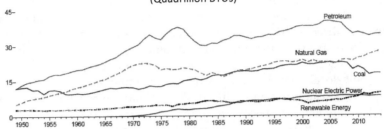

Source: US EIA, *Monthly Energy Review*, July 2015, Figure 1.3

tion. The share held by renewables had grown modestly to achieve 9 percent by 2011, more than three quarters of which was made up of hydro-electric generation and wood; wind, solar, and geothermal together contributed 1.2 percent of total primary energy (see figure 8-3). Globally the figures were not much better: 81 percent for hydrocarbons and 12 percent for renewables, with wind, solar and geothermal contributing less than one percent. And yet, the visionaries continue to assure us that a new energy nirvana can be achieved within a decade or two.[59]

Wind and solar

Wind and solar are the most-often cited alternative energy sources. To date, the delivery has been anaemic. British science writer Matt Ridley noted in March 2012: "To the nearest whole number, the percentage of the world's energy that comes from wind turbines today is: zero. Despite the regressive subsidy (pushing pensioners into fuel poverty while improving the wine cellars of grand estates), despite tearing rural communities apart, killing jobs, despoiling views, erecting pylons, felling forests, killing bats and eagles, causing industrial accidents, clogging motorways, polluting lakes in Inner Mongolia with the toxic and radioactive tailings from refining neodymium, a ton of which is in the average turbine – despite all this, the total energy generated each day by wind has yet to reach half a per cent worldwide. ... so how did the wind-farm scam fool so many policymakers? One answer is money. There were too many people with

---

59. Smil, *Energy Myths and Realities*, 44-54. Latest statistics can be gleaned from US EIA, *Monthly Energy Review*. See also Bryce, *Power Hungry*, and Glover and Economides, *Energy and Climate Wars*.

snouts in the trough."[60] Solar produces even less and comes with its own litany of negatives.

Voltaic cells and similar technologies that harness the energy of the sun directly may be effective on a small scale but cannot get the job done on a scale that makes a difference. Similarly, wind turbines can be used to capture kinetic energy in the atmosphere. Both can become local complements to more conventional energy sources, but large-scale solar and wind projects remain expensive and unreliable and require massive investments in transmission infrastructure to bring the electricity generated from areas with reliable sun or wind to major population centres. Both require huge subsidies to make the few large-scale projects now in place commercially feasible.[61]

Both wind and solar gobble up real estate, require complex land-use agreements, generate opposition from local and environmental interests, and are a blight on the country side. Windmills create noise pollution and have relatively short life spans before requiring refits; many older arrays stand as abandoned eyesores until some level of government takes on the task of dismantling them.[62] Ironically, from an environmental perspective, windmills kill wildlife at an alarming rate, particularly bats and migratory birds, and are certainly far more deadly than tailing ponds and other less desirable aspects of the mining industry featured prominently in modern ENGO campaigns.[63] Good locations for wind arrays usually

---

60. Matt Ridley, "The Winds of Change," *The Spectator*, March 3, 2012.

61. "The US Government Accountability Office (GAO) counted a whopping 641 programs in place at 130 federal agencies in 2010 to prop up windmill technology and underwrite solar panel manufacturers." "Global Warming Greed," Washington *Times*, March 14, 2012. Denmark and Germany, both of which have invested heavily in wind power, have learned that the unreliability of wind requires maintaining conventional back up and that there is no discernible impact on CO_2 reductions. Both wind and solar serve largely as symbols of good intentions rather than as reliable replacements.

62. See, for example, Tom Leonard, "Broken down and rusting, is this the future of Britain's 'wind rush'?" *Daily Mail*, March 19, 2012. The IPCC, in its *Special Report on Renewable Energies*, 96, envisages a future of windmills that rival the height of the CN Tower's observation deck (300 metres) with rotors 250 metres in diameter. The wingspan of the largest Boeing 747 is a mere 68.4 metres).

63. The Pembina Institute, for example, projects the loss of millions of birds as a result of oilsands development but dismisses those already being killed by windmills. See "Beneath the Surface: A review of key facts in the oilsands debate," January 28, 2013, 50, at *Pembina.org*. In a "Wind Factsheet," the Institute provides an enthusiastic endorsement of wind and dismisses bird kills as minor and less than the impact of tall buildings.

coincide with the main flyways of migrating raptors and other birds. The experience in trying to build a wind farm off Cape Cod, for example, has to date proven a demonstration project for the hypocrisy of champagne and brie environmentalists: alternative energy is good, but not in my backyard. Former Ontario Premier Dalton McGuinty's plan to bypass local planning authorities in order to facilitate placement of wind and solar facilities points to the problem that democratic decision-making poses for radical environmental solutions.[64] His successor was quick to remove this unpopular manifestation of regulatory overreach.

European governments were first out of the gate promoting wind and solar by offering huge subsidies and other incentives. All have found that performance has not even approached the original hype of ENGOs and industrial interests. From Denmark and Germany to Spain and Britain, government after government, after spending billions and subjecting consumers to ruinous rate hikes, have come to accept the folly of their original decisions and have quietly turned down the rhetoric, tried to turn off the spigots of government largesse, and licensed the building of more conventional generating facilities. *Investor's Business Daily* editorialized:

> The media aren't paying much attention, but in recent weeks Europe has decided to run, not walk, as fast as it can away from the economic menace of green energy. That's right, the same Europeans who used to chastise us for not signing the Kyoto climate change treaty, not passing a carbon tax and dooming the planet to catastrophic global warming. In Brussels last month, European leaders agreed to scrap per nation caps on carbon emissions. The EU countries – France, Germany, Italy and Spain – had promised a 40 percent reduction in emissions by 2030 (and 80 percent by 2050!). Now those caps won't apply to individual nations. Brussels calls this new policy 'flexibility.' Right. More like 'never mind,' and here's why: The new German economic minister, Sigmar Gabriel, says green energy mandates have become such an albatross around the neck of industry that they could lead to a 'deindustrialization' of Germany.[65]

---

64. For a detailed discussion of the folly of Ontario's wind policy, see Van Kooten, *Climate Change, Climate Science and Economics*, 375-81; Michael Trebilcock, "Speaking truth to 'wind' power," Submission to Ontario Legislative Committee Examining Bill C-150, CD Howe Institute, April 15, 2009.

65. "Europe Starts To Run, Not Walk, Away From Green Economics," *Investor's Business Daily*, February 5, 2014. See also Christopher Booker, "Ten years too late, it's good riddance to wind farms – one of the most dangerous delusions of our age," *Daily Mail*, October 31, 2012. By then, Britain had built 3,500 turbines and had hundreds more under construction before the government pulled the

The fundamental problem faced by both wind and solar is that, without development of massive and cost-effective storage capacity, they cannot provide a reliable base load. Neither wind nor solar is available on a 24/7 basis; both rely on the whims of nature. Few wind or solar arrays can deliver more than 25 percent of their rated or "installed" capacity over the course of a day, week, or month, and often deliver less. The farther a solar array is placed from the tropical zone, the less energy can be captured from the sun. Even in tropical regions, a daily average of more than 8 hours of direct sunlight to fuel an array is rare. Deserts are the best locations but are typically far from population centres. Wind varies even more erratically, with both high winds and light winds presenting problems. As a result, both solar and wind sources require a reliable base load generator of electricity to ensure a steady supply to a grid. Nuclear is ideally suited as a base-load generator as is hydro. Coal and gas can both perform this function; additionally, coal, gas, and hydro can be ramped up and down as dictated by demand.

Adding wind and solar to the grid enormously complicates the operator's ability to maintain a reliable supply. Many existing grids are already experiencing stress from having to deal with feed-ins of a growing mix of non-base load generation. Germany, which invested heavily in both wind and solar, can experience peaks of 12,000-14,000 megawatts of power from either on a sunny or windy day, and zero the next day. Because operators are required to buy any available solar or wind at premium prices, they may end up exporting solar and wind at a fraction of the acquisition cost to neighbouring countries.[66] In practice, this means that operators must ramp down low-cost sources – such as gas, coal, or hydro – and purchase high-cost wind or solar. The good news is that most wind or solar arrays have a useful life of only about 20 years, by which time they require either extensive refits or become very expensive scrap. Smil observes that "turning around the world's fossil-fuel-based energy system is a truly gargantuan task. ... Re-engineering and rebuilding the world's extensive energy infrastructure to ac-

---

plug. While the UK government ended subsidies to build new windmills, it continues to subsidize existing windmills, paying twice the market rate for land-based wind power and three times that rate for sea-based wind. Rowena Mason, "Wind farms get generous subsidies for another six years," *The Telegraph*, June 27, 2013.

66. Willem Post, "Wind Energy CO2 Emissions Reductions are Overstated," *The Energy Collective*, July 1, 2012.

commodate alternative energy – assuming such energy becomes economically and technologically feasible – is not a project of a decade or two – or five, for that matter. Replacing it with an equally extensive and reliable alternative based on renewable energy flows is a task that will require decades of expensive commitment. It is the work of generations of engineers."[67]

It is difficult to gain a realistic sense of the costs associated with either solar or wind projects. Industry associations such as the Canadian, American or European Wind Energy Associations or the Solar Industries Energy Association claim that they are already competitive with coal-fired and other facilities and that prices will decline further as more capacity is installed. At the same time, they acknowledge the critical role played by public policy incentives. Meanwhile, newspaper financial pages regularly report bankruptcies in both industries despite billions in subsidies.[68] The cost of photovoltaic cells and wind turbines may be declining, but the many other components of a functioning installation are not.[69] The US EIA 2013 outlook for 2018 provides a sobering, if optimistic, analysis of projected costs. The costs of producing solar energy are projected to be from 2-4 times those of modern natural gas plants, onshore wind 40 percent more, and offshore wind quadruple their costs. Taking account of their capacity versus their output, the differences in delivered costs are much higher (see table 8-1).

Smil notes that "the cost of electricity generated by residential solar systems in the United States has not changed dramatically since 2000. At that time the national mean was close to 40 US cents per kilowatt-hour, while the latest Solarbuzz data for 2012 show 28.91 cents per kilowatt-hour in sunny climates and 63.60 cents per

---

67.   "A Skeptic Looks at Alternative Energy," *IEEE Spectrum*, July 2012. For a comprehensive discussion of the economic and technical challenges of integrating wind energy into the grid, see Cornelis van Kooten, *Climate Change, Climate Science, and Economics*, 409-39.

68.   Trade in the RENIXX Index of the 30 largest renewable energy companies in the world bottomed out under 200 on May 16, 2012, having lost over 90 per cent of the Index's value since its start in 2008. Seventeen leading green-energy companies had filed for bankruptcy by that time; another 12 were teetering on the brink. Many of these companies were financed by taxpayers. "The worldwide crash of green energy companies," *The Hockeyschtick*, May 16, 2012. More than two years later, on August 21, 2014, more companies had gone bankrupt; the Index was now trading at 423.7 with a revamped cast of companies.

69.   Gordon Hughes, *Why is Wind Power So Expensive? An Economic Analysis*, Global Warming Policy Foundation, Report 7, 2012.

Table 8-1 – Projected comparative costs of new generating capacity in 2018

Plant type	Capacity factor (%)	Levelized capital cost	Fixed O&M	Variable O&M (including fuel)	Transmission investment	Total system levelized cost
US average leveled costs (2011 $/megawatthour) for plants entering service in 2018						
Dispatchable Technologies						
Conventional coal	85	65.7	4.1	29.2	1.2	100.1
Advanced Coal	85	84.1	6.8	30.7	1.2	123.0
Advanced Coal with CCS	85	88.4	8.8	37.2	1.2	135.5
Natural gas-fired						
Conventional Combined Cycle (CC)	87	15.8	1.7	48.4	1.2	135.5
Advanced Combined Cycle	87	17.4	2.0	45.0	1.2	67.1
Advanced CC with CCS	87	34.0	4.1	54.1	1.2	93.4
Conventional Combustion Turbine	30	44.2	2.7	80.0	3.4	130.3
Advanced Combustion Turbine	30	30.4	2.6	68.2	3.4	104.6
Advanced Nuclear	90	83.4	11.6	12.3	1.1	108.4
Geothermal	92	76.2	12.0	0.0	1.4	89.6
Biomass	83	53.2	14.3	42.3	1.2	111.0
Non-Dispatachable Technologies						
Wind	34	70.3	13.1	0.0	3.2	111.0
Wind– Offshore	37	193.4	22.4	0.0	3.2	86.6
Solar PV[1]	25	130.4	9.9	0.0	4.0	144.3
Solar Thermal	20	214.2	41.4	0.0	5.9	261.5
Hydro[2]	52	78.1	4.1	6.1	2.0	90.3

1   Costs are expressed in terms of AC power available to the grid for the installed capacity.
2   As modelled, hydro is assumed to have seasonal storage so that it can be dispatched within a season, but overall operation is limited by resources available by site and season.

Note: These results do not include targeted tax credits such as the production or investment tax credit available for some technologies, which could significantly affect the levelized cost estimate. For example, new solar thermal and PV plants are eligible to receive a 30-percent investment tax credit on capital expenditures if placed in service before the end of 2016, and 10 percent thereafter. New wind, geothermal, biomass, hydroelectric, and landfill gas plants eligible to receive either: (1) a $22 per MWh subsidy for technologies other than wind, geothermal and closed loop biomass) inflation-adjusted production tax credit over the plant's first ten years of service or (2) a 30-percent investment tax credit, if placed in service before the end of 2013 (or 2012, for wind only).

Source: US Energy Information Administration, Annual Energy Outlook 2013, December 2012, DOE/EIA-0383 (2012).

kilowatt-hour in cloudy ones. That's still far more expensive than using fossil fuels, which in the United States cost between 11 and 12 cents per kilowatt-hour in 2011."[70] To add insult to injury, most analyses of wind and solar arrays, taking into account materials, land area, transportation, distribution, and other factors, conclude that wind arrays generate more CO_2 over their lifespans than a modern, efficient gas plant.[71]

It took 30 years and billions of dollars to move from almost no wind-based electric energy to less than 3 percent of current global electricity consumption (US from 0.09 percent in 1990 to 3.5 percent in 2013, and Canada from a negligible amount to 2.12 percent in 2012).[72] Expecting that percentage to rise to 10 or even 20 percent in a decade is highly unrealistic. Smil points out: "Perhaps the most misunderstood aspect of energy transitions is their speed. Substituting one form of energy for another takes a long time. US nuclear generation began to deliver 10 percent of all electricity after 23 years of operation, and it took 38 years to reach a 20 percent share, which occurred in 1995. It has stayed around that mark ever since. Electricity generation by natural gas turbines took 45 years to reach 20 percent."[73] Solar numbers are even less impressive. The arrival of cheap natural gas as a result of the fracking revolution further undercuts the prospects of rapid additional wind and solar development. Europe is abandoning its push for wind and solar as the full costs have become more apparent.

Over time, both the technology and economics of solar and wind power are likely to improve. Neither technology nor economic efficiencies, however, can solve the fundamental limitations inherent in both: their intermittent and thus unreliable nature. Additionally, neither solar nor wind energy is sufficiently concentrated to match the energy intensity of nuclear and fossil fuels. As a result,

---

70. Vaclav Smil, "A Skeptic Looks at Alternative Energy," *IEEE Spectrum*, July 2012.

71. See, for example, Hughes, "Why Is Wind Power So Expensive?" 28-34. German environmental economist Joachim Weimann calculated that in order to save one ton of CO_2, Germany could either spend €5 on insulating an old building, €20 on investing in a new gas-fired power plant, or around €500 on photovoltaic arrays. The benefit to the climate was the same in all those scenarios. Alexander Neubacher and Catalina Schröder, "Germans Cough Up for Solar Subsidies," *Spiegel Online*, July 4, 2012.

72. U.S. Energy Information Administration, *Annual Energy Outlook 2014* and Natural Resources Canada, *Energy Markets Fact Book 2014-15*, 68.

73. Smil, "A Skeptic Looks at Alternative Energy."

both are likely to continue to remain expensive niche players, capable of delivering useful energy in limited circumstances.

Biofuels

The great biofuel scam of 2008 proved, once again, that there is no free lunch.[74] Heavily promoted by the nascent biofuel industry and by farmers eager to sell their corn at inflated or subsidized prices, biofuel turned out to be less than a saviour. The allure of selling corn and other biomass at premium or subsidized prices led to a major switch from growing food to growing fuel, leading to higher food prices, hardly the desired result, particularly in poorer countries. Carl Brehmer calculates that producing one gallon of ethanol requires about 22 pounds of corn containing about 10,560 calories, enough to feed one person for about four days. Producing 90 gallons of ethanol uses enough corn to sustain one person for an entire year. The US is currently producing 10.6 billion gallons of ethanol yearly, converting enough corn to feed 117 million people. According to the US Department of Agriculture, over 50 million people in the United States are in 'food-insecure households' because their families do not have sufficient funds to purchase adequate amounts of food.[75] While technologies may improve, at this point growing corn or similar crops and then manufacturing and distributing biofuels made from corn and other high-cellulose plants take more conventional energy than the energy content of the resulting fuel.[76] The overall carbon impact is larger, even if the CO_2 emitted *directly* by burning biofuels is lower.

Biofuels are produced from a range of plant materials and can replace either gasoline or diesel in transportation and other uses.

---

74. On the problems with biofuels in general, see Bryce, *Power Hungry*, 179-188. In 2008, demand for corn to produce ethanol to meet government mandates drove up the price for corn, an increase that was reflected throughout the food system given corn's central role in the food chain, from animal feed to starch, sweeteners, and edible oils.

75. "Why Do We Burn Our Food?" *Principia International*, September 7, 2012.

76. Cornell University researchers David Pimentel and Tad Patzek found that "ethanol production using corn grain required 29 per cent more fossil energy than the ethanol fuel produced. Ethanol production using switchgrass required 50 per cent more fossil energy than the ethanol fuel produced. Ethanol production using wood biomass required 57 per cent more fossil energy than the ethanol fuel produced. Biodiesel production using soybean required 27 per cent more fossil energy than the biodiesel fuel produced." "Ethanol Production Using Corn, Switch-grass, and Wood; Biodiesel Production Using Soybean and Sunflower," *Natural Resources Research* 14:1 (March 2005), Abstract.

The two principal biofuels are alcohols (ethanol, propanol, and butanol), and biodiesel. Ethanol is produced through a process of fermentation of crops such as corn, wheat, sugar cane, and sugar beet, and can be blended with gasoline. Biodiesel feedstocks include edible oils such as rapeseed, palm, and soybean oil but also waste products such as used cooking oil. These waste oils can make biodiesel a more attractive option since it can be produced without competing with food-grade feedstocks. Other second-generation pathways exist for ethanol and biodiesel based on alternative feedstocks; these are not yet commercially viable and are not expected to make a significant contribution to production in the near to medium term.

Table 8-2: Energy intensity of liquid fuels
(Megajoule/litre)

Regular gasoline	34.8	Diesel	38.6
Ethanol	23.5	Vegetable Oil	34.3
Gasohol (10% ethanol)	33.7	Biodiesel	35.1
E85 (85% ethanol)	33.1		

Despite the appeal of biofuels as a "natural" and renewable substitute for fossil fuels, none of them can compete with fossil fuels in energy intensity In practical terms, biofuels used as motor fuel require about 30 percent more fuel per mile or kilometer than conventional fuels (See table 8-2). As a means of reducing the carbon intensity of transport fuel, biofuels have proven to be a disappointment. Calculations based on EU experience indicate that reducing one tonne of CO_{2e} ranges from a low of US$165 for tallow-based biodiesel to US$400 for corn-based ethanol and US$1,100 for wheat-based ethanol.[77] Much more cost-effective results can be achieved by focusing on improving engine efficiencies.[78]

---

77. Rob Bailey, *The Trouble with Biofuels: Costs and Consequences of Expanding Biofuel Use in the United Kingdom*, Chatham House Papers, April 1, 2013, 15.

78. Henry Miller and Colin Carter, "Running on Empty," *Hoover Digest* no. 1, January 16, 2008. Further, as Bailey points out, "Biofuel policies create upward pressure on long-run food prices by increasing demand for agricultural commodities such as sugar, wheat, corn and edible oils. Biofuels produced from non-food agricultural feed-stocks can equally lead to higher food prices where they compete with food production for land, water or other inputs." *The Trouble with Biofuels*, 9. Miller and Carter point out that "an analysis by the Paris-based Organization for Economic Cooperation and Development suggested that replacing even 10 per cent of America's motor fuel with biofuels would require that about

Despite these limitations, ENGOs were originally attracted to biofuels because they appeared to offer greenhouse gas savings when compared to conventional petroleum products. As the feed-stock crops used in making biofuels grow, they remove carbon from the atmosphere through photosynthesis; when burned, these biofuels release the same carbon back into the atmosphere, suggesting a net carbon footprint of zero. In reality, a large number of other sources of emissions must also be taken into account – from chemical inputs and fertilizers to the energy needed to run farm machinery and refineries. Additionally, emissions from land-use change such as deforestation or drainage of peatland need to be taken into account for a full appreciation of their carbon footprint. Producing ethanol from biomass requires from 3-6 gallons of water per gallon; corn and similar crops are grown in areas that are water deficient, relying on aquifers and other finite sources. Locating ethanol plants closer to abundant water supplies would add to the cost – and CO_2 production – of transporting the biomass. From almost any perspective, current technology makes the use of biofuels as a substitute for fossil fuels a lose-lose situation.

Most OECD governments have found the appeal of biofuel mandates irresistible, satisfying not only ENGOs but also manufacturing and agricultural interests. Canadian federal and provincial governments provide up to US$ .40 per gallon in subsidies for biofuels. US subsidies include, in addition to various agricultural support programs, a US$ .51 per gallon tax credit for manufacturers as well as a US$ .54 a gallon import tariff. The EU provides as much as US$ 1.00 per gallon. US production of ethanol exceeds domestic demand and is exported to other countries, such as the UK, to meet their biofuel mandates. The EU reclassified E90 gasoline (70-90 percent ethanol) in 2012 to make it eligible for lower tariff treatment. Conventional engines can use fuel with up to 10 percent ethanol. Beyond that, they need significant modification to avoid high-cost damage.

In both Canada and the United States, most of the gasoline sold is a mixture of up to 10 percent ethanol and at least 90 percent petrol. The US Congress has increased the mandate, requiring that 36 billion gallons of ethanol be mixed with petrol by 2022, with the mixing ratio increased to 15/85. Estimates indicate that producing

a third of all the nation's cropland be devoted to oilseeds, cereals, and sugar crops. Achieving the 15 per cent goal would require the entire current US corn crop, a whopping 40 per cent of the world's corn supply."

this amount will require converting almost all of the US's 300 million acres of cropland to ethanol, while replacing just 15 percent of US oil demand with biofuels, 15 billion from corn, and 22 million gallons from other, yet-to-be developed sources.[79] In Canada, federal and provincial mandates vary from 5 to 10 percent of consumption. The EU is pushing to replace 10 percent of its reliance on conventional motor fuel with biofuels by 2020. Unlike North America, the EU relies much more on biodiesel than on alcohols.

Despite the mandates, biofuels are not proving economically feasible. Most OECD countries provide generous subsidies to both farmers and producers to encourage conversion to biofuels but, with few exceptions, ethanol producers in OECD countries are losing money and going bankrupt. Miller and Carter reveal that "major sugarcane-producing nations such as Brazil enjoy significant advantages over temperate-zone countries in producing ethanol, including ample agricultural land, warm climates amenable to vast sugarcane plantations, and on-site distilleries that can process cane immediately after harvest. At current world prices for sugar and corn, Brazilian ethanol production would remain competitive even if oil prices were to drop below $30 per barrel."[80]

Biomass can also be used to generate electricity or create steam to heat homes.[81] Waste wood products, for example, can be pelletized and used in much the same way as coal, often with the same generating facility. In order to comply with the Ontario government's determination to phase out coal-burning generating facilities, the Ontario Power Authority turned to wood pellets as an alternative. Municipal waste, much of which is biomass, can also be used to generate electricity. Sweden has long been a pioneer in converting municipal waste into energy. Modern mass-burn, gasification, and other processes reduce municipal waste to minimal residue while efficiently creating steam to drive turbines to generate

---

79. Max Schultz, "The Ethanol Bubble Pops in Iowa," *Wall Street Journal*, April 18, 2009.

80. Miller and Carter, "Running on Empty."

81. For an overview, see US EIA, "Biomass for Electricity Generation." Canada relies on biomass for nearly 5 per cent of its primary power, five times that of wind and more than ten times that of solar. Black liquor, the liquid residue from the paper-making process, is an important source of energy from biomass. Most pulp and paper plants now recover and process their residues into concentrates that can be used to supply the electrical energy needed to run the plant. Newer processes are able to produce enough surplus energy to feed it into the electric power grid for other users.

electricity. The residue itself has various applications, including as fill for road building.[82]

Unfortunately, if the objective is to reduce CO_2 emissions, this use of biomass will be disappointing. Modern fossil-fuel facilities are both more efficient and produce less CO_2. Wood fibre has many alternative applications that seem to make a better use of the Earth's forest resources. Diverting wood for use as fuel will inevitably drive up fibre prices. Burning current forests, one of the largest natural sinks for CO_2, as an alternative to fossil fuels, borders on the idiotic. Using biomass as a source of fuel to generate electricity only makes sense in areas where there is a plentiful supply of waste that needs to be disposed of in some manner or other.[83] The ability of politicians to pass off these kinds of projects as part of the solution to climate change, even to the applause of some ENGOs, speaks only to society's high levels of scientific illiteracy, including among leaders of the green movement.

Hydro-electricity

Greater reliance on water (e.g., by building more hydro-electric dams on rivers or capturing the energy potential of the ebb and flood of tides) and thermal-generated electricity are superficially very attractive. While capital costs are high, operating costs are relatively low. Hydro plants provide a good way to generate base-load electricity but are less flexible than fossil fuel-fired plants in providing electricity for periods of peak demand. As a result, hydro usual-

---

82. See Heather Hager, "Biomass Power," *Canadian Biomass Magazine,* March 2009; "Energy Update: Sweden's Waste-to-Energy Problem – Not Enough Garbage," September 30, 2012, at 21st Century Tech: A Look at Our Future. The city of Ottawa had a contract with Plasco Energy from 2005-15 to build a modern gasification plant to turn municipal waste into electricity. Plasco projected that for every tonne of municipal solid waste (with an average calorific value of 14,200 megajoules) processed with the Plasco Conversion System, the following would be produced: 1 kilowatt-hour of net electricity; 300 litres of water for reuse; 7 to 15 kilograms of metal (both ferrous and light metals like aluminium; 150 kilograms of slag (which can be used in place of quarried aggregates, such as sand, to mix with portland cement to make concrete or as aggregate in asphalt). Treena Hein, "Waste Not Want Not at Plasco," *Canadian Biomass Magazine,* May-June 2013. The contract was terminated in 2015 as Plasco sought protection from its creditors.

83. K. Holmgren, "Waste incineration in Swedish municipal energy systems," in V. Popov et al., eds., *Waste Management and the Environment* (WIT Press elibrary, 2012), and Liat Clark, "Sweden to import 800,000 tonnes of trash to burn for energy," *Wired,* October 29, 2012.

ly needs to be supplemented by more flexible gas, oil, or coal-fired generating capacity to deal with peak loads and other issues.

Hydro facilities tap the kinetic energy in flowing water to turn turbines that generate electricity. They can be built on various scales, from small, local facilities on small streams capable of generating a few mega-watts to major projects such as the Three Gorges Dam on the Yangtse River in China. Its 32 turbines can each generate 700 megawatts, each the equivalent of a medium-sized nuclear or thermal plant. The Hoover Dam in the United States has 19 turbines capable of generating a total of 2,080 MW. The province of Quebec, which relies almost exclusively on hydro-electric power, draws power from 60 hydro-generating stations with an installed capacity of 35,829 MW, which require 26 large reservoirs with combined storage capacity of 175 TWh plus 664 dams and 97 control structures. The average wholesale cost of a kilowatt hour was 2.09 cents at the end of 2012, among the lowest in the world.[84] The faster and larger the available flow of water, the more electricity can be generated. To ensure year-round generating capacity, most rivers need to be dammed to create a reservoir capable of providing a constant supply of water. As a result, many dam projects involve flooding hundreds of square kilometres of land, displacing whole towns and villages as well as local flora and fauna, and raising public concerns. The failure of a major hydro project can constitute a significant risk in populated areas. Remote areas far from population centres typically offer the best locations.

The kinetic energy in tides can similarly be used to drive turbines. Tidal energy has been used on a small scale as a source of power for centuries, but, because most tides are only a few feet high, there is often not enough energy for a major facility. Higher tides present formidable design and construction challenges. A major drawback of tidal energy is its low capacity factor and its inability to satisfy peak demand times because of the twice-daily cycle of the tides. Current tidal plants generate from 254 MW (Rance Tidal Power Station in France) to 20 MW (Annapolis Royal Generating Station in Nova Scotia). New turbine and construction technologies, however, have increased the potential for harnessing tidal power. Reaping that potential will require very high capital costs, e.g., for constructing a tidal barrage/bridge across an estuary. A number of these are on the drawing boards and are capable of generating as

---

84. HydroQuébec, "Hydropower Development: A Sustainable Solution for Present and Future Needs," at *Hydroforthefuture.com*.

much as 8 GW (Severn Estuary Barrage in the UK). Capital costs, however, are estimated to be in the billions, and maintenance costs are also expected to be high as are environmental impacts. To date, no large functioning tidal plant has been built.

The heat in thermal springs can also be used to drive turbines to generate electricity. Although geothermal plants are relatively inefficient because of the low heat content of thermal springs, they have been developed in various countries around the world. The US, for example, has a total installed capacity of 3,000 MW, but there is potential for much more.[85] Geothermal, however, is more suited to providing heat directly than to generating electricity, and that is how most installed capacity is used. Altogether, though, geothermal constitutes only a tiny slice of total global energy production. While there are no fuel costs, capital costs are high, placing geothermal on the expensive side of the energy spectrum.

Unfortunately, the number of suitable sites for building dams, constructing tidal arrays, or tapping thermal springs is limited. Many are often in remote locations and require major investments in transmission facilities. Long-distance transmission requires more powerful transmission lines in order to reduce loss of energy. And, similar to nuclear plants, hydro and thermal sites are controversial and attract widespread opposition from environmental groups. While hydro, thermal, and nuclear sources will undoubtedly form part of the future in generating electricity, the extent to which they can replace existing fossil-fuel-based plants in the foreseeable future is questionable. Smil concludes: "competing water uses, the unsuitability of many sites, seasonal fluctuations of flow, and the impossibility of converting water's kinetic energy with perfect efficiency – mean that the exploitable capacity will be a small fraction of the theoretical availability. ... [Further], not everything that is technically feasible is economically [or politically] acceptable."[86]

Nuclear

The most obvious and proven form of alternative energy is nuclear.[87] While generally more expensive than fossil fuels for generating

---

85. See International Geothermal Association, "Geothermal: A Natural Choice," at *Gtherm.net*.

86. Smil, *Energy Myths and Realities*, 122-3.

87. For more detailed discussion of nuclear, see Dieter Helm, "Nuclear Power: Climate Change and Energy Policy," in Helm and Hepburn, eds., *The Economics and Politics of Climate Change*, 263-83. Most reactors follow the original US design

electricity, nuclear energy has a fifty-year track record as a reliable, base-load generator, supplying 5.8 percent of primary energy globally from 436 power plants, most of them supplying from one to five giga-watts per plant, larger than most hydrocarbon facilities and much larger than wind and solar arrays, using a much smaller land area. Most plants now run at 90 percent or better capacity, making them ideal base-load generators. Coal-fired plants typically achieve 65 to 75 percent of capacity, hydro from 40 to 60 percent, and wind at best 25 percent.[88] This characteristic requires that electricity generated by nuclear plants be supplemented on the grid with fossil-fuel generating capacity that can be ramped up and down to handle peak loads.

Nuclear's high costs flow from the many safeguards that need to be built into plants, the cost of managing and storing nuclear waste, and the cost of satisfying multiple regulatory requirements. Many of these costs flow from the first two as well as from the opposition of environmental activists. As Smil points out, "as of January 1, 1971, the United States had some hundred codes and standards applicable to nuclear plant design and construction; by 1975, the number had surpassed 1,600; and by 1978, 1.3 new regulatory or statutory requirements, on average, were being imposed on the nuclear industry every working day."[89] Canadian and European regulations are equally detailed. Not surprisingly, few new plants were built after 1980 in North America. Most new plants since then have been built in developing countries. While the relative safety of nuclear energy, particularly newer plants, is well established, building new plants to replace current fossil-based electrical generating facilities continues to face high costs and major hurdles in overcoming local and environmentalist opposition.

Canada commissioned 23 commercial nuclear reactors between 1971 and 1992, all of the Candu type first developed at Chalk River, Ontario, by Atomic Energy of Canada. Three have been decommissioned, and seven older plants have been or are being refurbished to extend their operating lives. Most are or will be licensed to operate well into the 2020s and 2030s. Together, they provide 15 percent of Canada's electrical generating capacity. As in the United States,

---

developed to power nuclear submarines, using enriched uranium and water as a coolant.

88. Smil, *Energy Myths and Realities,* 40. See also table 8.1 above.

89. Smil, *Energy Myths and Realities,* 36.

costs arising from safety and other regulatory requirements have slowed Canada's nuclear program. Only one project – expansion of the Darlington facilities in Ontario with two additional reactors – is being considered.[90]

In the half century since the first generator went into operation in 1957, there have been three major incidents worldwide that have made the public wary of nuclear power: at Three Mile Island in Pennsylvania in 1979, at Chernobyl in Ukraine in 1986, and at Fukushima in Japan in 2011. While total fatalities were far fewer than those from other modern technologies, e.g., aviation, the spectre of long-term harm has made these incidents into powerful negative forces for the nuclear industry. In the Three Mile Island incident, safety features ensured that the partial meltdown of the core was wholly contained and resulted in no fatalities; nevertheless, the accident became the symbol of what could go wrong and has cast a long shadow. The stricken reactor was decommissioned, but the second reactor at the site is licensed to keep operating until at least 2034.[91] The Chernobyl incident proved that a poorly designed, built, and operated facility is vulnerable to a major accident; in retrospect, however, it also demonstrated that there are fewer long-term effects from modest radiation exposure than feared. An area covering a 30-km radius around the plant remains off-limits to humans but has become a large nature preserve in which both flora and fauna are thriving and showing no long-term negative effects.[92] Many of the ill effects attributed to the accident were more psychological (the fear factor) than physical. The event at Fukushima, the result of a 9.0 earthquake and a subsequent tsunami rather than a problem with the plant's design or operation, illustrated that location and post-

---

90.   A further dozen Candu reactors have been exported to South Korea (4), Romania (2), India (2), Pakistan (1), Argentina (1) and China (2), while India has built 13 Candu-derivative reactors. Candu reactors use heavy water (deuterium oxide) as a moderator and coolant, and are fuelled using natural uranium (as opposed to enriched uranium). Operating costs are lower because uranium does not have to be enriched, and reactor downtime is reduced due to shorter time needed for refuelling and maintenance. These savings are partially offset by the cost of producing deuterium oxide.

91.   See J. Samuel Walker, *Three Mile Island: A Nuclear Crisis in Historical Perspective* (Berkeley, CA: University of California Press, 2004) for a balanced assessment of the accident, its repercussions, and its longer-term impact on the nuclear industry in the United States.

92.   Eben Harrell and James Marson, "Apocalypse Today: Visiting Chernobyl, 25 Years Later," *Time,* April 26, 2011.

disaster responses need careful consideration. More people died from the Fukushima evacuation rather than from the effects of radiation.[93] Japan has now decided to refurbish the plant and put it back into operation.

Since few reactors have been built since 1980, most of the current generation of commercial reactors are more than thirty-five years old and have perhaps another 20 to 30 years of useful life. In the intervening years, there have been significant advances in nuclear technology. As the World Nuclear Association points out: "An international task force is developing six nuclear reactor technologies for deployment between 2020 and 2030. Four are fast neutron reactors. All of these operate at higher temperatures than today's reactors. In particular, four are designated for hydrogen production. All six systems represent advances in sustainability, safety, economics, reliability and proliferation-resistance."[94] Scientists and engineers are also working on thorium reactors, which use a more abundant fuel source and raise fewer safety and proliferation concerns.[95] Regardless of climate change and GHG emissions concerns, governments will need to address how best to replace aging facilities. Climate change considerations, however, will be a further, major factor in the debate over the future role of nuclear.

Most experts see nuclear reactors as critical suppliers of base-load energy in sufficient quantities to replace hydrocarbons. Opposition to nuclear energy, however, continues to be strong, not only among environmentalists but also more generally due to safety and cost concerns. Both governments and the nuclear industry will face an uphill battle in convincing their populations that nuclear is critical to maintaining current levels of energy availability, let alone expanding that availability in developing countries. Given the difficulties experienced with renewables such as wind and solar, environmental activists such as Patrick Moore, one of the founders of Greenpeace, and James Lovelock, inventor of the Gaia hypothesis, have rethought their original opposition and parted with many of

---

93. Lawrence Solomon, "Evacuation a Worse Killer than Radiation," *Financial Post,* September 22, 2012. See also "Special Report: The Fallout from Fukushima," *New Scientist,* March 9, 2012 for a series of articles examining the aftermath of the accident.

94. "Generation IV Nuclear Reactors," at *WorldNuclear.org.*

95. See "Thorium," at *WorldNuclear.org.* India has already built an operating thorium reactor and is home to the largest known reserves of thorium.

their colleagues by insisting that there is no solution to GHG-induced climate change that does not include nuclear energy.[96]

Electric cars

Trucks, trains, buses, and automobiles are the mainstays of modern surface transportation systems. Attempts to wean these means of transportation off hydrocarbons and replace them with electricity provide a further example of silo thinking. Some progress has been made, but at significant costs. Regulatory requirements and con-sumer preferences have steadily improved fuel performance and engine efficiency.[97] Hybrid and electric cars are making inroads but remain a tiny part of the total. The cost premiums are still signifi-cant and explain why hybrid sales remain modest. Electric cars still use energy that must be generated somewhere and distributed to consumers. A wholesale change to electric cars would require major infrastructure investments in a distribution network as well as an expansion of electrical generating capacity from coal, gas, hydro, or nuclear. Electrical generating capacity in North America is already showing signs of strain due to growing demand; strategies to re-duce consumption are not working.

Over time, with the development of appropriate, widespread infrastructure and better battery technologies, it may be possible to replace liquid fuel engines in light-duty vehicles, but diesel and jet fuel will be needed to power heavy-duty vehicles, trains, boats, and airplanes as well as heavy-duty equipment for mining, construction, and similar applications. It is difficult to envisage electric motors capable of generating that kind of power on a sustained basis from energy stored in batteries. The great benefit of hydrocarbons is that they provide a very concentrated, portable form of energy rivalled only by nuclear energy. The energy density of gasoline assures 80kw hours per kilogram, compared to one kw for the latest lithi-um-ion battery.[98]

---

96. See George Mack, "From Greenpeace Dove to Nuclear Power Phoenix," *The En-ergy Report,* September 29, 2011, and James Lovelock, "Nuclear Energy for the Twenty-first Century," Speech to the International Conference in Paris, 21-22 March 2005, at jameslovelock.org/page12. html.

97. See Michael Hart, "Potholes and Paperwork: Improving Cross-Border Integra-tion and Regulation of the Automotive Industry," *Commentary* No. 286, C.D. Howe Institute, April, 2009.

98. Bryce, *Power Hungry,* 190-91.

A successful all-electric car will require smaller and lighter vehicles; increased battery capacity; faster recharging; reduced battery weight; extended battery life as well as enhanced capacity to recharge over time; and a dense generating and distribution network. Reliance on wind and solar ignores the fact that most electric cars will be recharged at home overnight. As consumers have made clear, electric cars will need to demonstrate that they are able to compete in both cost and performance against cars equipped with steadily improving internal combustion engines as well as with partially self-charging hybrid cars. The most optimistic scenarios suggest that it will take at least two or three fleet turnover cycles (average car life is now nine years) before hybrid cars occupy more than a niche position; the prospect for all-electric cars is much cloudier.

As with the early investors in solar, wind, and biofuels, the electric car industry has not fared well. Tesla has succeeded in carving out a high-end niche market for its luxury all electrics, but the more competitively priced Nissan Leaf is still losing money. Its high price means consumers only break even after 8-9 years. Fisker has gone out of business. The surge in investment in battery and other related industries has been followed by bankruptcies: recent startups A123 Systems, Ener 1, Coda, Better Place, and Think Global all benefitted from government soft loans and subsidies; by 2013 they were all out of business.

Finally, environmentalists like to claim that reduced CO_2 emissions can be accomplished through more efficient use of existing energy sources. There is some truth to this claim. Indeed, energy use per unit of economic output has steadily decreased over the years as greater efficiencies have been achieved. Ironically, the same innovation and ingenuity producing energy efficiencies per unit of output also lead to new energy-consuming devices. As well, the population continues to increase and prosperity to spread. The net result is that humans use energy much more efficiently than 60 years ago but use more of it. Even then, as Figure 8-4 illustrates, US CO_2 emissions – as a proxy for energy use – have only doubled over the past 60 years, even as economic output has increased eight-fold. Under normal conditions, this would be considered a human achievement. For environmentalists, however, this is now considered a problem requiring government intervention.

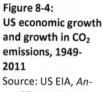

Figure 8-4:
US economic growth
and growth in CO₂
emissions, 1949-
2011
Source: US EIA, *Annual Energy Review 2012*, Figure 11.1.

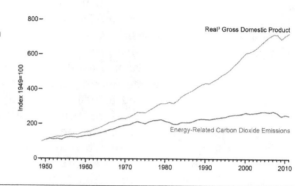

The stark choices imposed by energy reality

No matter how much alarmists and their supporters huff and puff about the dangers of fossil fossils and the impending catastrophe of run-away global warming, a large segment of the population will continue to resist their policy preferences unless it becomes clear that there are viable and cost-effective alternatives. The market behaviour of ordinary citizens will continue for the foreseeable future to resist the reordering of the economy on ideological grounds. Hybrid cars, for example, which have now been available for more than a decade, still command less then 2 percent of the market and, despite their price premiums, are sold by their makers as loss leaders to showcase their "green" credentials. American-made hybrids, such as the GM Volt or Ford Fusion, depend on large federal consumer subsidies and even then are proving to be a commercial bust. Convincing consumers to switch to more energy-efficient light bulbs and less powerful home appliances, to turn off the lights when they leave a room, to wear more sweaters, and to turn the thermostat down – or up as the case may be – has at best a marginal impact on total consumption. Only massive subsidies and command- and-control policies and programs can *begin* to achieve the reductions in fossil fuel use required to make even a modest dent in the modelled impact of CO_2 on the climate. Such subsidies and command-and-control policies would result in major distortions in efficiency and productivity and lead inevitably to a significant decline in standards of living.

Reaching a global reduction of 50 percent in greenhouse gas emissions within a generation will require much more than the relatively modest goal of stabilizing emissions in industrialized countries at 6-10 percent below 1990 levels by 2012, Kyoto's unachieved goal. Many alarmists think that an 80 percent reduction by mid-century is both desirable and achievable. Only the most radical steps would make that possible. The *Wall Street Journal* concluded:

> Currently, alternative sources – wind, solar, biomass, hydroelectric and geothermal – provide less than 7 percent of yearly domestic [US] consumption. Throw out hydro and geothermal, and it's only 4 percent. For the foreseeable future, renewables simply cannot provide the scale and volume of energy needed to meet growing US demand, which is expected to increase by 20 percent over the next two decades. Even with colossal taxpayer subsidies, renewables probably can't even slow the rate of growth of carbon-based fuel consumption, much less replace it. ... Environmentalists love the idea of milking Mother Nature for power, but they hate the hardware needed to make it work: huge windmills, acres of solar panels, high-voltage transmission lines to connect them to the places where people live. Of course, they still totally, absolutely, wholeheartedly support green energy – as long as it gets built where someone else goes yachting.[99]

In the face of the problems arising from both carbon reduction strategies and alternative fuels, the favourite response of the alarmist community is that all this may be true, but that something must be done. In fact, when the solutions are worse than the problem, it is best not to do anything. Invoking the precautionary principle is also not much help. As Roy Spencer points out, "The precautionary principle is a guiding philosophy that unrealistically assumes we can have benefits with no risks."[100] To take the kind of draconian steps being called for to change the fundamental underpinnings of modern civilization on the whim that the theory of anthropogenic global warming may over time prove to have more explanatory power than it does now is the height of irresponsibility. Talk of tipping points and of runaway warming has no basis in science but does underline that the alarmist community is becoming desperate and more strident in its calls for action.

The lack of reality in the movement's push for a fossil-free energy future is well illustrated by the Obama administration's continuing war on coal. The United States currently generates nearly forty percent of its electrical energy from coal. Over the past 12 years, the EPA has waged a cynical effort to make coal-based generation uneconomi-

---

99. "Blowhards," *Wall Street Journal*, January 24, 2009.

100. Roy Spencer, *Climate of Confusion*, 142. The full implications and intellectual legerdemain of the "precautionary principle" are explored in the next chapter.

cal by imposing new standards that few power facilities can meet on a cost-effective basis. The goal is clear: destroy the economics of coal generation in order to make wind and solar generation look more attractive. As noted earlier, most power companies are opting for gas conversion but fear that a war on gas may not be far off.

Unfazed by the Supreme Court's June 2015 decision disallowing 2012 regulations on the basis that the costs were astronomically out of line with any conceivable benefits,[101] the President announced even more draconian measures at the beginning of August 2015.[102] Using all the verbal pyrotechnics and deceits that have become his trademark, Obama claimed that his plan will deliver public health benefits, create tens of thousands of new jobs, drive renewed investment in renewable technologies, reduce future energy bills, and continue American climate change leadership. As the above analysis makes clear, each claim is patently untrue.[103] Not mentioned in his announcement was the impact these measures would have on climate. The reason is clear: even committed IPCC climatologists are not prepared to say that they will have anything other than a marginal impact.[104]

The new EPA rules will require a 32 per cent cut in power-plant carbon dioxide emissions by 2030 from 2005 levels, an increase from the 30 per cent target proposed last year at an estimated annual cost by 2030 of $8.4 billion. Such estimates are typically off by a factor of two or more. The burden will lie with the states which must each provide compliance plans by 2018, i.e., a year after Obama leaves office. A number of states have already indicated that they will not comply, and Washington lawyers have already set to work preparing legal challenges. The plan can also be rescinded January 21, 2017 with the stroke of a pen by the next president or disallowed by the Congress. As the *Wall Street Journal* put it, "Mr. Obama's argument

---

101. In *Michigan v. the EPA*, the Court ruled that the EPA must consider costs in imposing regulations on an industry. Justice Antonin Scalia, writing for the majority, wrote that it is irrational and inappropriate to "to impose billions of dollars in economic costs in return for a few dollars in health or environmental benefits."

102. See "Announcing Obama's new 'Carbon Pollution' plan," at *WattsUpWithThat*, August 3, 2015.

103. Julian Morris provides a point-by-point rebuttal of Obama's claims in "Obama's clean calamity," *Financial Post*, August 5, 2015.

104. Judith Curry calculates that the earlier Obama commitment to reduce US emissions by 28 per cent by 2030 might reduce global temperatures by 0.03°C by 2100. *Climate Etc*, August 3, 2015.

is that climate change is too important to abide by relics like the rule of law or self-government."[105] Paul Albaugh, channelling Winston Churchill, argues that "socialism is a philosophy of failure, the creed of ignorance and the gospel of envy. Its inherent virtue is the equal sharing of misery. And make no mistake: Obama's plan has little or nothing to do with climate and everything to do with his social justice worldview, in which he wants to handicap the US ostensibly to the benefit of the rest of the world."[106] Perhaps, but it is also about political theater in preparation for the Paris Conference of the Parties to the UNFCCC at the end of 2015, at which time the world will adopt a global carbon reduction plan. Or not. Time will tell, but as discussed in chapter 13, this kind of bombast has to date proven the direct inverse of actual achievements.

Rarely discussed in considering the feasibility of making the transition to a low carbon economy is the reality that transitions necessarily take a very long time. Globally, trillions of dollars are tied up in infrastructure predicated on the current mix of energy sources. Electric transmission grids, for example, have been built to distribute electricity from multiple sources as efficiently as possible with a minimal loss of power. The longer the transmission lines, the greater the loss of power. Maintaining the integrity of the grids and ensuring a steady flow of uninterrupted electricity to all customers have become increasingly complex challenges to engineers, particularly considering the government mandates to include feed-in from retail producers of solar and wind power as well as from larger solar and wind arrays, none of which provide a steady flow of electricity. Similarly, fossil-fuel powered transportation is made easy and affordable because of the build-up of huge, widespread distribution systems. With proper maintenance, the life cycle of this infrastructure is rarely less than thirty years and often as many as fifty years or more. The cost of prematurely retiring purpose-built infrastructure and replacing it with new, untried infrastructure and distribution regimes is rarely taken into account in calls for transitioning to a low-carbon economy.[107]

Putting aside whether there is a pressing need for emission reductions, the only viable long-term approach to reducing depend-

---

105. "Climate Change Putsch," *Wall Street Journal*, August 3, 2015.

106. Paul Albaugh, "Obama's Clean Power Plan is Costly Political Theater," *The Patriot Post*, August 4, 2015.

107. See the discussion in Smil, *Energy Myths and Realities*, 133-49.

ence on fossil fuels for electricity generation is to start planning now for a transition in two steps: converting many current fossil-fuel generators to high efficiency gas generators, such as super critical gas plants, and replacing them as their useful life comes to an end with third and fourth generation nuclear facilities. Step one will stabilize emissions, while step two will reduce them. With large-scale take-up of the nuclear option, new technology and designs will become economically viable. These designs will also reduce the major issue of waste disposal because there will be much less of it. Governments will need to screw up their courage and convince alarmists and environmentalists to accept this two-step, long-term transition as the best-case scenario.[108]

The preceding survey of alternative energy sources above also indicates that, if there is to be a response to climate change, it must be focused on adaptation. The IPCC itself admits that even if governments pursue aggressive carbon reduction strategies, the impact on global temperatures will be marginal for many decades, if not centuries. The theory on which the IPCC and affiliated scientists rely holds that the amount of CO_2 already in the atmosphere, as well as additional emissions during the transition to a low-carbon global economy, will remain in the atmosphere for hundreds of years and continue to affect the climate adversely. In addition, the theory maintains that warming already in the pipeline as, for example, in increased ocean heat content, will ensure continued global warming for many decades into the future. Optimism about containing the anticipated global temperature increase since the start of the Industrial Revolution to 2°C by limiting emissions and holding atmospheric concentrations to 550 ppm has no basis in science but rather originated in Europe as a convenient political target.

Adaptation has the advantage of being focused on much more realistic time horizons. Rather than pursuing policies to address possible adverse impacts decades and even centuries into the future, adaptation strategies address problems that are either already apparent or can be anticipated on the basis of real world evidence rather than on model projections. Adaptation is a strategy focused on local issues and sharply reduces the need for international and national action.

There are those, particularly in the business community, who assume that reducing GHG emissions to stable levels is the end of

---

108. For a full discussion of this approach, see Bryce, *Power Hungry*, 207-82.

the story. But to AGW alarmists, that is only the beginning. Based on the theory that they promote, alarmists assert that anthropogenic atmospheric CO_2 lasts several generations and that the foreseeable impact of stabilizing global CO_2 emissions to levels equivalent to 1990, 2000, or even 2010, will only delay global warming at the margins. In their scenarios, it will be several centuries before the climate will be stabilized at optimal levels and the damage done by fossil-fuel based civilization finally neutralized. Buying into the AGW hypothesis means accepting the whole story and all of AGW's alleged consequences. AGW enthusiasts in fact do not have a solution. What they want is for mankind to expiate its guilt by taking measures that will eradicate modern civilization and reduce the planet to a biosphere in which the human species is at most one of many and no more important than any other.

The truth is that there is only one reliable way in which a fifty percent or more reduction in CO_2 levels by mid-century can be achieved, and that is through a drastic reduction in global population. From the onset of the Industrial Revolution, the rise in global population and the rise in total CO_2 emissions have moved in lock step. Gains in efficiencies have been offset by the expansion of industrial activity to ever more countries. The current world population of 7 billion people emits about 8.6 billion metric tons of CO_2 annually, a carbon intensity of about 1.25 metric tons per capita, a number that has remained fairly constant for over a century.[109] With a projected increase of the world's population to 9 billion by 2050, reaching a target of 4.3 billion metric tons of annual CO_2 emissions, i.e., a fifty percent reduction in current levels will – superficially – require one of the following:

- a reduction in global population to fewer than 3.5 billion people;[110]

---

109. This figure, of course, does not take into account the very low levels of carbon intensity in developing countries (under 1 ton per person per year) and the high levels in industrialized countries (from 20 tons in the United States to 6.0 in France). As a result, intensity reductions are much more important for the industrialized – and much more daunting – than for developing countries and will require radical changes in life styles in the advanced economies.

110. A major reduction in total global population, of course, has been a goal of the leaders of the alarmist movement for more than a generation, expressed in a willingness to take draconian measures. What such measures might entail is never explained, but it never seems to include their own demise. South African philosopher David Benatar, for example, in *Better Never to Have Been* (New York: Oxford University Press, 2006), argues that for the good of the planet, humans should stop procreating and become extinct. Apparently, he justifies his own

- a major reduction in global economic activity, by limiting industrial output, air and surface transportation, food production, heating and air conditioning, and other energy-dependent activities to levels prevailing in the 1940s, leading to major increases in global poverty and reductions in living standards;
- a reduction in per capita energy intensity to less than .5 metric tons by vastly expanding modern technological efficiencies or reducing per capita consumption to less than half of current levels;
- a miracle breakthrough in developing and implementing economically feasible alternative, non carbon-based sources of energy; or
- some combination of all four.[111]

It is difficult to imagine what it would take to reach an 80 percent reduction. Mad-men like Bill McKibben want to go even farther: zero emissions as soon as possible and geo-engineering so that atmospheric CO_2 can be reduced to 350 ppm as quickly as possible. As far as McKibben is concerned, we have already reached Climate Armageddon despite the inconvenient fact that even alarmist scientists admit that there has been no warming since at least 1998.

Finally, there is the inconvenient fact that both warming and atmospheric CO_2 are, on balance, beneficial. In their zeal to advance the alarmist cause, environmentalists have sought to demonize both and thus to erase all thought that there may be a silver lining for their cloud of doom. Evidence for the benefits of both warming and CO_2 is not difficult to document. Princeton's William Happer, for example, points out that:

> The 'green revolution' has increased crop yields around the world. Part of this wonderful development is due to improved crop varieties, better use of mineral fertilizers, herbicides, etc. But no small part of the yield improvement has come from increased atmospheric levels of CO_2. Plants photosynthesize more carbohydrates when they have more CO_2. Plants are also more drought-tolerant with more CO_2, because they need not 'inhale' as much air to get the CO_2 needed for photosynthesis.

---

continued existence by the need to spread his message. Journalist Alan Weisman explores similar themes in *The World Without US* (Toronto: Harper Canada, 2008) and *Countdown: Our Last, Best Hope for a Future on Earth* (Boston: Little Brown and Co., 2014).

111. These ideas, known as the Kaya Identity, are fully developed in Yoichi Kaya and Keiichi Yokobori, eds., *Environment, Energy, and Economy: Strategies for Sustainability* (Tokyo: United Nations University, 1997).

At the same time, the plants need not 'exhale' as much water vapour when they are using air enriched in CO_2. Plants decrease the number of stomata or air pores on their leaf surfaces in response to increasing atmospheric levels of CO_2. They are adapted to changing CO_2 levels and they prefer higher levels than those we have at present.[112]

Other benefits include longer growing seasons, a result which adds to the prospect of more successful and larger harvests. More generally, history attests that civilization blooms during periods of relative warming and declines during periods of relative cooling. Fossil fuels and technology have made the extremes of weather less and less challenging to most people in developed countries. The transformation of Washington, DC, for example, from a sleepy backwater to a thriving global capital was materially assisted by the widespread adoption of air conditioning; the same is true of large parts of the US South.[113] Modern humans may be preoccupied with climate, but their experience of weather's normal extremes is waning as a result of technology and prosperity.

It is little wonder that the alarmist community is opposed to cost-benefit analysis of the problem and to many of the proposed solutions. Instead, they hype alarm. Without an alarmist message, the prospect of imposing their preferred solutions is slim. For adherents of the religion of global warming, human emissions of GHGs are the prime drivers of a new and dramatic type of climate change that will inexorably result in a significant warming over the next century and lead to catastrophe for all of life on earth. As far back as the 1992 UN Rio Earth Conference, this belief has been used as a legitimizing myth for a gamut of interconnected political agendas focused on fossil fuels, trade, transportation, economic growth, and global corporations. The language used tends to be authoritarian and religious in character. It is not an accident that its most celebrated advocate is a failed seminarian, onetime politician, serial exaggerator, and now full-time activist, former US Vice-President Al Gore. For him, and for many others, the myth has become an article of secular faith exhibiting all the characteristics of a pre-modern religion, above all demanding sacrifice to the earth, or Gaia.

---

112. William Happer, Testimony before the US Senate Environment and Public Works Committee, February 25, 2009.

113 See David Brinkley's charming account of Washington before air conditioning and its role in the city's transformation. *Washington Goes to War* (New York: Ballantine Books, 1988).

If the past is any guide to the future, technological development will open up new, as yet unproven, solutions to future energy needs, from the ethane trapped in deep-sea natural gas hydrates to the energy contained in every atom. Environmentalists exhibit a curious mix of emotions about this: a naïve faith in solutions that have proven technically, economically, or commercially impracticable and an equally strong scepticism about future technological developments, a scepticism that leads them to advocate immediate draconian solutions. At this point a "clean" energy future involves a very unstable structure built on a foundation of questionable assumptions. The failure of any one of these assumptions will bring down the whole house of cards. Much of the discussion betrays an astounding level of scientific, technological, and economic illiteracy, particularly within the political class and among journalists. Their unwillingness to deal with scientific, energy, and economic realities provides fertile ground for the purveyors of doom and gloom.

Until renewable sources of energy become economically viable on a sufficiently large scale to replace the many uses of fossil-based energy on which modern economies and life-styles rely, the only way to reduce fossil fuel-based CO_2 emissions is through significant life-style changes and reductions in economic activity, not only in the already developed countries but also in developing countries. As Philip Stott notes, UN and national efforts to reduce fossil fuel use are "a serious threat to everyone, but especially to the 1.6 billion people in the less-developed world who have no access to any modern form of energy. The twin curses of water poverty and energy poverty remain the real scandals. By contrast, the political imposition on the rest of the world of our Northern, self-indulgent ecochondria about 'global warming' could prove to be a neo-colonialism too far."[114]

٭٭٭٭٭

---

114. Stott, "Global Warming as Myth," *SPPIblog*, February 4, 2010.

9 | The Economics of Climate Change Policy

Striving for sustainable development and an equitable world must be central features of any study of this kind.[1]

Bert Bolin, IPCC's first Chair

One has to free oneself from the illusion that international climate policy is environmental policy. Instead, climate change policy is about how we redistribute de facto *the world's wealth.* [2]

Ottmar Edenhofer, IPCC senior official

Assessing risks, costs, and benefits

Taking action on climate change requires that governments consider not only the risks and consequences of climate change but also its benefits. They will need to compare the costs, risks, and feasibility of mitigation and of adaptation measures. What public policy initiatives are indicated by the evidence being advanced? What competing public policy initiatives will have to be rejected or postponed in order to devote finite resources to the issues raised by climate change? As Bjørn Lomborg has demonstrated, climate change came in dead last when a group of leading economists evaluated the costs and benefits of various programs to help gauge how governments can achieve the most good with limited resources in addressing a range of global policy challenges such as communicable diseases,

---

1. Bert Bolin, *A History of the Science and Politics of Climate Change: The Role of the Intergovernmental Panel* (Cambridge: Cambridge University Press, 2008), 118.

2. Interview with *Neue Zürcher Zeitung,* November 14, 2010. Edenhofer is the chief economist of Germany's Potsdam Institute for Climate Impact Research, chair of Working Group 3 of the fifth assessment report of the IPCC, and an energetic green activist.

education, corruption, migration, malnutrition and hunger, and access to clean water. Subsequent reiterations of the process have refined some of the issues and sharpened the cost-benefit analysis, but the conclusions remain the same: the world would be a better place if the money already being spent on climate change issues was being used to fight malnutrition, improve public health, increase access to clean water, and strengthen reforestation, all areas in which government policies have proven to be effective.[3]

To clarify the public policy issues presented by the alleged threat of anthropogenic global warming, it is worth reviewing the extent of the harm that could arise from an increase in global temperature reaching 2°C above pre-industrial levels by the end of the 21st century. Such an increase would require a rate of warming more than three times that experienced since the end of the Little Ice Age, a rate that is highly unlikely given growing appreciation of the impact of such factors as coupled atmospheric-oceanic circulation cycles and solar activity. Peer-reviewed scientific analysis of the climate system's sensitivity to rising levels of atmospheric CO_2 is steadily pushing that rate downward. Nevertheless, for the sake of argument, what are the risks – and benefits – of a 2°C increase? Are there solutions that lie within known human experience? Is there public support to pursue those solutions?

Warming

The warming of the atmosphere by a degree or two over the course of a century presents no significant direct harm and in many ways may be beneficial. As already noted, historically periods of warming have been beneficial to humans, flora, and fauna alike. If the GHG hypothesis is correct, its principal effect will be at higher latitudes at night and in winter, i.e., in reducing heat loss to the upper atmosphere and out into space. Warmer winters and warmer nights will generally extend growing seasons and increase harvests. A whole industry has emerged to model the negative impacts of global warming with generous research funds flowing to those who can show alarming impacts. In most instances negative impacts are assumed rather than demonstrated, while significant evidence of the benefits of a warmer world are ignored.

---

3.   The report of the first Copenhagen Consensus meeting came out in 2004 (*Global Crises, Global Solutions*, Cambridge University Press). Lomborg repeated the process in 2008 and 2012, with the latest report published in October 2013 (*Global Problems, Smart Solutions*, Cambridge University Press).

Higher levels of atmospheric CO₂

CO_2 is not a pollutant; it is essential to all forms of life. In and of themselves, therefore, the higher levels of atmospheric CO_2 feared by the alarmist community will be beneficial. Increased levels have already had a marked greening effect on the planet. It would require levels many times higher than those of today before human or animal health might be adversely affected. It is only the alleged impact of CO_2 on global temperature that may be of concern and then only if the feedback sensitivity is high and positive. As discussed in earlier chapters, in the absence of high sensitivity leading to significant feedbacks, the increase would be less than 0.5°C, and that could even be nullified by other climatic factors.

Increased precipitation

Everything else being equal, a warmer climate will probably mean a slightly wetter climate, but distribution patterns of precipitation may differ, creating possible habitat pressures on some species and problems of fresh water availability in some poorer regions of the world. The alarmist community routinely underrates the adaptability of most species, including humans, and pays insufficient attention to the prospect of innovation and technological developments. Most flora and fauna have survived far more dramatic natural climate change over historical and geologic time. A marginally warmer, wetter world would, on balance, bring more benefits than problems.

Availability of fresh water

Over the course of the 20th century rapid population growth has made access to fresh water a matter of concern. It is alleged that global warming has exacerbated this problem by reducing meltwater from disappearing low-altitude glaciers and by increasing drought in some areas while increasing rainfall in others. These highly speculative impacts are largely a matter of conflating economic growth, population pressures, and climate change. The hydrological cycle will not cease in a warmer world but potable water may not always be found in the most convenient places. Technology, however, can address this issue more easily and reliably than massive changes in energy use. It is morally repugnant to deprive poor people in developing countries of the energy they need to develop their economies so that they can adopt modern sanitation technologies, including up-to-date water distribution systems.

Increases in extreme weather

It has become an article of faith among alarmist scientists that warmer temperatures will lead to more extremes, from heat waves and floods to hurricanes and tornadoes. There are no observational data to substantiate this claim. Indeed, the historical record indicates that colder temperatures lead to more extreme weather patterns.[4] This makes sense in physical terms. Storms are the product of temperature differentials; a warmer climate at higher latitudes would reduce these differentials.

Higher sea levels

Most of the rise in sea levels resulting from the melting of water stored in the last, Ice Age icecap and mountain glaciers has already taken place. Some further modest increase may result from the thermal expansion of sea surface water; ocean depths will remain a frigid 4°C. The threat of major melting of the Greenland and Antarctic ice caps is remote and, even at accelerated melting rates, would take thousands of years. The prospect of further sea-level increases in the foreseeable future, therefore, is limited to 3-4 millimeters per year for a total of no more than 30-40 centimeters over a century, and probably as little as the ca. 20 cm. increase experienced over the 20th century. We know from past experience that it is well within the scope of human ingenuity to address any resulting issues of coastal erosion and similar problems taking place over a period of many years. Again, as poor countries gain access to modern energy sources and develop their economies, they too can afford to take these steps.

Pressures on global food supplies

Over the past century global food supplies have increased ten-fold while population has increased six-fold. The rate of population growth is slowing to the point that projections of global population increases have been moved downward several times. The ability to increase available food supplies even further will be enhanced under somewhat warmer, wetter conditions, with higher levels of CO_2, and with the application of new technologies such as genetic modification. The opposition of environmentalists to genetic modification and other technologies should not be allowed to obscure tech-

---

4.   See, for example, Brian Fagan, *The Little Ice Age: How Climate Made History 1300-1850* (New York: Basic Books, 2000) and *The Long Summer: How Climate Changed Civilization* (New York: Basic Books, 2004).

nology's critical role in addressing food and other issues. Reducing fossil fuel use – as both an important source of energy in agriculture and as a feedstock for fertilizers and pesticides – will create significant pressures on the ability to maintain adequate food supplies for a still currently growing population.

Pressures on flora and fauna

As has already happened since the end of the last ice age and since the depth of the Little Ice Age, flora and fauna adapt to a changing climate by shrinking or expanding their ranges. For most species, a warmer and wetter climate is better than a colder and drier one, and under warmer and wetter conditions, most species will tend to expand their range. Habitat loss and change are primarily due to factors other than climate change and will not disappear in a "post-carbon" world. Building massive solar- and wind-generating facilities *will* have major habitat implications for many species.

Modest increases in premature deaths

Warmer weather will more than balance any increases in warm-weather induced premature deaths with larger decreases in premature deaths from colder weather. Neither scenario, however, will be of great significance, given the human capacity to adapt to changing climatic conditions through technology and other factors. As the developing world becomes more prosperous, it too will be able to adopt modern cooling and heating technologies.

Increases in vector-borne diseases

While often mentioned in the alarmist literature, it is rarely explained how this problem would expand with warmer temperatures. Progress in understanding the basis for many vector-born diseases and the development of both preventative and therapeutic strategies have largely eradicated such diseases as malaria, bubonic plague, and yellow fever in developed countries. The prevalence of those diseases in warmer climates is less a matter of climate than of inadequate public health measures.

More generally, economist Richard Tol, in reviewing the impact of climate change from 1900-2050, concludes that its impact has been beneficial and will continue to be beneficial until well into the 21st century or until global temperatures rise 3°C above pre-industrial levels, or 2.2°C above current levels. In Tol's view, the rise in temperature, ambient CO_2 levels, and other changes over the past century and a half have already contributed to a 1.4 percent increase in global welfare and will continue to do so for many years to come.[5] Matt Ridley, after reading Tol's paper, concludes that "climate change has done more good than harm so far and is likely to continue doing so for most of this century. This is not some barmy, right-wing fantasy; it is the consensus of expert opinion. Yet almost nobody seems to know this. ... Good news is no news, which is why the mainstream media largely ignore all studies showing net benefits of climate change. ... There are many likely effects of climate change: positive and negative, economic and ecological, humanitarian and financial. And if you aggregate them all, the overall effect is positive today – and likely to stay positive until around 2080."[6] In sum, effects will only turn negative in the unlikely event that the IPCC prognosis is correct, and then only towards the end of the 21st century, by which time increasing prosperity and new technologies should postpone negative impacts further into the future to a time when the cyclical nature of the climate system may well have ushered in entirely different climatic conditions.

Problems that *might* emerge as a result of global warming seem to be fully manageable on the basis of routine adaptation and problem-solving and are hardly in need of radical programs and policies that could undercut the basis of modern prosperity. Indeed, problems that are exacerbated by poverty will become easier to address as the developing world continues on the path to greater prosperity. The past century of mild global warming correlated with an exponential increase in global well-being. It is difficult to understand why a further, *modest*, and *gradual* increase in global temperatures would have catastrophic results. Concern about the marginal climate change experienced so far only makes sense for those who find it a convenient horse to ride towards a future transformed by radical public policy measures.

Enter the precautionary principle

As discussed in chapter two, post-industrial society has developed an aversion to risk that our ancestors would find puzzling. The *public* perception of risk is often much greater than that of experts. Governments have responded to activist demands for risk ameliora-

---

5.   Richard Tol, "Climate Change," in Bjørn Lomborg, ed., *How Much Have Global Problems Cost the World? A Scorecard from 1900 to 2050* (Cambridge: Cambridge University Press, 2013), 117-30.

6.   Matt Ridley, "Why climate change is good for the world," *The Spectator*, October 19, 2013.

tion with a torrent of regulatory requirements governing all aspects of life, from food and health to transportation and industry. No government wants to be caught in the politically uncomfortable position of having ignored a risk when disaster strikes. Nevertheless, as philosopher Roger Scruton cautions: "regulations imposed by the state have side effects that often worsen what they aim to cure."[7] Regardless of the limited prophylactic effect of much public regulation, OECD governments are now spending billions regulating risk, and private enterprises have been saddled with an increasingly costly burden as they implement more and more requirements. *Individual responsibility has given way to social responsibility.* Sociologist Frank Furedi adds:

> We live in an era where problems associated with uncertainty and risk are amplified and, through our imagination, mutate swiftly into existential threats. Consequently, it is rare that unexpected natural events are treated as just that. Rather, they are swiftly dramatized and transformed into a threat to human survival. ... Policies designed to deal with threats are increasingly based on feelings and intuition rather than on evidence or facts. ... One of the principal accomplishment[s] of precautionary culture is the *normalization of irresponsibility.*[8] [emphasis added]

Environmental activists have been among the most successful in raising public perceptions of risk and in convincing governments to act. From pesticides to species diversity, environmentalists have used alarm and exaggeration as a powerful tool to raise awareness and move governments to legislate or regulate. Two potent weapons in their arsenal are the so-called precautionary principle (PP) and international agreements. The Rio Summit in 1992, for example, declared: "in order to protect the environment, the precautionary approach shall be widely applied by States according to their capabilities. Where there are threats of serious or irreversible damage, lack of full scientific certainty shall not be used as a reason for postponing cost-effective measures to prevent environmental degradation."[9] Similar wording can be found in an increasing number of international environmental agreements, including the UN Framework Convention on Climate Change, the Convention on Biological Diversity, the Montreal Protocol on Substances that Deplete the

---

7.   Roger Scruton, *How to Think Seriously about the Planet: The Case for an Environmental Conservatism* (Oxford: Oxford University Press, 2012), 1.

8.   Frank Furedi, "Fear is Key to Irresponsibility," *The Australian,* October 9, 2010.

9.   Principle 15 of the Rio Declaration on Environment and Development.

Ozone Layer, the Cartagena Protocol on Biosafety, and the Stockholm Convention on Persistent Organic Pollutants. The PP has also been enshrined in domestic legislation. The government of Canada, for example, intones in its 1999 *Environmental Protection Act* that it "is committed to implementing the precautionary principle that, where there are threats of serious or irreversible damage, lack of full scientific certainty shall not be used as a reason for postponing cost-effective measures to prevent environmental degradation."[10] The idea of "full scientific certainty" is a political rather than a scientific statement. The best that science can do is point to the preponderance of evidence.

An obsession with risk and uncertainty has now been the preoccupation of two generations of activist scientists and ENGOs.[11] To them, the development of nuclear weapons, for example, was the result of a society that did not place sufficient controls on the risks flowing from irresponsible developments in science and technology. Their perspective was well-captured by the initial reports of the Club of Rome: *Limits to Growth* (1972) and *Mankind at the Turning Point* (1974). Both reports were prepared by scientists and technocrats who believed that "the rapid succession of crises which are currently engulfing the entire globe are the clearest indication that humanity is at a turning point in its historical evolution. ... *Our scientifically conducted analysis of the long-term world development based on all available data points out quite clearly that ... a passive course leads to disaster.*"[12] [emphasis in original] To such a mindset, the precautionary principle seems both necessary and obvious, and its shortcomings and disadvantages inconsequential. As Dutch chemist Jaap Hanekamp points out, from this perspective, "science and technology need to be assessed on a continuous basis in order to keep a firm grip on its development."[13] The precautionary principle provides that grip.

Over the past thirty years the emergence of the precautionary principle in European law and practice has subtly and importantly changed the nature of earlier, more balanced approaches consistent,

---

10.   Preamble, para 6, Environmental Protection Act (CEPA 1999).

11.   See the discussion in Jeffrey Foss, *Beyond Environmentalism: A Philosophy of Nature* (Hoboken, NJ: John Wiley & Sons, 2009), 15-19.

12.   Mihaljo Mesarovic and Eduard Pestel, *Mankind at the Turning Point: The Second Report of the Club of Rome* (Boston: Dutton, 1974), vii.

13.   Jaap C. Hanekamp, "Neither Acceptable nor Certain – Cold War Antics for 21st Century Precautionary Culture," *Erasmus Law Review*, 2:2 (2009), 222.

for example, with the rules of the WTO. Rather than a preference for evidence-based policy making, there is increasing resort to policy-based evidence making. The requirement to demonstrate the possibility of serious harm, an issue requiring *expert judgment* based on scientific or other evidence-based investigation, changed first to a need to demonstrate only a *perception* of possible risk, i.e., a political judgment, and then to a requirement to demonstrate the *absence* of harm – a logical impossibility. This subtle shift in the burden of proof has had tremendous implications for governments' regulation of risk, including in response to claims of threatened harm to the environment. Marchant and Mossman conclude: "The precautionary principle may well be the most innovative, pervasive, and significant new concept in environmental policy over the last quarter century. It may also be the most reckless, arbitrary, and ill-advised, ... [due in large measure to] its potential for overregulation of insignificant or even nonexistent risks, its disregard for scientific evidence, and its failure to adequately consider the economic cost and risk-risk trade-offs inherent in risk regulation."[14]

EU experience has been instructive. Perhaps well-intentioned originally, the precautionary principle has evolved into a concept having so many interpretations and applications that it has lost all capacity to guide governments in addressing difficult, risk-related regulatory challenges. Marchant and Mossman find that the EU experience in applying the precautionary principle shows a record of arbitrary and inconsistent decisions, applying it where it makes little sense and ignoring it where it does.[15] The absence of clear criteria to guide decision-makers leaves the door wide open to highly politicized and arbitrary decisions. At the same time it is wholly consistent with the political and social history of postwar Europe. After two devastating world wars, European elites see politics as a means of social and institutional transformation along statist lines rather than as a matter of brokering competing interests.

---

14. Gary E. Marchant and Kenneth L. Mossman, *Arbitrary and Capricious: The Precautionary Principle in the European Union Courts* (Washington: AEI Press, 2004), 1.

15. Marchant and Mossman, 27-43. In their view, it "provides an open invitation for arbitrary and unreasonable decisions by both regulators and judges," 65. Cass Sunstein, who is more sympathetic to its intent, concludes that without much greater definition and clarity, the precautionary principle risks perverse and paralyzing regulatory action. Sunstein, *Laws of Fear: Beyond the Precautionary Principle* (Cambridge: Cambridge University Press, 2005), 34.

The EU Commission, with its ever-growing enthusiasm for directives and enforcers, may have been in the vanguard, but other OECD countries have followed suit. Between 1993 and 2013, the US federal government issued 81,883 rules (an average of ten a day). In fiscal 2011-12, Congress passed 29 laws, while the Administration issued 3,708 rules. The annual cost to the US economy of this regulatory burden is estimated to have reached $1.8 trillion.[16] The courts have played a central role in this expansion because American litigiousness has become a weapon in the hands of one special interest after another, stretching Congress's legislative intent in ever more directions. The latest twist under the Obama administration has been sue-and-settle cases. A special interest is encouraged by officials to sue; once it does, officials will negotiate a judicially supervised settlement and enlarge their agency's mandate. The Environmental Protection Agency has proven particularly adept at this technique.[17] Canada is not far behind, with federal, provincial, and municipal officials vying to outdo each other in keeping Canadians "safe". Canadians have experienced the precautionary principle's effect, for example, in the new municipal enthusiasm for banning the use of pesticides in maintaining weed-free residential lawns, pesticides licensed for use by the federal government and still available to farmers and golf course operators.[18]

Given the exponential growth of the regulatory state, OECD members have devoted considerable resources over the past twenty years to an ambitious program of reviewing members' regulatory practices, sharing information, and developing and promoting best practices. In all jurisdictions studied, legislatures have played a minor role in this expansion. Instead, the complex task of governing modern societies has been appropriated by officials relying on the courts and interpretations of existing statutory instruments. Booker and North conclude: "What this was creating ... was what came to be known as the 'health and safety culture', centred on the 'precautionary princi-

---

16. Clyde Wayne Crews, *Ten Thousand Commandments: An Annual Snapshot of the Federal Regulatory State* (Washington, Competitive Enterprise Institute, 2013), 2.

17. See US Chamber of Commerce, *Sue and Settle: Regulating Behind Closed Doors* (Washington: US Chamber of Commerce, 2013).

18. In a 2001 survey the Fraser Institute indicated that between 1975 and 1999, over 117,000 new federal and provincial regulations were enacted, an average of 4,700 a year or nearly 13 a day. See Laura Jones and Stephen Graf, "Canada's Regulatory Burden: How Many Regulations? At What Cost?" *Fraser Forum* (August 2001). Unfortunately, the Institute has not done updates similar to those done by Crews for the United States.

ple', whereby it was always necessary to eliminate the faintest degree of risk, however remote and imaginary that possibility might be. ... If flawed science was on offer, [officials] fell for it."[19] Sociologist Ulrich Beck argues that scientific and technological advances have created a new society, one that involves a much deeper consciousness of risk and a need for political institutions – sub-national, national and supra-national – to accommodate them.[20]

Philosopher Steve Fuller suggests that "the precautionary principle can look quite short-sighted, as it places too much trust in today's science, overlooking science's long-term tendency to shift its ground, often as a result of a massive reinterpretation of data, which in turn leads to new projections."[21] The continuing application of the precautionary principle promotes stasis rather than experimentation and creativity and may deny the fruits of innovation to both present and future generations. Most research and technological developments involve the possibility of both risk and benefit. The institutionalization of the precautionary principle, however, is leading to the denial of funding and even of permission for research projects that could lead to important and beneficial new insights and developments.

Environmentalists have been among the most enthusiastic proponents of this regulatory expansion based on the precautionary principle and bureaucratic empire-building. Their only criticism is that officials have been too cautious in their definition and application of the precautionary principle. According to environmentalists, the best definition was provided by a group of self-appointed guardians of the environment in the so-called Wingspread Statement, adopted at the Wingspread Conference Center in Racine, Wisconsin in 1998: "Where an activity raises threats of harm to the environment or human health, precautionary measures should be taken even if some cause and effect relationships are not fully established scientifically. In this context the proponent of an activity, rather than the public, bears the burden of proof." In this statement, the burden of proof shifts from demonstrating risk to demonstrating

---

19. Christopher Booker and Richard North, *Scared to Death: From BSE to Global Warming: Why Scares Are Costing Us the Earth* (London: Continuum Books, 2007), 156. The book provides a number of case studies of premature or excessive regulatory actions based on zealous application of the precautionary principle.

20. Ulrich Beck, *Risk Society: Towards a New Modernity* (London: Sage, 1992).

21. Steve Fuller, "The Proactionary Principle: Between No Caution and Precaution," *Breakthrough Institute*, August 8, 2013.

the absence of risk, an impossible burden. This activist sleight of hand replaces the need to consider real risks with the requirement to avoid *perceptions* of risk. It is wholly consistent with the postmodern culture of precaution. As Hanekamp points out, "in contemporary post-modern society the goal of affluence yields to that of life-term ... safety. ... in economically and industrially highly developed societies, diverse regulation of a mainly precautionary nature has found its way into many areas."[22] From this perspective, scientific progress becomes less a matter of *objective, evidence-based* enquiry and more a matter of *socially acceptable* enquiry, as advanced by the proponents of post-normal science.

Most human activity entails some risk; the essence of risk-related regulation is to assess the extent of risks, benefits, and costs and develop rules about appropriate use. Changing the burden of proof, of course, opens the door to using public fear and political perceptions of the moment to drive regulatory action, thus reducing the scope for governments to resist the pressures of single-interest groups. The absence of any reference to cost-benefit analysis or evidence-based risk assessment in the Wingspread Statement is both deliberate and telling. For the environmental movement, alarm advances its agenda while evidence undermines it. Scruton observes that "alarms turn problems into emergencies, and so bring the ordinary politics of compromise to a halt. ... People who pursue a politics of top-down control ... find emergencies extremely useful."[23] Fear and ignorance thus become the drivers of regulations implemented by an ever-growing army of health and safety inspectors, environmental protection officers, and social workers, each armed with their own rules, procedures, and interests and, inevitably, mounting societal costs.

Against this background, the global warming movement's enthusiasm for the precautionary principle is not difficult to understand. In this context, the emphasis of the alarmist camp on irreversible and catastrophic harm – concerns that are beyond rational discussion – is an imperative. "Great emergencies require top-down solutions. They can be met only by mobilizing society as a whole, and establishing a command structure that will unite the people around a single goal."[24] Australian climate scientist Robert Carter

---

22. Hanekamp, "Neither Acceptable nor Certain," 242.

23. Scruton, *How to Think Seriously about the Planet*, 39.

24. Scruton, *How to Think Seriously about the Planet*, 82.

similarly suggests: "Increasingly the world's press and politicians have come to treat IPCC utterances as if they were scribed in stone by Moses. This is a reflection, first, of superb marketing by the IPCC and its supporting cast of influential environmental and scientific organizations; second, of strong media bias towards alarmist news stories in general, and global warming political correctness in particular; and, third, of a lack of legislators and senior bureaucrats possessed of a sound knowledge of even elementary science, coupled with a similar lack of science appreciation throughout the wider electorate."[25]

For some people, risk is a matter of choice best left to each individual: nature is neither malign nor benign, fragile or robust; human experience is one of adapting to changing circumstances. For many environmentalists, however, risk is a social issue, particularly in an overcrowded world where humans have disturbed fragile nature and placed the planet in peril.[26] To them, Scruton notes, "nature is precariously balanced and jeopardized by the hubris of risk takers."[27] Scruton sees politics as either the collective pursuit of an egalitarian goal or as a free association of individuals rather than as a social project aimed at imposing the choices of some on the whole.[28]

The need for cost-benefit analysis

Alarmists have long insisted that there is no time or need for sober economic analysis of the costs and benefits of the measures they assert are required *now*. Surprisingly, governments have shown little interest in cost-benefit analysis. Only reluctantly have they sought to justify demand-led measures by undertaking such analyses, and the results have been far from credible. As the authors of a report prepared for the US Congressional Research Service candidly point out: "It is difficult (and some would consider it unwise) to project costs up to the year 2030, much less beyond. The already tenuous assumption that current regulatory standards will remain constant becomes more unrealistic as time goes forward, and other unfore-

---

25. Robert M. Carter, "Knock, Knock: Where is the Evidence for Dangerous Human-Caused Global Warming?" *Economic Analysis & Policy* 38:2 (September 2008), 178.

26. This is one of the themes explored by Jeffrey Foss in *Beyond Environmentalism*. As he points out, nature's fragility is a *value*, rather than a *fact*, and that makes all the difference in how we approach environmental issues.

27. Scruton, *How to Think Seriously about the Planet*, 81.

28. Scruton, *How to Think Seriously about the Planet*, 97.

seen events (such as technological breakthroughs) loom as critical issues which cannot be modelled. Hence, long-term cost projections are at best speculative, and should be viewed with attentive scepticism. The finer and more detailed the estimate presented, the greater the scepticism should be."[29]

Environmentalists are generally suspicious of cost/benefit and other economic analyses which only lead to delay and to the danger of reaching tipping points that could cause irreparable harm.[30] They see climate change as a problem going beyond the tools available to economists and are thus prepared to take draconian steps and spend enormous amounts to mitigate climate change, which may not become acute, if at all, until well into the future. This approach is neither rational nor necessary. Tol points out by way of example that "more sober people would recognize greenhouse gas emissions as an externality. It is an externality that is global, pervasive, long-term, and uncertain – but even though the scale and complexity of this externality [are] unprecedented, economic theory is well equipped for such problems – and advice based on rigorous economic analysis is anyway preferred to wishy-washy thinking."[31]

Environmentalists assume that today's generation, with today's resources and technology, should address issues that may have an impact on their grandchildren, great-grandchildren, or even farther into the future. Based on the patterns of the past 150 years, future generations are likely to be substantially better off than today's generation and will have access to technology and resources to address problems more effectively and at lower cost than is possible today. As Nobel laureate-in-economics Gary Becker argues, "we don't have to sabotage today's thriving economy to insure ourselves against environmental upheaval. ... Where we set the discount rate for future environmental damage – if it is discounted at all – will make an enormous difference in what we spend today. Does it make sense to impose steep taxes on emissions and pay for carbon

---

29. Larry Parker and Brent D. Yacobucci, *Climate Change: Costs and Benefits of the Cap-and-Trade Provisions of H.R. 2454*, Document R40809, Congressional Research Service, September 14, 2009.

30. See, for example, Frank Ackerman, "Critique of Cost-Benefit Analysis, and Alternative Approaches to Decision-Making," A report to Friends of the Earth England, Wales and Northern Ireland, January 2008. In his view, cost-benefit analysis is a flawed procedure, which should not be central to public policy decisions on climate change or other issues.

31. Richard S. J. Tol, "Targets for Climate Change Policy: An Overview," *Journal of Economic Dynamics & Control*, 37:5 (May, 2013), 913.

sequestration? It depends on the costs, many years from now, of ameliorating environmental damage."[32] Similar to the critical role of climate sensitivity to the science of climate change, the choice of the discount rate makes all the difference in economic assessments of climate change policy.

Governments and firms routinely use discount rates to estimate the present or current value of future costs and benefits. A 1.4 per cent discount rate, for example, means that $100 today is worth $25 a hundred years from now, and $6 a hundred years later. Increase the discount rate to 3 per cent, and that same $100 would be worth only $5 a century later and 30 cents two centuries later. Depending on assumed market conditions and time frames, governments routinely rely on 5-7 per cent discount rates, i.e., they heavily discount the future. A very low discount rate assumes that future benefits are equivalent to today's benefits, strengthening the case for immediate action to secure those benefits; a high discount rate indicates that future values are relatively insignificant to today's generation and that their cost should be assumed by future generations. Higher discount rates also reflect the fact that future generations are likely to be better off and able to pay for their own benefits.

*****

Economists began showing professional interest in the global warming issue at the beginning of the 1990s. In the July, 1991, issue of *The Economic Journal*, for example, editor David Greenaway introduced articles by William Cline, William Nordhaus, and David Pearce examining a number of economic themes, all three accepting the basic outlines of the issue as set out in the first report of the IPCC (1990). Cline concluded in his review of the science that "many scientific uncertainties remain about the greenhouse effect. However, uncertainty is not necessarily grounds for policy inaction. Indeed, if policymakers are risk averse, they should attach a higher weight to the upper-bound warming and damage estimates than to the lower-bound estimates, so that a wider uncertainty spread around the central expected values might even be the basis for greater action."[33] He urged economists to become engaged and

---

32. Gary S. Becker, "An Economist Looks at Global Warming," *Hoover Digest*, April 2007. In economics, externality refers to the positive or negative impact of an economic activity that is experienced by third parties with no direct interest in the activity, for example, pollution.

33. Cline, "Scientific Basis for the Greenhouse Effect," *The Economic Journal* 101:4 (July 1991), 916. Cline expanded the article the following year into a full-length

complement the scientific work of the IPCC with their own assessments of the economic implications of climate change.

Most, but not all, economists have followed Cline's lead in accepting that the science is settled and that their role is to look at the economic policy implications. But David Henderson, a former chief economist at the OECD, has written a number of sharp critiques, particularly of government-employed economists, for their failure to think independently and critically about climate change and about related dimensions of what he first called "global salvationism." He cautions: "given the huge complexity of the climate system and the large gaps in present knowledge, the unsurprising existence of a range of views among the scientific upholders, and the extent of professional doubts and dissent, generalized references to a 'scientific consensus' are out of place."[34] Ross McKitrick has gone farther, collaborating with applied mathematician Chris Essex to write a blistering assessment of the shortcomings of the science.[35] They remain in the minority.

Cline points to one of the most challenging aspects of analyzing the economics of climate change: the need to project far into the future. Economic analysis of climate policy requires projections not only of the anticipated state of the climate thirty, fifty, or even a hundred years into the future, but also of the state of the economy, population growth, energy use, and developments in technology. As we have seen, climate projections have raised considerable controversy about both methodology and assumptions. In his 1998 generally optimistic treatment of climate change, Hoover Institution economist Thomas Gale Moore cautions against taking economic and other projections too seriously: "Economists are poor soothsayers and often over- or under-estimate growth. Accurate forecasts for a long period are not possible. Not only are we unable to predict the economic future but technology can change greatly. ... We can project, however, that future generations will have better technology at their disposal; that they will be wealthier; and that they will live

book, *The Economics of Global Warming* (Washington: Institute for International Economics, 1992).

34. Henderson, "Economists and Climate Science: A Critique," *World Economics* 10:1 (2009), 65. See also "Economics, Climate Change Issues, and Global Salvationism," *Economic Education Bulletin* XLV:6 (June 2005).

35. McKitrick and Essex, *Taken by Storm: The Troubled Science, Policy and Politics of Global Warming* (Toronto: Key Porter, 2002)

longer. They will certainly be in a better position to deal with any adverse changes in the climate than is mankind today."[36]

The IPCC relies on "scenarios" or "story lines" for both its climate and economic projections, many of which have been found to be wildly unrealistic. Most economic analyses of public policy are limited to projections ten or fifteen years into the future, and even then they often prove to be way off base. There is, for example, a long and not very stellar record of projections about anticipated resource depletion and population growth. Economist Julian Simon has demonstrated that all pessimistic projections make the same mistake: they underestimate human ingenuity. As he wrote:

> In the short run, all resources are limited. ... The longer run, however, is a different story. The standard of living has risen along with the size of the world's population since the beginning of recorded time. There is no convincing economic reason why these trends toward a better life should not continue indefinitely. ... The most important benefit of population size is the increase it brings to the stock of useful knowledge. Minds matter economically as much as, or more than, hands or mouths. Progress is limited largely by the availability of trained workers.[37]

Much discussion of the economics of climate change and of environmental issues more generally is characterized by the pervasive pessimism first advanced by Thomas Malthus in the early 19th century: not only is climate change characterized as a "wicked" problem, but alarmists acknowledge few upsides. From this perspective, the middle of the 20th century offered the best of all climates, and it has been downhill from the 1980s on. Malthus has been proven wrong time and again, as have his modern-day followers, and yet the media and much popular imagination continue to rely on a pessimistic view of the future. In that sense, widely held views about the alarming state of the world and the need for urgent government

---

36. Thomas Gale Moore, *Climate of Fear: Why We Shouldn't Worry About Global Warming* (Washington: Cato Institute, 1998), 143.

37. Julian Simon, *The Ultimate Resource 2* (Princeton: Princeton University Press, 1996), 12. See also Wilfred Beckerman, *In Defence of Economic Growth* (London: Jonathan Cape, 1975), for a thorough debunking of the many claims of the environmental doomsday cult. Beckerman revisited the issues 20 years later and found that, if anything, the claims of the cult had become even more fatuous and intellectually bankrupt. Wilfred Beckerman, *Small is Stupid: Blowing the Whistle on Green* (London: Duckworth Publishing, 1995). Both books are also available in US editions: *Two Cheers for the Affluent Society* (New York: St. Martin's Press, 1975), and *Through Green-Colored Glasses: Environmentalism Reconsidered* (Washington: Cato Institute, 1996).

action are not new, just misleading and unhelpful.[38] Today, as Jeffrey Foss observes, "environmentalism has become the most shared ideology of the legions of people discontented with the state of the world." For them, "the word fragile is almost always used to describe the environment."[39] Unfortunately, governments are willing to cater to these concerns and support policy responses that will prove ruinously expensive but have little if any prophylactic effect on nature, which remains indifferent to human impacts and will continue evolving as it has for the past four-and-a half billion years, often in more destructive ways than humans could ever contemplate, from volcanoes and earthquakes to hurricanes and tornadoes.

Serious analysis of climate change policy requires a realistic assessment of trends in population, economic growth, and energy use. That kind of assessment must take into account possible technological and political developments. It is not difficult to appreciate that the farther analysts look into the future, the more problematic projections become. Looking back only 30 years, one generation, it would have been difficult in 1985 to anticipate the many technological and political developments in the offing, from the revolution in information technology to the fall of the Soviet Union and the emergence of China, India, Brazil, and others as major players in the global economy. The availability of energy is even more volatile and subject to constant technological developments. Two generations ago, back to 1960, it would have been even more challenging. In climate change policy scenarios that look to the end of the 21st century and into the 22nd century, the uncertainties limiting the analysis are multiplied by the structural uncertainties of future population and economic growth, as well as energy, technological, and political developments.[40]

---

38. Simon and some of his fellow analysts have prepared a number of studies on the state of the world to counter the pessimistic views of the environmental movement. See Simon, ed., *The State of Humanity* (Oxford: Blackwell, 1995); Ronald Bailey, ed., *The True State of the Planet* (New York: Free Press, 1995); and Indur M. Goklany, *The Improving State of the World* (Washington: Cato Institute, 2007).

39. Foss, *Beyond Environmentalism*, 51 and 38.

40. See an early discussion of the many problems with this kind of forecasting and the false allure of computer-based models in Wilfred Beckerman, "Economists, Scientists, and Environmental Catastrophe," *Oxford Economic Papers,* New Series, 24:3 (November 1972), 327-344. Beckerman was responding to Donella Meadows, et al., *The Limits to Growth* (New York: New American Library, 1972), and The Ecologist, *A Blueprint for Survival* (London: Penguin, 1972). Both were the work of scientists using computers to forecast the future on the basis of sim-

Harvard economist Martin Weitzman offers a further twist to interpreting the economics of catastrophic AGW. He argues that the implications of structural uncertainties in the case of "wicked" problems, such as climate change, render traditional economic analysis inadequate. He writes: "The economic uniqueness of the climate-change problem is not just that today's decisions have diffi-cult-to-reverse impacts that will be felt very far out into the future, thereby straining the concept of time discounting and placing a heavy burden on the choice of an interest rate. ... Much more unset-tling for an application of (present discounted) expected utility analysis are the unknowns: deep structural uncertainty in the sci-ence coupled with an economic inability to evaluate meaningfully the catastrophic losses from disastrous temperature changes."[41]

Weitzman's Dismal Theorem led to considerable discussion among environmental economists. He is probably right when he ar-gues that in cases in which potential outcomes involve the deaths of hundreds of millions of people at least three or four generations into the future as well as the destruction of the world economy, it makes little sense to argue about discount rates and present value. Alarm-ists have used his Theorem to argue that, given the potential magni-tude of the problem, we must act now regardless of the cost. In such circumstances cost-benefit analyses are irrelevant. Lawrence Solo-mon places this kind of irrational anxiety into its proper perspec-tive: "There is no shortage of low-probability, high-consequence perils that could befall us at any time. If we must indulge a human need to worry about the unknown, let's at least pick fears worthy of trembling over, that at least offer some plausibility – no matter how remote – that they can befall us in future."[42] He places future global warming at the very bottom of this kind of anxiety.

Yale's William Nordhaus points out that Weitzman's analysis, while theoretically interesting, breaks down with some real-world examples. The probability of an asteroid, for example, wiping out

---

plistic linear projections. Their appreciation of economics, to put it kindly, was problematic.

41. Martin Weitzman "On modelling and Interpreting the Economics of Cata-strophic Climate Change," *Review of Economics and Statistics,* 91:1 (2009), 1. Weit-zman overstates the extent of scientific uncertainty. Based on paleoclimatologi-cal research, scientists are well aware of the boundaries within which global climate operates, boundaries that are acknowledged by all but the most extreme doomsayers.

42. Lawrence Solomon, "Let's Play Chicken Little," *Financial Post,* September 23, 2013.

much of civilization is extremely low but not impossible. Taking steps now, at astronomical costs, would not make much sense given that low probability. The utility of public policy for addressing climate change, in his view, can be subjected to meaningful economic analysis despite the uncertainties in both the science and the economics. He writes: "The results of the Dismal Theorem are important in emphasizing that we must always be cautious in our assumptions about specific functional forms in empirical research – whether those concern the utility functions or the probability distributions. There are indeed deep uncertainties about virtually every aspect of the natural and social sciences of climate change. But these uncertainties can only be resolved by continued careful analysis of data and theories."[43] It should be noted that Nordhaus is a respected environmental economist but is also a firm believer in anthropogenic climate change and its potentially catastrophic consequences.[44]

Given the complexity of the issues that must be included, Nordhaus and other environmental economists rely on Integrated Assessment Models (IAMs) for their analytical work. Climate change IAMs are tools that allow analysts to integrate different types of information into a coherent framework that can then be used by researchers and policy makers.[45] They can be used, for example, to examine key interactions between the climate system and anticipated socio-economic developments. They are not forecasts; they can provide, however, a framework for understanding climate change within a broader socio-economic context, providing policy researchers with insights into the pros and cons of different approaches to dealing with climate change. Parson and Fisher-Vanden explain: "integrated assessment models seek to combine knowledge from multiple disciplines in formal integrated representations; inform policy-making, structure knowledge, and prioritize key uncer-

---

43. "An Analysis of the Dismal Theorem," Cowles Foundation Discussion Paper No. 1686, January 2009.

44. Nordhaus's views are well represented in an exchange of views in the pages of the *Wall Street Journal*, in January and February of 2012, between two groups of prominent scientists and economists. Not satisfied with that exchange, Nordhaus contributed a further criticism, "Why the Global Warming Sceptics are Wrong," in the *New York Review of Books*, March 22, 2012. It in turn led to further discussion with three of the sceptics in the April 26, 2012 issue of the *Review*: "In the Climate Casino: An Exchange."

45. Modern IAMs are more sophisticated versions of the model built at MIT by Donella and Dennis Meadows in the late 1960s to provide the input into their 1972 report for the Club of Rome, *Limits to Growth*.

tainties; and advance knowledge of broad system linkages and feedbacks, particularly between socio-economic and biophysical processes. They may combine simplified representations of the socio-economic determinants of greenhouse gas emissions, the atmosphere and oceans, impacts on human activities and ecosystems, and potential policies and responses."[46] The aim is to develop insights going beyond science or economics that can be used by policy makers.

IAMs built by environmental economists such as Nordhaus or Richard Tol are mathematical computer models based on explicit assumptions about how the modelled system behaves. They allow a researcher to calculate the consequences of different assumptions and to interrelate many factors simultaneously. They use mathematical equations to model demographic, political, and economic variables that affect greenhouse gas emissions scenarios. Various equations link one period to the next, for example, an equation that adds emissions of CO_2 in one period to the stock already in the atmosphere. There may also be equations that add investment in that period to capital stock minus the depreciation of capital stock. Investments in capital stock will be a decision variable, as will investments in technologies that reduce emissions per dollar of GDP. Other decision variables will include consumption and thereby CO_2 emissions. IAMs are not directly linked to climate models, are much less complex, and can be run on most desktop computers equipped with the right program.[47]

IAMs require the modellers to make assumptions about future population and economic growth in both developed and developing countries and on that basis to assess future global welfare or probable energy consumption levels. They are thus constrained by the quality and character of the assumptions and data that underlie the model. They also rely on highly simplified representations of socio-economic and physical relationships. Models are constructed in different ways and rely on different assumptions about the interrelationships of the many factors that need to be taken into account. As they grow more complex, the range of assumptions also increases. Not all the models include the same variables nor do they model

---

46.　Edward A. Parson and Karen Fisher-Vanden, "Integrated Assessment Models of Global Climate Change," *Annual Review of Energy and the Environment* 22 (1997), abstract.

47.　I am indebted to Cornelis Van Kooten for some clarifications on the nature and use of IAMs.

their interrelationships in the same way. Like so much in climate science, IAMs remain relatively primitive; at this stage, they are more helpful in pointing to areas that need further research than in providing policy makers with credible policy advice. Michael Mastrandrea concludes, "the most important information to be gleaned from IAM efforts is not the specific numerical results of a particular modelling analysis but broader insights into the general structure of the policy challenge posed by climate change that come from examining results across models and understanding the relative importance of differences in assumptions that drive the results."[48] To date, it has not yet been possible to integrate IAMs with climate models in order to develop more dynamic results that provide for the interaction of both socio-economic and physical developments.[49]

The UK's Stern Review

Prior to 2005, the only known official economic analysis was prepared by the US Council of Economic Advisors during the George H.W. Bush administration. Under pressure from his chief environmental adviser, William Reilly, to support the UN's climate crusade, Bush sought advice from his closest political advisers, including his chief of staff, John Sununu, head of the Office of Management and Budget, Richard Darman, his science adviser, Allan Bromley, and the head of the Council of Economic Advisors, Michael Boskin. The four of them were well versed in both the science and economics, and all were of the view that there was no urgency. Instead, there was need for much better data and analysis than had been proffered to date.[50] In their 1990 *Annual Report*, the Council of Economic Advisors pointed out: "The United States is taking a leadership role in in-

48.   Michael Mastrandrea, "Representation of Climate Impacts in Integrated Assessment Models," Workshop Proceedings, *Assessing the Benefits of Avoided Climate Change: Cost-Benefit Analysis and Beyond*, Pew Center for Global Climate Change, May 2010, 97.

49.   For a full discussion of what is involved, see Paul N. Edwards, *A Vast Machine: Computers, Climate Data, and the Politics of Global Warming* (Cambridge: MIT Press, 2010). On their limitations, see Orrin Pilkey & Linda Pilkey-Jarvis, *Useless Arithmetic: Why Environmental Scientists Can't Predict the Future* (New York: Columbia University Press, 2007), and J. Scott Armstrong, Kesten Green, and Willie Soon, "Research on Forecasting for the Manmade Global Warming Alarm," *Energy & Environment* 22:8 (December 2011), 1092.

50.   Rupert Darwall, *The Age of Global Warming: A History* (London: Quartet Books, 2013), 136-46.

ternational efforts to reduce scientific and economic uncertainties about global climate change and to build a common understanding about all aspects of the climate change issue from the basic Earth science, to impacts on human activities, to potential response strategies. The data now available on the economic costs of reducing greenhouse gas emissions suggest that it may be as important to improve understanding of the economics of global warming as it is to improve current ability to predict warming itself."[51] Bush ultimately agreed to sign the 1992 UN Framework Convention on Climate Change at Rio de Janeiro, but the concern expressed by his advisers about the lack of serious economic analysis of climate change and of potential policy responses was not addressed by the administrations of either of his successors.

Early in 2005, the British House of Lords Select Committee on Economic Affairs, including former Chancellors of the Exchequer Nigel Lawson and Norman Lamont, and led by former Energy Secretary John Wakeham, held hearings into the economics of climate change; the committee was astounded to learn that the government had yet to conduct an enquiry into the economic implications of climate change for Britain. They were underwhelmed by both the scientific and economic analysis of the IPCC[52] and drew attention to many areas of the science's uncertainty and the need for more research. In their conclusions, the Lords expressed concern that "the links between projected economic change in the world economy and climate change have not been as rigorously explored as they should have been by the IPCC. We believe the complex interactions between world economic growth and climate change need additional scrutiny at the international level, and that the UK Government has a role to play in ensuring that this happens."[53]

---

51. Michael J. Boskin, Richard L. Schmalensee, and John B. Taylor, *The Annual Report of the Council of Economic Advisors* (Washington, 1990), 223. The Congressional Budget Office of the US Congress later published an economic report suggesting that a phased approach to carbon taxes would have a 1 to 2 per cent impact on total Gross National Expenditure. Alan D. Hecht and Dennis Tirpak, "Framework agreement on climate change: a scientific and policy history," *Climatic Change* 29 (1995), 389.

52. Economic analysis was the responsibility of IPCC WG3. In its first three assessment reports, WG3 had basically outlined the economic issues that governments would need to address but had not engaged in much serious analysis. That lacuna was finally addressed in the fourth and fifth assessment reports with predictable results.

53. House of Lords, Select Committee on Economic Affairs, "The Economics of Climate Change," 2nd Report of Session 2005-06, paragraph 145.

Shortly after publication of the Lords' report, the Labour government responded to their criticism of UK climate policy, and by
implication the work of the IPCC, of which it was one of the prime
sponsors, by appointing Sir Nicholas Stern to organize a thorough
cost-benefit analysis of climate change policy.[54] His team produced
a 700-page report (popularly referred to as the Stern Review) that
met all the expectations of the UK government, if not of its critics.
The Review starts off by assuming that climate change is the mother
of all market failures, presenting a unique challenge to economic
analysts and policy makers.[55] The concept of market failure, one of
the few economic ideas embraced with any enthusiasm by environmentalists, only applies to the extent that emissions of carbon
dioxide have a significant malign impact on the atmosphere; if the
effect is minor or even benign, little or no market failure exists.

The Review points to the potential impacts of climate change on
water resources, food production, health, and the environment as
compelling reasons for justifying environmental taxes and other
measures to minimize the economic and social disruptions that
would otherwise flow from climate change. It concludes that the benefits of strong, immediate action far outweigh the costs of not acting
or delaying action. Stern's team calculated that, without action, the
overall costs of climate change would be equivalent to losing at least
5 per cent of global GDP each year, now and forever.[56] Taking into
account a wider range of risks and impacts could increase this esti-

---

54.   Stern was, at the time, second permanent secretary of the Treasury and head of
      the British government's economic service. He had previously served as chief
      economist at the World Bank and has since pursued a career as an academic
      economist and climate activist.

55.   "The evidence shows that ignoring climate change will eventually damage eco-
      nomic growth. Our actions over the coming few decades could create risks of
      major disruption to economic and social activity, later in this century and in the
      next, on a scale similar to those associated with the great wars and the economic
      depression of the first half of the 20th century," *Economics of Climate Change,*
      "Executive Summary," ii. First released as an e-document by the UK govern-
      ment and then published as Nicholas Stern, et al., *The Economics of Climate
      Change* (Cambridge: Cambridge University Press, 2007). Stern published a
      shorter, more accessible version as *The Global Deal: Climate Change and the Crea-
      tion of a New Era of Progress and Prosperity* (New York: Public Affairs, 2009).

56.   Stern arrived at this number by assuming negligible damage now but massive
      damage later and by calculating the present value of this future damage on the
      basis of a very low discount rate, giving the impression that current damage is
      already 5 per cent of global GDP, one of the many statistical sleights of hand
      throughout the Review.

mate to 20 per cent of GDP or even more. The Review further indicated that devoting as little as one per cent of global GDP per annum to climate mitigation policies would avoid the worst effects of climate change by stabilizing GHG emissions at an atmospheric equilibrium of 500-550 ppm CO_2 equivalent (CO_{2e}) and at a temperature rise of no more than 2°C above pre-industrial levels.[57] The low-balling of costs was particularly risible in view of the lamentable results of the UK government's ten-year record in trying to reduce emissions. Costs had been astronomical, while results were minimal. As Oxford economist Dieter Helm concludes: "The happy political message that we can deal with climate change without affecting our standard of living – which is a key implicit message from the Stern Report on which politicians have publicly focused – and do so in a sustainable way, turns out, unfortunately, to be wrong."[58]

The Review projects the impact of climate change some two hundred years into the future. It recommends that today's and future generations devote part of their income to mitigation measures in order to avoid damage that will not have its full impact until well into the future, by which time, based on the Review's own estimates, incomes will have grown at least seven-fold, and perhaps as much as twelve-fold. It would be remarkable for even the best analysts to consider themselves competent to anticipate the many scientific, technological, economic, cultural, demographic, and political developments of the next two hundred years. It would be even more remarkable if they could convince their contemporaries to address and assume the costs arising from policy problems that would not begin to have a significant impact until well into the 22nd century. Such is the political economy of climate change.

Stern's team drew heavily on dramatic, non-peer-reviewed alarmist literature. The Review was presented as an independent assessment by an academic economist; in fact, it is a political document, the work of a group of UK officials led by the head of the UK's economic service. Unlike the IPCC, which in each of its assessment reports analyzes a wide range of possible future scenarios for economic and population growth, GHG emissions, and their impact on climate and global welfare, the Stern Review relies on on-

---

57. CO_2equivalent refers to the cumulative impact of increased CO_2 and other anthropogenic greenhouse gases, such as nitrous oxide, ozone, chlorofluorocarbons, and methane.

58. Dieter Helm, "Climate-change policy: why has so little been achieved?" *Oxford Review of Economic Policy* 24:2 (2008), 228.

ly one scenario, and an outlier at that. It was not peer-reviewed and achieved its dramatic conclusions by drawing on the most alarmist views of future climate and its impacts and on the least realistic assessments of the ability of clean energies to replace fossil-fuel based energy. It projects that global population will triple by 2150 to 21 billion, despite the widely held view among demographers that it would likely reach a maximum of 9-10 billion in the second half of the 21st century and then decline. The result is a much higher estimate of future emissions and climatic impact, including global temperatures rising by up to 10°C, twice the highest IPCC estimate.

The most controversial aspect of the Review was Stern's choice of a 1.4 percent discount rate as the basis for costing future damage and current policy action, based on a 0.1 percent time preference and an assumption of a 1.3 percent annual growth in per capita consumption. By choosing a low discount rate, Stern inflated the cost of future damage and deflated current costs. Stern later admitted that a 2.8 percent rate would have been more realistic, thus reducing his damage estimate of 5 percent per annum to only 1.4 percent per annum.[59] He justified his low discount rate as a matter of intergenerational ethics,[60] an argument that has stirred up considerable discussion among economists and ethicists, particularly given his own projections that average incomes will rise substantially over his projected, two-century time span.[61]

The Review was well received by its client, the UK government, by the IPCC community as a whole, and by the media. From their perspective, Stern had provided everything needed to dispel criticism: a strong affirmation of the rising danger of climate change, a ringing call to action, and confirmation that remedial action would cost significantly less than failure to act. The world of ENGOs similarly embraced the Review, although with the caveat that the report was perhaps too conservative and not alarmist enough. Franck

---

59. Stern, "Technical Annex to Postscript to Stern Review," Table PA.3.

60. The Review's extreme view of intergenerational ethics originated in the advice of philosopher John Broome, a professor of moral philosophy at Oxford and the author of a number of books on the ethics of economics, intergenerational justice, and cost of climate change. Broome later explained his view in an article in *Scientific American:* "The Ethics of Climate Change," June 2008, 96-102. He later served as one of the lead authors of IPCC Working Group 3's 2014 assessment report.

61. For a full, measured discussion of the economics of climate change and mitigation measures, see van Kooten, *Climate Change, Climate Science, and Economics,* particularly chapters 6-8.

Ackermann of the Global Development and Environment Institute at Tufts University, for example, concluded his assessment of the Review and its critics for the Friends of the Earth as follows: "The Stern Review is far from being the last word on every aspect of the economics of climate change – but it is much less wrong than the analyses that preceded it. It has decisively laid to rest the notion that standard economic methods somehow counsel timidity in the face of global crisis."[62]

Not everyone agreed, particularly economists, many of whom had previously analyzed some of the preferred policy choices and had indicated the range of costs associated with no action, with adaptation, and with various mitigation provisions, particularly carbon taxes and cap-and-trade schemes. None of their estimates came close to those put forward by Stern and his team.[63] It is not surprising that Stern's work was met by a storm of criticism, even from economists generally sympathetic to the idea of catastrophic anthropogenic climate change and the need for early action.

Richard Tol pointed to the many internal inconsistencies in the Stern Review. The Review, for example, asserted that Africa would experience famine, disease, and other problems but at the same time would become much richer. A richer Africa should be able to import food during periods of scarcity and spend more on public health, agricultural modernization, and other issues. Similarly, a richer world could afford to adapt to a gradual rise in sea level and thus avoid the many horrors Stern predicted. Such inconsistencies are the product of relying exclusively on the PAGE model for damage assessments rather than of considering a wider range of model results.[64] Tol concludes that "the Stern Review is very selective in

---

62. Frank Ackerman, "Debating Climate Economics: The Stern Review vs. Its Critics," at www.ase.tufts.edu.

63. Peter Lilley notes that Stanford University's *Energy Modelling Forum* looked at 21 economic models of the costs of climate change mitigation and found the average of the range of estimates to be 2.2 per cent of GDP, rising to 6.9 per cent by 2100. One per cent is emphasized in Stern's conclusions; the body of the Review, however, indicates that it is one per cent, +/- 3 per cent, a much wider range. Stern has subsequently indicated that he now believes 2 per cent is the appropriate figure. See Peter Lilley, *What Is Wrong With Stern: The Failings of the Stern Review of the Economics of Climate Change* (London: Global Warming Policy Foundation, 2012), Report 9, at 20. See also *Climate Change* 123:3-4 (April 2014), which provides a series of articles resulting from the *Energy Modelling Forum's* comparative assessments.

64. The PAGE (Policy Analysis of Greenhouse Effect) model is an IAM conceived by Chris Hope, a business economist at Cambridge University. It can be used to

the studies it quotes on the impacts of climate change. The selection bias is not random but emphasizes the most pessimistic studies. The discount rate used is lower than the official recommendations by HM Treasury. Results are occasionally misinterpreted. The report claims that a cost-benefit analysis was done, but none was carried out. The Stern Review can therefore be dismissed as alarmist and incompetent."[65]

Other mainstream environmental economists, while generally sympathetic to Stern's goals, were critical of both his methods and his conclusions. Nordhaus found that the Review's unambiguous conclusions about the need for urgent and immediate action would not survive the substitution of assumptions consistent with the real interest rates and savings rates of today's marketplace. "So the central questions about global warming policy – how much, how fast, and how costly – remain open."[66] His colleague Robert Mendelsohn, while acknowledging Stern's contribution to the literature, concludes that policy makers will need both better science and better economics before committing hundreds of billions of dollars to mitigation measures. Until such time, the economics of climate change still indicate that the costs of mitigation at this stage far outweigh the benefits.[67]

Economists less persuaded of the urgency of the climate issue were even more critical. Climate scientist Robert Carter and economist Ian Byatt organized a long critique of both the science and the

---

project future increases in global mean temperature (GMT), the economic costs of damages caused by climate change, the economic costs of mitigation policies, and the overall impact of adaptation measures (including costs of adaptation measures and reduction in damage costs that result from adaptation). It uses simple equations to simulate the results from more complex specialized scientific and economic models, while accounting for the uncertainty that exists around climate change. It works with eight world regions, ten time periods to the year 2200, four sectors (sea level rise, market, nonmarket and major discontinuities), and is able to examine the impacts of climate change, as well as the costs of mitigation and adaptation. See "Overview" at *Climatecolab.com*.

65. Richard S.J. Tol, "The Stern Review of the Economics of Climate Change: A Comment," *Energy & Environment* 17:6 (2006), 979-80.

66. William Nordhaus, "Critical Assumptions in the Stern Review of Climate Change," *Science* 317 (13 July 2007), 202. See also Nordhaus, *A Question of Balance* (New Haven: Yale University Press, 2008), for a more complete discussion of the economics of climate change.

67. Robert O. Mendelsohn, "A Critique of the Stern Report," *Regulation*, Winter 2006/07, 42-6.

economics of the Review.[68] The scientific critique tracks the arguments outlined in chapters four to seven above, noting the extent to which Stern and his colleagues selected some of the most alarmist conclusions from the literature, including much from non-peer-reviewed sources. It concludes that the "Review fails to present an accurate picture of scientific understanding of climate change issues, and will reinforce ill-informed alarm about climate change among the general public, the bureaucracy and the body politic."[69]

The economic half of the critique provides a companion assessment of the Review's shortcomings, from its many questionable assumptions to its faulty handling of data, its failure to take account of the wider literature on the economics of climate change, its reliance on model-based speculations, and its choice of remedial instruments. The authors point out that "the Stern Review has understated [prospective costs], probably by a wide margin. The combination of projected benefits that are pitched too high and projected costs that are pitched too low has led to a seriously unbalanced presentation of policy alternatives. ... [In discounting the future], critical issues are not fully explored, the bias towards immediate and far-reaching actions to reduce emissions is reinforced, and the risks and problems that would arise from following the Review's prescriptions for policy are not faced."[70] The authors conclude that the Review greatly understates the extent of uncertainty of both the science and the economics of climate change and relies to an excessive extent on selected and biased sources, making the Review a vehicle for "speculative alarmism. ... [rather] than a basis for informed and responsible policies."[71]

One of the more telling criticisms the authors make is that Stern, as well as the IPCC before him, relies on market-exchange rates (MERs) rather than purchasing power parity rates (PPPRs). MERs, much favoured by development activists, tend to distort the differences between rich and poor and have a significant inflationary ef-

---

68. "The Stern Review: A Dual Critique," *World Economics* 7:4 (Oct.-Dec. 2006) – Part I: The Science (Robert M. Carter, C.R. de Freitas, Indur Goklany, David Holland and Richard Lindzen); Part II: Economic Aspects (Ian Byatt, Ian Castles, Indur M. Goklany, David Henderson, Nigel Lawson, Ross McKitrick, Julian Morris, Alan Peacock, Colin Robinson and Robert Skidelsky).

69. "Dual Critique,"194.

70. "Dual Critique," 200.

71. "Dual Critique," 224.

fect when projected far into the future.[72] Thus in constructing scenarios about the future impact of global warming, the use of MERs
tends to project unreliable levels of future global consumption, energy use, and damage, and thus paints a much more dramatic picture than would result from the use of PPPRs. Despite the broadly
acknowledged credibility of this criticism, the UN, the World Bank,
and other international organizations continue to use MERs in their
forecasts when it suits their purposes.

British politician Peter Lilley has also provided a comprehensive critique, finding the Review to be duplicitous and misleading,
"achieved by verbal virtuosity combined with statistical sleight of
hand," well outside the mainstream of economic analysis, based on
inappropriate comparisons, conflating future and current costs and
grounded in a peculiar view of intergenerational ethics. [73]

Stern subsequently responded to his critics. In a June 2008 article, Stern increased the estimate for the annual cost of achieving
stabilization between 500 and 550 ppm CO_{2e} to 2 percent of GDP to
account for faster than expected climate change; rather than toning
down his assessment, Stern chose to go farther down the most
alarmist road despite the fact that in the preceding decade the planet had failed to warm. By this time, the alarmist community had begun to subtly change the language from anxiety about global warming to alarm about climate change. A few years later, the focus shifted to the impact of anthropogenic climate change on extreme
weather, a wholly untestable proposition, as is the use of the concept of "climate weirding." Nevertheless, Stern felt fully vindicated
when the SPM of the IPCC's 2007 Fourth Assessment Report concluded that the need for action was even more urgent than his Review had assumed.

---

72. See the discussion in Aynsley Kellow, *Science and Public Policy: The Virtuous Cor-
    ruption of Virtual Environmental Science* (Cheltenham, UK: Edward Elgar, 2007),
    50-53.

73. Stern's estimates of the harm unchecked global warming will have on human-
    kind are ten to twenty times the average of those in the economic literature he
    reviewed. He reached this conclusion without any original research of his own,
    other than the projections of the PAGE 2002 IAM. As Lilley concludes: "the Re-
    view's estimates of the cost of reducing emissions to acceptable levels are as op-
    timistic as his estimates of the likely scale and damage of global warming are
    pessimistic." (Lilley, *What Is Wrong With Stern,* 23-4). Lilley points out that there
    is a curiously incestuous relationship between the IPCC's 2007 fourth Assess-
    ment Report and the 2006 Stern Report. Although not peer-reviewed and not
    published until past the deadline for inclusion, the IPCC cites the Stern Review
    26 times, including on matters of science (Lilley, *What Is Wrong With Stern,* 17).

Given its many shortcomings, the Review has not had the influence its authors had anticipated. Its function was largely to reinforce the policy choices already made by the then Labour government in Britain. In short, it was official economics. It also found favour among the alarmist community, but most professional economists who took the trouble to read it agreed with Australia's independent Productivity Commission, which saw it "as much an exercise in advocacy as it is an economic analysis of climate."[74] Nevertheless, Stern succeeded in drawing more attention to the need for serious economic analysis of climate policy. It is not hard to appreciate that economists' perspectives on the science of climate change as well as their views on the benefits of public economic intervention heavily influenced their assessments of the Review in particular and of the economics of climate change more generally. Conservative or market-oriented economists such as David Henderson, Ian Byatt, and Ross McKitrick fell on one side and liberal or progressive economists such as Joseph Stiglitz,[75] Jeffrey Sachs,[76] and Paul Krugman[77] on the other. Public policy specialists can be excused for throwing up their hands and bemoaning the lack of objective, evidence-based analysis. The fact that the analysis is so often value-laden, however, whether the issue is the science of climate change or the economics of mitigation, indicates that ultimately the issue is a matter of politics, to be determined by governments on the basis of their best assessment of both the science and the economics. Arriving at such an assessment requires a willingness to listen to the views of more than just the advocates of alarm.

---

74.   Australia, Productivity Commission, "The Stern Review: An Assessment of Its Methodology," January 2008, x.

75.   Joseph Stiglitz, Stern's predecessor as chief economist at the World Bank, has strongly supported action on climate change. See, for example, "Turning tides: The climate change message is finally getting through; it's time for political leaders to move beyond mere rhetoric and act," *The Guardian*, February 10, 2007.

76.   Jeffrey Sachs now heads the Earth Institute at Columbia University, an institute dedicated to all issues related to sustainable development, including climate change. See "The Dangerous New Era Of Climate Change," *Economy Watch*, August 2, 2012.

77.   Princeton's Paul Krugman has never met a left-wing cause that he will not support and has used his New York *Times* columns to castigate all those who fail to take the climate change "crisis" as seriously as he does. See Krugman, "Betraying the Planet," New York *Times*, June 29, 2009, and "Building a Green Economy," New York *Times Magazine*, April 7, 2010.

Australia's Garnaut Review

Only one other government sought to make its own assessment of the costs of climate change mitigation. In 2007, then Australian opposition leader Kevin Rudd, together with the Australian states and territories, commissioned economics professor Ross Garnaut to lead a study on the impact of climate change on Australia. Rudd became prime minister later that year and confirmed the participation of the federal government. Garnaut delivered his report in September 2008 as well as a revised version in 2011.[78]

Like Cline, Stern, Nordhaus, and other environmental economists, Garnaut took the view that "the outsider to climate science has no rational choice but to accept that, on the balance of probabilities, the mainstream science is right in pointing to high risks from unmitigated climate change." Also like Stern, Garnaut took a more alarmist view than the IPCC, assuming a 5°C increase in global temperatures by the end of the 21st century but concluding that taking mitigation action now would not bring significant benefits until the 22nd and 23rd centuries. Nevertheless, he was confident that "a high proportion of Australians are prepared to pay for mitigation in higher goods and services prices." (*Final Report*, xviii)

Garnaut recommended that the government promote a global effort to achieve a CO_{2e} level of 450 ppm. Australia's contribution would require an ambitious 25 percent reduction from 2000 levels by 2020, and a 90 percent reduction by 2050.[79] In the event that the global target of 450 ppm could not be achieved, Australia should agree to a 550 CO_{2e} level, requiring a more realistic reduction of 10 percent by 2020, and an 80 percent reduction by 2050. Should UN negotiations fail, Australia should still reduce its emissions by 5 percent by 2020, based on 2000 levels.

In modelling the economic and physical impacts, the Report's principal focus was on Australia's commitment to achieving a 550 ppm global stabilization objective by 2050 and avoiding a 5°C temperature increase by the beginning of the 21st century if no action

---

78. *Final Report on Climate Change Review*, September 2008 at Garnautreview.org. Similar to the Stern Review, the 2008 Report is almost 700 pages long, relies on modelling, and seeks to project a high level of scientific and economic sophistication in its pages.

79. During the Kyoto Protocol negotiations, the Howard government had agreed to a commitment to keep Australian emissions to no more than an 8 per cent rise beyond 1990 levels. Even then, the government had concluded that this could not be achieved and had not ratified the Protocol.

was taken – that is, an even higher estimate than the IPCC range. Garnaut concluded that this could be achieved because, in his view, "Australian material living standards are likely to grow strongly through the 21st century, with or without mitigation" (*Final Report*, 565). His modelling results found that "mitigation [would cut] the growth rate over the next half century, [but] lift it somewhat in the last decades;" GDP at the end of the century would be "higher with 550 mitigation than without" (*Final Report*, 245). Following Garnaut's policy advice would mean that the present generation of Australians would experience lower growth for the next 40 or so years but that the next (and later) generations would benefit from higher growth. By the end of the 21st century, Australian GDP would have grown 700 percent by pursuing his recommendations. The unanswered question is why today's generation should assume significant costs on behalf of future generations who are modelled to be seven-times better off.

In Australia Garnaut's Report met with even more criticism than Stern's did in the UK. Environmentalists, as has become a familiar pattern, condemned Garnaut for lack of ambition, particularly by offering a second-best option. However, Australia's very active and articulate community of climate sceptics found little to redeem the high cost of the Report. Economist Tim Curtin summed up their views succinctly:

> The Garnaut Report's stringent emission reduction targets stand or fall on the validity of the climate science of the IPCC. ... But even if this science proves correct, the Report's unsound economic cost-benefit analysis results in policy proposals that impose inordinate costs now for uncertain benefits far in the future. It is much more certain that by 2100 the Report will have taken its place alongside Malthus (1799), Jevons (1865), Ehrlich (1970), and the Club of Rome (Meadows at al. 1972) for being as spectacularly wrong as these eminent "scribblers" were with their equally fanciful predictions.[80]

Des Moore, a former deputy treasurer in the Australian government, concluded in his assessment of the Garnaut Report:

> The uncertainties about mainstream science and the extent of dissent are so large that they rule out any application of the so-called precautionary principle. I also conclude that, even if it were accepted that temperatures will increase over time, the large uncertainties about the timing and extent of the alleged mitigating action said to be needed

---

80. Tim Curtin, "The Contradictions of the Garnaut Report," *Quadrant Online*, January 1, 2009.

suggests that no case exists for governments to start a comprehensive program now to reduce greenhouse gas emissions. ... My assessment of the published economic modelling, and the potential availability of alternative technology, leads me to conclude that there is no substantive basis for urgent action by Australia, let alone the world, to reduce greenhouse gas emissions.[81]

Nevertheless, Australia's then government adopted the basic policy thrust of the Garnaut Report and used it to justify its acceptance of the Kyoto Protocol, activist participation in the work of the IPCC and the annual conferences of the Parties to the UN Framework Convention on Climate Change, and the imposition of a carbon tax on major emitters. Garnaut's assessment of the willingness of Australians to adopt mitigation measures proved to be wrong. Climate change policies added to the increasing unpopularity of the Labour Government and led to its defeat by the Liberal-led Coalition in the September 7, 2013, federal election. The new prime minister, Tony Abbott, campaigned on withdrawing the carbon tax and initiating a thorough review of Australian climate policy. Garnaut's advice has not been sought by the new government.

IPCC economic analysis of mitigation

For the first decade or two of concern about anthropogenic climate change, serious economic analysis and integrated assessments were limited to the specialist literature and were not an integral part of public policy discussions. The only widely available official analysis was done by the IPCC in the Reports of Working Group 3 (WG3). It addressed mitigation in each of its five Assessments. The first Report (1990), chaired by US Assistant Secretary of State for Oceans and International Environmental and Scientific Affairs, Fred Bernthal, focused on the policy issues that governments would need to consider in negotiating a climate treaty and implementing it in domestic policy. This first report relied to some extent on modelling and scenarios, but its conclusions and recommendations were relatively modest in comparison with the more baroque and alarmist tone of later reports. Its main findings included:

- climate change is a global issue requiring a cooperative global response;
- industrialized countries have a special responsibility given their much larger emissions and capacity;

---

81. Des Moore, "Global Warming and Uncertainty – What is the appropriate response?" Address to Economic Society (Australia), October 2, 2009, 2 and 4.

- growing developing country emissions need to be factored into any solutions;
- global economic development will increasingly have to take account of environmental impacts;
- mitigation and adaptation strategies need to be part of an integrated approach;
- the potential seriousness of consequences points to the need to begin response measures in the face of significant uncertainties; and
- public education is essential to promote awareness and to guide positive practices.

The report concluded that these factors should lead to a flexible and progressive approach.[82]

In preparation for AR3 (2001), the IPCC commissioned an expert group to develop emissions scenarios in order to provide a more integrated basis for both the scientific and mitigation assessments. Scenario development had been part of global systems modelling since the 1970s when computer modelling was first introduced. The new IPCC scenarios were not completed in time to be fully incorporated into the third assessment but became the basis for much of the analysis that went into WG3's fourth assessment report (2007). The approved scenarios were based on four story-lines, each developed to describe the interrelationships among the forces driving emissions and their evolution. The resulting set of 40 scenarios covered a wide range of the imagined demographic, economic, and technological driving forces that would determine levels of future greenhouse gases and other emissions.

Each scenario represents a specific quantification of one of the four story-lines. The four story-lines are then run through different integrated assessment models developed by both private and public modelling groups to produce the outcomes that are the heart of the IPCC exercise. In effect, each run represents a triple layer of assumptions: possible demographic, economic, and technological developments over the period being modelled, i.e., until the end of the 21st century; modelled levels of emissions resulting from each scenario; and the impact of these emissions (or forcings) on the evolving state of the climate. The extent of uncertainty flowing from this exercise cannot be overstated due to model inadequacies, parameter uncer-

---

82. IPCC, Working Group 3, *Climate Change: The IPCC Response Strategies* (1990), at IPCC.ch.

tainties, and climate system internal variability. With different story-lines and assumptions, totally different results could be modelled that would be as plausible as those used by the IPCC.[83]

The results, as synthesized by the IPCC, bristle with tables, charts, and schematics to provide the impression of a high level of expertise and precision in the projections. In reality, the results are far from scientific forecasts. Rather, they are stories that the IPCC asserts are based on plausible developments over the modelled time period.[84] Not everyone would agree with their plausibility, particularly for those scenarios that involve rapid economic and global population growth which, not surprisingly, produce the most alarming climate results. Also not surprisingly, the scenarios based on full national implementation of the IPCC's preferred policies produce the desired climate outcomes.

Based on this extensive exercise in modelling, the Report of AR4 WG3 was able to adopt a much more aggressive and less nuanced approach. Gone were the caution and uncertainty of the first report. Co-chaired by Bert Metz of the Netherlands and Ogunlade Davidson of Sierra Leone, the 2007 Report is full of notes, charts, and images to guide the reader and to leave no doubt that while climate change is a serious matter, solutions are clear, feasible, and cost effective. Much of that confidence draws on assertions by environmental groups rather than on assessments found in the scholarly literature. Its *Summary for Policymakers* provides a confident assessment of what can and must be done by governments, most of them involving a high level of international and national regulatory action.[85] With proper management, the Report concludes, these measures would solve the climate crisis and would even strengthen the global economy by making it more equitable. The scenario with the greatest reduction in greenhouse gases resulted in a reduction of global GDP of less than three percent in 2030, measured against the background of rising income levels over the same period. The same scenario would trim av-

---

83. On the limits of IPCC scenario building, see G. Cornelis van Kooten, *Climate Change, Climate Science, and Economics: Prospects for an Alternative Energy Future* (New York: Springer, 2013), 102-09.

84. IPCC, "Emission Scenarios and Storylines," Technical Summary, *Special Report on Emission Scenarios.*

85. The *Summary* is a modest 23 pages of relatively straightforward, if perhaps controversial, language. The underlying report, however, is dense and runs to more than 800 pages of small print.

erage annual GDP growth rate by less than 0.125 percentage points in 2030.[86]

Richard Tol is among the few environmental economists to have read the full 2007 report. He concludes: "the WG3 report did not attract the same scrutiny [as the WG2 report]. This could create the impression that WG3 wrote a sound report. That impression would be false. Just as WG2 appears to have systematically overstated the negative impacts of climate change, WG3 appears to have systematically understated the negative impacts of greenhouse gas emission reduction."[87] He points out, for example, that WG3 Chapter 11 suggests that climate policy could stimulate economic growth and create jobs. These claims are supported by grey or advocacy literature only, not by peer-reviewed studies, as alleged. Similarly, the Report systematically ignores studies and findings that do not support its bullish conclusions and distorts others in order to boost its case. It leaves much to be desired as a report surveying the scholarly literature, cherry-picking what helps, ignoring what does not, and relying on grey literature for some of its more controversial findings.

Similar to the reports of Working Groups 1 and 2, the *Summary for Policy Makers*, the only pages likely to be read by the public and policy makers, draws conclusions that are not supported or that are even contradicted by the detailed analysis in the main report. The report should be read in the context explained by Ottmar Edenhofer of the Potsdam Climate Institute in Germany. Edenhofer was a lead author of the 2007 WG3 report and was promoted to chair of the Group, serving as editor of its 2014 report.

> [I]t's a big mistake to discuss climate policy separately from the major themes of globalization. The climate summit in Cancun [2010] at the end of the month is not a climate conference, but one of the largest economic conferences since the Second World War. ...Developed countries have basically expropriated the atmosphere of the world community. But one must say clearly that *we redistribute de facto the world's wealth by climate policy.* ... One has to free oneself from the illusion that international climate policy is environmental policy. This has almost nothing to do with environmental policy anymore, with problems such as deforestation or the ozone hole.[88] [emphasis added]

---

86. IPCC Fourth Assessment Report: *Climate Change 2007*, Report of Working Group 3, *Summary for Policymakers*, Table SPM 4.

87. "Richard Tol on Working Group 3 of IPCC," *Klimazwiebel*, February 28, 2010.

88. "IPCC Official: Climate Policy is Redistributing The World's Wealth," *WattsUp WithThat*, November 18, 2010.

Edenhofer clearly followed his own advice in delivering WG3's 2014 report and inserted a paragraph early in the *Summary for Policy Makers* encapsulating the IPCC's guiding philosophy: "Limiting the effects of climate change is necessary to achieve sustainable development and equity, including poverty eradication. ... Consequently, a comprehensive assessment of climate policies involves going beyond a focus on mitigation and adaptation policies alone to examine development pathways more broadly, along with their determinants."[89]

The three working groups apparently devoted limited attention to ensuring that the reports of the three groups were reasonably consistent with each other. WG1 claimed that its confidence in the malign impact of human impacts on the climate had reached 95 percent, even as its estimate of the range of that impact had widened, a discrepancy most commentators found bizarre and inconsistent with emerging research on the climate system's sensitivity to greenhouse gases. WG2 concluded that the impact of anthropogenic climate change in the near future was such that governments should devote resources as much to adaptation as to mitigation. Not so WG3. It fell back on the familiar mantra of looming tipping points and limited time. For more than 30 years, the UNEP and the IPCC have warned that humanity has less than 10 to 15 years to act in order to solve the climate crisis. The deadlines have come and gone, and global temperatures have not risen but fluctuated within a narrow band well within historical experience at local, regional, and global levels.

WG3's 2014 report, again based on extensive modelling of climate, energy use, and socio-economic factors, has no more credibility than the 2001 and 2007 reports. This time around, the modellers relied less on the story lines used in 2007 and instead developed "representative concentration pathways." They collected about 900 mitigation scenarios based on IAMs. The report continues to emphasize mitigation rather than adaptation and argues that in order to achieve a stable temperature of no more than 2°C above pre-industrial levels by 2100, governments need to take more aggressive steps to scale back on fossil fuel use by adopting a four-step approach: phase out coal-fired, electrical generation as quickly as possible; rely on gas-fired generation as a bridging technology for the immediate future; phase in nuclear generation to replace both as soon as possible; and continue to subsidize and increase wind and

---

89. IPCC, AR5, WG3, *Summary for Policy Makers*, 5.

solar facilities. But achieving these goals would require significant investments in the order of four percent of global GDP per year through 2030 in order to complete the transition to a largely decarbonized world. The report also notes that western diets need to become more sustainable by reducing meat consumption. If governments fail to meet these goals, the report maintains, global temperatures will rise by up to 4.7°C by the end of the century, leading to irreversible and dramatic damage to the planet.

The original draft of WG3's 2014 report created considerable unrest among both authors and governments. Before its release, Richard Tol asked that his name be removed from the list of authors because he concluded that both in tone and content the report was more alarmist than the facts and the analysis warranted.[90] Government representatives apparently took note and, at the final session to approve the *Summary for Policy Makers,* toned it down to such an extent that a number of authors, including lead author Edenhofer, claimed it had lost its sense of urgency.[91] The much less influential, but more alarmist, *Technical Summary* survived intact.

Even in its diluted form, WG3's report remains true to the broader agenda central to the UN as a whole but only incidental to climate change. The real issue is sustainable development, a vague concept that has been a central UN theme for more than thirty years. Its full implementation would involve extensive social engineering and major changes in global governance and the global economy in order to create a more "just" world patterned on the UN's image. This is what activists have in mind when they hold up signs calling for "climate justice." Analysts who read IPCC reports from the more limited perspective of an alleged climate crisis miss the broader context in which this agenda is being pursued. That approach is also clear in both the Stern Review and the Garnaut Report.

At the dawn of the environmentalist era, Barbara Ward and René Dubos asserted that "the two worlds of man – the biosphere of his inheritance, the technosphere of his creation – are out of balance, indeed potentially in deep conflict. And man is in the middle. This is the hinge of history at which we stand, the door of the future opening on to a crisis more sudden, more global, more inescapable, and more bewildering than any ever encountered by the human

---

90. See Cheryl K. Chumley, "UN climate author withdraws because the report has become 'too alarmist'," Washington *Times,* March 27, 2014.

91. Nick Miller, "IPCC report summary censored by governments around the world," Sydney *Morning Herald,* April 15, 2014.

species and one which will take decisive shape within the life span of children who are already born."[92] They see technology as a threat to the biosphere and thus to humans, rather than as a potential instrument for meeting human needs and aspirations. Ward and Dubos do not celebrate human achievement but characterize it as a threat to the survival of the planet. Similar to Edward Goldsmith,[93] E.O. Wilson, and others who believe that mankind has evolved beyond its ecological niche, the core concern from this perspective is that there are too many humans, all of them seeking to enjoy the fruits of modern technology. Ehrlich, Goldsmith, Wilson, Ward, and the rest of the utopians who want to save the planet from the human scourge are all of one mind: to keep the Earth green, the world's population must be severely reduced and governed on progressive principles, i.e., by an enlightened cadre of experts capable of providing centralized direction.[94]

Ward and Dubos were at the vanguard of a movement that has become more and more aggressive over the years. Hailed as humanitarians by a fawning media, they and their followers are in fact profoundly anti-human in both their diagnosis and prognosis of the issues facing mankind. Robert Zubrin echoes their words but turns

---

92. Barbara Ward and René Dubos, *Only One Earth: The Care and Maintenance of a Small Planet* (New York: WW Norton, 1972), 12. See also the religious tone of the conclusions, 220.

93. Edward Goldsmith founded the UK journal, the *Ecologist,* in 1970 with an editorial that set out its commitment to deep ecology. In Goldsmith's view, the widespread adoption of agriculture moved humans beyond their natural ecological niche as hunter-gatherers, increased their numbers, and led to the widespread development of civilization, the results of which were cataclysmic for the biosphere. Similar to the Club of Rome, he characterized modern humans as parasites on a fragile planet. The planet would only endure by reducing humanity's presence and ecological footprint. In 1972, the *Ecologist* published *A Blueprint for Survival* (London: Penguin, 1972), a statement by 36 eminent British academics largely drawn from the sciences, expanding on this dismal view of man. It was well received by media elites and sold well. Similar to reports from the Club of Rome and other utopian pessimists, Goldsmith's dire prognostications have not worn well outside of the environmental movement.

94. For a full discussion of this theme, see James Delingpole, *Watermelons: The Green Movement's True Colors* (New York: Publius, 2011). Similar to its contemporary counterparts, peace and feminist studies, environmental studies emerged from the swamp of left-wing theories that have mushroomed and infected many traditional fields of study since the 1970s, from literature to history. Roger Scruton points to the pervasive influence of Marxism as the intellectual underpinning of the new secular religion of the academy. *Thinkers of the New Left* (Burnt Mill, UK: Longman, 1985).

their perspective on its head: "Humanity thus stands at a crossroads, facing a choice between two very different visions of the future. On one side stands anti-humanism which, disregarding its repeated prior refutations, continues to postulate a world of limited supplies, whose fixed constraints demand ever-tighter controls upon human aspirations. On the other side stand those who believe in the power of unfettered creativity to invent unbounded resources, and so, rather than regret human freedom, insist upon it. The contest between these two views will determine our fate."[95]

Superficially, the struggle is portrayed as a matter of science, drawing on the continuing, if misplaced and exaggerated, authority that it enjoys. In realty, climate change provides a proxy for an ideological, even cultic struggle, and the goal is power, the power to transform the world on the basis of an alluring utopian vision. We turn now to an assessment of the movement's cultic dimensions.

&&&&&

95. Robert Zubrin, *Merchants of Despair: Radical Environmentalists, Criminal Pseudo-Scientists, and the Fatal Cult of Antihumanism* (New York: Encounter Books, 2012), 252. See also Wesley J. Smith, *The War on Humans* (Seattle, WA: Discovery Institute) and Foss, *Beyond Environmentalism*, particularly 57-75.

10 | Baptists, Bootleggers, and Opportunists

If we allow science to become politicized, then we are lost. We will enter the Internet version of the dark ages, an era of shifting fears and wild prejudices, transmitted to people who don't know any better. That's not a good future for the human race. That's our past. So it's time to abandon the religion of environmentalism, and return to the science of environmentalism, and base our public policy decisions firmly on that.[1]

Novelist-physician Michael Crichton

In science, refuting an accepted belief is celebrated as an advance in knowledge; in religion it is condemned as heresy.

Physicist Robert Park, University of Maryland[2]

Ever since Rachel Carson published her call to action in 1962, the world has been inundated by one book after another warning us about the impending demise of the planet, civilization, and the natural world. Shelves in university libraries are groaning under their weight, and used bookstores flog them for as little as a penny each. The numbers of trees sacrificed to this voracious god are incalculable. All these calls to action have four things in common: 1) their fears are grounded in the ever-growing human population of the planet and its impact on a perceived fragile biosphere; 2) every prophecy of doom relies on a linear extrapolation into the future of contemporary circumstances, without regard to technological or

---

1. Michael Crichton, "Environmentalism as Religion," Commonwealth Club, San Francisco, September 15, 2003, at *Crichton-official.com*.
2. Bob Park, *What's New*, October 3, 2008.

other developments; 3) virtually every doom-laden prognostication from the first generation or two has proven wrong, not just as a matter of timing or detail, but spectacularly wrong; and 4) the solution lies in more central planning, more regulation, more government, and less individual freedom, choice, and responsibility. Newer versions may need more time before their own improbable assumptions and wild, often computer-driven, projections are disproven, but their defects are already being exposed. And yet, the public appetite for this kind of doom remains insatiable. The safest, richest, healthiest generation in recorded history apparently lives in constant fear that its good fortune will soon end unless people repent and change their ways, not at the margins, but radically, and messiahs of one kind or another are appointed to take charge and set things right.[3]

Within this context, concern about global warming or climate change has become central to fashionable opinion. Those prepared to raise doubts or to question any of its vital points are ostracized from polite society and characterized as sceptics or, even worse, deniers. Those most firmly attached to the dogma rely on expressions of opprobrium because many are woefully under-informed about the science of climate change and the technology and economics of mitigation. Even such icons of the movement as Al Gore, Bill McKibben, and David Suzuki regularly display an appalling level of ignorance about the basic concepts involved. All have abandoned the long-held principle that science is grounded in evidence and reproducibility. Not surprisingly, they refuse to debate sceptics. Even scientists committed to the dogma regularly exhibit a narrow understanding of the science as a whole and ignore the fact that there is little real-world evidence to substantiate the central tenets of the official science. Politicians who have embraced global warming as the cause of their lives would be embarrassed if they had to explain the basic concepts to a high school science class. Their advisers prepare briefs summarizing the issues but can rarely explain the science or recognize that it is far from settled. Earnest Sunday sermons admonishing the faithful to join in efforts to save the planet come out of the mouths of preachers who would find reading even one of

---

3.   See Ron Bailey, "Earth Day, Then and Now," *Reason Magazine,* May, 2000, for a review of the many pessimistic prognostications marking the first celebration of Earth Day in 1970 and their fate since then. For an example of their anger and dismay, see Christopher Manes, *Green Rage: Radical Environmentalism and the Unmaking of Civilization* (Boston: Little Brown, 1990).

the IPCC's *Summaries for Policy Makers* a major challenge. Social scientists write learned papers on what must and can be done without ever having examined the science justifying the draconian steps they advocate. Opportunistic business leaders mine this latest anxiety for profit and urge politicians to help pave the way with regulations and subsidies; others rail against the cost and waste of the same regulations and subsidies. The media breathlessly report every cry of alarm from the committed but rarely find space or time to explain that much of this alarm is speculative, rarely based on anything other than computer modelling, and no more credible than earlier, thoroughly discredited alarms.

When a sceptic is impertinent enough to raise inconvenient questions, polite society falls back on the formulaic answer that the science is settled and that panels of experts have reached consensus on all the issues that matter. More than two millennia ago, Aristotle concluded that the claim of consensus is one of the weakest forms of argument. The public discourse about climate change is replete with Aristotle's other twelve logical fallacies, including arguments from authority and those dependent upon ambiguous language, red herrings, false equivalence, or *ad hominem* criticisms. The ignorant but convinced will fall back on the always useful dodge that the issue is too complicated and should be left to the experts. This may be true for some technically demanding aspects, but when the "experts" warn us that the whole of human society must be remade to address climate change, people may want to become better informed.

The totalitarian nature of environmentalism as ideology is also evident in the zeal with which critics are demonized. In the alarmist view, the issue is of such importance that dissent needs to be rooted out and punished. Thus, James Hansen called for the chief executives of oil companies to be prosecuted and jailed, telling *The Guardian*: "When you are in that kind of position, as the CEO of one of the primary players who have been putting out misinformation even via organizations that affect what gets into school textbooks, then I think that's a crime."[4] Hansen may have gotten the idea from Suzuki, who had told McGill University students a few months earlier: "What I would challenge you to do is to put a lot of effort into trying to see whether there's a legal way of throwing our so-called

---

4.   Ed Pilkington, "Put oil firm chiefs on trial, says leading climate change scientist," *The Guardian*, June 23, 2008.

leaders into jail because what they're doing is a criminal act."[5] Ironically, while Hansen was prepared to see his critics dragged through the courts, he was adamant that as a public employee, the government had no right to place limits on his right to criticize his own employer. Similarly, President Obama's science adviser, John Holdren, asserts that "the extent of unfounded scepticism about the disruption of global climate by human-produced greenhouse gases is not just regrettable, it is dangerous. It has delayed – and continues to delay – the development of the political consensus that will be needed if society is to embrace remedies commensurate with the challenge."[6] The Boston *Globe's* Helen Goodman put the final point on the issue for the alarmist camp, opining that "I would like to say we're at a point where global warming is impossible to deny. Let's just say that global warming deniers are now on a par with Holocaust deniers."[7] The new intolerance that marks this age of political correctness extends aggressively to those who insist on evidence and reason rather than faith and authority. Those who dissent from climate orthodoxy are dismissed as either wicked or stupid and are considered unworthy of being engaged in reasoned discussion.

Climate change alarmism is a belief system underpinning a political agenda. A wide range of interests have hitched their future to that agenda and need to be evaluated accordingly. Patrick Moore, one of the founders of Greenpeace who became more and more disenchanted with the movement as it departed from evidence-based science in pursuit of its political and ideological objectives, points to the "powerful convergence of interests among key elites that support the climate 'narrative.' Environmentalists spread fear and raise donations; politicians appear to be saving the Earth from doom; the media have a field day with sensation and conflict; science institutions raise billions in grants, create whole new departments, and stoke a feeding frenzy of scary scenarios; business wants to look green, and get huge public subsidies for projects that would other-

---

5. Craig Offman, "Jail politicians who ignore climate science: Suzuki," *National Post*, February 7, 2008. Suzuki likes to characterize his critics as shills for big corporations while happily accepting donations from dozens of corporations, including oil companies, for his own foundation.

6. Holdren, "Convincing the Climate-Change Sceptics," Boston *Globe*, August 4, 2008. Holdren has a long history of association with radical environmental causes.

7. Ellen Goodman, "No change in political climate," Boston *Globe*, February 9, 2007.

wise be economic losers, such as wind farms and solar arrays. ...
The Left sees climate change as a perfect means to redistribute
wealth from industrial countries to the developing world and the
UN bureaucracy."[8]

Two reasons in particular point to the weakness of the case for
action as recommended by alarmists. First, unlike other areas of science, in which uncertainty, misconduct, and error are acknowledged as a regrettable but normal part of human endeavour, core
climate scientists are quick to deny either error or uncertainty; rather, with any sign of criticism, they circle the wagons and launch *ad
hominem* attacks on those who criticize the basic tenets of the alarmist movement. Second, none of the mainstream scientists have dis-
owned the lies and distortions proffered by Gore and other prophets of doom. Not even the ideological rantings of a Naomi Klein
warrant a disavowal from movement leaders.[9] Instead, these Jeremiahs are hailed as upholders of the science. Both problems suggest
that there is something seriously wrong at the centre of the catastrophic climate change movement. Robert Tracinski points out that
"a theory with this many holes in it would have been thrown out
long ago, if not for the fact that it conveniently serves the political
function of indicting fossil fuels as a planet-destroying evil and allowing radical environmentalists to put a modern, scientific face on
their primitivist crusade to shut down industrial civilization."[10]

Notwithstanding the many serious questions raised about the
anthropogenic global warming hypothesis and about the solutions
on offer, the fact remains that there are many scientists who believe
it is a serious problem, are prepared to support the hypothesis, and
demand that governments take remedial action. At the same time,
there are other scientists, equally well-credentialed, who are confident that climate change is largely a natural phenomenon and that
human influence is marginal and non-catastrophic. Given that we
live in an age in which the voice of experts is very powerful, the argument from authority remains one of the most effective instru-

---

8.   Patrick Moore, "Why I am a Climate Change Skeptic," *Heartlander*, March 15,
     2015.

9.   See Naomi Klein, *This Changes Everything: Capitalism vs. the Climate* (New York:
     Simon and Schuster, 2014). The New York *Times* characterized her arguments as
     hopeful and sophisticated, "the most momentous and contentious environmen-
     tal book since 'Silent Spring'." Rob Nixon, "Naomi Klein's 'This Changes Every-
     thing'," November 6, 2014.

10. Robert Tracinski, "The End of an Illusion," April 4, 2013, at *RealClearPolitics*.

ments available to the alarmist community. In Richard Lindzen's words: "Most arguments about global warming boil down to science versus authority. For much of the public, authority will generally win since they do not wish to deal with science. For a basically political movement, as the global warming issue most certainly is, an important task is to co-opt the sources of authority. This, the global warming movement has done with great success."[11] Having gained control of the commanding public heights of the issue, from government environment and meteorology departments to some of the leading science journals and the two key UN agencies (UNEP and WMO), the "experts" have resorted to demonizing their critics, no matter how false the charges. So-called deniers of the AGW hypothesis are fighting an up-hill battle. And yet, their numbers are growing, and the claims of the alarmist community have become ever more shrill. Deconstructing alarmist ideology provides a key to understanding the nature of the debate and of its support.

The religious appeal of climate alarmism

Some analysts of environmental alarmism have observed that much of it operates as a secular religious cult and that this cultic dimension helps to explain its widespread appeal.[12] Philosopher of science Jeffrey Foss explains that "environmental science conceives and expresses humankind's relationship to nature in a manner that is – as a matter of observable fact – religious." Much of it is expressed in religious rather than scientific terms, relying on value statements rather than testable hypotheses. It "prophesies an environmental *apocalypse*. It tells us that the reason we confront apocalypse is our own environmental *sinfulness*. Our sin is one of *impurity*. We have fouled a pure, 'pristine' nature with our dirty household and industrial wastes. The apocalypse will take the form of an environmental *backlash*, a payback for our sins. ... environmental scientists tell people

---

11. Richard Lindzen, "Climate Alarm: What We Are Up Against, and What to Do," 2nd International Conference on Climate Change, Heartland Institute, March 8, 2009.

12. According to Al Gore, the global-warming challenge provides "a compelling moral purpose; a shared and unifying cause; the thrill of being forced by circumstances to put aside the pettiness and conflict that so often stifle the restless human need for transcendence. ... Those who are now suffering from a loss of meaning in their lives will find hope." *An Inconvenient Truth: The Planetary Emergency of Global Warming and What We Can Do About It* (New York: Rodale Books, 2006), 11.

what they must do to be blameless before nature."[13] [emphasis in original] Historian William Cronon adds: "Environmentalism is unusual among political movements in offering practical moral guidance about virtually every aspect of daily life, so that followers are often drawn into a realm of mindfulness and meditative attentiveness that at least potentially touches every personal choice and action. Environmentalism, in short, grasps with ultimate questions at every scale of human existence, from the cosmic to the quotidian, from the apocalyptic to the mundane. More than most other human endeavours, this is precisely what religions aspire to do."[14] Environmentalism relies more on a will to believe than on evidence and reason and rejects that its claims can be subjected to logical criticism. It partakes of the sensibilities of the Romantic era and the Gnosticism of the New Age movement: nature is sacred and worthy of worship and sacrifice.

Environmentalism – or ecologism – as religion is part of a broader pattern in contemporary life which rejects traditional values, religious or otherwise. Historian Anna Bramwell writes: "Ecologists believe in the essential harmony of nature. But it is a harmony to which man may have to be sacrificed. Ecologists are not man-centred or anthropocentric in their loyalties. Therefore they do not have to see nature's harmony as especially protective towards or favouring mankind. ... There is a conflict between the desire to accept nature's harmonious order and a need to avert catastrophe because ecologists are apocalyptical, but know that man has caused the impending apocalypse by his actions. Ecologists are the saved."[15] This phenomenon is steadily advancing on both sides of the Atlantic as well as in Oceania, perhaps more so in western Europe than elsewhere. The authorities in developing countries are right in concluding that the issues associated with climate change in particular and environmentalism more generally raise concerns peculiar to prosperous societies. The principal stake for developing countries is to extract rents from the wealthy countries and to ensure that whatever measures are adopted do not inflict collateral damage on themselves. To them, sustainable development is an economic concept,

---

13. Jeffrey Foss, *Beyond Environmentalism: A Philosophy of Nature* (Hoboken, NJ: John Wiley & Sons, 2009), 177.

14. William Cronon, foreword to Thomas R. Dunlap, *Faith in Nature: Environmentalism as Religious Quest* (Seattle: University of Washington Press, 2004), xii.

15. Anna Bramwell, *Ecology in the 20th Century: A History* (New Haven: Yale University Press, 1989), 16.

while in the North it has religious appeal, a difference being shamelessly exploited by the UN and its progressive allies. Over the years, as environmentalism's religious appeal has grown in Western countries, the populations of developing countries have begun to suspect that environmentalism is little more than a new form of colonialism. Rising concerns among environmentalists about growing populations and limited resources do little to alleviate those suspicions. There is no more disturbing image for rich-country environmentalists than the prospect of the poor in developing countries attaining the living standards prevalent in industrialized countries.[16]

As early as 1990, Robert Nelson, who had spent more than two decades as an economist at the US Department of the Interior, observed that "environmental policy-making often turns out to be a battlefield for religious conflict. Rather than rational policy analysis, the making of natural resource and environmental policy in the United States has become an exercise in theological controversy. ... Increasingly, the environmental policy analyst must address matters not only of physical science, economics, and other conventional policy subjects, but of theology as well."[17] In a later book, Nelson characterizes contemporary environmentalism as "Calvinism minus God," a populist faith, "more successful at attracting followers and influencing ... government policy, than in the formulation of a logically developed body of thought."[18] That lack of coherence is well illustrated by the fact that a movement that originated in a revolt against experts and authority and against the products of science and technology, now insists that when it comes to climate change, science is king. The science they worship, of course, is the perverted version discussed in earlier chapters, a post-normal science based on a carefully selected set of convenient precepts and militantly defended from critics pointing to inconvenient observational evidence.

Despite the claim that we need to transition as quickly as possible to a fossil-free energy world, for example, that transition is slowed by the actions of, yes, environmentalists, who have never

16. Developing country concerns are fully explored in Paul Driessen, *Eco-Imperialism: Green Power, Black Death* (Bellevue, WA: Free Enterprise Press, 2003-04). He describes the many perverse consequences of applying western environmental standards to developing countries.

17. Robert H. Nelson, "Unoriginal Sin: The Judeo-Christian Roots of Ecotheology," *Policy Review* 53 (Summer, 1990), 52.

18. Robert H. Nelson, *The New Holy Wars* (University Park, PA: Pennsylvania State University Press, 2009), xx.

seen an industrial-scale energy project that they like, regardless of the energy source. Renewables' advocate Amory Lovins, for example, argues that "future power plants should be small-scale facilities that can be effectively networked by smart grids, thus being made safer because they'd be more diversified, i.e., panels on rooftops, parking lot surfaces, and such."[19] The splenetic Paul Ehrlich famously said that "giving society cheap, abundant energy at this point would be the moral equivalent of giving an idiot child a machine gun."[20] But as Steve Stein remarks in response, "In addition to preservationists and minimalists, yet another environmental faction has curbed its enthusiasm for renewable energy. This is the no-growth contingent, who see population growth and resource development as weaknesses of the capitalist system."[21]

Philosopher Roger Scruton points out that humans "are religious beings, with a need to submit to divine imperatives and to find comfort in the community of believers."[22] In that vein, Joseph Bottum observes: "The new elite class of America is the old one: America's Mainline Protestant Christians, in both the glory and the annoyingness of their moral confidence and spiritual certainty. They just stripped out the Christianity along the way. ... When we recognize their origins in Mainline Protestantism, we can discern some of the ways in which they see the world and themselves. They are, for the most part, politically liberal, preferring that government rather than private associations ... address social concerns. They remain puritanical and highly judgmental ..., and like all Puritans they are willing to use law to compel behaviour they think right."[23]

---

19. Quoted in Steve Stein, "The Environmentalist's Dilemma: Making the perfect the enemy of the good," *Policy Review* 174, August 1, 2012.

20. Paul Ehrlich, "An Ecologist's Perspective on Nuclear Power," Federation of American Scientists, *Public Interest Report* 28:5-6 (May-June 1975), 5.

21. Stein, "The Environmentalist's Dilemma," *Policy Review* 174, August 1, 2012.

22. Roger Scruton, *The West and the Rest: Globalization and the Terrorist Threat* (London: Bloomsbury Academic, 2003), 1.

23. Joseph Bottum, "The Post-Protestant Ethic and Spirit of America," *The Bradley Lecture* at the American Enterprise Institute, February 22, 2014. The lecture was excerpted from Bottum's highly original book, *An Anxious Age: The Post-Protestant Ethic and the Spirit of America* (New York: Image, 2014), in which he provides a compelling tour through the religious, intellectual, cultural, and sociological evolution of today's American ruling class. In a review of Bottum's book, David Goldman adds: "post-Protestant secular religion ... gained force and staying power by recasting the old Mainline Protestantism in the form of catechistic worldly categories: anti-racism, anti-gender discrimination, anti-inequality, and so forth. What sustains the heirs of the now-defunct Protestant

Joel Garreau adds that "beyond influencing – one might even say colonizing – Christianity, the ecological movement can increasingly be seen as something of a religion in and of itself."[24] The same phenomenon is at play in other western countries.

The appeal of post-Christian secular religion for many people – whether rooted in environmentalism or other belief systems – is that they have become sceptical of institutional Christianity but seek the comfort, certainty, and direction that religion can provide in their daily lives. Environmentalism offers a fully secular version of all the characteristics of more traditional transcendent belief systems: the need to avoid disaster by turning from our sinful ways and by following a path of righteousness leading to harmony between man and nature. In a much-read speech delivered shortly before his death, novelist-physician Michael Crichton reached the same conclusion: "If you look carefully, you see that environmentalism is in fact a perfect 21st century remapping of traditional Judeo-Christian beliefs and myths."[25] In a June 2007 essay, John Brignell similarly lays out in exquisite detail the extent to which the global warming faith parallels the basic tenets of other religions, including the role of faith, sacrifice and ritual, sin and absolution, confession and salvation, infidels and apostates, prophecy and divination, and contradictions and irrationality.[26]

Nelson explains that environmentalism, "like most of the secular religions of the modern age, to a significant extent rework[s] in a new language the earlier messages of Jewish and Christian faiths ... by recasting ... traditional Jewish and Christian understandings in a new (ostensibly more modern and scientific) vocabulary – thus disguising the origins, lending them greater authority in an age that gives greater public legitimacy to scientific methods than the read-

---

consensus ... is a sense of the sacred, but one that seeks the security of personal salvation through assuming the right stance on social and political issues. Precisely because the new secular religion permeates into the pores of everyday life, it sustains the certitude of salvation and a self-perpetuating spiritual aura." "The Rise of Secular Religion," *The American Interest*, March 17, 2014.

24. Joel Garreau, "Environmentalism as Religion," *The New Atlantis*, Summer 2010, 66. Garreau is the Lincoln Professor of Law, Culture, and Values at Arizona State University.

25. Michael Crichton, "Environmentalism as Religion," speech to the Commonwealth Club of San Francisco, September 15, 2003.

26. John Brignell, "Global Warming as Religion and not Science," at number-watch.co.uk/religion.htm.

ing of the Christian Bible."[27] Nelson believes that environmentalism's appeal to many on the left stems from the fact that they see "the story of history as one of decline from an earlier existence in true harmony with nature. ...[Environmentalism] follows Marx in locating the fall of man in history; in this case, it is the arrival of agriculture and an organized society. ... In the gospel of [David] Foreman [founder of Earth First], if it was the wilderness that created man, man has now rebelled against his primitive naturalness and fallen into sin."[28] He adds that "the mainstream of American environmentalism is significantly defined by its opposition to economic religion, and is thus part of a wider reaction against what many now see as the excesses of modern optimism."[29]

Bill McKibben in *The End of Nature,* writes of a severe "crisis of belief" in the current era but asserts that "many people, including me, have overcome it to a greater or lesser degree by locating God in nature."[30] Garreau adds: "As climate change literally transforms the heavens above us, faith-based environmentalism increasingly sports saints, sins, prophets, predictions, heretics, demons, sacraments, and rituals. Chief among its holy men is Al Gore – who, according to his supporters, was crucified in the 2000 election, then rose from the political dead and ascended to heaven twice – not only as a Nobel deity, but as an Academy Awards angel."[31]

French social philosopher Pascal Bruckner, commenting on the rising level of anxiety in Western life, characterizes environmentalism as "the sole truly original force of the past half-century. [It] has challenged the goals of progress and raised the question of its limits. It has awakened our sensitivity to nature, emphasized the effects of climate change, pointed out the exhaustion of fossil fuels. Onto this collective credo has been grafted a whole apocalyptic scenography that has already been tried out with communism, and that borrows from Gnosticism as much as from medieval forms of messianism. Cataclysm is part of the basic tool-kit of Green critical analysis, and

---

27. Nelson, *The New Holy Wars,* xxi.

28. Nelson, "Unoriginal Sin," 55.

29. Nelson, *The New Holy Wars,* 3.

30. McKibben, *The End of Nature,* revised edition (New York: Random House, 2006), 61.

31. Garreau, "Environmentalism as Religion." It is worth recalling that Gore spent a year at Vanderbilt Divinity School in order to explore spiritual issues and consider his future.

prophets of decay and decomposition abound. They beat the drums of panic and call upon us to expiate our sins before it is too late."[32]

Some of the leading critics of the demands of the environmental movement have been economists, including Julian Simon, Wilfred Beckerman, David Henderson, and Ross McKitrick, practitioners of what Nelson calls economic religion. He characterizes the clash between environmental and economic values to be among the defining characteristics of our age: a preference for evidence and reason versus that of emotion and authority rooted in the irrational canons of post-normal science.[33] Famed Indo-British development economist Deepak Lal has added his own voice. "The religious nature of the movement is further supported by its failure even to admit that its predictions have been wrong, and to continue making the same assertions based on its world view despite evidence to the contrary. ... Environmentalists do not respect the 'evidence' even if it is incontrovertible. ... The ecologists, as much as the religious fundamentalists, have launched an attack on modernity [and] are seeking to use transnational (rather than national) institutions to pursue their goals."[34]

⁂

The willingness of progressive Christians to adopt many of the tenets of environmentalism as their own betrays a disturbing shallowness in their religious understanding. It is hard, for example, to reconcile the anthropocentric nature of Christianity – man as the steward and centre of creation whose sinful nature is redeemed through the sacrifice of Jesus Christ – with the anti-human core of environmentalism, which holds that man is the despoiler of nature and posits the ideal as nature without man. As papal scholar George Weigel astutely observes: "I doubt that ... there is much 'common ground' to be found [by traditional Christians] with 'creation care' folks who implicitly worship Gaia rather than the God of

32. Pascal Bruckner, "Against Environmental Panic," *The Chronicle of Higher Education*, June 17, 2013.

33. Robert H. Nelson, *Economics as Religion: From Samuelson to Chicago and Beyond* (University Park, PA: Pennsylvania State University Press, 2001).

34. Deepak Lal, "Ecofundamentalism," *International Affairs* 71:3 (1995), 522, 523, 524. *The Economist* observes that "academic disciplines are often separated by gulfs of mutual incomprehension, but the deepest and widest may be the one that separates most economists from most environmentalists. ... [It] is not so much disagreement about facts as disagreement about how to think" about the world. "Never the Twain Shall Meet," *The Economist*, February 2, 2002, 74.

the Bible. Gaia worship ... carries with it an anthropology that treats the human person as a kind of anthropollutant, which leads in short order to a eugenic 'morality'."[35] In a similar vein, Michael Novak writes: "where people are poor, environmental conditions tend to be abysmal; and if the twentieth century proved anything, it was that the best way to end poverty isn't red – the color of socialism – but blue, the color of liberty, personal initiative, and enterprise. ... Blue Environmentalism, therefore, stands for the spreading of those institutions of empowerment that promote private property and creativity. It is not the natural endowment God gave the poor that is currently at fault, but the inadequacy of political systems and social institutions that fail to nurture and support it."[36]

Weigel and Novak were writing before Pope Francis decided to join the full flowering of the green/red environmentalism crusade in 2015 with his encyclical *Laudato Si'*.[37] As Joel Kotkin points out:

> With Francis's pontifical blessing, the greens have now found a spiritual hook that goes beyond the familiar bastions of the academy, bureaucracy, and the media and reaches right into the homes and hearts of more than a billion practicing Catholics. ... What makes the Pope's position so important – after all, the world is rejecting his views on such things as gay marriage and abortion – is how it jibes with the world view of some of the secular world's best-funded, influential, and powerful forces. In contrast to both Socialist and capitalist thought, both the Pope and the greens are suspicious about economic growth itself, and seem to regard material progress as aggression against the health of the planet.[38]

Ignoring the carefully worded teachings of his predecessor on environmental issues, Francis enthusiastically jumped on board the alarmist bandwagon. Catholic commentators have gingerly tried to parse the pope's message and square it with traditional Church teachings.[39] Non-Catholic commentators have not been so kind. Putting aside the applause from such usually atheist, anti-life, and anti-family organs as the New York *Times*, most critics found the encyclical to be an incoherent, internally contradictory collection of

---

35. George Weigel, "Replies," *First Things* 252 (April 2015), 13.

36. Michael Novak, "Blue Environmentalism," *Patheos*, July 2015. Most of the article was adapted from material originally written in 2003.

37. *Laudato Si' (Praise Be to You), On Care for Our Common Home*, June 17, 2015.

38. Joel Kotkin, "Green Pope Goes Medieval on Planet," *Daily Beast*, July 13, 2015.

39. See, for example, Philip Booth, "Pope Francis is Unduly Pessimistic about the World," *Catholic Herald*, June 18, 2015.

thoughts on things theological, moral, environmental, economic, and scientific. On the first two, the Pope may be on solid ground, but on the other three, he is clearly out of his depth but willing to align the church with the anti-humanism of environmentalism, the destructive economics of statism, and the dubious science of the IPCC. Mixing all five together leads to a confusing and unpalatable brew. As Jesuit scholar James Schall sees it, "the Pope's ... plea for 'humility' in this document would seem to apply here concerning scientific questions that are at best hypotheses subject to change."[40] It is equally problematic that the climate warmer basket also comes encrusted with economic and political values that are at odds with Church teaching, from population control to world government. As Peter Foster notes, "apart from the religious references, Pope Francis' climate Encyclical ... could have come from any branch of the UN, any environmental NGO, or the World Economic Forum. This is hardly surprising since they all promote Global Salvationism, which is based on projections of doom to be countered by morally charged, UN-centric, globally-governed sustainable development and corporate social responsibility."[41]

From the beginning of his papacy, Francis has been focused on the plight of the world's poor. Not surprisingly, concern for the poor is central to *Laudato Si'*, but in hitching his concern for the poor to the climate change crusade, the pope reveals an appalling level of economic illiteracy – a fault he shares with many religious leaders. Like them, he focuses on the issue of distribution, but without understanding the importance of production. The models of development that the pope criticizes are the very ones that are leading to falling rates of poverty, reduced inequality, and fewer deaths due to natural disasters. The most hopeful sign of the past thirty years is that countries from India and China to Chile and Brazil have learned what it takes to produce and distribute wealth: private property rights, free markets, and entrepreneurship – an economy in which those who participate can, by their talents, work, and ingenuity, earn a decent living.

The pope seems to prefer the failed and discredited approach of statism, centered on the UN. He foresees the need for a global bureaucracy of unprecedented size and power, a technocracy on ster-

---

40. James V. Schall, "Concerning the 'Ecological' Path to Salvation," *Catholic World Report*, June 21, 2015.

41. Peter Foster, "The Pope's eco-mmunist manifesto," *Financial Post*, June 18, 2015.

oids.[42] In this context, he endorses the Earth Charter, the brainchild of Maurice Strong and Mikhail Gorbachev, two of the world's leading advocates of statist sustainable development. It would be difficult to find a movement that has done more damage to the prospects of the world's poor than the UN's sustainable development campaign in all of its ramifications. Despite billions spent in the UN's top-down approach to development, little of that has trickled down to the poor. Rather, as scholars such as William Easterly, Surjit Bhalla, Elinor Ostrom, and Deepak Lal have shown, the most effective development tool has been the gradual abandonment of the *dirigiste* economics favoured by the UN and the embrace by more and more developing-country governments of open markets. As Lal demonstrates in *Poverty and Progress,* the combination of anti-capitalism, corporate social responsibility, central planning, environmentalism, and the other core tenets of sustainable development, has been lethal for the world's poor. Yet elite opinion in the West maintains that the path to progress lies in more of the same. In Lal's view, "the world is at a strange pass. Instead of rejoicing in what has been one of mankind's most amazing achievements over the last three decades – the spread of economic progress around the world, which is gradually eliminating the ancient scourge of mass poverty – we hear wails of doom and gloom in the West, not least from those who see this progress as threatening the very survival of Spaceship Earth." He adds that putting "a limit on the use of fossil fuels without adequate economically viable alternatives is to condemn the Third World to perpetual poverty."[43]

For most people, Australian Cardinal George Pell's take on global warming in 2012 was much more in keeping with traditional Church teaching. Pope Francis ignored Pell's counsel in favour of such icons of the church of global warming as Ban Ki-Moon, Al Gore, Jeffrey Sachs, and Joachim Schellnhuber. In his 2012 lecture, Pell clearly points to the anti-humanism of environmentalism, its incompatibility with Biblical teachings, and the tortured science it promotes. He also cautioned:

---

42. See R.R. Reno, "Laudato Si'" *First Things,* August 2015.

43. Deepak Lal, *Poverty and Progress: Realities and Myths About Global Poverty* (Washington: Cato Institute, 2013), end of chapter 11 and beginning of chapter 10, kindle edition. Lal notes that Argentina, the pope's home country, remains mired in the legacy of Juan Peron and Raul Prebisch. Francis' Argentinean origins may be one explanation for his lack of understanding of markets.

Theologians do not have too much to contribute on AGW except, perhaps, to note the ubiquity of the 'religious gene' and point out regressions into pseudo-religion or rudimentary semi-religious enthusiasms. ... the appeal must be to the evidence, not to any consensus, whatever the levels of confusion or self-interested coercion. First of all we need adequate scientific explanations as a basis for our economic estimates. We also need history, philosophy, even theology and many will use, perhaps create, mythologies. But most importantly we need to distinguish which is which.[44]

Pell maintains his more traditional perspective, and told an interviewer following the release of *Laudato Si'* that the papal encyclical has "many, many interesting elements. There are parts of it which are beautiful. But the church has no particular expertise in science. The church has got no mandate from the Lord to pronounce on scientific matters. We believe in the autonomy of science."[45]

The pope seems clearly to have regressed into pseudo-religion as peddled by his secular, atheist advisers in his quest to address an age-old question: social justice. Kotkin concludes: "Social justice may be an important value, but it is dubious that the Church's credibility will be well served by a neo-feudal alliance dominated by those who abhor the Church's other core values such as family, the sanctity of human life and some degree of social prudence."[46] The pope has now aligned himself with the rest of progressives by embracing environmentalism as the key to global salvation. Anglican bishop Peter Forster and Lord Bernard Donoughue conclude: "Overall, the encyclical strikes us as well-meaning but somewhat naïve. ... The environmental and especially the energy policies advocated in the encyclical are more likely to hinder than to advance this great cause. ... To regard economic growth as somehow evil, and fossil fuels as pollutants, will only serve to increase the very poverty that he seeks to reduce."[47]

⚜⚜⚜⚜⚜

Back in 1967, historian Lynn White, Jr. introduced a controversial dimension to environmentalism's cultic character by blaming

---

44. George Cardinal Pell, "One Christian Perspective on Climate Change," 2011 Annual Lecture, Global Warming Policy Foundation, October 26, 2011.

45. Steve Doughty, "Senior cardinal breaks ranks by questioning the Pope's authority on climate change," *Mail Online*, July 23, 2014.

46. Joel Kotkin, "Green Pope Goes Medieval on Planet," *Daily Beast*, July 13, 2015.

47. "The Papal Encyclical: A Critical Christian Response," Global Warming Policy Foundation, Briefing 20, 2015, 1 and 7.

Christianity for the Earth's ecological crisis, arguing that "whereas older pagan creeds gave a cyclical account of time, Christianity presumed a teleological direction to history, and with it the possibility of progress. This belief in progress was inherent in modern science, which, wedded to technology, made possible the Industrial Revolution. Thus was the power to control nature achieved by a civilization that had inherited the license to exploit it." White believed that only a new religious sensibility could save mankind, a religion that provided a new balance between man and nature. "Since the roots of our trouble are so largely religious, the remedy must also be essentially religious."[48] White's thesis has met with more than success. It would have been hard for him to imagine nearly fifty years ago the extent to which urban elites have adopted environmentalism as a critical part of their belief systems and the extent to which mainline Protestantism – and now the pope – has accommodated environmentalism in a vain attempt to keep the pews filled. To many of the remaining faithful, the new pagan pantheism embraced by their leadership is either bewildering or an abomination. Some have sought refuge in less flexible denominations or have decided that formal, institutional expressions of their faith can be limited to baptisms, marriages, funerals, Christmas, and Easter.

White's thesis generated wide discussion, much of it favourable, particularly among ecologists and other upholders of the doomsday thesis, but some also critical. Richard John Neuhaus, for example, in his 1971 book *In Defense of People,* provided one of the earliest indictments of the rise of the mellifluous 'theology of ecology' then emerging in the western world.[49] Other Christian writers agreed, condemning the eco movement's attempt to subvert or supplant their religion. "We too want to clean up pollution in nature," *Christianity Today* demurred, "but not by polluting men's souls with a revived paganism."[50] Thomas Sieger Derr subjected White's thesis to a searching analysis and concluded that it had been appropriated by

---

48. Lynn White, Jr., "The Historical Roots of Our Environmental Crisis," *Science* 155:3767 (March, 10 1967), 1203-07.

49. Richard John Neuhaus, *In Defense of People* (New York: Macmillan, 1971). At the time, Neuhaus was a Lutheran pastor. He later converted to the Catholic Church and became a major voice in neo-conservatism, dedicated to bringing an orthodox – Catholic, Protestant, and Jewish – religious perspective to the Public Square, to which end he established the monthly journal, *First Things*.

50. As quoted in Joel Garreau, "Environmentalism as Religion," *The New Atlantis,* Summer 2010, 64.

those who see Christianity in a much darker light than White did. He notes that "the irony is surely bitter, then, that 'Historical Roots' has been embraced by an environmental movement deeply tinged with elitist, antidemocratic values; by ecologists ready to sacrifice millions to starvation and to institute totalitarian methods to keep the population down; by romantic technology haters ready to abandon centuries of civilization; and by various anti-Christian types with a mixture of personal motives."[51] Nevertheless, much of mainline Protestantism has felt compelled to integrate an ecological dimension into its worship, often conflicting with earlier core beliefs.

Environmentalism is thus religious in the sense that it makes fundamental truth claims about the nature of the universe and man's place in it. These claims arise out of belief rather than reason. As Foss sees it, "environmentalists profess that environmental health must take precedence over … other values. In other words, popular environmentalism tends to treat environmental health as a *transcendent objective,*"[52] (emphasis in original), a value that trumps evidence and reason. Their concern for nature brooks no limits. As a result, environmentalists are vulnerable to the charge that, if their views are based on belief, then they cannot be advanced as science; if their claims are scientific, they do not require faith and are subject to the normal strictures of scientific practice, including evidence and falsifiability.

In modern secular society, belief is a personal matter, not a matter for the state to legislate and enforce. One of the greatest triumphs of western democracies is the wide acceptance of the separation of church and state: freedom to practice the religion of one's choice without interference from the political authorities. That freedom is rapidly eroding in the face of progressives demanding a special place for their own belief systems organized around the increasingly pervasive canons of sustainability. A generation ago, Christianity was banned from public schools and diminished in universi-

---

51. Derr, "Religion's Responsibility for the Ecological Crisis: An Argument Run Amok," *Worldview* 18:1 (January 1, 1975), 45. The churches and their leaders are not monolithic in their message. Many of the mainline churches have joined the dominant cultural chorus, but there are some articulate voices, mainly originating from within more traditional and evangelical groups, presenting well-documented and argued reports that question this perspective. Among them are the statements and reports put out by the Cornwall Alliance. Jeffrey Foss provides a thorough dissection of White's thesis and its problems in *Beyond Environmentalism*, 169-77.

52. Foss, *Beyond Environmentalism*, 45.

ties to an object of sociological interest. In its place, civic leaders and educational specialists have fully integrated environmental sustainability into the school curriculum as the core of a new civic religion. On university campuses, sustainability programs and faculties have become the new religion departments.[53] Some 695 US university presidents have signed on to the American College and University Presidents' Climate Commitment in which they agreed that they are: "deeply concerned about the unprecedented scale and speed of global warming and its potential for large-scale, adverse health, social, economic and ecological effects. We recognize the scientific consensus that global warming is real and is largely being caused by humans. We further recognize the need to reduce the global emission of greenhouse gases by 80 per cent by mid-century at the latest, in order to avert the worst impacts of global warming and to re-establish the more stable climatic conditions that have made human progress over the last 10,000 years possible."[54]

Few of these presidents are even modestly informed on the science, but all are prepared to sign a statement accommodating New Age sensibilities. They preside over institutions that militantly dismiss Plato's distinction between knowledge and opinion and insist that there are no universal values other than the doctrinal shibboleths of the dominant culture. Like many members of the elite class of liberal progressives, academic administrators have long abandoned the virtues of evidence and empiricism in favour of cultural relevance and political correctness. The modern university no longer prizes diversity of thought, critical enquiry, and speculative thinking, nor does it seek to preserve and enlarge the accumulated intellectual capital of the past.[55] At the same time, the leaders of the modern university would be scandalized by Peterson's and Wood's assessment:

---

53. Rachelle Peterson and Peter W. Wood, *Sustainability: Higher Education's New Fundamentalism* (New York: National Association of Scholars, 2015), provide a detailed assessment of the extent to which the religion of sustainability has permeated American higher education, particularly at the most prestigious colleges and universities.

54. Statement of commitment at presidentsclimatecommitment.org. As of March 31, 2015, the statement carried 695 signatures. The Statement is a project of *Second Nature*, an ENGO founded by John and Teresa Heinz Kerry.

55. See Roger Scruton, "The End of the University," *First Things* 252 (April 2015), 25-30.

[The academic sustainability movement is] cultivating a susceptibility in today's students for the allures of command economies and undemocratic forms of political control. At its heart, sustainability is opposed to freedom. It offers students an imaginary world where important decisions about how to use resources will be made by properly credentialed experts, not by citizens making their own choices. The anti-consumerist impulse in this vision marches side-by-side with a wish for authoritarian control. Sustainability advocates are never too clear on exactly what regime they would like to install to bring about sustainatopia, but they are united in the belief that leaving people free to govern themselves can only create a tragedy of the commons.[56]

Like all elites throughout history, modern progressives will brook no questioning of their moral certainty and its universal application. There is no room for humility, for dissent, or for the idea that other issues might take precedence over "saving the planet," particularly given global salvationism's dubious premises. Progressive liberalism, of which sustainability is a central tenet, now has the distinction of becoming the closest thing to an established religion in much of Western society, aggressively supported by the state, inculcated from an early age in the young, elevated to the heights of moral behaviour on university campuses, embraced by religious leaders, embedded in legislation and regulation, and part of activist political campaigns for many more statist interventions.[57]

The politics of the climate change movement

Are attitudes towards climate change policy a matter not only of one's religious sensibilities but also of one's politics? Are those of one political persuasion more easily convinced of the need for action than those of another? The fact that so many people are prepared to accept the climate change mantra suggests that perhaps something more than science, or even religion, is involved. Harping on climate change and insisting on the need for extreme measures to address the issue suggest an ideological or opportunistic predisposition favouring the solutions on offer rather than a conviction

---

56. Peterson and Wood, *Sustainability: Higher Education's New Fundamentalism*, 228-9.

57. See Andrew Montford and John Shade, "Climate Control: Brainwashing in Schools," Global Warming Policy Foundation, Report 14, 2014; Richard Gray, "Scaring Kids for Gaia," *Quadrant*, April 17, 2014, for an account of the efforts by the educational establishment in Australia to indoctrinate children at school; and Ian Plimer, *How to get expelled from school: A guide to climate change for pupils, parents and punters* (Ballan, Australia: Connor Court, 2012).

that there is a real problem.[58] Jonathan Adler of the Case Western Reserve University School of Law observes: "Those who are risk averse, or place a high value on avoiding anthropogenic disruption of natural systems, or place a relatively low value on economic growth may be more predisposed to support costly controls on greenhouse gas emissions. In contrast, those who place a high value on individual liberty and property rights, or are more suspicious of government regulation, or believe that economic growth is more important than a pristine environment to human prosperity may be more reluctant to endorse emission control measures. Similarly, those who have a utilitarian preference for the maximization of net human welfare may come to different policy conclusions than those who believe that certain actions necessarily violate rights or otherwise constitute 'wrongs' worthy of redress."[59] Indeed, it is sometimes difficult to tell whether climate change policy is called for because of concerns about climate or because it provides a politically convenient reason to pursue particular policies. Certainly the conviction that there is need for a wholesale reordering of the way national economies function and interact with one another betrays a strong disposition in favour of statist, interventionist, progressive solutions, regardless of the problem. It is not surprising, therefore, to find that people on the political left are more easily persuaded of the need for an aggressive climate policy than are those on the political right.[60]

All but the most committed ideological analyses of climate change and mitigation policy indicate that at least until the end of the 21st century, the impact of any climatic changes will be more beneficial than malign and that the costs of effective mitigation measures will be much higher than the long-term benefits that might result. This perspective is even shared by many of those who accept the IPCC view that in the long run climate change could lead to catastrophic results.[61] Only outliers, such as the Stern review, see

---

58. See the discussion in Cardinal Pell, "One Christian Perspective on Climate Change."

59. Jonathan Adler, "Taking Property Rights Seriously: The Case of Climate Change," Working Paper 08-16, Case Research Paper Series in Legal Studies, Case Western Reserve University School of Law, 297.

60. See, for example, the work done by Yale's Dan Kahan, "Climate Science Communication and the *Measurement Problem*," *Advances in Political Psychology* 36:S1 (February 2015), 1-43.

61. See, for example, Martin L. Parry (ed.), "An Assessment of the Global Effects of Climate Change under SRES Emissions and Socio-Economic Scenarios," *Global*

the costs of delay in adopting mitigation measures to be so high as to warrant immediate, drastic measures, a view suggesting that there is a strong disposition among those who are concerned about climate change to pursue interventionist policies, regardless of the costs.

Those who express the greatest concern about the threat of a climate "crisis" are adamant that the answer lies in centrally planned, highly interventionist mitigation measures rather than in more local adaptation solutions, and they dismiss any analysis that points to the opposite conclusion. They roundly reject, for example, Indur Goklany's view that "the world can best combat climate change and advance well-being, particularly of the world's most vulnerable populations, by reducing present-day vulnerabilities to climate-sensitive problems that could be exacerbated by climate change rather than through overly aggressive GHG reductions."[62] The UN process does express a need for rich countries to pay for adaptive measures in developing countries but more in the interest of distributive justice and sustainable development than out of any commitment to adaptation. Most of the scenarios the IPCC uses for its forecasts of the impact of future climate conditions foresee a much wealthier world, including in developing countries. Despite these positive scenarios, the IPCC's leadership persists in the view that today's taxpayers in advanced economies need to pay for mitigation and possible adaptation measures that might make more sense at the time that they are actually needed.

The left has thus appropriated the post-normal science of climate change for its own purposes and insists that all other perspectives are tainted by special interests and ideology. Similar to environmentalism's success in "colonizing" mainline Protestantism, Roger Scruton argues that the left has colonized environmentalism. In his words:

> Environmentalism has all the hallmarks of a left-wing cause: a class of victims (future generations), an enlightened vanguard who fights for them (the eco-warriors), powerful philistines who exploit them (the capitalists), and endless opportunities to express resentment against

Environmental Change, Special Issue 14:1 (2004), 1–99. Most of the contributors to this study also contributed to IPCC reports. See also Richard Tol, "The economic impact of climate change in the 20th and 21st centuries," *Climatic Change* 117 (2013), 795–808.

62. Indur M. Goklany, "What to Do about Climate Change," *Policy Analysis* 609 (Washington: Cato Institute, February 5, 2008), 1.

the successful, the wealthy, and the West. The style too is leftist: the environmentalist is young, dishevelled, socially disreputable, his mind focused on higher things; the opponent is dull, middle aged, smartly dressed, and usually American. The cause is designed to recruit the intellectuals, with facts and theories carelessly bandied about, and activism encouraged. Environmentalism is something you *join*, and for many young people it has the quasi-redemptive and identity-bestowing character of the twentieth-century revolutions. It has its military wing, in Greenpeace and other activist organizations, and also its intense committees, its *odium theologicum* [the intense hatred often exhibited in theological controversies] and its campaigning journals.[63]

To many in the environmental movement, Marxism and its many variants and mutations have achieved a sacral character and require total commitment. This becomes evident in surveying the movement's founding works. Stephen Schneider's first book, for example, *The Genesis Strategy: Climate and Global Survival* (1976), offered a sweeping plan to reorganize global governance and the world's economy along statist lines to meet the purported threats of catastrophic climate change and overpopulation. At the time, Schneider was 31, had been narrowly trained in mechanical engineering and plasma physics, and had little formal knowledge of economics, history, or politics. Similarly, Paul Ehrlich's *Population Bomb* (1968) first built a catastrophic scenario of overpopulation and depleting resources and then offered utopian, collectivist solutions based on central planning and global governance. The environment is the cause, and the reorganization of the global political and economic order along statist, *dirigiste* lines is both the means to save the planet and also the ultimate goal. At the time he wrote his first jeremiad, Ehrlich was 36 with a narrow education as an entomologist. These religious sentiments, first expressed by Ehrlich and others, now flourish in universities, and Scruton characterizes their proponents as parasites on societies that they attempt to overturn. David Henderson similarly argues that climate alarmism is part of a broader movement that "combines alarmist visions and diagnoses with confidently radical collectivist prescriptions for the world."[64] Henderson finds that the most pernicious dimension of global salvationism is the willingness of otherwise sensible people to buy into

---

63. Roger Scruton, "Conservatism Means Conservationist," *Modern Age* 49:4 (Fall 2007), 351.

64. David Henderson, "Economics, Climate Change Issues, and Global Salvationism," *Economic Education Bulletin,* XLV:6 (June 2005), 2.

its fundamental precepts. Rajendra Pachauri, for 13 years chair of the IPCC but now disgraced by a sex scandal, set out his own commitment to global salvationism in his self-serving letter of resignation: "For me the protection of Planet Earth, the survival of all species and sustainability of our ecosystems is more than a mission. It is my religion and my *dharma* [the path of righteousness in the Hindu faith]."[65]

The climategate e-mails made the left's claims about the integrity of the science even more hollow than they manifestly always were. Political scientist Ivan Kennealy finds "these behind-the-scenes discussions among leading global-warming exponents ... remarkable both in their candour and in their sheer contempt for scientific objectivity. There can be little doubt after even a casual perusal that the scientific case for global warming and the policy that springs from it are based upon a volatile combination of political ideology, unapologetic mendacity, and simmering contempt for even the best-intentioned disagreement."[66] Richard Lindzen adds:

> The fact that the focus of climate alarm keeps changing (from global cooling to global warming to climate change to extreme weather to ocean acidification to ......) is suggestive of an agenda in search of a scientific rationale. Given the destructive, expensive and corrupting nature of the proposed or, alas, implemented) policies (as well as their demonstrable irrelevance to climate) leaves one with a disturbing view of the proposed agenda. It would appear that the privileged members of the global society regard as dogma that the rest of humanity is a blight on the planet, and all effort should be devoted to preventing their economic improvement and development. If this selfish and short-sighted view is what the privileged regard as morality, then God help us all.[67]

Those on the left are always quick to argue that the presence of an externality justifies their preference for political action of one kind or another. But, as Nobel laureate in economics, Ronald Coase, observes,

> the existence of 'externalities' does not imply that there is a *prima facie* case for governmental intervention, if by this statement is meant that, when we find 'externalities,' there is a presumption that governmental

65. February 24, 2015, at ipcc.ch/pdf/ar5/150224_Patchy_letter.pdf.

66. Ivan Kennealy, "The Climate E-mails and the Politics of Science," *New Atlantis*, November 24, 2009.

67. Richard Lindzen, "Submission to UK Parliament Select Committee," December 20, 2013.

> intervention (taxation or regulation) is called for rather than the other courses of action which could be taken (including inaction, the abandonment of earlier governmental action, or the facilitating of market transactions) ... The fact that governmental intervention also has its costs makes it very likely that most 'externalities' should be allowed to continue if the value of production is to be maximized. ... The ubiquitous nature of 'externalities' suggests to me that there is a *prima facie* case against intervention.[68]

It might also be added that those on the left are more prone to identify externalities to justify policy measures than others. The carbon dioxide that results from burning fossil fuels is only an externality if one considers it to be a pollutant, which fact in turn assumes a high climate sensitivity, a wholly unproven hypothesis.[69]

The alacrity with which climate change policy advocates dismiss market-based solutions is a further indication of their ideological predispositions. The use of property rights in addressing enviromental issues, is dismissed as evading the problem because "everyone knows" that capitalism, markets, and property rights are sources of environmental problems in the first place. As Adler points out, "Property-rights systems take advantage of the dispersed knowledge possessed by individuals about their own circumstances and subjective value preferences, as well as the availability of, and demand for, resources. At the same time property-rights systems preserve a relatively large sphere of individual autonomy and reinforce notions of personal responsibility."[70] Rather than acknowledging the benefits of property rights in addressing the "tragedy of the commons," environmentalists prefer social control, apparently convinced that social control leads to more certain outcomes than property rights. The evidence favouring social control is not convincing. The worst environmental impacts from industrialization were experienced in the Soviet Empire as a result of central planning and social control. China is learning from this experience: it is increasing private ownership and property rights in order to overcome problems stemming from cen-

---

68. R. H. Coase, *The Firm, the Market, and the Law* (Chicago: University of Chicago Press, 1988), 24, 26.

69. Economist Paul Heyne counsels, "Ardent environmentalists need to discover and acknowledge that the same limitations which made economic central planning impossible will make it impossible to establish a comprehensive system of central environmental planning." Heyne, "Economics, Ethics, and Ecology," in Roger Meiners and Bruce Yandle, eds., *Taking the Environment Seriously* (Lanham, MD: Rowman and Littlefield, 1993), 47.

70. Adler, "Taking Property Rights Seriously," 199.

tral planning. The same experience holds true in mitigating the environmental impacts of resource exploitation, whether from mining or forestry or fisheries.

Ideologies that rely on central direction and command-and-control mechanisms pose a seductive appeal to many people. Wolfgang Kasper observes that it is not difficult to understand the appeal of central planning and control to scientists and engineers, "who view the spontaneous coordination of actions in the market as disorderly chaos. ... Many do not seem to comprehend the working of the invisible hand. They prefer instead some high-minded, well-informed authority to sort out all necessary information prior to any action, and to control all subsequent actions."[71] Many are naïve when it comes to political and economic systems and are reluctant to see debate about alternatives based on socio-economic and political considerations. Such scientists lean towards global governance rather than towards open markets. They believe that environmental agreements should trump trade agreements. They view the command-and-control Kyoto Protocol as the future and the market-oriented WTO as the past. Additionally, as Steven Hayward points out, their "preference for soft despotism has become more concrete with the increasing panic over global warming in the past few years. Several environmental authors now argue openly that democracy itself is the obstacle and needs to be abandoned."[72]

The extent to which climate alarmism is rooted in left-wing ideologies and infected with cultic tones and anti-human sentiments becomes clear upon reading some of the speeches and articles of its most fervent advocates. They all exhibit, in Derr's words, "a somewhat murky antipathy to modern technological civilization as the destroyer of a purer, cleaner, more 'natural' life, a life where virtue dwelt before the great degeneration set in. The global warming campaign is the leading edge of an environmentalism that goes far beyond mere pollution control and indicts the global economy for its machines, its agribusiness, its massive movements of goods, and

---

71. "The Political Economy of Global Warming, Rent Seeking and Freedom," in Paul Reiter, et al., *Civil Society Report on Climate Change* (London: International Policy Press, 2007), 90. Kasper's observation is borne out by surveys of members of scientific organizations. See Matthew Nisbet, *Climate Shift: Clear Vision for the Next Decade of Public Debate*, American University School of Communication, Spring 2011, 61-80.

72. Steven F. Hayward, "All the Leaves are Brown," *Claremont Review of Books*, Winter 2008.

above all its growing population."[73] For many of its adherents, salvation lies in a return to nature and a much diminished role for man and industrial civilization. This dark perspective is not grounded in the work of scientists presenting the results of careful, qualified, evidence-based research but is premised on the teachings of a cult bent on imposing its belief system on a wider world, regardless of the consequences, and woe to those are prepared to oppose it. They are people looking for a crisis and manufacturing evidence to justify draconian measures. Scruton points out that "when a radical Left movement becomes discredited, there is seldom an act of penitence. There is rather a sideways migration to another movement with the same emotional structure. During the '70s and '80s, therefore, as the reality of communism could no longer be denied, people began to migrate from red to green."[74] Their solutions are all nostrums dug out of the bag of discarded left-wing notions of earlier eras, seasoned with a religious overlay calling for repentance and asceticism: "drive less, buy less, walk lightly upon the earth."[75]

The radical roots of environmental alarmism

As clearly religious, statist, and anti-human as modern environmentalism has become, it was not always that way. Originally, environmentalism was rooted in a desire to *conserve* nature for the enjoyment of man. It was anthropocentric in orientation, concerned with the ability of mankind to continue to benefit from the natural world. Based on the Biblical commandment to act as stewards of the earth, early environmentalists were primarily concerned with maintaining forests, wetlands, and other pristine locations for the enjoyment of future generations. The movement to set aside parkland and wilderness in the late 19th century was spearheaded by the same people who founded the Sierra Club and the Audubon Society. Their heirs today are not the Sierra Club and the Audubon Society, but Ducks Unlimited, a private group that buys wetlands as a trust in order to set them aside and ensure breeding grounds and a supply of ducks for hunters and wetlands for nature lovers.

---

73. Thomas Sieger Derr, "Strange Science," *First Things*, November 2004. This sour perspective on modern culture is fully laid out in Neil Postman, *Technopoly: The Surrender of Culture to Technology* (New York: Random House, 1992).

74. Roger Sruton, "A Righter Shade of Green," *The American Conservative*, July 16, 2007.

75. Brett Stephens, "Global Warming as Mass Neurosis," *Wall Street Journal*, July 1, 2008.

Environmentalism also has roots in 19th century Romanticism. The English poet William Wordsworth is often viewed as an exemplar of the Romantics' reverence for the wild and untamed and their uneasiness about the impact of industrialization on nature. He had his counterparts in Germany, France, the United States, and elsewhere. More widespread, however, was the impact of prosperity on attitudes towards nature. It is only with affluence that people can begin to take a romantic attitude towards nature. Affluence made it possible for ordinary people to enjoy home gardens or to pursue tourism in remote, "unspoiled" places, two 19th century popular pastimes that altered attitudes towards nature and fuelled the conserver movement.[76] Foss observes that Romanticism may have died by the end of the 19th century, but it lives on in the "sensibilities of industrialized city dwellers ... [who] accept the beauty of nature unquestioningly," sensibilities that explain their "rapid and unquestioning acceptance of the need to rescue nature from human science and technology."[77]

With affluence it is also possible for people to become preoccupied with anxieties and neuroses that earlier generations could not have imagined. Philip Stott observes that "the richer we get, the more neurotic we become. Hypochondriacs worry constantly about their bodily health, and they see every little twinge, however trivial, as evidence of a serious, and often terminal, condition. Ecochondriacs are fundamentally the same, with ecochondriasis being the unrealistic and persistent belief, or fear, that the Earth, and thus we, are suffering from one critical sickness after another, despite the fact that the Earth is the toughest of old boots and life goes on – indeed, is improving for many people."[78]

The rapid growth in prosperity in the second half of the 20th century gave rise to a new kind of attitude to nature, this time rooted in anti-materialism and a disdain for modern industrial society. It rejected the anthropocentrism of earlier generations and replaced

---

76. See David Peterson del Mar, *Environmentalism* (London: Pearson Longman, 2006), 5-30. Bramwell explores the roots of environmentalism in earlier cultural, philosophical, religious, and ideological movements in *Ecology in the 20th Century*. See also David Pepper, *Modern Environmentalism: An Introduction* (New York: Routledge, 1996), John McCormick, *The Global Environmental Movement* (New York: John Wiley & Sons, 1995), and Jeffrey Foss, *Beyond Environmentalism*.

77. Foss, *Beyond Environmentalism*, 197.

78. Philip Stott, "Ecochondriasis," at *Parliament of Things*, a website that is now defunct.

it with a new anti-humanism rooted in reverence for a romanticized nature in the raw. The traditional Christian idea of man as steward of the natural world was replaced by man as the spoiler of that world. Freeman Dyson points to the fundamental difference in values evident among the old (humanist) and the new (naturalist) environmental movements: "Naturalists believe that nature knows best. For them the highest value is to respect the natural order of things. ... Nature knows best, and anything we do to improve upon Nature will only bring trouble. The humanist ethic begins with the belief that humans are an essential part of nature.... For humanists, the highest value is harmonious coexistence between humans and nature. The greatest evils are poverty, underdevelopment, unemployment, disease and hunger, all the conditions that deprive people of opportunities and limit their freedoms. The humanist ethic accepts an increase of carbon dioxide in the atmosphere as a small price to pay, if world-wide industrial development can alleviate the miseries of the poorer half of humanity."[79]

James Lovelock's idea of the planet as Gaia, a self-regulating organism in danger of being overwhelmed by a parasite, man, became one of many radical new perspectives on nature, many of them tied to utopian views of how life on earth should be reorganized in order to place limits on the parasite and its damaging impact. This new kind of environmentalism is largely an urban phenomenon and is embraced by people who are far removed from the reality of nature and whose understanding of the natural world is largely abstract and romantic. It is not surprising that the scientific leaders of global warming alarmism put their faith in models rather than in observation and field work. Some of their harshest critics are those scientists who have spent their careers studying nature close up, often at great personal discomfort. Michael Crichton adds: "People who live in nature are not romantic about it at all. They may hold spiritual beliefs about the world around them, they may have a sense of the unity of nature or the aliveness of all things, but they still kill the animals and uproot the plants in order to eat, to live. If they don't, they will die."[80]

Craig Loehle argues that environmentalism in general and climate change, in particular, are beset with categorical thinking, a

---

79. Dyson, "Heretical Thoughts about Science and Society," *Edge* 219 (August 9, 2007).

80. Michael Crichton, "Environmentalism as Religion."

mode of thinking that gained in popularity in the 1960s, particularly in education. It devalued the benefits of the critical, contextual thinking that was part of a classical education (including language skills and literature, history, science, and music), and replaced it with fads and ideas steeped in political values. Loehle suggests that "a premise in categorical thinking about the environment that goes back before the current debate is that natural is good and artificial is bad, where artificial means anything affected by humans. In the case of nature this means that wilderness is good and trees planted in rows are bad ... . The categorical mindset means that any touch by humans ruins the wilderness, so humans in the US are being progressively excluded from [the] wilderness ... that they are supposed to value so highly."[81] A good example of categorical thinking is the knee-jerk opposition of some environmentalists to many of the benefits of science, such as the bizarre campaigns against vaccination or the chlorination of drinking water. Vaccination against common childhood illnesses and the addition of chlorine to municipal water systems marked two of the great advances in public health, eradicating many common diseases and leading to no known secondary effects. To an environmentalist, however, neither vaccines nor chlorine are "natural" but are products of industry and, therefore, must be banned. This kind of silliness led Patrick Moore, one of the founders of Greenpeace, to resign. Moore concluded that the movement had strayed too far from its scientific roots and had become an intellectually bankrupt movement of the left opposed to a myriad of useful and life-saving products and technologies.[82]

Finally, environmentalists are heirs to the Progressive movement of the opening years of the 20th century. From muckrakers like Upton Sinclair to politicians like Teddy Roosevelt and Woodrow Wilson, progressives sought to address the social ills of rapid industrialization through state-centred reforms. The development of the regulatory welfare state in the middle years of the 20th century was the solution offered by the progressive movement as a middle way between the excesses of both capitalism and socialism. Seeing governments, rather than individuals and voluntary associations, as the key organizers of modern life, the progressives opened the way for a quantum leap in the role of the state. Modern environmentalism

---

81. "Categorical thinking and the climate debate," *WattsUpWithThat*, March, 2013.

82. Patrick Moore, "Why I Left Greenpeace," *Wall Street Journal*, April 22, 2008.

similarly sees a large role for the state and for state-centric regulations and is wary of private initiatives and property rights.

By the 1960s, the emerging breed of environmentalists concluded that conserving wilderness and wildlife was not enough. Public
attention needed to be focused on a wider spectrum of concerns –
from air and water pollution to solid waste disposal, dwindling energy resources, radiation, pesticide poisoning (particularly as described in Carson's *Silent Spring*), and noise pollution. For each of
these concerns, the answers lay in regulation and a massive increase
in state intervention. The vehicles for advancing the agenda were,
first, venerable conservationist groups such as the Sierra Club and
the Audubon Society, and soon a swarm of new groups: Earth First,
Greenpeace, the Worldwide Fund for Nature, Environmental Defense, the WorldWatch Institute, Earth Liberation Front, Natural Resource Defense Council, the Nature Conservancy, Rainforest Action
Network, the Pembina Institute, the David Suzuki Foundation, and
more, each competing for funds by advancing its message through
exaggeration and some mendacity. [83]

Their message fell on receptive ears. By the end of the 1980s,
many of their initial concerns had been addressed by various government programs and regulations. In Europe, North America, Japan, and Oceania, powerful new bureaucracies engaged the issues
and, in many cases, became part of the movement. But support for a
cleaner world was widespread and was rewarded with measurable
and visible success. As a result, the world is not only safer and more
prosperous but also cleaner. The relentless pessimism of the environmental movement, however, refuses to accept that human material circumstances have improved. Their pessimism reminds us of
Thomas Babington Macaulay's criticism in 1830 of the Malthusians
of his day: "On what principle is it that, when we see nothing but
improvement behind us, we are to expect nothing but deterioration
before us?" [84]

Additionally, from the perspective of the environmental movement, all these problems were being addressed within the framework of market-based economies and democratic polities. The new

---

83. For a robust criticism of the takeover of environmentalism by radicals from a
     veteran of field-based biology, see Jim Steele, *Landscapes and Cycles: An Environ-
     mentalist's Journey to Climate Skepticism* (self-published, ISBN: 1490 390189, 2013).
     See also Elizabeth Nickson, *Eco-fascists: How Radical Conservationists are Destroy-
     ing Our Natural Heritage* (New York: HarperCollins, 2012).

84. *Edinburgh Review*, January, 1830, in a review of Southey's *Colloquies on Society*.

movement had a broader goal: to "save" the planet and to do so on the basis of radical political and economic change. Its leaders believed that continued industrial development and economic growth were incompatible with the health of the biosphere.[85] Enter global warming! Here was a problem of truly planetary proportions that they insisted could only be solved on the basis of fundamental political and economic change. A new ideology was born.

It is ironic that the new breed of environmentalists is not only disdainful of materialism and capitalism but even more so of rational science. The very scientific process of discovery to whose authority the climate alarmist community appeals has been systematically rejected by many of their environmental activist supporters. From a reverence for organic food and a rejection of pesticides to astrology and psycho-healing, post-modern environmentalism has rejected the tenets and benefits of two centuries of scientific discovery and application. European governments, in particular, have become captive of the environmentalists' reverence for the "precautionary principle," a concept that is rooted in risk aversion and is disdainful of the careful balancing of scientifically established risks and benefits. *The Economist,* in one of its more sober articles, advised that "predictions of ecological doom, including recent ones, have such a terrible track record that people should take them with pinches of salt instead of lapping them up with relish. For reasons of their own, pressure groups, journalists and fame-seekers will no doubt continue to peddle ecological catastrophes at an undiminishing speed. … Environmentalists are quick to accuse their opponents in business of having vested interests. But their own incomes, their fame and their very existence can depend on supporting the most alarming versions of every environmental scare."[86]

The perils of scepticism

Global climate does change and there is room for debate about what drives change, but global warming advocates in the science commu-

---

85. In 1984, Lester Brown and the Worldwatch Institute began an annual volume on the *State of the World,* premised on the conceit that, despite years of ever more stringent environmental regulations, the biosphere was under ever greater threat. Ronald Bailey (*The True State of the Planet*) and Bjørn Lomborg (*The Skeptical Environmentalist*), among others, demonstrated that Brown and others like him were misrepresenting the facts on a monumental scale.

86. "Plenty of Gloom: Environmental Scares – Forecasters of Scarcity and Doom Are Not Only Invariably Wrong. They Think that Being Wrong Proves them Right." *The Economist,* December 20, 1997-January 3, 1998, 19-21.

nity have undermined their case by their willingness to become associated with the most extreme political and activist agendas, their unwarranted *ad hominem* attacks on other scientists, and their calls to punish sceptics. Attacking scientists for the sources of their funding, for example, often alleged by weak association, has become a common form of criticism. Funding from ExxonMobil, for example, is bad, but if it comes from BP, activist foundations, or, particularly, government, it is good. It is all part of a movement that has badly discredited evidence-based scientific enquiry and vigorous discussion. British philosopher Edward Skidelsky warns that "the extension of the 'denier' tag to group after group is a development that should alarm all liberal-minded people. One of the great achievements of the Enlightenment – the liberation of historical and scientific enquiry from dogma – is quietly being reversed."[87] And perhaps not so quietly.

Danish statistician Bjørn Lomborg has become one of the most passionate critics of this new ideology and, as a result, one of the most vilified individuals on the planet. Surprisingly, Lomborg is an environmentalist of the old school and remains convinced that manmade CO_2 may be an important contributor to global warming. Nevertheless, he is sceptical that the problem is as large or as catastrophic as it is made out to be and that the proposed mitigation strategies are worth their high cost. He believes solutions can be found within the existing political and economic order and can be implemented on a gradual basis.

Lomborg was initially drawn to study environmental issues because he became sceptical of the many outlandish claims made by environmentalists, both about problems and their solutions, and he learned that global warming was one of the most outlandish. Using his considerable skill as a statistician, he found that many of the claims made by environmentalists did not add up. The result has been a series of books and articles questioning the claims of environmental lobbies and advocates. His first book, *The Skeptical Environmentalist*, exposed the hyperbole and hysteria that fuel the environmentalist lobby as a whole. In his own words:

> We're defiling our Earth, we're told. Our resources are running out. Our air and water are more and more polluted. The planet's species are becoming extinct, we're paving over nature, decimating the biosphere. The problem is that this litany doesn't seem to be backed up by facts. When I

87.　Edward Skidelsky, "The tyranny of denial," *Prospect Magazine*, February 2010.

set out to check it against the data from reliable sources – the UN, the World Bank, the OECD, etc. – a different picture emerged. We're not running out of energy or natural resources. There is ever more food, and fewer people are starving. In 1900, the average [global] life expectancy was 30 years; today it is 67. We have reduced poverty more in the past 50 years than we did in the preceding 500. Air pollution in the industrialized world has declined – in London the air has never been cleaner since medieval times."[88]

Lomborg's next publication (as editor), *Global Crises, Global Solutions: the Copenhagen Consensus,* brought together Nobel prize winners to think through what global problems could be solved with an injection of US $50 billion. Global warming came in dead last. His third, *Cool It,* explored the global warming hysteria and subjected all the claims about catastrophic impacts to cool, rational analysis. Lomborg's latest book (again as editor), *Solutions for the World's Biggest Problems: Costs and Benefits,* returns to the theme of his second and explores what it would take to solve some of the more challenging global problems. As the dust cover summarizes:

> The world has many pressing problems. Thanks to the efforts of governments, NGOs, and individual activists there is no shortage of ideas for resolving them. However, even if all governments were willing to spend more money on solving the problems, we cannot do it all at once. We have to prioritize; and in order to do this we need a better sense of the costs and benefits of each 'solution.' This book offers a rigorous overview of twenty-three of the world's biggest problems relating to the environment, governance, economics, and health and population. Leading economists provide a short survey of the state-of-the-art analysis and sketch out policy solutions for which they provide cost-benefit ratios.

Lomborg is right. The global warming agenda is but one of many problems the world faces, and one of the least pressing. Even if the hypothesis is correct, the likelihood of anything more than a 1-2° C increase in average temperature over the course of a century or more falls well within previous human experience and, as discussed in earlier chapters, will present as many benefits as problems. Given the time scale, such problems as exist are readily addressed through adaptive strategies. Urgent action on a large scale makes no public

---

88. Bjørn Lomborg, "Smearing a Skeptic," *Wall Street Journal,* January 23, 2003. His original motivation was to demonstrate that economist Julian Simon was wrong in his optimistic assessment of the state of the planet. His research, however, demonstrated that there had been significant progress in improving the environment and that Simon was right that human ingenuity could prove a powerful weapon in addressing environmental and resource problems.

policy sense. Economic analyses show that it will be far more expensive to cut carbon dioxide emissions radically than to increase resilience and pay the costs of adaptation to marginally increased temperatures.

♦♦♦♦♦

To analysts not trained in the details of the abstruse scientific issues that figure in the debate among physicists, astronomers, meteorologists, mathematicians, paleoclimatologists, and others, the discussions can at times be bewildering. Fortunately, the alarmist community has provided a helpful guide to sorting out who is probably right. Time and again, the most dedicated proponents of the AGW hypothesis, rather than sharing data and discussing methodologies, engage in invective and *ad hominem* attacks. It may well be true that certain sceptics are associated with conservative think tanks supported by business interests. Many others are wholly independent researchers. At the same time, many alarmists are associated with think tanks and centres dependent on government, foundations, ENGOs, and, yes, business support. The issue is not the source of funding but the integrity of researchers, the quality of their research, and the resulting evidence. Under circumstances in which one group is trying to demonstrate the validity of a highly controversial hypothesis with critical implications, it is even more important that those advancing the hypothesis do so on the basis of the best evidence and refrain from arguments from authority, *ad hominem* attacks, reliance on questionable data, and recourse to totalitarian tactics. If they were as confident of their conclusions as they claim, they would be prepared to enter into the kinds of searching discussions pursued by Albert Einstein and his contemporaries on quantum mechanics and other game-changing theories. Instead, they hide behind official science and seek to discredit their critics.

The intolerance shown to climate change sceptics is not unique but has become a growing, if unwelcome, feature of today's popular discourse, particularly on university campuses. Demographer Joel Kotkin, for example, observes:

> Climate change is just one manifestation of the new authoritarian view in academia. On many college campuses, 'speech codes' have become an increasingly popular way to control thought .... Like medieval dons, our academic worthies concentrate their fire on those whose views – say on social issues – offend the new canon. ... A remarkable 96 percent of presidential campaign donations from the nation's Ivy League

faculty and staff in 2012 went to Obama, a margin more reminiscent of Soviet Russia than a properly functioning pluralistic academy.[89]

In smearing sceptical scientists and other commentators, the alarmists have the full support of some political leaders. In February 2015, for example, US President Obama used his personal website to initiate an ugly campaign to "call out the climate change deniers," encouraging his supporters to go after anyone who dissents from the official view on climate change.[90] The New York *Times* and Washington *Post* took the hint and warmed up an ancient story that Willie Soon, a scientist associated with the Harvard-Smithsonian Center for Astrophysics, had benefited from funding from the Charles G. Koch Charitable Foundation and from Southern Company, a Georgia-based energy company. Like most universities, Harvard solicits financial support from many sources, as it had in this case, and Soon's work had benefited from those grants. Soon's real sin, however, was that he is a principled critic of the official science, maintaining that the main driver of climate change is the sun, a view shared by many solar scientists as well as by other scientists.[91]

Four of Obama's fellow Democrats followed up. First, Arizona Congressman Raul Grijalva wrote letters to the presidents of seven universities demanding that they clarify all funding and expenses of seven climate scientists who, in Rich Lowry's colourful phrase, had expressed "impure thoughts" on climate change – Robert Balling, John Christy, Judith Curry, Stephen Hayward, David Legates, Richard Lindzen, and Roger Pielke, Jr. – and had dared to testify to that effect before congressional committees. Not to be outdone, Senators Ed Markey of Massachusetts, Barbara Boxer of California, and Sheldon Whitehouse of Rhode Island, minority members of the Senate Committee on Environment and Public Works, sent letters to the heads of 100 universities, corporations, foundations, and other institutions demanding to know whether "they are funding scientific studies designed to confuse the public and avoid taking action to cut carbon pollution, and whether the funded scientists fail to disclose the sources of their funding in scientific publications or in tes-

---

89.   "The Spread of 'Debate is Over' Syndrome," Orange County *Register*, March 21, 2014.

90.   The campaign can be found at barackobama.com/climate-change-deniers/#/.

91.   Justin Gillis and John Schwartz, "Deeper Ties to Corporate Cash for Doubtful Climate Researcher," New York *Times*, February 21, 2015 and Terrence McCoy, "Things just got very hot for climate deniers' favorite scientist," Washington *Post*, February 23, 2015.

timony to legislators." To their credit, ten other members of the Committee, led by its chair, Senator James Inhofe of Oklahoma, responded by sending letters to the same addressees, apologizing for the first letter and urging them to "continue to support scientific inquiry and discovery, and protect academic freedom despite efforts to chill free speech."[92] The irony is, as Henry Payne pointed out in the *National Review*, that "the overwhelming majority of climate research funding comes from the federal government and left-wing foundations. And while the energy industry funds both sides of the climate debate, the government/foundation monies go only toward research that advances the warming regulatory agenda. With a clear public-policy outcome in mind, the government/foundation gravy train is a much greater threat to scientific integrity."[93] Richard Lindzen agrees, writing in the *Wall Street Journal* that "Mr. Grijalva's letters should help clarify for many the essentially political nature of the alarms over the climate, and the damage it is doing to science, the environment and the well-being of the world's poorest."[94]

Charles Krauthammer warned a few years ago that "the left is entering a new phase of ideological agitation – no longer trying to win the debate but stopping debate altogether, banishing from public discourse any and all opposition."[95] Similarly, British pundit Melanie Phillips sees the unwillingness of alarmists to engage their critics on the merits of their scientific research as part of the larger crisis of modern liberalism. She observes that "liberalism is all about shaking off the shackles of religious authority to make everywhere a better place: ending poverty and oppression, eradicating prejudice, turning swords into ploughshares (aka international law) and creating the brotherhood of man, all based on the rule of reason. Liberalism is thus a utopian project. But, like all utopias, it is impossible to achieve. And so, having now realized this, the west is simply giving up on liberalism, resorting instead to deeply illiberal, pessimistic doctrines which are hostile to freedom, progress and humanity itself." She adds in another column: "And so the warmists continue

92. Grijalva letters at democrats.naturalresources.house.gov/documents. Markey letters at markey.senate.gov/news/press-release. Rich Lowry, "A Shameful Climate Witch Hunt," *RealClearPolitics*, February 28, 2015.

93. Henry Payne, "Global Warming: Follow the Money," *National Review*, February 25, 2015.

94. "The Political Assault on Climate Skeptics," March 5, 2015.

95. "Thought police on patrol: Charles Krauthammer," Washington *Post*, April 11, 2014. See also Tod Lindberg, "Left 3.0," *Policy Review*, February/March 2013.

to make their ever more ludicrous claims, exhibiting that total absence of insight characterizing all fanatics – who, by definition, are totally incapable of grasping quite what a ludicrous spectacle they present every time they open their mouths."[96]

A good deal of the tenacity with which the alarmist community maintains its views – and resists open discussion – can be explained by the sheer amount of money involved. At the beginning of the first Bush administration, US federal support for research into the global warming hypothesis amounted to less than $200 million. After James Hansen's congressional testimony in 1988, it rapidly escalated, reached $2 billion a year by the end of the first Bush administration, and has continued to climb ever since. European, Canadian, Japanese, and Australian governments add more, as do state and provincial governments, private firms, foundations, and ENGOs. Global warming has become a multi-billion dollar industry that cannot afford to have its underlying premises subjected to searching, open scrutiny. It has become a matter of survival not only to the hundreds of scientists whose research budgets are tied to advancing alarmist stories but also to a growing range of industries ready to cash in on mitigation strategies, from wind and solar to biofuel and nuclear energy.[97]

Other large economic interests are also involved. Government subsidies to the ethanol, corn, solar, wind, and other industries, in Canada, the United States, Europe, Australia and elsewhere, have

---

96. Melanie Phillips, "How the liberal West has lost the plot," December 17, 2013, and "Yes! The science is settled! Even MORE sunshine can be extracted from cucumbers!" October 7, 2013, at her weblog, embooks.com/blog. See also Brendan O'Neill, "Global warming: the chilling effect on free speech," *Spiked*, October 6, 2006, and Mark Steyn, "The slow death of free speech," *National Post*, April 18, 2014, for more detailed expositions of the same theme.

97. Given the large number of governments involved, it is difficult to come to a precise figure for the amounts that have been spent on climate change. Suffice it to say that there is more than enough money to provide for a cadre of highly motivated scientists, industries, and other advocates. In the United States alone, some 18 federal agencies are involved in climate change programs and policies. The Congressional Research Service estimates that from fiscal 2001 to 2012, these agencies were authorized to spend US $111.4 billion on climate-related projects. See Jane Leggett, Richard K. Lattanzio, and Emily Brunner, "Federal Climate Change Funding from FY2008 to FY2014," Congressional Research Service Report 7-5700, September 13, 2013. State governments similarly authorize climate change-related spending, as do national and sub-national governments in Canada, Europe, Japan, Australia, and elsewhere. Foundations and industry have added billions more. See also William N. Butos and Thomas J. McQuade, "Causes and Consequences of the Climate Science Boom," forthcoming in *The Independent Review*.

led to sophisticated lobbying by energy and other interests to keep the funds flowing. Duke Energy is only one of many energy companies that have become big boosters of alternative energy and major contributors to ENGOs in an effort to keep the pressure on governments to open the spigots.[98] As part of his 2009 stimulus package, President Obama promised a minimum of $15 billion a year for the next ten years for research on alternative energy. Money on this scale creates passionately committed interests. The idea that industry is lined up on one side of the debate is ludicrous. Instead, there is increasing evidence of a gargantuan Baptist-bootlegger alliance to push through such policies as cap and trade.

Environmental NGOs and activists

Lomborg's audacity lay in pointing out the essential nonsense of the environmentalist doomsday litany. It earned him the opprobrium of all the politically correct, including, particularly, such media favourites as Ehrlich, Schneider, Gore, Worldwatch Institute founder Lester Brown, Earth First! founder David Foreman, and the leading environmentalist lobbying groups, such as Greenpeace, the Worldwide Fund for Nature, the Club of Rome, the Union of Concerned Scientists, and Friends of the Earth. Lomborg exposed them for what they are: a collection of special interests having to hype doom in order to survive financially and politically. Unable to demonstrate that Lomborg is wrong, they resort to *ad hominem* attacks in an effort to discredit the depth and quality of his research.[99] They claim he does not understand. Unfortunately for them, he understands only too well and continues to be a thorn in their sides.

The fatal flaw in the predictions of environmental doomsters is their dismissal of human ingenuity and its role in solving problems.

---

98. The CEO of Massey Energy, Don Blankenship, candidly explains that "in the case of big businesses, it is not the grip of a medieval superstition, nor is it credulity that is driving them. Instead it is what motivates most business people: profit and fear of government retaliation. Or maybe worse: hope for government favouritism." "In the Name of Climate Change," *American Thinker*, July 28, 2009.

99. *Scientific American* lent its pages to a condescending, sneering review of Lomborg's first book, *The Skeptical Environmentalist*, by Stephen Schneider, John Holdren, John Bongaarts, and Thomas Lovejoy. ("Misleading Math About the Earth," January 2002, 61-71). The review led to wide comment in the media. Lomborg replied in the May, 2002 issue. John Rennie, *Scientific American's* editor-in-chief, added a tendentious note which illustrated that the magazine had become little more than a house organ for environmental advocacy.

From Malthus to Ehrlich, Wilson, and Suzuki, they have simple-mindedly extended observations from the behaviour of insects, plants, and animals to that of human societies.[100] Not surprisingly, their predictions fall flat. From famines to epidemics, the very science and technology they denigrate have reduced such issues to manageable problems. As with Marxism, the claim of environmentalists that they have the support of science has proven hollow. To environmentalists, man is subordinate to nature and subject to its limitations; to their critics, man transcends nature, can deal with it, and can use it to expand his prospects and possibilities, while nature remains indifferent, doing what nature has done for 4.5 billion years: change. The static view of nature that animates most environmentalists has no basis in science.

The original scientific quest in the 1970s to reach a better understanding of the drivers of climate change focused on the threat of cooling and a new ice age. By the 1990s it had become wholly captive of ideological environmentalism and of official science. Scientific breakthroughs expanding understanding of such issues as the role of the sun, clouds, greenhouse gases, and ocean thermal and circulation cycles became secondary to the more important ideological goal of using global warming as a driver of broader social and economic change. Increasing focus on a minor driver of climate change, the greenhouse effect of CO_2, provided the perfect wedge issue. The environmental movement is hostile to the material well-being made possible by modern industry. Oil, in particular, has become a symbol of the excesses of modern consumerism, and the large corporations controlling the extraction and distribution of oil are perfect targets for the anti-capitalism that animates much of environmental alarmism. Oil is also the basis of the chemical revolution and all the wonder products that have made possible the medical, agricultural, and materials revolutions enriching human lives today. To current environmental ideologues, it would be hard to find a more perfect target.

Having placed all their eggs in the AGW basket, the alarmist community has become ever more persistent in defence of its goals.

---

100. Every generation since Malthus has produced people prepared to propagate the same fears. Modern Malthusianism can be traced back to books produced in the 1940s: Fairfield Osborn's *Our Plundered Planet* and William Vogt's *Road to Survival*. See Pierre Desrochers and Christine Hoffbauer, "The Post War Intellectual Roots of the Population Bomb," *The Electronic Journal of Sustainable Development* 1:3 (2009).

It is not surprising that as the evidence to support the AGW hypothesis fails to materialize and the gargantuan costs of decarbonizing the economy – in both economic and broader terms – continue to increase, that support is beginning to wane. But, as political scientist Roger Pielke, Jr. argued in 2009, the shrillness is symptomatic of a looming collapse. "Policy makers and their political advisers (some trained as scientists) can no longer avoid the reality that targets for stabilization such as 450 ppm (or even less realistic targets) are simply not achievable with the approach to climate change that has been at the focus of policy for over a decade. Policies that are obviously fictional and fantasy are frequently subject to a rapid collapse."[101] Steven Hayward agreed; he noted a year earlier that "opinion surveys show that the public isn't jumping on the global warming bandwagon, despite a multi-million dollar marketing campaign and full-scale media hysteria. More broadly there are signs that 'green fatigue' is setting in. Magazine publishers recently reported that their special Earth Day 'green' issues generated the lowest newsstand sales of all issues published in 2008."[102]

Pielke and Hayward may have been somewhat premature in their assessments, but the experience since the collapse of the 2009 Copenhagen climate conference has not been encouraging for climate change alarmists. Their hold on European fears and imagination remains formidable, but opposition is now sufficient to allow politicians to begin the painful process of dismantling some of the more damaging programs they implemented and the unrealistic expectations they created. The election of Barack Obama in the United States created new hope and optimism among alarmists, but his delivery has been more rhetorical than effective, and the Congress is now more adamant than ever in its opposition to large-scale legislation. As a result, the Obama administration has been limited to executive action and scare campaigns. Some of his initiatives have imposed costs and burdens on the economy but on a scale that, hopefully, can still be managed and undone. Governments in Canada, Australia, Japan, and Russia have placed climate change on the back burner. Selected sub-national jurisdictions in Canada and the United States, on the other hand, continue to impose costly burdens on their citizens in vain attempts to alter fundamental climate patterns.

---

101. Roger A. Pielke, Jr., "The Collapse of Climate Policy and the Sustainability of Climate Science," Sciencepolicy.colorado.edu, February 7th, 2009.

102. Steven F. Hayward, "All the Leaves are Brown," *Claremont Review* 9:1 (Winter 2008/09), 2.

The environmentalist goal of mitigating global climate change by devising means to effect much more centralized control, at a global level, over all economic activity is looking more and more utopian as time goes by.

It is conventional wisdom among environmentalists that they are at a disadvantage in the propaganda wars because they are seriously outspent by opposing forces. In reality, think tanks and activist groups *supporting* global warming restrictions raise and spend far more money than think tanks and activist groups *opposing* global warming restrictions. Drexel University sociologist Robert Brulle, for example, maintains that a "counter-movement involv[ing] a large number of organizations, including conservative think tanks, advocacy groups, trade associations and conservative foundations, with strong links to sympathetic media outlets and conservative politicians," is determined to confuse people. Their goal, "in the face of massive scientific evidence of anthropogenic climate change, [is] to manipulate and mislead the public over the nature of climate science and the threat posed by climate change."[103]

Towards that end, Brulle's so-called "counter-movement" raises about $900 million a year in the United States for a range of causes, including $64 million in "dark" money provided by private donors who wish to remain anonymous. The effort of these more conservative organizations is more than offset by the energy expended by the hundreds of local, national, and international ENGOs united in their campaign to fight global warming through the Climate Action Network. The CAN is divided into national, state, and provincial chapters, each with its own website, all part of a carefully orchestrated campaign to give the impression that millions of concerned citizens support the crusade to stop global warming, even though polling consistently shows diminishing public support. Brulle himself found in an earlier paper that there were more than 6,500 national and 20,000 local ENGOs in the United States, with some 20-30 million members and more than $5 billion in annual revenue.[104] Whatever Brulle thinks, the money and the media are more committed to alarmism than to the message of sceptics.

---

103. Robert J. Brulle, "Institutionalizing delay: foundation funding and the creation of U.S. climate change, counter-movement organizations," *Climatic Change* 122 (2014), 692.

104. "The US Environmental Movement," in K. Gould and T. Lewis, eds., *20 Lessons in Environmental Sociology* (Los Angeles: Roxbury Press, 2008).

James Taylor reports that alarmists and their media allies presented "Brulle's [2012] paper as 'proof' that money drives the global
warming debate and the money is heavily skewed in favour of sceptics. ... Giving the global warming activists every benefit of the
doubt, no more than $90 million of conservative think tank money
addresses global warming, and no more than $68 million supports
conservative think-tank efforts opposing global warming activism.
This $68 million is counterbalanced by $22 million for conservative
think tank efforts *supporting* global warming activism [emphasis
added]. That leaves a net of merely $46 million among 91 conserva-
tive think-tanks opposing global warming activism."[105] American
University's Matthew Nisbet found numbers similar to Brulle's but
placed them in the context of the amount of money spent by environmental groups and their allies. He reports: "Overall, in 2009, the
most recent year for which data [are] available, the major conservative think tanks, advocacy groups and industry associations took in
a total of $907 million in revenue, spent $787 million on all pro-
gram-related activities, and spent an estimated $259 million specific
to climate change and energy policy. In comparison, the national
environmental groups took in $1.7 billion in revenue, spent $1.4 billion on program activities, and spent an estimated $394 million on
climate change and energy-specific activities."[106] In addition, governments spend millions "educating" the public on official science,
and schools and many churches have become mainstays of ENGO
propaganda.

The US Senate Committee on Energy and Public Works held
hearings on the rising role of ENGOs, and the then Republican Minority concluded that over the past fifty years they have grown
from loosely organized college kids to a billion dollar industry em-
ploying lobbyists, lawyers, and public relations professionals commanding huge budgets. No veteran of the early movement would
recognize the sophisticated organizations making up current advocacy groups. They have moved far beyond the independent, grass-

---

105. James Taylor, "'Dark Money' Funds To Promote Global Warming Alarmism
     Dwarf Warming 'Denier' Research," *Forbes,* February 2, 2014.

106. Matthew Nisbet, "Climate Shift: Clear Vision for the Next Decade of Public De-
     bate," American University School of Communication, Spring 2011, iii. See pag-
     es 1-29 for a detailed breakdown of these numbers. Pages 30 to 47 break down
     the millions more devoted to environmental causes by US foundations. Often
     the original sources of these funds are successful entrepreneurs, including those
     representing gas, coal, and oil interests. Current earnings of ENGOs also come
     from investments in capitalist enterprises.

roots, and citizen-funded groups of the 1960s and 1970s. What many still perceive as a small and disparate movement is, in fact, a large and well-oiled machine receiving its funding from a handful of super-rich, liberal donors operating behind the anonymity of foundations and charities. Similar to other lobby groups, employees move in and out of environmental regulatory agencies and remain part of the network of activists.[107] Another report found that its members are disproportionately white, college-educated, and well off. "From Earth Day 1970 until today ... the majority of the people directing, staffing, and even volunteering at green groups have not only been white men, but they also hail from wealthier households with elite educational pedigrees."[108]

Modern environmentalism's evolution into large, professionally-run organizations has enhanced its ability to become formidable lobbyists and advocates penetrating to the very heart of capitalism and convincing the corporate sector to cooperate.[109] Money has become critical to its operations: the salaries of the professionals who manage these organizations rival those of senior business executives. Unlike business executives, however, who seek to increase value *for* their shareholders, ENGO executives seek to increase the value of the organization *from* their members with constant appeals for more money to pursue their mission. It is a very successful business model. Volunteers may still exist at the local level, but nationally and internationally ENGOs have become wholly professional and formidable fund raisers. If one scare does not work out or the facts do not pan out, there are always other alarms to raise. Much of ENGO funding comes from members, foundations, and wealthy donors who, having made their fortunes in industry, including energy and other supposedly suspect sources of wealth, now lavish it on the

---

107. "The Chain of Environmental Command," US Senate Committee on Environment and Public Works, Minority Report, July 30, 2014. For an earlier and much more detailed study, see Ron Arnold, *Undue Influence: Wealthy Foundations, Grant-Driven Environmental Groups, and Zealous Bureaucrats That Control Your Future* (Bellevue, WA: The Free Enterprise Press, 1999).

108. Brentin Mock, "New report expounds on old problem: Lack of diversity in green groups," *Grist*, July 28, 2014. The report, *The Green Insiders Club*, is available at diversegreen.org/report.

109. For a discussion of this evolution in the United States, see Christopher J. Bosso, *Environment, Inc.: From Grassroots to Beltway* (Lawrence, KS: University Press of Kansas, 2005). He concludes that "the environmental movement has evolved into a mature and very typical American interest group community, albeit with an impressive array of policy niches and potential forms of activism." 157.

opponents of capitalism. The "dark" money raised by ENGOs also dwarfs that raised by conservative groups, with environmental groups raising much more from anonymous donors than do conservative ones. It appears that, in reality, the two sides in the public discussion, at least in the United States and probably elsewhere, dispose of substantial financial resources, but the alarmist side has not only more money to spend but can also count on the considerable resources of the state and its many tentacles.

Some ENGO resources also now come directly from corporate donors who hope to gain public applause for their "green" generosity. Such benevolence can turn out to be a mixed blessing. Physicist Norman Rogers reports, for example, that "a few years ago, the natural gas industry gave $25 million to the Sierra Club for their 'beyond coal' campaign that is trying to destroy the coal industry. The natural gas people thought that the Sierra Club through its influence over the government would kill the coal industry, thereby helping the alternative fuel, natural gas. The natural gas industry did not understand that you can't buy off ideological fanatics. The Sierra Club later turned on its benefactor and launched an attack on fracking. The Sierra Club is an important church in the carbon cult."[110] One is reminded of Lenin's astute observation: "The capitalists will sell us the rope with which we will hang them."

The media and climate alarmism

The means by which a large share of the public has been convinced of the dangers of global warming are not subtle. The principal agents have been:

- Reports from the United Nations, particularly the IPCC, reinforced by reports from all its other agencies, from UNEP and the WMO, to the World Bank and the WHO;
- Similar reports, often prepared by the same people, issued by national agencies, such as the *US National Climate Change Assessment*, issued by the US Global Change Research Program in 2014;[111]

---

110. Norman Rogers, "American Geophysical Union Scraps Science, Now Faith Based," *American Thinker*, June 29, 2013.

111. It was released with much pomp and circumstance by the White House in May, 2014, but was so over-the-top that its influence proved to be short-lived. The report can be accessed at nca2014.globalchange.gov/report.

- Uncritical acceptance of these reports by national bureaucracies attracted to the opportunities they create for expanding the role of government through regulations and supportive programs;
- Their reinforcement by a steady drumbeat from progressives in the academy, churches, schools, the world of entertainment, and elsewhere about the deteriorating state of the planet and the need to re-organize society along statist lines in order to stave off further disaster;
- Incessant lobbying, press releases, and funding appeals by EN-GOs and allied scientists;
- Further lobbying by business and related groups insisting that they have solutions to the problems allegedly created by AGW and that those solutions can be implemented at minimal cost with a little help from the public purse and government regulators; and
- The obligingly one-sided circulation of alarmist climate information by the media.

The combined alarmist activities of all seven can only be described as a propaganda campaign now deeply embedded in popular culture, at least in developed countries. Because all these many interest groups communicate with the public primarily through the gatekeepers of the press, it is the press that carries the prime responsibility for the unbalanced state of public discussion and opinion on global warming. Sociologist Alison Anderson points out that "the media play a crucial role in framing the scientific, economic, social and political dimensions through giving voice to some viewpoints while suppressing others, and legitimating certain truth-claims as reasonable and credible."[112] It was the media that from the outset characterized the IPCC as a uniquely reliable and authoritative voice on the science of climate change, the objective voice of 2,500 of the world's top climate scientists. A little bit of digging would have told them it was nothing of the sort. The media breathlessly, and uncritically, report every study that claims things "are worse than we thought," when in fact the study is often little more than conclusions teased out of a computer model. It is the media that ignore studies and reports based on observational evidence as well as studies that contradict claims made by IPCC-affiliated scientists. As Christopher Booker has concluded: "As the story continued

---

112. Alison Anderson, "Media, Politics and Climate Change: Towards a New Research Agenda," *Sociology Compass* 3:2 (2009): 166.

to unfold, most journalists remained so locked into the narrative that, with honourable exceptions, they abandoned any attempt to exercise their supposedly ingrained professional scepticism, continuing to accept the claims of the 'environmentalist' lobby at their face value, however absurd."[113] Not until the last few years have the media reported on the extent to which the weather has failed to match alarmist claims and predictions, including, as of the summer of 2015, the 18-year hiatus in global warming.

Alarmism is generally consistent with the outlook of most journalists: progressive, interventionist, and concerned. Many media outlets are quick to criticize *FoxNews* and *The National Post* for their conservative perspective but insist that their own progressive outlook is, in fact, neutral. Even supposedly moderate media outlets have serious problems covering climate change and other hot topics in anything approaching a neutral fashion. Claire Parkinson observes that "both journalists and experts would serve the public better if they would be slower to hype a story without stating (or reiterating) the reasonable qualifiers. Sadly in today's society that might not be feasible for reporters and researchers intent on maintaining their competitive edge."[114] Alternate views in editorials and opinion columns do not mitigate the repetition of alarmist stories, few of which have any credible basis in science or reality. Peter Glover and Michael Economides observe that "the media have a long history of taking up 'end is nigh' scaremongering. It's good for ratings. We have had a litany of warnings that 'billions could die' when AIDS, Avian flu, SARS, Ebola, mad cow disease, the millennium bug hit the headlines. When they didn't, of course, media alarmists shrugged, claimed they 'simply report the facts' and moved on to warn about the next looming disaster."[115] The use of weasel words helps to salvage the conscience of some of the better informed. Something "may," "could," "possibly," or "likely" happen or be modelled or projected. Melanie Phillips summarizes: "The way global warming is being reported by the science press is a scandal. In selecting only those claims that support a prejudice and disregarding evidence that these claims are false, it is betraying the

---

113. Christopher Booker, *The Real Global Warming Disaster* (London: Continuum, 2009), 330.

114. Claire Parkinson, *Coming Climate Crisis: Consider the Past, Beware the Big Fix* (Lanham MD: Rowman and Littlefield, 2010), 283-4.

115. Peter C. Glover and Michael Economides, "The Real Climate Deniers," *The American Spectator*, April 3, 2009.

basic principles of scientific inquiry and has become instead an arm of ideological propaganda."[116] David Whitehouse, argues that

> an essential component of the scientific enterprise is the science journalist, and there as the saying goes, we have a problem. ... The spectrum of stories being covered has narrowed to a worrying degree. Many survive as a science journalist just by paying attention to press releases and reproducing them, as long as others do the same. ... Too many who profess to practice journalism are the product of fashionable science communication courses that have sprung up in the past fifteen years. It's my view that this has resulted in many journalists being supporters of, and not reporters of, science. ... Journalists should portray where the weight of evidence lies, but that is the least they should do. ... They should criticize, highlight errors, make a counterbalancing case if it will stand up, but don't censor, even by elimination, don't be complacent and say the science is settled in areas that are still contentious.[117]

The prevailing view among environmental alarmists and activists is that the media provide a falsely balanced perspective on global warming, giving sceptics equal or more space and time than those who uphold the consensus. Thus, Mike Shanahan of the International Institute for Environment and Development (IIED) maintains that "journalistic balance ... arises from the media's need to appear unbiased and tell a story from two sides. And in news terms, conflict sells more than consensus. For years, journalists have been 'balancing' science with scepticism, offsetting evidence with emotion. By ignoring the overwhelming scientific consensus, ... this effectively instils bias. It serves to confuse and misinform the public and has helped to delay action to address climate change."[118] If true, such reporting would be a welcome development and provide the basis for more informed and balanced public discussion. In reality, as study after study has shown, the opposite is true. Nisbet, for example, examined 1,862 climate change stories in 2009-10 at the New York *Times*, Washington *Post*, *Wall Street Journal*, CNN, and *Politico*. Eight out of ten news articles reflected the IPCC perspective. Without the *Wall Street Journal* and *Politico*, the ratio was nine out of ten. Even at the conservative *Wall Street Journal*, where opinion pages

---

116. Melanie Phillips, "The global warming scam," April 25, 2005, at melaniephillips.com/diary/archives/001153.html

117. David Whitehouse, "The Pathetic State of Science Journalism," *Huffington Post*, December 6, 2011. Whitehouse is a former BBC science reporter.

118. Mike Shanahan, "Talking about a revolution: climate change and the media," An IIED Briefing, December 2009, at pubs.iied.org/pdfs/17029IIED.pdf.

were evenly split, news reporting reflected the IPCC perspective eight out of ten times.[119]

In a similar study, Reiner Grundmann and Mike Scott, using computer-based searches of newspapers in the United States, Germany, the UK, and France, conclude that: "All countries show a dominance of advocates over sceptical voices for the period 2000-2010. Sceptics are much more visible in the US and France compared to Germany and the UK. Al Gore is the dominant reference point across all countries and there is a dominance of political actors over scientists within the group of advocates."[120] Another study based on an e-mail survey of 170 professional science journalists from Germany, India, Switzerland, the UK, and the United States, representing five types of leading news outlets, tested their agreement with four statements describing the IPCC perspective. The authors conclude: "In spite of different national and editorial contexts, journalists display a broad consensus. First, the journalists largely agree to all four statements of the IPCC consensus. Second, they agree on the assessment of climate change sceptics: Their contributions are seen as hardly scientifically proven. Third, journalists agree that sceptics should be given the chance to make their points, provided that what they say is critically assessed. Most of the journalists do not want to provide sceptics with space equal to [that] granted to other voices."[121]

To those new to the environmental beat, help is close at hand. As Christopher Alleva learned, the Society of Environmental Journalists (SEJ) is the answer. Founded in 1989, "the association is considered an indispensable resource among many reporters. The SEJ proclaims their mission to be the creation of a formal network of reporters that write about environmental issues. To that end, they maintain a website, run a listserv and send out regular email alerts to coordinate the coverage and make sure that no one deviates from the story template and action line. To reinforce this, they regularly

---

119. Nisbet, "Climate Shift," v-vi. Pages 48-60 provide the details behind these numbers.

120. Reiner Grundmann and Mike Scott, "Disputed climate science in the media: Do countries matter?" *Public Understanding of Science* 23 (2014), 232. Similar findings were reported in a study sponsored by the Reuters Institute for the Study of Journalism (RISJ): James Painter, *Poles Apart: The international reporting of climate scepticism* (London: RISJ, 2011).

121. Michael Brüggemann and Sven Engesser, "Between Consensus and Denial: Climate Journalists as Interpretative Community," *Science Communication* 36:4 (2014), 419.

conduct conferences and workshops teaching propaganda writing techniques and holding indoctrination seminars."[122] In January of 2007, SEJ published the helpful *Climate change: A guide to the information and disinformation,* a remarkably one-sided guide that could have been written by the IPCC or any of the big ENGOs.[123]

Conclusion

Walter Russell Mead acerbically concludes that "environmentalists will only be able to help the world when they grow up. And they will only grow up when the rest of the world – and especially the mainstream press and serious writers and thinkers – start holding them to serious, grown-up standards. Screwy but superficially appealing ideas like the Kyoto Protocol should be mercilessly criticized, and all their flawed assumptions and wishful thinking be held up for the whole world to see – when they are first proposed and debated, not after twenty years of uncritical praise ending in failure. The green agenda and the environmental movement are victims of 'social promotion'; their self-esteem has been stoked and their grades inflated – and nobody has ever explained the hard facts of life, or equipped them with the skills needed for actual, as opposed to virtual, success."[124]

And yet, ordinary people continue to resist the collective forces of this formidable array of advocates. Polling indicates a kind of wary general support that evaporates with questions about what specific sacrifices people are prepared to make. Politicians, who have keen noses for both the superficial appeal of these zeitgeist movements and an innate suspicion of their longevity, learn how to play them to their advantage. Democracy, with all its faults, still works. The movement can fool some of the people all the time, but it cannot fool all of them, or at least not for long. Time is running out, and the movement will soon need another scare if it wants to keep people alarmed. Climate change has had its run, and alarmist scientists will soon need to return to real science to earn their living. At the same time, elite opinion remains committed to the climate alarm narrative which, in turn, may lead to continued mischief at global, national, and local levels, mischief that will be difficult to re-

---

122. "Global Warming Propaganda Factory," *American Thinker,* August 3, 2007.

123. The guide is available at the Society's website, sej.org/initiatives/climate-change/overview.

124. Walter Russell Mead, "Kyoto Fraud Revealed," October 14, 2010, *The American Interest.*

verse and will embed unwarranted and costly policies into the body politic.

We now turn to a discussion of the movement's political trajectory, tracing its early successes at the UN and with national governments and the gradual decline of its appeal. Over the years, the enormity of the project has begun to sink in even as probing questions about the science remain unanswered. By the second decade of the 21st century it had reached a familiar status in the pantheon of other utopian projects: stasis, neither waxing nor waning. Two questions remain: one, how long will it take before climate alarmism sinks into the same oblivion enjoyed by earlier UN projects, from the New International Economic Order project of the 1960s and 1970s to the succession of feel-good world conferences in the 1990s; and two, how much damage will the movement wreak on the economies and polities of western nations and developing countries, and on the integrity of science before it has run its course.

11 | Building Global Consciousness

I am not going to rest easy until I have articulated in every possible forum the need to bring about major structural changes in economic growth and development. That's the real issue. Climate change is just a part of it. ... For me the protection of Planet Earth, the survival of all species and sustainability of our ecosystems is more than a mission. It is my religion and my dharma.[1]

Rajendra Pachauri, IPCC's third Chair

No matter if the science is all phony ... climate change provides the greatest opportunity to bring about justice and equality in the world.

Christine Stewart, Canadian Minister of the Environment, 1997-99[2]

Anxious scientists and the UN find each other

Anxiety about dangerous anthropogenic global warming originated with a small group of scientists in both Europe and North America involved in the emerging discipline of climate science. While for some, the search for greater understanding of the drivers of climate change was purely scientific, for others it involved a political dimension from the start. For the latter group, the need for climate change policy and international cooperation was the product of a cascade of events flowing from human interaction with nature, an age-old phenomenon that, according to them, had reached crisis proportions by the second half of the 20th century due to the grow-

---

1. Letter of Resignation to UN Secretary General Ban Ki-Moon, February 24, 2015 and Gabrielle Walker, "The Team Captain," *Nature*, December 19, 2007, 1153. Pachauri remained consistent from the beginning to the end of his tenure.

2. Quoted in Peter Stockland, "Social engineers sniffed out among greenhouse gases," *Calgary Herald*, December 15, 1998.

ing number of humans and their technological prowess. To them, the minor changes experienced in global climate over the course of the 20th century seemed outside the boundaries of nature, and they were determined to find a human dimension to explain those changes. Additionally, they believed that the rapid pace of post-war economic growth had accelerated the internationalization of re-source exploitation, further increasing the "malign" human impact on earth systems, including the climate system.

Governments, then, needed to develop an international regime to address the crisis by forcing mankind to reduce its numbers and adopt less environmentally destructive lifestyles. These scientists soon found common cause with other scientists who shared their anxieties about human impacts on natural systems. To ameliorate the problems they thought they had identified, all these scientists and their supporters needed a political process that would lead to the necessary changes in the way humans interacted with nature. By the 1970s these concerns had coalesced into a political movement and had found in the United Nations a potent vehicle to address their anxieties. At the same time UN leaders and like-minded pro-gressive politicians recognized in the environmental movement, particularly in its climate dimension, a powerful narrative through which to promote their common interests in a world that was focus-ing on the progressive goals of social justice and global equity.

With the establishment of the IPCC in 1988, the UN, scientists, and the climate change movement more generally had passed a criti-cal hurdle in their political goal of using climate change as the path to broader global socio-economic reforms. A good idea of what these utopians had in mind had been provided by an earlier, progressive global initiative sponsored by the UN: the World Commission on the Environment and Development chaired by former Norwegian prime minister Gro Harlem Brundtland. Canada's Maurice Strong was among those who had called for such a Commission, helped to or-ganize it, and participated as a member. One of his collaborators, Jim MacNeill, moved from his OECD assignment to act as the Commis-sion's Secretary General and as principal author of its report.[3] Swe-

---

3. MacNeill recalls his role in "Brundtland Revisited," *Open Canada,* Canadian In-ternational Council, February 3, 2013. Rupert Darwall notes: "One nation stood above all the rest: seemingly every organ of the Canadian government, its prov-inces and territories and a multitude of societies, students and individuals – two hundred and seven in all – were mobilized and made it their business to be in-volved [in the work of the Commission]." *The Age of Global Warming: A History* (London: Quartet, 2013), 97.

den's Bert Bolin, the lead climate movement scientist soon to be named first chair of the IPCC, was recruited to help in preparing the Brundtland report and ensuring that it fit well within the story line that he and his fellow scientists had developed over the previous 20 years.[4] The Commission duly reported with an ambitiously progressive document, *Our Common Future* (1987), outlining a bold global agenda for change.

Not surprisingly, the Commission found that both development and the environment were in dire straits and that governments needed to become much more active in ensuring *sustainable* development, the nebulous concept destined to become a mainstay of UNspeak.[5] Sustainable development provided a new rationale for central planning and for top-down control. As Peterson and Wood described it in their report for the US National Association of Scholars:

> The goals of the sustainability movement ... go far beyond ensuring clean air and water and protecting vulnerable plants and animals. As an ideology, sustainability takes aim at economic and political liberty. ... Sustainability's alternative to economic liberty is a regime of far-reaching regulation that controls virtually every aspect of energy, industry, personal consumption, waste, food, and transportation. Sustainability's alternative to political liberty is control vested in agencies and panels run by experts insulated from elections or other expressions of popular will. ... Sustainability thus combines an environmental theme with an economic call to arms and a recipe for harsh and often non-democratic forms of political control. ... There is in these demands something that borders on totalitarianism.[6]

---

4. In his history of the IPCC, Bolin considered this link essential to achieving success on the climate project. See *A History of the Science and Politics of Climate Change* (Cambridge: Cambridge University Press, 2007), 40. Bolin had by now fully morphed from a scientist into a political activist and would put his stamp on the work of the IPCC. It is hard to tell whether his science served his ideology or vice versa.

5. Rupert Darwall concludes that sustainable development was "a doctrine in search of scientific authentication. As a political ideology, Marxism always claimed to be derived from scientific analysis. By contrast, sustainable development was an ideology, developed from a political formula, in search of science." *The Age of Global Warming* (London: Quartet, 2013), 98. For a dissection of the intellectual problems with the concept, see Wilfred Beckerman, "'Sustainable Development': Is it a Useful Concept?" *Environmental Values* 3 (1994), 191-209, and "How Would you Like your 'Sustainability', Sir? Weak or Strong? A Reply to my Critics," *Environmental Values* 4 (1995), 169-79.

6. Rachelle Peterson and Peter W. Wood, *Sustainability: Higher Education's New Fundamentalism* (New York: National Association of Scholars, 2015), 14, 17, 18.

A greater role for the UN and its many organs went without speaking. In her foreword, Brundtland appealed to "citizens groups, to non-governmental organizations, to educational institutions, and to the scientific community. They have all played indispensable roles in the creation of public awareness and political change in the past. They will play a crucial part in putting the world onto sustainable development paths, in laying the groundwork for *Our Common Future*."[7]

According to MacNeill, the Commission's most important role lay in changing the focus of attention from addressing symptoms of environmental degradation to dealing with its underlying causes. He observes that "we found that the environmental protection agenda that nations adopted before and after Stockholm [the 1972 Stockholm Human Environment Conference] tackled only the *symptoms* of environmental degradation; it completely ignored the *sources*. The sources were to be found not in our air, soil, and waters, which were the focus of the environmental-protection agenda, but in a whole range of perverse public policies, especially our dominant fiscal and tax policies, our energy policies, and our trade, industry, agriculture, and other policies."[8] It is an astute point. The focus of environmental activism after the Brundtland Report was less one of solving specific problems such as air or water quality but rather the much more ambitious goal of remaking the world so that it operated on *progressive* ethical, social, economic, and ecological principles, requiring state regulation of every aspect of life having an impact on the environment. From this perspective, the environment was no longer a *consideration* in addressing a range of public policy problems, but the *central principle* around which life should be organized and directed by the state. In its own words *Our Common Future* "serves notice that the time has come for a marriage of economy and ecology, so that governments and their people can take responsibility not just for environmental damage, but for the policies that cause the damage. Some of these policies threaten the survival of the human race. They can be changed. But we must act

---

Peterson and Wood provide a detailed assessment of the evolution of the concept and its increasing hold on the academic world.

7. "Chairman's Foreword," Report of the World Commission on Environment and Development, *Our Common Future*, United Nations, 1987.

8. MacNeill, "Brundtland Revisited." Ever the environmental pessimist, MacNeill believes that in 2013, "the journey to a more sustainable world is barely underway, even though we have made a significant amount of progress."

now."[9] It was a perspective clearly shared by the scientists active in moving the climate agenda forward. French philosopher Pascal Bruckner notes sardonically in his study of modern apocalyptic movements: "Save the world, we hear everywhere: save it from capitalism, from science, from consumerism, from materialism. Above all, we have to save the world from its self-proclaimed saviours who brandish the threat of great chaos in order to impose their lethal impulses."[10]

The year 1987 also provided climate scientists with a powerful example of what could be done if they raised sufficient alarm. That year, governments agreed to the Montreal Protocol, which committed them to phasing out the use of chlorofluorocarbons (CFCs). This group of chemicals had proven very useful as refrigerants, solvents, propellants, and in other applications. Because these chemicals are relatively stable, early consensus had been that their release into the atmosphere was not a problem. By the late 1970s, however, some atmospheric physicists had concluded that not only were CFCs greenhouse gases contributing to global warming but also that their stability meant that they could reach the stratosphere – the atmospheric layer 8-15 to 50-60 kilometres above the Earth's surface. At that height, CFCs' reaction to ultraviolet radiation released chemicals that reacted with ozone, destroying the very thin layer of atmospheric ozone that plays a critical role in shielding the Earth's surface from the ultraviolet segment of the sun's radiation. In the sardonic view of Fred Singer, the environmentalists had found "at last, an industrial chemical – and produced by big bad DuPont and their ilk. What a cause!"[11]

Not all scientists agreed that CFCs presented an urgent problem. Knowledge of the chemistry of the upper atmosphere, for example, was still in its infancy, and it was not clear whether the ozone hole over the Antarctic was part of a natural cycle or a worrisome new development; nor did atmospheric chemists fully understand the forces at work over the polar regions that might lead to changes in the level

---

9. Back Cover, printed version of *Our Common Future.*

10. Pascal Bruckner, *The Fanaticism of the Apocalypse* (Malden, MA: Polity Press, 2013), 185.

11. Singer, "My adventures in the ozone layer," *National Review,* June 1989. Singer's opposition, grounded in his experience with the issue as an EPA and NASA official and as an atmospheric scientist, added to the motivations for environmentalists to take every opportunity to discredit his sceptical approach to politicized science.

of ozone. Much of the evidence favouring regulation came from models rather than from more reliable observational studies. As with climate change, data and measuring challenges remained immense.[12] Nevertheless, governments were convinced they needed to act, reassured by the fact that industry could use substitute chemicals without ozone-depleting characteristics and do so at a reasonable cost. A number of companies viewed the issue as more of an opportunity to market new products than as a problem and withdrew their opposition once this became clear. Phasing out CFCs, therefore, could be done without causing significant political or economic collateral damage. As a result, governments succeeded in short order in negotiating an international instrument to ensure that everyone was on the same page.[13]

For global warming interests, this early success provided a powerful incentive to press on and convince governments to act on climate change, not always acknowledging that the two issues were only superficially similar. Climate change science was much more complex and controversial, as were the solutions. Nevertheless, as EPA officials Alan Hecht and Dennis Tirpak point out, "this issue helped to build a non-governmental environmental constituency in the US which became very well organized in the subsequent climate negotiations, and to enhance the reputation of UNEP, which coordinated the negotiations of the Vienna Convention. This issue also served to forge an important alliance between the EPA and the Department of State that was crucial in resolving many domestic political issues and in implementing US foreign policy."[14]

---

12. Subsequent scientific literature, which is now of little media or activist interest, bears out that the science was incomplete and the problem less urgent than imagined in the 1980s. See Will Happer, "The Ozone Hole Debacle from an Insider," *Icecap*, August 5, 2010, and Quirin Schiermeier, "Chemists poke holes in ozone theory," *Nature* 449, 382-383 (27 September 2007). Matt Ridley concludes that "you do not have to dig far to find evidence that the ozone hole was never nearly as dangerous as some people said, that it is not necessarily healing yet and that it might not have been caused mainly by CFCs anyway." "The ozone hole was exaggerated as a problem," *The Times*, September 25, 2014.

13. The 1987 Montreal Protocol is part of the 1985 Vienna Convention on the Protection of the Ozone Layer and is managed by the parties under the auspices of the UNEP. It has been amended six times as scientists and governments have gained greater understanding of the problem the Protocol seeks to address.

14. Hecht and Tirpak, "Framework agreement on climate change: a scientific and policy history." *Climatic Change* 29 (1995), 377. There is significant literature examining the Montreal Protocol as an important precedent for efforts to address climate change. See, for example, Cass Sunstein, "Of Montreal and Kyoto: A

The die was now cast for a more concerted campaign to convince governments to tackle the GHG problem and to do so with the same vigour and dispatch as they had displayed on CFCs. The next few years saw a flurry of conferences and seminars involving scientists, officials, and ministers trying to build a broad consensus for policy measures to address the looming threat of global warming: the Toronto Conference (June 1988), the Ottawa Conference (February 1989), the Tata Conference (February 1989), the Hague Conference and Declaration (March 1989), the Noordwijk Ministerial Conference (November 1989), the Cairo Compact (December 1989), the Bergen Conference (May 1990), and the Second World Climate Conference (November 1990).[15] Consistent with the long-established UN practice of using frequent meetings to move an issue forward, even if only by a few inches at a time, frequent meetings helped to create a climate of urgency, wear down resistance, make what seemed radical at one point commonplace after many repetitions, and provide allied domestic interests with additional ammunition to put political pressure on reluctant governments.[16] At the same time, experience with the New International Economic Order in the 1970s had taught international bureaucrats that governments' willingness to engage in negotiations as framed by the "international community" did not necessarily translate into enforceable commitments. As Loren Cass points out: "The headline normative debates masked the complexity of implementation. Norm affirmation under political pressure did not ensure compliance because there were insufficient incentives to force compliance."[17]

Holding meetings in different countries helped to strengthen host government commitments. The 1988 Toronto Conference, for example, was opened by Canadian Prime Minister Brian Mulroney, who saw in environmental issues an opportunity to offset the opprobrium he had earned from left-wing activists everywhere with his 1985 initiative to negotiate a free-trade agreement with the Unit-

---

Tale of Two Protocols," *Harvard International Law Review* 31:1 (2007), 1-65; and Richard J. Smith, "The Road to a Climate Change Agreement runs through Montreal," Peterson Institute for International Economics, *Policy Brief* PB 10-21, August 2010.

15. UN, "The international response to climate change," Climate Information Sheet 17.

16. See the discussion in Loren Cass, *The Failures of American and European Climate Policy* (Albany, NY: State University of New York, 2006), 49-52.

17. Cass, *The Failures of American and European Climate Policy,* 51.

ed States. The ubiquitous Jim MacNeill was a member of the conference's steering committee. Mulroney had already convinced Ronald Reagan to take steps to resolve the cross-border acid-rain problem. The negotiation of the Montreal Protocol in Canada was another feather in his cap. The Toronto Conference allowed him to further burnish his environmental credentials by hosting an important climate conference and ensuring that Canada would be seen to be taking the lead in any subsequent negotiations.[18] The conference was scheduled immediately after the G-8 meeting in Toronto that year in the hope that some ministers would stay and attend the climate conference. For the first time, delegates at the conference came up with a specific target: governments needed to take steps to reduce CO_2 emissions by 20 per cent by 2005.[19]

These meetings saw the transition from discussions among scientists considering a *scientific* issue to meetings among government officials focused on a *policy* problem. While sponsored by the UN and its organs, the flurry of seminars and conferences from the early 1960s through the mid-1980s lacked full participation from all UN members and did not benefit from a formal policy goal. Most of the participants were scientists, invited by the organizers. Even government-based scientists came without a policy mandate. At the same time, organizers recognized that the complexity of climate science went beyond the expertise of the atmospheric specialists who had dominated the meetings to date. Climate was more complicated than that, and, before governments could be persuaded to act, the case had to be grounded in a wider assessment of the science as well as the economics. Scientists needed to make an unassailable case based on three elements: the physical basis of continued anthropogenic warming; the negative impacts of that warming on both humans and the biosphere; and the steps that could be taken to mitigate warming. A broader base of scientific knowledge would make the case for action more credible. To that end, as discussed in chapter five, the UN created the IPCC and gave it a mandate to research the *human* dimension of climate change, including its physical and socio-economic impacts, and the range of possible measures that governments could take to mitigate the human impact on climate change.

18. See Mulroney's remarks on the occasion of being honoured as Canada's Greenest Prime Minister by Corporate Knights, April 20, 2006.

19. Daniel Bodansky, "The History of the Global Climate Change Regime," in Urs Luterbach and Detlef F. Sprinz, eds., *International Relations and Global Climate Change* (Cambridge, MA: MIT Press, 2001), 27.

At the same time UN members agreed to initiate negotiations on a Framework Convention on Climate Change (UNFCCC), with the goal of completing negotiations in time for the Convention to be opened for signature at the Earth Summit, scheduled to be held in Rio de Janeiro, Brazil, in June, 1992. Twenty years after the first UN Environmental Conference, environmental activists believed the time had come to take the international environmental agenda to a new level, with the climate change "crisis" sharing centre stage with the biodiversity "crisis."[20] The two crises had come of age and were now ready for prime time, not only in the world of environmental activists but as part of national political agendas. Bolin, John Houghton, and their fellow scientists had convinced themselves and environmental activists that the minute rise in global temperatures from 1977-88 – less than half a degree Centigrade, a number teased out of the highly manipulated temperature series recently produced by the teams at the Climatic Research Unit at the University of East Anglia and NASA's Goddard Institute for Space Studies – was the prelude to a steady rise in the years to come, reaching catastrophic levels by the middle of the next century unless governments took drastic steps to curb modern man's appetite for the energy locked up in fossil fuels and to forgo the benefits that came from exploiting that energy. They had also convinced themselves of the wholly un-scientific idea that their views represented the consensus of climate scientists and that only a few cranks and sceptics believed otherwise. It was a tall order, particularly since they had nothing but their models to make the case and no viable alternatives for replacing the energy from fossil fuels.

The UN, the progressive agenda, and the climate "crisis"

Up until 1988, the development of an international scientific "con-sensus" had been driven largely by three interests: 1) scientists who believed that they had identified a serious problem needing not on-ly more research and funding but also urgent political action; 2) en-vironmentalists who were convinced that human interference with nature was destroying the planet; and 3) the UN system whose

---

20. Similar to concern about climate change, concern about species diversity pro-vides an equally vague issue similarly harnessed to address various aspects of the progressive agenda. It too is a product of speculative science grounded in computer models, with little observational science to give it substance. While environmentalists claim that thousands of species become extinct every year, the number of *documented* cases is far less dramatic.

leaders believed that scientists had identified an issue requiring government action and cooperation among all nations and one that dictated a large role for the UN and its agencies. The scientists' and environmentalists' interests were clear, the UN's less so, until it was realized that the climate crisis offered a golden opportunity to make significant progress on its goal of global governance based on a progressive agenda of transformative social engineering. This was already clear to Sonia Boehmer-Christiansen in 1994: "Scientists naturally prefer to experiment with mathematical models for the Earth's various systems free of responsibility for policy. Uncertainty is their security. Indeed, some can already be seen withdrawing from policy involvement [in climate policy discussions]. For others, including the chairman of the IPCC, global warming has become the justification for a crusade against materialism and for a 'new organizing principle' – the preservation of the Earth."[21]

The 1972 UN Conference on the Human Environment in Stockholm may have touched on climate only tangentially, but it set in train the second theme of the post-security, post-1960s UN. The first theme had been launched in Geneva at the first UN Conference on Trade and Development (UNCTAD – 1964), followed by the first UN Development Decade, the quest for a New International Economic Order, the Charter of Economic Rights and Duties of States, and a series of other declarations, programs, and agencies, such as the UN Industrial Development Organization (UNIDO).[22] All were dedicated to propagating and advancing the global salvationist doc-

---

21. Sonia Boehmer-Christiansen, "A scientific agenda for climate policy?" *Nature* 372: 6505 (December 1, 1994), 400. IPCC Chair Bolin, protesting in a letter to *Nature*, called Boehmer-Christiansen naïve: volume 374 (March 16, 1995), 208. With a background in both science and international relations, she was at the time with the respected Science and Technology Policy Research Unit (SPRU) at the University of Sussex and, judging by the Climategate e-mails, was already regarded with suspicion by the global warming movement.

22. Canadian economist Harry Johnson described the NIEO as a classic example of progressive thinking: "first, that there is something fundamentally wrong with the existing system of international economic relations, which needs to be corrected by a change in the system or order; second, that that something wrong is blameable on the past and present policies of the advanced Western countries, which have been blatantly immoral and should atone for their guilt by accepting the obligations of the new international economic order; and third, that the change in the international order indicated requires a massive shift of political power from the major countries to the voting assembly of the United Nations." "The New International Economic Order," *Woodward Court Lecture*, The University of Chicago, October 5, 1976. It is not difficult to imagine how this could be reshaped into a description of sustainable development.

trine – one resolution, conference, commission, seminar, program, and document at a time. Government leaders and their officials actively participated in the discussions as if they were offering real solutions to serious problems rather than engaging in the building of a series of evermore elaborate Potemkin villages. Ministers' patience – in both developed and developing countries – gradually wore thin, and while some of these agencies continue to exist and pour out studies and hold meetings, ministers ceased attending meetings by the mid-1980s, and the resources devoted to UNCTAD, UNIDO, and similar UN development efforts have gradually dwindled. In their place, a newer version of the doctrine has emerged: rather than establishing a new international economic order, the focus has shifted to advancing sustainable development.

German official Herman Ott identifies the link between the current climate agenda and the earlier economic agenda: "Suffice it to say that the [climate agenda] involves nothing less than a technological and social revolution within the next 100 years – the conscious development of a global society that has outgrown its fossil-fuel resource base. It hardly needs to be said that complex questions of international and intergenerational equity loom large behind almost every aspect of the problem. It is no coincidence that many of the arguments familiar from the debate on a 'new international economic order' in the 1970s have been resuscitated in recent years."[23] In place of UNCTAD and UNIDO, we now have UNEP, WMO and particularly the IPCC and the UNFCCC as the central actors in the UN's efforts to remake the world along progressive lines, efforts that continue to this day.

Economist David Henderson finds serious fault with senior economic officials in capitals who sit idly by as the UN's salvationist doctrine continues to beguile governments. He notes that "they have failed, among other things, to wake up to, and try to do something about, the growing influence of anti-business and anti-market NGOs, the interventionist and anti-market line taken by most international agencies, the uncritical endorsement by their own governments of questionable notions such as 'sustainable development,' 'social exclusion,' and 'corporate social responsibility,' and – in particular – the substantial and continuing erosion of freedom of con-

---

23. "Climate Change: An Important Foreign Policy Issue," *International Affairs* 77:2 (April 2001), 278. Ott was an official at the German Foreign Office as a policy planner in 2000-01. He is now a Green Party politician associated with the Wuppertal Institute for Climate, Environment and Energy in Germany.

tract through intrusive laws and regulations."[24] Steven Bernstein, however, argues that following the Brundtland Report, the UN and member governments succeeded in crafting a compromise that integrated environmentalism into the liberal architecture of post-war economic institutions, allowing markets to take the lead in addressing environmental and development issues.[25] It is an intriguing argument but not one that recommends itself to anyone well-versed in the evolution of post-war economic institutions, particularly the GATT/WTO regime.[26]

An international organization focused on environmental and social issues may not have been the original plan of those governments who came together in 1945 to establish the United Nations with a view to avoiding the excesses and tensions that had led to two world wars and a global recession. It soon became clear, however, that, for many social democratic governments, the UN could also be a forum for advancing the progressive view of how the world should be organized, politically, economically, and socially. With the influx of many members from the newly independent, developing world and the UN commitment to speak to their needs from a progressive perspective, the die was cast for the UN's steady march towards its version of utopia. That march has had two main features: alarmism about the present and future course of events unless changes are made in member policies and practices; and the conviction that solutions are available but that they require collectivist action by governments and the international community to be effective. The system does not provide for competing analyses or resolutions crafted on the basis of different political values and per-

---

24. David Henderson, "Economics, Climate Change Issues, and Global Salvationism," *Economic Education Bulletin* XLV:6 (June 2005), 5. See also Henderson, *The Role of Business in the Modern World: Progress, Pressures, and Prospects for the Market Economy* (Washington: Competitive Enterprise Institute, 2004).

25. Steven Bernstein, *The Compromise of Liberal Environmentalism* (New York: Columbia University Press, 2001).

26. The GATT, firmly under the control of member governments, managed to stay largely aloof from the UN's machinations for most of the postwar years. The IMF and World Bank proved less successful as their roles gradually evolved into becoming the bankers of last resort for bankrupt developing countries. The latter two boasted large bureaucracies generating their own agenda; the GATT did not. Since the turn of the century, however, all three have become increasingly involved in the broader UN progressive agenda, as discussed further below, including the WTO which now boasts a large bureaucracy of its own. All bureaucracies require expanding agendas to thrive.

spectives. Climate change discussions at the UN provide a classic example of how the UN operates.

The UN learned from its experience pursuing the NIEO that the adoption of resolutions, the way in which it had typically moved the agenda forward, did not have the same impact as treaties and similar instruments. In the environmental area, therefore, the emphasis shifted to negotiating formal multilateral environmental agreements (MEAs). While some legal scholars insist that UN resolutions have the force of international law, their practical effect on the policies of member states is much less than the provisions of more formal treaties that are opened for signature and subject to ratification. The case can be made, however, that the experience with MEAs is as much a matter of diluting the force of international treaty law as it is of raising the bar above the plethora of resolutions adopted by the UN General Assembly and many other UN bodies. Many countries have signed on to MEAs; few of them have fully implemented their provisions into law or lived up to them in their domestic policies. MEAs, similar to UNGA resolutions, have become matters of rhetoric or, perhaps, moral pledges rather than of legal, enforceable commitments.[27]

At many meetings and conferences, the UN relies on consensus to move the agenda along. The psychology of group dynamics makes it difficult for a state to register a formal objection. Decisions at international meetings rarely involve a recorded vote because a vote would require member governments formally to express an opinion rather than to accept the broad consensus of the meeting as perceived by the chair. A further UN convention helping to move the agenda along is that developing countries need not accept the same level of obligation as do developed countries nor make more than token financial contributions to the implementation of UN programs. As a result, the Secretariat can count on a block of at least 150 countries who will support the agenda and help bring it into force, knowing that financial obligations will only affect the 40 or so developed countries participating in the discussion. Given that the

---

27. See the discussion in Jeremy Rabkin, *Law Without Nations: Why Constitutional Government Requires Sovereign States* (Princeton: Princeton University Press, 2007). Rabkin is dismissive of the many claims of international lawyers and makes a strong case that without appropriate representation and police powers, as are enjoyed by nation states, the claims of international law are generally hollow. For the progressive perspective in full bloom, see Philippe Sands, *Lawless World: America and the Making and Breaking of Global Rules* (London: Penguin Books, 2005).

EU and its 28 members are often represented by the Commission, enlarging its "competence," Canada, the United States, Japan, Australia, New Zealand, Norway, and a few of the more advanced developing countries (e.g., Singapore, South Korea, Turkey, Chile, and Mexico) often find themselves relatively isolated from "the will" of the international community. It is not difficult to understand, therefore, why the UN's progressive agenda moves steadily in the same direction, as it has for more than five decades, with only a few occasional setbacks. Ministers may not always be aware of what is happening at these meetings, but, even if they were, it is unlikely that they would want their delegation to be isolated. Foreign ministry officials will certainly counsel against voting no to progressive resolutions, eager as they typically are to remain in step with the rest of the so-called international community.

The development of the EU from a six-member customs union into a 28-member economic and quasi-political union provides a further example of the progressive agenda at work and helps to explains why the EU is now regularly part of the progressive wing at these international gatherings. Policy is brokered in Brussels by a Commission that is committed to a similar agenda for similar reasons: it is the most efficient path to power for a non-elected elite. Additionally, international agreements can add to the competence of the Commission, i.e., the subject areas for which the Commission, rather than the member states, has lead responsibility. There are EU member governments who may not always share the agenda and occasionally manage to slow it down, but the direction of change remains the same. On climate change, the Poles, Czechs, and a few others have been laggards. The UK at the beginning was an enthusiastic subscriber led by Margaret Thatcher, a commitment she learned to regret, both in general and specific terms. Once Tony Blair and Labour replaced the Conservatives, the UK became even more enthusiastic. Only in the last few years has the Liberal-Conservative Coalition developed some doubts, not only about climate change but also about the extent to which Commission enthusiasm for multilateralism and the climate agenda has dragged the UK into obligations that some now regret.

The EU is a social democratic institution. Socialism is built into its very fabric, as is readily apparent when one studies the *acquis communautaire*, the complex regulatory regime that all acceding states must incorporate into their domestic laws and institutions before they can become members. Daniel Hannan, a UK member of

the European Parliament, aptly points out that the modern upholders of liberty are to be found in the Anglosphere from which the liberalism of John Locke emerged. That liberalism is not to be found
on the continent and explains the increasing discomfort of the British electorate with the European model.[28]

Progressive politics at the global level is a rather incestuous
business. Socialists International (SI), for example, the voice of some
170 socialist, labour, and social democratic national political parties
around the world, works hand in glove with the United Nations in
advancing progressive causes. Its Commission for a Sustainable
World Society (CSWS), for example, was established to articulate
how the world of progressive politics could help facilitate the resolution of global environmental concerns and their global governance. The CSWS focuses on supporting and contributing to UN climate change conferences and the global effort to reduce global
warming.[29] There is very little to distinguish the statements and
principles of Socialists International from UN resolutions. Sustainable development and the redistribution of wealth are central organizing principles for both. SI's leaders meet regularly in New York at
UN headquarters with direct access and input to policy. Its members include national governments, national leaders, and former national leaders wearing their party hats. Progressive groups such as
the Global Leadership for Climate Action and the International Crisis Group involve former senior members of social-democratic governments engaged with other global progressive leaders such as
George Soros, Maurice Strong, Mikhail Gorbachev, Carol Browner,
Achim Steiner, and Mark Malloch Brown in reinforcing and advancing progressive causes. The image of thousands of progressives
speaking with one voice is misleading, given the number of times
that the same names keep popping up supporting the same cause
under a different name. The linkages are endless and incestuous.[30]

---

28. Daniel Hannan, *Inventing Freedom: How the English-Speaking Peoples Made the Modern World* (New York: Broadside Books, 2013).

29. Socialists International, "Progressive Politics for a Fairer World," at Socialistinternational.org. See also its statement of principles.

30. The OECD, which for years acted as a brake on the ambitions of the UN by providing more sober and thoughtful analyses of world problems, has over the last two decades morphed into another branch of Socialist International. In 2010 the secretary-general of the OECD, Mexico's Angel Gurria, hosted a meeting of SI's Council at OECD headquarters. The Managing Director of the IMF is tradi-tionally a European, and most have been social democrats. The head of the World Bank is traditionally an American appointment but, with the exception of

The Global Legislators Organization for a Balanced Environment (GLOBE) is another international organization with many national chapters, including in Canada and the United States. It supports national parliamentarians in developing and advancing common legislative responses to the major challenges pursued in the name of sustainable development, including climate change. At the beginning of 2013, GLOBE released its 3rd Climate Legislation Study, prepared largely by the Grantham Institute at the London School of Economics, headed by now Lord Nicholas Stern. GLOBE's purpose is to provide a forum for legislators to share experiences in developing, passing, and overseeing the implementation of climate change legislation and in supporting legislators in advancing that agenda.[31] "Without the burden of formal governmental negotiating positions, legislators have the freedom to push the boundaries of what can be politically achieved. GLOBE's vision is to create a critical mass of legislators within each of the parliaments of the major economies who can agree to common legislative responses to major global environmental challenges and can demonstrate to leaders that there is cross-party support for more ambitious action. Its goal is to ensure that all major government policy decisions are consistent with climate change goals."[32] Christiana Figueres, the current executive secretary of the UN Framework Convention on Climate Change, gave a stirring speech at GLOBE's 2013 forum urging legislators to act in support of the UN's work on climate change. Dennis Ambler concludes: "The UN network, fed by the climate change agenda, is infinite .... It is like a vast mycelium with threads spreading over the globe, springing up new bodies on a regular basis, but growing from the same root. Most of the people who started the process,

---

Robert Zoellick, has in recent years been staffed by social democrats. The GATT's first four Directors General were technocrats, but, with its evolution into the WTO, its last four Directors General have been former political figures from social democratic parties. The current DG, Roberto Azevedo of Brazil, is a technocrat from Brazil. Members of SI now occupy the commanding heights of global governance, leaving little room for alternative points of view. Many progressive politicians accept that it will ultimately be necessary to establish a world government with power to tax and regulate. They downplay this objective when in office but play it up when meeting as NGOs.

31.   See its website, globeinternational.org, for an overview of its activities.

32.   Quoted in Dennis Ambler, *The United (Socialist) Nations – Progress on Global Governance, Climate Change, Sustainable Development and Bio-Diversity*, SPPI Original Paper, February 17, 2011, 19.

such as Strong, are still in place and they survey the results of their social engineering with great satisfaction."[33]

The climate change agenda now permeates the work of virtually all international organizations, and, even as national governments are losing their enthusiasm for some of the more radical ideas advanced by activists, international organizations and their supporters remain firmly committed. Similar to national governments run by social democrats, the UN faces the dilemma that its appetite is larger than its reach. Liberals and social democrats alike are convinced that they can create a better world and provide better results than can be achieved through private-sector and voluntary initiatives. History has long since belied this claim. Competence, efficiency, and delivery have never been hallmarks of public programs, whether delivered at the local, national, or international level. The image of municipal workers leaning on their shovels while one fills a pothole rings a universal bell. UN pretensions that it can deliver a more prosperous, sustainable, equitable, and just world are as hollow today as they were fifty years ago.

Since the early success of the US space program and the release of the first iconic image of Earth from space, progressives have penned many a hymn of praise to man's technological and planning prowess, particularly when promoted by government. They also saw in the space program a metaphor for the need to govern the planet more justly and efficiently through greater central planning. Adlai Stevenson, US ambassador to the UN, former Democratic candidate for the presidency and progressive icon, told the UN in 1965: "We travel together, passengers on a little space ship, dependent on its vulnerable reserves of air and soil; all committed for our safety to its security and peace; preserved from annihilation only by the care, the work, and, I will say, the love we give our fragile craft. We cannot maintain it half fortunate, half miserable, half confident, half despairing, half slave – to the ancient enemies of man – half free in a liberation of resources undreamed of until this day. No craft, no crew can travel safely with such vast contradictions. On their resolution depends the survival of us all."[34]

His friend and fellow progressive Barbara Ward followed up the next year with an extended essay, *Spaceship Earth*. In her words,

---

33. Ambler, *The United (Socialist) Nations*, 33.

34. Adlai Stevenson, Speech to the Economic and Social Council of the United Nations, Geneva, Switzerland, July 9, 1965. Barbara Ward wrote the speech.

"the most rational way of considering the whole [human] race to-day is to see it as the ship's crew of a single spaceship on which all of us, with a remarkable combination of security and vulnerability, are making our pilgrimage through infinity." She added, "in short, we have become a single human community."[35] To continue the journey safely, she argued, we will need a captain and rational rules of behaviour. As economist Gary North pointed out at the time, "the analogy of spaceship earth is more than an analogy; it is a call to religious commitment. The call is to faith in centralized planning. ... The thing which strikes me as ironic is that the language of the spaceship involves a chain of command approach to the solution of human problems. Those humanitarian intellectuals who decry the petty military dictatorships in underdeveloped nations want to impose a massive system of command over the whole earth. That is what the call to world government implies."[36]

American literary historian and environmentalist Leo Marx, in a much-quoted 1970 article in *Science*, bluntly summed up what by then had become one of the core beliefs of the modern progressive movement: "On ecological grounds, the case for world government is beyond argument. Meanwhile, we have no choice but to use the nation-states as political instruments for coping with the rapid deterioration of the physical world we inhabit."[37] Since then, many have become impatient with the slow pace with which democracies address environmental problems and the willingness of politicians to make compromises with those who do not share the progressive view of the brave new world order required to save the planet. For some, the threat of climate change provides a particularly strong argument for more efficient and centralized forms of government.

Recent expressions of the need to provide a global authority to address the ills of the planet can be found throughout the environmental and climate change literature. All are predicated on what the *Wall Street Journal's* Holman Jenkins characterizes as "the idea that humanity might take charge of earth's atmosphere through some supreme triumph of the global regulatory state over democracy,

---

35.   Barbara Ward, *Spaceship Earth* (New York: Columbia University Press, 1966), 14-5. This essay became the inspiration for her next book prepared for the 1972 Stockholm Conference, *Only One Earth*.

36.   Gary North, "The Mythology of Spaceship Earth," *The Freeman*, November 1, 1969.

37.   Leo Marx, "American Institutions and Ecological Ideals," *Science* 170 (November 27, 1970), 945.

sovereignty, nationalism and political self-interest, the very facts of political human nature."[38]

Two Australians, a physician and a lawyer, give voice to the claim that only a new authoritarian form of government headed by scientific experts can save the planet. In their words, they "confront the reader with problems of such magnitude that issues of personal liberty pale into insignificance. ... [and] present the case against democracy, showing how freedom and liberalism have the potential to propagate environmental tyranny far greater than any threat posed by the former Soviet Union." They see themselves as "critics, on ecological grounds, of the capitalist economic system and existing authoritarian systems [and] argue that even the allegedly more environmentally preferable liberal democratic societies fail to provide humanity with ecologically sustainable structures. ... Humanity will have to trade its liberty to live as it wishes in favour of a system where survival is paramount. Perhaps this choice should not be put for democratic approval, or humanity will elect to live as it wishes." Shearman and Smith insist that "liberal democracy – the meshing of liberalism and democracy – is the core ideology responsible for the environmental crisis. ... It has become a matter of real politics intrinsically enmeshed with capitalism, and it is virtually impossible to separate the effects of each."[39] Much of their book is rooted in the usual left-wing, utopian bin of discredited ideas and is readily countered with the work of Popper, Hayek, and other more grounded thinkers, who emphasize the cognitive limits of human knowledge, no matter how expert, particularly when moving from the general to the specific.[40]

Shearman and Smith are not eccentric outliers. Similar views, if not always as boldly and fully articulated, can be found throughout

---

38. Holman Jenkins, "Personal Score-Settling Is the New Climate Agenda," *Wall Street Journal*, February 28, 2014.

39. David Shearman and Joseph Wayne Smith, *The Climate Change Challenge and the Failure of Democracy* (London: Praeger, 2007), 2, 4, and 12. The book is part of a series sponsored by the Pell Center for International Relations and Public Policy and was published in a serious, scholarly series of books on political science by Praeger. The book fits well within the tradition of left-wing, anti-human diatribes from Francis Galton (eugenics) to Paul Ehrlich (ecology). Shearman has penned a number of similar books and was a contributor to the third and fourth IPCC reports.

40. See, for example, Karl Popper, *The Open Society and its Enemies* (London: Routledge 1945), which focused largely on Plato, Hegel, and Marx, as well as Friedrich Hayek, *The Road to Serfdom* (London: Routledge, 1944).

environmental and alarmist literature by both natural and social scientists. Two German academics associated with the Wuppertal Institute for Climate, Environment and Energy, for example, maintain that "national welfare is no longer an effective frame of reference for enlightened foreign policy; it must be extended to encompass the common welfare of a world society. ... Perhaps what is required 60 years after the founding of the United Nations is a new attempt to establish a genuinely sustainable global order. ... Without an environmental organization with legal powers to combat global ecological crises and without an international social politics concerned with justice, there will never be peace, for rich and poor alike."[41] Their sentiment is reflected in the papal encyclical, *Laudato Si'*, in which Pope Francis calls for "stronger and more efficiently organized international institutions, with functionaries who are appointed fairly by agreement among national governments, and empowered to impose sanctions. ... What is needed is a politics which is farsighted and capable of a new, integral and interdisciplinary approach to handling the different aspects of the crisis. ... Interdependence obliges us to think of *one world with a common plan*."[42] (emphasis in original)

On this side of the Atlantic, a group of biologists, including such alarmist icons as Paul Ehrlich and Michael Oppenheimer, recently provided a more subtle version of this mind-set:

> Government policies are needed when people's behaviours fail to deliver the public good. Those policies will be most effective if they can stimulate long-term changes in beliefs and norms, creating and reinforcing the behaviours needed to solidify and extend the public good. ... In much of the world, there is growing awareness that we face potentially catastrophic global environmental problems and that significant shifts in policies, technologies, and behaviours will be required to address them. Therefore, many people are primed to accept solutions that evoke social norms involving our shared responsibility to the environment and to other people, and many policymakers are searching for policies that can have long-term impacts on behaviour and environmental outcomes.[43]

---

41. Wolfgang Sachs and Hermann E. Ott, "A New Foreign Policy Agenda: Environmental Politics is Resource Politics is Peace Politics," *Limits to Growth* (Spring 2007), 22.

42. *Laudato Si' (Praise Be to You), On Care for Our Common Home,* June 17, 2015, paras 175, 197, and 164.

43. Ann P. Kinzig et al., "Social Norms and Global Environmental Challenges: The Complex Interaction of Behaviours, Values, and Policy," *BioScience* 63:3 (2013), 164.

The politicization of climate change science in the late 1980s rapidly became more than a matter of finding the most effective way to address a specific problem. Rather, climate change became the central organizing principle for a much more ambitious movement. By 2011 UN Secretary General Ban Ki-Moon was urging world business leaders to consolidate and organize and to embrace UN-style economic innovation to save the planet. He told the denizens of Davos: "We need a revolution. Climate change is also showing us that the old model is more than obsolete." He called the current economic model a recipe for "national disaster" and said: "We are running out of time. Time to tackle climate change, time to ensure sustainable ... growth."[44] In 2015, the pope joined the bandwagon, seemingly inspired by the words and ideas of the global progressive movement.

The UN and one-world religion

In 2009 the UN named April 22nd – already known as Earth Day in much of the world – Mother Earth Day, thus establishing worship of the goddess Earth as its official religion. "One would suppose [that] the UN would be committed to cultural and religious diversity, eschewing the establishment of a global religion, given the fact it has representatives from countries around the world. But as it turns out, the UN has long had an interest in a global government and universal religion."[45] The 2009 resolution, introduced by Bolivia's Evo Morales, marked more than two decades in which the UN had been actively engaged in promoting global religious harmony, coincident with its efforts to promote global salvationism and sustainable development.[46]

By the early 1990s, as the UN's ambitions grew, it faced the reality that the membership in local UN associations and World Federalist chapters was aging and the numbers declining. A new way was needed to appeal to a broader audience than representatives from member governments. The UN found the answer in religion,

---

44. "Ban Ki-Moon: World's economic model is 'environmental suicide'," *The Guardian*, January 28, 2011.

45. Fay Voshell, "The UN and One World Worship," *American Thinker*, April 24, 2011.

46. The full expression of these religious sentiments can be found in the *Universal Declaration of the Rights of Mother Earth*, adopted at the World People's Conference on Climate Change and the Rights of Mother Earth, April 22, 2010 in Cochabamba, Bolivia.

not the traditional kind, but in the growing number of progressives within and outside existing faith communities who were looking for ways to make their spiritual life more relevant to the modern age. Senior UN officials correctly perceived that religion had been one of the most divisive forces on the planet, the basis of wars throughout history. They also realized that religious leaders from many faiths were committed to peace and that there should be a way to bridge this gap and at the same time promote UN values. To that end, the UN declared 1993 to be the Year of Inter-religious Co-operation and Understanding.

To make progress, it would be necessary to work with religious leaders whose appreciation of their own faith was sufficiently flexible to provide for such a movement. The growing influence of progressive ideas within many faith communities suggested that the time was ripe. Former UN assistant secretary-general Robert Muller had long dreamed of this moment. As he wrote in 1982: "The world's major religions in the end all want the same thing, even though they were born in different places and circumstances on this planet. What the world needs today is a convergence of the different religions in the search for and definition of the cosmic or divine laws which regulate our behaviour on this planet."[47] Existing organizations, such as the World Council of Churches, were not considered up to the task because they were insufficiently comprehensive in their membership. Indeed, one of the UN's advisers, the adaptable and controversial Swiss Catholic theologian Hans Küng, wrote that "traditional religions have an ethical obligation to cease to exist ... there will be no room for religious diversity because if ethics is to function for the wellbeing of all, it must be indivisible. The Undivided World increasingly needs an undivided ethic."[48] In an ethically progressive world, there would be no room for Christians, Jews, Buddhists, or Moslems, but only for adherents of the UN-sponsored, one-world religion of peace, brotherhood, and sustainable development.

---

47. Robert Muller, *New Genesis: Shaping a Global Spirituality* (Garden City, NJ: Doubleday, 1983), 126-7. He was a UN Assistant Secretary-General for nearly 40 years until his retirement in 1986 and was well known for his ideas about world government, world peace, and spirituality. See his blog at goodmorningworld.org for insight into his many ideas on world government and international peace and harmony.

48. Küng, *Global Responsibility: In Search of a World Ethic* (Crossroads, 1991), as quoted by Voshell, "The UN and One World Worship."

For a start, UNESCO commissioned Küng to create a "Declaration of a Global Ethic." Küng's effort became the focus of a meeting of the World Parliament of Religions, held in Chicago in September 1993 on the centenary of the first such parliament held there in 1893. Muller gave the opening address, and Gerald Barney of the Millennium Institute gave the keynote address on the state of the environment. Barney is interested in the development of a "mutually enhancing relationship between humans and Earth." A deep ecologist, he is a firm believer in integrating sustainable development into contemporary religion. His most important contribution to the movement was the pessimistic *Global 2000 Report* commissioned by President Jimmy Carter.[49] Barney urged the delegates to work together to "(a) create the religious, social, and economic conditions necessary to stop the growth of human population, (b) reduce the use of resources and disposal capacity by the wealthiest, (c) assure civil order, education, and health services for people everywhere, (d) preserve soils and species everywhere, (e) double agricultural yields while reducing both agricultural dependence on energy and agricultural damage to the environment, (f) convert from carbon dioxide-emitting energy sources to renewable, non-polluting energy sources that are affordable even to the poor, (g) cut sharply the emissions of other greenhouse gases, (h) stop immediately the emissions of the chemicals destroying the ozone layer, and (i) bring equity between nations and peoples of the North and South."[50]

The Parliament has met subsequently in Cape Town, South Africa (1999), Barcelona, Spain (2004), and Melbourne, Australia (2009). The next meeting in Brussels, Belgium was scheduled for 2015 but has now been postponed to 2017 for lack of local funding.[51] At the

---

49. Barney is one of the co-authors of *Threshold 2000: Critical Issues and Spiritual Values for a Global Age* (Ada, MI: CoNexus Press, 1998), as well as *Christian Theology and The Future of the Creation*, Proceedings of the Holden Village Conference, December 3-7, 1989, and similar reports and studies. He was the study director of *The Global 2000 Report to the President*, three volumes, US Government Printing Office (1980).

50. Gerald O. Barney with Jane Blewett and Kristen R. Barney, *Global 2000 Revisited: What Shall We Do?* (Arlington, VA: Millennium Institute, 1993). Georgetown's Leo D. Lefebure, commented: "For me and many other participants, Barney's address was the most powerful presentation of the entire Parliament. The speech crystallized the aim of the event and set forth a clear and compelling agenda for inter-religious cooperation for the sake of all life on the planet." "Global Encounter," *Christian Century*, September 22-29, 1993, 887.

51. See the website of the Council of the World Parliament of Religions, headquartered in Chicago, at Parliamentofreligions.org.

1993 meeting, the participants adopted Küng's "Declaration Toward a Global Ethic"[52] as an expression of their own ambitions. In 2007, the Parliament spun off the Universal Forum of Cultures to explore religion and culture and to broaden its appeal to contemporary society.

UN leaders also looked hopefully at an effort by the Episcopalian bishop of California, William Swing, to bring together religious leaders from around the world in a United Religions Initiative (URI). A staunch champion of progressive causes, Swing has worked tirelessly to build a world-wide movement dedicated to "a world at peace, sustained by engaged and interconnected communities committed to respect for diversity, nonviolent resolution of conflict and social, political, economic and environmental justice."[53] In 1997, Swing brought nearly 200 religious leaders from around the world together at Stanford University. He hoped that this "spiritual United Nations," as some have referred to it, would be a world assembly for humanity's myriad spiritual traditions. URI cooperates with the Council of the World Parliament of Religions but prefers its more grass-roots approach to that of the Council.

Concurrently, the ubiquitous Maurice Strong had been building a network to craft and sponsor an appropriate creed and catechism – an Earth Charter – for the new world religion or ethic. Strong told the 1992 Earth Summit that "the responsibility of each human being today is to choose between the force of darkness and the force of light. We must therefore transform our attitudes and values, and adopt a renewed respect for the superior law of Divine Nature."[54] The Earth Charter would put this superior law into words.

The idea for such an initiative originated during the work of the Brundtland Commission and initially resulted in the document adopted at the 1992 Rio Summit: Agenda 21. While Strong took great pride in Agenda 21, he recognized that its length – 500 pages – and detail went well beyond the kind of document required to create a groundswell of popular support for a world ethic. In its place, he began thinking about a shorter and more appealing document. He founded a new group, the Earth Council, and recruited Mikhail Gorbachev and his group, Green Cross International, to sponsor a

---

52. The Declaration was prepared by some 200 scholars representing many of the world's religions and was signed by 143 religious leaders at the Parliament.

53. Vision statement at www.uri.org/about_uri.

54. Maurice Strong, as quoted in "In their own words," at SPPI, March 10, 2008.

new effort. Strong and Gorbachev had found that, despite their different backgrounds, they saw eye to eye on many issues. They consulted widely and recruited many to help out and by 2000 had succeeded in crafting their Charter. Thousands had contributed to its drafting. It was discussed at the 1999 World Parliament of Religions. The final text was approved at a meeting of the Earth Charter Commission at UNESCO headquarters in Paris in March 2000. The official launch was on 29 June 2000 in a ceremony at the Peace Palace in The Hague, Netherlands.[55]

According to its authors, the Charter reflects the concerns and aspirations expressed at the seven UN Summit meetings held during the 1990s on the environment, human rights, population, children, women, social development, and the city. There are calls for universal health care; education; disarmament; the equitable distribution of wealth; elimination of discrimination in all its forms, such as that based on race, colour, sex, sexual orientation, religion, language, and national, ethnic or social origin; participatory democracy; freedom of expression, assembly, and dissent; debt relief for developing nations; gender equality and equity as prerequisites to sustainable development; and the participation of women in all aspects of economic, political, civil, social and cultural life as full and equal partners. Similar in conception to Küng's global ethic, the Charter conveys a more optimistic tone without the agonizing evident in Küng's declaration. It finds an echo in the tone and wording of the 2015 papal encyclical, *Laudato Si'*.

The Charter's founders believe that the document, while not formally adopted by governments, has become "widely recognized as a global consensus statement on the meaning of sustainability, the challenge and vision of sustainable development, and the principles by which sustainable development is to be achieved. ... [It] continues to grow in international stature as a source of inspiration for action, an educational framework, and an international soft law document, as well as a reference document for the development of policy, legislation, and international standards and agreements."[56] Ultimately, its founders hope it will take its place alongside the 1948

---

55. The Dutch government had provided the funding for much of the work of the Earth Charter Commission. The Charter can be found at the website of the Earth Charter Commission, www.earthcharterinaction.org. Steven Rockefeller, emeritus professor of religion at Middlebury College, is widely credited with being the principal intellectual resource person for the movement.

56. As expressed in the official history of the Charter at Earthcharterinaction.org.

Universal Declaration of Human Rights and will have a similar impact.[57] As formidable as such an accomplishment would be, the UN is strongly committed to a Charter that clearly reflects the views of those who see the UN as the principal agent for the development of a world government with a one-world ethic or religion.

Thomas Sieger Derr found it all a bit much. He noted that its "tone, in sharp contrast to UN documents, breathed a quasi-religious spirit, an overarching pantheism that not everyone could share, echoing the Gaia hypothesis that earth is, in effect, a single organism deserving of the name of a goddess – here, 'Mother Earth,' a name deliberately chosen for such symbolism."[58] With the adoption of Mother Earth Day, the aims of its founders are a step closer to realization. Nevertheless, Derr finds that the Charter's strong anti-anthropocentric tones, reflective of the anti-humanism that colours much New Age religion and deep ecology, will not appeal to developing countries which look to the UN to promote their developmental aspirations and which view the sustainability agenda as at best a distraction from that goal, appealing more to dissidents and utopians in the developed world than to the citizens of developing countries. Derr adds: "There is also, undoubtedly, a kind of neo-paganism among many Charter supporters, whose antipathy to modern society in all its aspects, from industrial to religious, has led them back to a radical pre-modernism, a pan-religiousness that appears to be some (partly imagined) basic form of religious life before the destructive divisiveness of the historic religions appeared."[59] In summary the UN has opened, with its one world religious initiative, a veritable Pandora's box.

Framing a policy response

By the early 1990s, anxious climate scientists and their supporters in the environmental and progressive movements could look back on two major achievements: they had clearly placed climate change at

---

57. Then Canadian Minister of External Affairs, Lester Pearson, viewed the Declaration of Human Rights with some misgivings and along with US Secretary of State John Foster Dulles feared it would become the source of much mischief. They were not wrong. See the discussion in Michael Hart, *From Pride to Influence: Towards a New Canadian Foreign Policy* (Vancouver: UBC Press, 2005), 41 and 350.

58. Thomas Sieger Derr, "The Earth Charter and the United Nations," *Religion and Liberty,* March/April 2001, 5-6.

59. Derr, "The Earth Charter and the United Nations," 8.

the top of the UN agenda as an integral part of its efforts to pursue sustainable development, and they had concluded that governmental funding for more research was no longer a sufficient response to the awareness and anxiety that had been raised among political leaders in the major economies. Governments now needed to act or, at least, to be seen to act, both internationally and domestically. Internationally they needed to develop agreements mandating cooperative approaches, and domestically they needed to be prepared to incorporate internationally agreed goals into domestic measures. Less clear was what specific measures would be required to address the problems. Policy experts indicated two basic approaches: mitigation and adaptation. As Australian climate scientist David Evans put it: "If human emissions of CO_2 are causing a major planetary problem, then there are only two plausible solutions: wait and adapt, or regulate and reduce."[60] From the perspective of all those pushing for action to address climate change, mitigation is the preferred and only acceptable option. For them, mitigation provides the key to the global economic transformation they have long sought. Again, in Evans' words: "To regulate CO_2 emissions effectively and fairly you must regulate nearly all energy use, and thus most of the economy, in every nation of the world. The regulating class promotes the dual beliefs that the 'problem' of global warming is very scary and that it is caused by human emissions of CO_2. *The only solution they offer just happens to be complete regulation of the whole world's economy.*"[61] (emphasis in original) Mitigation often forces policy makers to choose between two equally unattractive alternatives: "do nothing and face potentially cataclysmic consequences or act expeditiously and risk destroying the global economic infrastructure."[62]

The choice of mitigation over adaptation also admirably lends itself to the model manipulation that has become the trademark of alarmist scientists. Their models can generate the kinds of numbers that can form the basis for negotiation. Abstract ideas are difficult to negotiate and implement. Hard numbers – no matter how fictitious once examined more closely – give the illusion of negotiating real disciplines. None of the numbers that have become integral to the

60. David Evans, "Climate Coup – The Politics," *JoanneNova.com*, March 19, 2012.

61. David Evans, "Climate Coup – The Politics."

62. Timothy O'Donnell, "Of loaded dice and heated arguments: Putting the Hansen-Michaels global warming debate in context," *Social Epistemology* 14:2/3 (2000), 114.

negotiations bear close examination since they are all the result of speculative science and model crunching, from the impact of GHGs on the global average temperature (itself a suspect number), to the amount of GHGs added to the atmosphere through human agency. Mitigation is also the option with which governments have had relatively limited experience. Adaptation to and preparedness for climate extremes and natural disasters are well-known areas of public policy, and it is not difficult to think through what more might be required if some of the projections advanced by climate scientists turn out to be true. Costs might be high, but with good forward planning this could be managed, particularly since needs might not become apparent for decades or not at all. Most infrastructure is built on the basis of a 40 to 50 year time horizon. Factoring potential impacts of climate change into that kind of planning cycle is well within the experience of both governments and private industry.

Adaptation, however, is of little interest to the zealots – scientific or otherwise – pushing for a response to the threat of climate change. Adaptation is largely a matter for local governments – state, provincial, and municipal. For federal states, the government's role is largely a matter of facilitation and financial support. For unitary states, central governments may play a larger role. A role for international agencies can be foreseen in the need for assistance in developing countries and perhaps for information sharing. None of this requires international treaties and programs. The climate change lobby, however, is interested in a global response to a global problem, one that will provide a basis for a quantum leap in the march toward greater global governance and central planning. Nothing else will do.

Activist climate scientists, environmentalists, and various progressive groups have identified industrialization, overpopulation, and economic growth as the sources of the climate crisis, and they want governments to frame their policy responses on that basis. As Jim MacNeill pointed out, the great contribution of the Brundtland Commission was to change the focus from efforts to address symptoms to an attack on root causes. Mitigation, therefore, has become the focus of policy discussion while adaptation has been pushed into the background. Given the central role assumed by the UN system in framing the issue and in pushing governments to respond, primary responsibility for policy development resides in foreign ministries and environmental agencies. Neither have much experience in devising policies addressing an issue of this scale and complexity, nor do they command the analytical and programmatic human resources that may be required. Other agencies may, of course, need to become involved, but the driving force behind policy development would be the capacity to participate in international discussions.

Most public policy originates in problems and opportunities identified at the local level and requires a response at the appropriate level of government. However, as governments have taken on more and more responsibilities, foreign ministries have had to expand their range of expertise to deal with the foreign-relations aspects of domestic policies, often in large part to ride herd on the interests of domestic departments in dealing with their counterparts in foreign countries and in addressing the need, if any, for international agreements. As a result, foreign ministries have gained some capacity to deal with a wide range of domestic issues, although at a relatively superficial level, relying on domestic departments to provide them with more detailed and knowledgeable policy support. In few countries do foreign ministries have large political constituencies. Their legitimacy within the hierarchy of political and bureaucratic interests lies in maintaining prestige and responding to the international needs of domestic departments.

Setting society's goals within the confines of its own frontiers is largely under the control of the government of the day, subject to the vagaries of democratic politics. Their delivery in the context of foreign policy is more complicated because it is critically dependent on factors and circumstances that are beyond any government's control. More often than not, implementing foreign policy goals requires the cooperation and good will of other states. Governments pursue joint projects with other governments because they believe they can solve problems or advance goals that cannot be resolved through domestic or unilateral measures alone. Done well, foreign policy can make a critical contribution to the security, prosperity, and well-being of a country's citizens, just as a well-ordered domestic policy framework is essential to strengthening a government's hand in meeting its foreign policy objectives. Nevertheless, aside from matters of war and peace, it is difficult for foreign ministry officials to convince governments to act unless they have a domestic partner capable of harnessing a significant domestic political constituency. Unless and until climate change should become an issue with a strong domestic constituency, it would be difficult to advance the UN agenda much beyond what had been accomplished by the early 1990s, i.e., rhetorical expressions of good will and the promise of some future action.

In addition to the traditional functions of national governments, such as defence and foreign relations, governments expand their mandates in two ways: by providing services sought by a significant domestic constituency, such as income redistribution, or by responding to anxieties, such as food safety or air pollution. Given the growing costs of supplying their populations with services, modern governments have learned that they can increase their responsibilities by promising that they can reduce risk, advance health, improve the environment, or solve any of the numerous perceived social and environmental problems that have become preoccupations of the modern state. Much of this activity requires constituencies devoted to keeping populations alarmed and media that thrive on alarm stories. None of the alarms need to be true; all they need is enough cogency to spread anxiety in the electorate and push governments towards a perceived political need to respond. In the case of climate change, the required anxiety level might not have been reached by the early 1990s, but among international elites a campaign begun with the 1972 Stockholm Conference had matured enough to convince governments of the need to, at least, appear to act.

As a result, unlike many domestic environmental issues, such as air and water pollution or habitat loss, climate change emerged as a policy problem without a significant domestic constituency but, nevertheless, one requiring governments to respond to a call from the UN. From the 1972 Stockholm conference through the 1992 Rio Earth Summit, governments had participated in meetings at UNEP, WMO, and other agencies, most of them of a technical nature and largely staffed by science specialists from responsible domestic agencies. These discussions were either largely technical – advances in weather forecasting and development of standardized data – or at a high level of abstraction. Reading UN reports and resolutions on climate and other environmental issues can be a mind-numbing experience. Nevertheless, they served their purpose in spreading alarm among progressive elites. Repetitive, full of high-minded ideas, and short on concrete details and proposals, they largely served to reinforce the faith and advance the agenda.

Most OECD economies established environmental agencies and ministries in the 1970s to address the increasing environmental consciousness of their electorates. The division of responsibility for environmental matters posed particular challenges for federal countries like Canada, Australia, and the United States, as well as for an economic alliance such as the European Union. Since most environmental issues need to be addressed at the local level, the federal or central government's role is thus a matter of coordinating policies, perhaps setting standards, and using the power of the purse to create national policy. When, however, an issue becomes a matter of international discussion and negotiation, state and provincial governments must yield to the superior jurisdiction. Environmental and foreign ministry officials then quickly discover that they have mutual interests.

For many governments, therefore, the first concrete manifestation of the need for a serious policy response to the climate issue came with the call at the 1988 Toronto Conference for the negotiation of an international framework agreement to address climate change by reducing GHG emissions. The call had been carefully orchestrated with the support of a few governments committed to an activist response to this newly identified problem, particularly the Commission of the European Union and some of its more progressive member governments. UK officials, initially among the prime movers of the climate agenda, had become less enthusiastic as the enormity of what some had in mind became clear and as more agencies needed to become involved. Germany was not yet ready to exercise leadership. US officials from the State Department and various specialized science agencies had been among the most enthusiastic proponents at the technical meetings but were now being held back by the White House and economic agencies. Canada remained one of the more enthusiastic proponents of action. Australia did not share that enthusiasm. Developing countries were interested, but only if the required action did not involve any commitments on their part *and* if the action included financial assistance of one kind or another.

Despite initial lukewarm support among the major economies that would need to finance the burden of mitigation, the UN pressed on and organized an intergovernmental committee to prepare for negotiations at the beginning of 1991 with a goal of concluding them in time for the Rio Summit. As Loren Cass points out, the UN and similar institutions "facilitated a socialization process that built momentum toward international negotiations to mitigate the threat of climate change. ... The international framing of the problem provided a tool for domestic advocates to influence" governments to adopt both domestic and international mitigation measures. "The origins of the norm [requiring states to accept domestic emission reduction commitments] were in the work of NGOs and scientists."[63] The norm would be pursued most vigorously by the EU, led by Germany, Denmark, and the Netherlands.

By the middle of 1992, in time for Rio, delegates had succeeded in crafting a UN Framework Convention on Climate Change (UNFCCC) setting out the parameters within which governments would negotiate more specific commitments. It built on an approach similar to that addressing the ozone depletion issue: a framework convention providing for the subsequent negotiation of protocols containing specific commitments. As described on the UNFCCC website, "the ultimate objective of the Convention is to stabilize greenhouse gas concentrations 'at a level that would prevent dangerous anthropogenic (human induced) interference with the climate system.' It states that 'such a level should be achieved within a time-frame sufficient to allow ecosystems to adapt naturally to climate change, to ensure that food production is not threatened, and to enable economic development to proceed in a sustainable manner'."[64]

It was a shrewd step, providing a much firmer platform on which to build more expansive policies. In Rupert Darwall's words, global warming had by this time become "embedded in a pre-existing ideology, built on the belief of imminent planetary catastrophe – which many scientists subscribed to – with a UN infrastructure to support it and a cadre of influential political personages to propagate it."[65] Nevertheless, given the extent of scientific uncertainty, the complexity of the issues, the lack of a widespread sense of crisis, and the reluctance of less developed countries to assume any burdens, it is surprising that the issue had reached the point at which governments were prepared to enter into a framework convention, as they did at the Rio Earth Summit in 1992. Translating this framework into concrete commitments, however, was a much more daunting challenge. Activists were convinced that climate change was a major issue, requiring an urgent response to a pending disaster that could be avoided by deploying existing technologies at reasonable cost. There are grounds for suspecting that many of the leaders of the movement were well aware that they were engaging in both exaggeration and dissembling. For many, the means

---

63. Cass, *The Failures of American and European Climate Policy*, 50.

64. UNFCCC, "First steps to a safer future: Introducing The United Nations Framework Convention on Climate Change," at UNFCC.int.

65. Darwall, *The Age of Global Warming*, 99.

justified the end: a more ordered, just, and equitable world controlled by high-minded experts. But could they build a strong enough base at the domestic level to convince political leaders to do more than endorse symbolic acts?

Expanding the political base favouring climate action

In the four years between the 1988 Toronto Conference on the Changing Atmosphere and the 1992 Rio Earth Summit, alarmist scientists and their allies at the United Nations, in national bureaucracies, and in environmental nongovernmental organizations (ENGOs) had succeeded in transforming an issue preoccupying environmentalists, a few scientists, and a handful of sympathetic politicians to one commanding sufficient numbers among the public to move more politicians to embrace the issue and pursue a cooperative, multilateral, course of action. This had been no simple matter. Since the Second World War, multilateralism had become an integral part of international statecraft. Virtually all of it had been directed towards three widely shared goals: enhancing security, increasing prosperity and economic development, and strengthening cooperation on such technical issues as telecommunications, food and agriculture production, and weather forecasting. But the negotiation of multilateral environmental regimes required not only agreement on some scientifically challenging issues; it also required a willingness to restrict various types of economic activity, potentially leading to reduced economic welfare. The record by 1990 of successful multilateral environmental agreements that included reasonably enforceable provisions was negligible. Those that were more than constructions of rhetorical commitments only addressed limited issues, such as the 1975 Convention on International Trade in Endangered Species (CITES) and the 1987 Vienna Convention for the Protection of the Ozone Layer. The first of these was based on a widely shared desire to end the indiscriminate slaughter of such charismatic fauna as whales, tigers, elephants, and black rhinos. The second worked because those most affected by restrictions on the use of chlorofluorocarbons (CFCs) could rely on reliable substitutes for most applications.[66]

---

66. See Fen Osler Hampson with Michael Hart, *Multilateral Negotiations: Lessons from Arms Control, Trade and the Environment* (Baltimore: Johns Hopkins University Press, 1995), generally and 255-77 on the ozone accords. See also Scott Barrett, *Environment and Statecraft: The Strategy of Environmental Treaty-Making* (Oxford: Oxford University Press, 2002). By 2000, there were well over 150 treaties

The kind of social engineering envisaged by the UN and the crisis-oriented scientists could not be achieved either by national action alone or by reliance on existing rules of customary international law.[67] Luterbacher and Sprinz, however, suggest that "two major cooperative problems emerge at the international level concerning the environment, in general, and climate change, in particular: 1) international cooperation is often needed to achieve a collective good and to create a particular institutional framework to keep free-riding from occurring. ... [and] 2) international cooperation often consists of enforcing rules of mutual restriction, such as the reduction of GHG emissions. ... The question of international cooperation is complicated further by the fact that the two categories of collaboration ... can often *not* be separated in the analysis of concrete situations. The creation of an international climate change regime involves both the creation of a public good and the establishment of rules for mutual restriction in order to avoid a mutually detrimental outcome."[68]

Three critical elements need to be satisfied in negotiating multilateral environmental agreements: 1) the development of transnational networks of knowledge-based or epistemic communities that succeed in gaining political influence on the basis of authoritative knowledge; 2) a sufficient sense of widespread public alarm to provide political leaders with a perceived crisis that needs their attention; and 3) a credible range of measures that, if widely adopted, will ameliorate the problem at reasonable cost. For climate change, the first had been largely achieved by the late 1980s but would be reinforced with the reports of the IPCC, which underscored the politically desired authority of this epistemic community.[69]

---

on the UNEP Register of international treaties in the field of the environment. Few of them, however, contain enforceable provisions.

67. As international law scholar Sir Ian Brownlie had demonstrated in "A Survey of International Customary Rules of Environmental Protection," *Natural Resources* 13 (1973), 179-89.

68. Urs Luterbacher and Detlef Sprinz, *International Relations and Global Climate Change* (Cambridge, MA: MIT Press, 2001), 13-4.

69. For a discussion of the role of epistemic communities in environmental agreements, see Peter M. Haas, "Obtaining International Environmental Protection through Epistemic Consensus," *Millennium: Journal of International Studies* 19:3 (1990), 347-63, including the extensive body of literature discussed. For a more theoretical discussion of the difficulties encountered in negotiating a multilateral environmental agreement, see Oran Young, "The politics of international regime formation: managing natural resources and the environment," *International Organization* 43:3 (Summer 1989), 349-75, and Hugh Ward, et al., "March-

Creating a widely shared and sustained perception of alarm is more difficult to achieve. Effective levels of alarm require a wider public conviction of the authority of the consensus science, the potentially catastrophic impact of further global warming, and the human role in pushing climate outside the envelope of natural variation and historical experience. The factor that helped to make the ozone issue resonate with a wider public was the fear that a thinning ozone layer would allow more ultra-violet rays to penetrate the atmosphere, leading to widespread – and deadly – skin cancer. As a matter of fact, the thinning was limited to the polar regions, threatening only sun-bathing penguins and polar bears, but instilling fear of a dangerous cancer was a potent way to convince the public to support action. Nearly forty years after that alarm was first raised, there is little evidence of adverse human health effects from ozone depletion at the poles.[70] Consequently, much more work was required to convince those other than environmentalists of the need to adopt significant changes in life-styles and standards of living in order to ameliorate the alleged climate crisis.

By the early 1990s, the UN, climate scientists, and the environmental movement had convinced themselves that mitigation measures were the responsible way to address the climate crisis and that central to successful mitigation would be significant reductions in energy usage, particularly energy based on fossil fuels. As discussed in chapter eight, this third element would prove to be the Achilles heel of climate change policy, something which some had already sensed at an earlier stage. John Holdren, a movement pioneer and a colleague of both Stephen Schneider and Paul Ehrlich, had pointed out in 1975 that, "one is always reading that the sorts of drastic changes proposed here are infeasible, impractical, and unrealistic, particularly in the economic and political sense. I believe that the alternative of proceeding along our present course is *physically* impractical as well as socially unacceptable. Can our political and economic scientists devise ways to bring the world's institutions into line with physical reality in time?" Holdren and his colleagues

---

ing at the Pace of the Slowest: A Model of International Climate-Change Negotiations," *Political Studies* 49 (2001), 438-61.

70. Medical literature on the health effects of ozone depletion indicates that most of it is speculative. A search of PubMed and the Cochrane Collaboration found no articles that linked changing patterns of disease with ozone depletion. See, for example, F.R. de Gruijl, "Skin cancer and solar UV radiation," *European Journal of Cancer* 35:4 (December 1999), 2003-09.

knew full well that their prescription of transformative social engineering was radical and would require a central global authority to coerce humankind to repent, change its ways and, most importantly, reduce global population to a more sustainable level.[71] Like the prophets of old, the new generation of Jeremiahs asked too much and lacked a practical sense of what would be involved. According to the new prophets, mankind needed to repent and believe: they would then understand and accept what was required, allowing governments to exercise the political will to make it happen.

Regardless of the practical limits of proposed solutions, further progress on action to mitigate climate change required that the discussion of global warming migrate from science journals, laboratories, and international conferences to the political arena. Scientists would need to become activists in a political campaign and, in the process, adopt the language and techniques of politics. Activist scientists would also need to make common cause with others who agreed that only transformative social engineering and world governance could lead to a sustainable, just society of fewer people that would place more sustainable demands on a fragile planet. As their campaign progressed, it would become more and more difficult to determine whether concern about climate change drove the need for transformative change or whether the progressive movement had found in climate change a compelling basis for pursuing its transformative objectives. In either case, it proved to be a match made in heaven. The climate change campaign might not have been conceived as a political movement, but by the beginning of the 1990s it had clearly become a movement of the left, championed by political

---

71. "Energy and Resource Program," University of California at Berkeley (1975), as quoted in Stephen Schneider, *The Genesis Strategy* (New York: Plenum Press, 1976), 289-90. Holdren was responding to the negative reviews by economists of books such as Ehrlich's *Population Bomb* and the Club of Rome's *Limits to Growth*. Like Ehrlich, Schneider and other movement pioneers, Holdren's core concern was that of population, particularly in developing countries. Without bringing world population down to what the movement considered sustainable numbers, the environmental crisis could not be solved. Holdren and Ehrlich wrote in 1969 that "it cannot be emphasized enough that if the population control measures are not initiated immediately and effectively, all the technology man can bring to bear will not fend off the misery to come." Paul R. Ehrlich and John P. Holdren, "Population and Panaceas: A Technological Perspective," *BioScience* 19:12 (December 1969), 1070-1. At the time, world population was approaching 3.5 billion. Ehrlich and Holdren believed that there was not enough food for that many people and that technological advances could not erase the gap between hungry mouths and the world's food supply, a prognostication that has proven spectacularly wrong.

parties and leaders on the sinister side of the political spectrum and resisted by those on the right.[72]

Spreading alarm about the malign impacts of global warming was made more difficult, however, by the scientists' awareness that any significant adverse effects were unlikely to become significant in less than half a century into the future, if at all, while the costs of mitigation would need to be borne immediately. It was critical, therefore, that the alarm be raised, even if by a less than truthful campaign to bring the media onside through a steady diet of alarm stories.[73] It was especially important to demonstrate that adverse impacts were already evident. Stephen Schneider admitted as much in his first alarmist book, *The Genesis Strategy*: "Most of the crucial issues of human survival that will confront humanity over the next few decades will call for ethical and political value judgments – *decisions on how to act in the face of uncertainties*. In few cases will these decisions be based on issues clear enough to be decided easily by an input of scientific truths comprehensible to only a handful of specialists."[74] [emphasis in original] Those who disagreed were gambling with the planet's future. The decades have passed, the uncertainties remain, and the planet has failed to warm, even while the calls to repent have become ever more strident.

Former New Hampshire governor John Sununu, who experienced the beginning of this campaign when he served as chief of staff (1989-91) to the first President Bush, observes that "the alarmists have learned well from the past. They saw what motivates policy makers is not necessarily just hard science, but a well-orchestrated symphony of effort. Their approach is calculated and deliberate. ... They have used that strategy to execute an orchestrated agenda over the last two decades: announce a disaster; cherry pick some results; back it up with computer modelling; proclaim a consensus; stifle the opposition; take over the process and control

---

72. See the discussion, for example, in James Delingpole, *Watermelons: The Green Movement's True Colors* (New York: Publius, 2011).

73. The perceived need to sustain alarm on the basis of less-than-truthful claims continues, as demonstrated by the egregious exaggerations in the 2014 US National Climate Change Assessment issued by the Obama administration. As Paul (Chip) Knappenberger characterizes it: "The National Climate Assessment is a political call to action document meant for the president's left-leaning constituency. What pretence of scientific support that decorates it quickly falls away under a close and critical inspection." "National Climate Assessment Report Raises False Alarm," Washington *Times*, May 8, 2014.

74. Schneider, *The Genesis Strategy*, xiv.

the funding; and roll the policy makers."[75] In doing so, as writer J.R. Dunn agrees, science would "become entwined and infected with ideology. ... Environmentalist Greens needed a threat – one that menaced not only technological civilization, but life on earth itself. They have promoted an endless parade of such threats since the 1960s – overpopulation, pollution, runaway nuclear power, and global cooling – only to see them shrivel like old balloons. They required a menace that was overwhelming, long-term, and not easily disproven. With global warming, the climatologists gave them one. In exchange for sky-high funding, millennial scientists ... continually inflated the nature and extent of the CO_2 threat by using the sleaziest methods available, as we now know."[76]

Australian scientist Brian Tucker further points out that "fear is as hardwired into human thought processes as the temptations of utopia, and both have influenced social development throughout human history."[77] Environmentalism draws on both fear and the utopian impulse for its support, as first became evident during the economically turbulent 1970s when a sufficient number of people were prepared to accept the pessimistic prognostications of the new generation of prophets of doom. Concern, however, declined during the more prosperous 1980s. Environmentalism then gained new life in the early 1990s in the years around the Rio Earth Summit and then declined again. The pattern repeated itself in the lead up to the Copenhagen Climate Summit in 2009 and then again declined. Fear can be a potent mobilizer of public concern, but its staying power is limited unless reinforced by evidence that the alleged problem is real and will affect people's well-being within their lifetimes. Fear of calamities that do not materialize for two or three generations is very difficult to sustain.

The internet became one of the most effective tools for spreading alarm, but it did not have its full impact until the turn of the century. The Y2K scare and widespread concern over flu pandemics

---

75. "The Politics of Global Warming," Remarks delivered at the 2009 International Conference on Climate Change, March 10, 2009.

76. J.R. Dunn, "Global Warming Fraud and the Future of Science," *American Thinker*, November 29, 2009.

77. Brian Tucker, "Science friction: the politics of global warming," *The National Interest* 49 (Fall 1997), 78-9. Tucker was a senior official at Australia's Commonwealth Scientific and Industrial Research Organisation (CSIRO) and led its scientific support for developing a national response to the threat of human-induced climate change. For Tucker, the science had to be rigorous and the uncertainties not underestimated.

in the opening years of the 21st century demonstrated the web's potential, but exaggerated responses to both problems also bred complacency and a suspicion that both alarmists and governments were crying wolf too soon and too often. Yet for climate alarmists, the web had become a critical tool in organizing support and spreading alarm, even if it would also serve to help sceptics in defusing and debunking it and thus provide the basis for a much more informed debate than had been possible earlier. In the early 1990s, however, alarmist scientists and their allies had had to rely on more traditional ways of spreading their anxiety, including demonstrations, the popular media, books, and similar techniques.

What Sununu characterized as a "symphony of effort" started soon after the 1988 Toronto Conference. Bill McKibben initiated his career in 1989 with *The End of Nature,* a book focused on the horrors of global warming.[78] A number of environmental groups sponsored and published alarmist reports. Greenpeace International, for example, published a report organized by its science director, Jeremy Leggett, which included contributions from such familiar names as Stephen Schneider, George Woodwell, and Amory Lovins.[79] A year earlier, Schneider had prepared a full-length popular treatment of his concerns for the Sierra Club, making up for his earlier anxiety about a new ice age by describing the many disastrous consequences that would flow from just one degree of warming, let alone more.[80] Francesca Lyman, an environmental journalist, prepared *The Greenhouse Trap* for the World Resources Institute.[81] Martin Ince, another science journalist, prepared *The Rising Seas* for Earthscan,[82] and activist Lynn Edgerton wrote *The Rising Tide* for the Natural Resources Defense Council.[83] Paul Ehrlich contributed the foreword

---

78. *The End of Nature* (New York: Anchor, 1989) had limited success when first published but was re-issued in 1999 and again in 2006 to cash in on the author's popularity among a new generation of worriers.

79. *Global warming: the Greenpeace report* (Oxford: Oxford University Press, 1990).

80. *Global warming: are we entering the greenhouse century?* (San Francisco: Sierra Club Books, c1989).

81. *The Greenhouse Trap: what we're doing to the atmosphere and how we can slow global warming* (Boston: Beacon Press, c1990).

82. *The Rising Seas* (London: Earthscan, c1990). Earthscan is the publishing arm of the International Institute for Environment and Development (IIED) in London.

83. *The Rising Tide: Global Warming and World Sea Levels,* foreword by George M. Woodwell (Washington, D.C.: Island Press, c1991). Edgerton was an attorney with the Natural Resources Defense Council.

to a collection of essays about the impact of climate change.[84] Lester Brown, who had been emoting about the disastrous state of the planet since his first book in 1963 (*Man, Land and Food*), had added global warming to his catalogue of horrors in his annual *State of the World* series (1984-2001).[85] Other books would follow as more and more journalists and activists discovered that global warming alarm books were enjoying good sales and maintaining media attention.

Media interest in global warming picked up in 1988 following the Toronto Conference and the first signs of political interest in the issue. It was further stimulated by the weather. The summer of 1988 was hot and dry in the eastern part of North America, typical of an extended la Niña pattern. It was not unprecedented except to gullible reporters, none of whom were around when similar hot weather had persisted much longer in the 1930s, particularly in 1936. Nevertheless, within the frame of interest in "unprecedented" global warming, the summer of 1988 turned out to be a bonanza. For leading English-language newspapers around the world, media coverage increased for the next couple of years but could not be sustained until the end of the decade, when it picked up again as a result of the Kyoto discussions, complemented by the giant El Niño of 1997-98, which did send global temperatures to levels rivalling those of the 1930s. The US heat index, however, based on the best available records on the planet, showed only a modest increase (see Figure 11-1). Media coverage became more sustained after 2004 when the Kyoto Protocol came into force and the IPCC had succeeded in raising the alarm further with its 2001 report and its iconic – and fraudulent – hockey-stick graph.

By the end of the 1980s, an endless supply of alarmist research papers was being made available to feed the media appetite for alarm. The number of scientists active in researching climate-related issues had increased, as had funding, stimulating enterprising academics and publishers to establish more outlets for publishing their

---

84.   Richard L. Wyman, ed., *Global Climate Change and Life on Earth*, foreword by Paul Ehrlich (New York: Routledge, 1991). Contributions came from a wide range of environmental organizations, including the World Wildlife Fund, the National Audubon Society, and the Environmental Defense Fund, as well as from academics and government officials.

85.   Since 1963, Brown has written some fifty books bemoaning the state of the planet and also founded, first, the Worldwatch Institute (1974) and then the Earth Policy Institute (2001) to provide his followers with books and papers raising global consciousness about the plight of the planet.

Figure 11-1: US annual heat wave index, 1895-2014

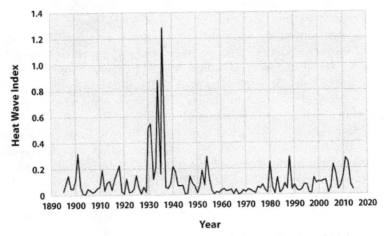

Source: www.epa.gov/climatechange/images/indicator_downloads/high-low-temps-download1-2015.png

research. *Theoretic and Applied Climatology* had been the first specialized journal in the field (from 1948). It was followed by *Climatic Change* (1978), founded and edited by Stephen Schneider as an outlet for modellers and those concerned about human-induced climate change. Others followed: *the International Journal on Climatology* (1981), *Climate Dynamics* (1986), *Journal of Climate* (1988), *Global Environmental Change* (1990), *Mitigation and Adaptation Strategies for Global Change* (1996), *Climate Policy* (2000), *Nature Climate Change* (2007) and *Carbon and Climate Law Review* (2007). As the issue heated up, less specialized journals were also prepared to accept articles. By the 1990s hundreds of climate-related papers were looking for a publisher every year, with their university or institutional public relations departments scrambling to interest the media in reporting the innovative angle of each paper. Most of the papers, of course, do not offer much that is truly innovative but, in the competition for funds and reputation, efforts to make a paper newsworthy have become a critical part of the game.[86] Even so, until the final years of the 20th

---

86. William Butos and Thomas McQuade document the "boom in climate science, sustained by massive levels of funding by government entities, whose scientific direction is set by an extra-scientific organization, the IPCC, which has emerged as a 'big player' in the scientific arena, championing the hypothesis of anthropogenic global warming." "Causes and Consequences of the Climate Science Boom," forthcoming in the *Independent Review*.

century, media attention to climate change remained sporadic. Cross-country polling confirmed that public awareness had grown significantly by this time: about a third of people surveyed worried a great deal, and up to another third worried somewhat. These worries, however, did not translate into anything beyond vague support for some unspecified action and, when compared to other, more immediate concerns, climate change fell far down the list and below other environmental issues. The public's understanding of the issue ranged from vague to superficial.[87]

More recent polling also indicates that there are diminishing returns for fear-based campaigns. Low-information individuals are most easily influenced by such campaigns, but the fear dissipates if the alarm does not materialize. Repeated claims in the media that changing climate is having a devastating impact fall on deaf ears when people perceive that their own experience belies the alarm. Polling indicates that support for action on climate declines among more informed voters.[88] To the chagrin of climate alarmists, better educated people are less inclined to accept the alarm stories, indicating that they understand the basic science, whereas people who admit that they do not understand the science are more easily convinced that there is need for concern about the prospect of catastrophic climate change.

By the late 1980s, movement leaders had concluded that the option that worked best in raising the political profile of climate change was to appeal directly to political leaders through the UN, seeking to convince them on the basis of principle and hope that their generally meagre understanding of science would carry the day. This campaign resulted in success sufficient enough to translate the UN preparatory work into a modest framework agreement. Going to the next step, however, was a larger challenge. As Kathryn Harrison and Lisa Sundstrom learned in their comparative analysis

---

87. See, for example, Irene Lorenzoni and Nick Pidgeon, "Public Views on Climate Change: European and USA Perspectives," *Climatic Change* 77 (2006), 73-95. Lack of understanding among the public is not difficult to understand given simplistic media coverage.

88. See, for example, Paul M. Kellstedt, Sammy Zahran, and Arnold Vedlitz, "Personal Efficacy, the Information Environment, and Attitudes Toward Global Warming and Climate Change in the United States," *Risk Analysis* 28:1 (2008), 113-26, and Robert J. Brulle, Jason Carmichael, and J. Craig Jenkins, "Shifting public opinion on climate change: an empirical assessment of factors influencing concern over climate change in the U.S., 2002–2010," *Climatic Change* 114:2 (September 2012), 169-88.

of climate change policy in western Europe, North America, Japan, and Australia, "policy-makers' ideational commitments can be fragile in the face of persistent political and institutional obstacles."[89] Nevertheless, by the early 1990s, activist scientists had succeeded in making global warming an integral part of what Brian Tucker characterizes as "the international environmental culture wars... with activist Cassandras and conservative Pollyannas both trying to marshal the authority of science as justification for their views."[90] The campaign had more success in western Europe than anywhere else, in large part because institutional factors helped to sustain initial political commitments.

♦♦♦♦♦

89. Kathryn Harrison and Lisa McIntosh Sundstrom, "The Comparative Politics of Climate Change," *Global Environmental Politics* 7:4 (November 2007), 16. Like the hundreds of social scientists worldwide who have latched onto climate change policy as their area of specialization, Harrison and Sundstrom indicate no critical awareness of the weakness of the science and mask their limited understanding in social science jargon.

90. "Science friction: the politics of global warming," 78.

12 | National Interests vs. Global Norms

The urge to save humanity is almost always only a false-face for the urge to rule it. Power is what all messiahs really seek: not the chance to serve.[1]

H.L. Mencken

The role of climate science remains to put the problem-facts on the table and to identify options for appropriate solutions. The role of politics is then to mobilize the will of the citizens with the aim of implementing decisions that are based on science.

Hans-Joachim Schellnhuber, German climate scientist-activist[2]

By the end of the 1980s, understanding of climate change had markedly increased because billions of dollars had been poured into government, university, and even private laboratories and think tanks to study the science and its policy implications as well as to subsidize technology to replace fossil-fuel-based energy. All this activity only confirmed the complexity of the issues and the extent to which scientific understanding of the drivers of climate change remained incomplete. Nevertheless, between the 1988 Toronto Conference and the 1992 Rio Earth Summit, climate change morphed

---

1. Henry Louis Mencken, *Minority report: H. L. Mencken's Notebooks* (New York: Alfred A. Knopf, 1956), 247.

2. Quoted by Peter Heller, "Schellnhuber and Ruling the World," *Frankfurter Allgemeine Zeitung,* translated at *NoTricksZone.com,* July 15, 2012. Schellnhuber is the founder of the Potsdam Institute for Climate Impact Research and has served as a principal adviser to German Chancellor Angela Merkel and the EU Commission on climate change issues and as a coordinating lead author for the IPCC. He also advised Pope Francis on the preparation of the June 17, 2015 encyclical, *Laudato Si' (Praise Be to You).*

from a scientific quest into a political and diplomatic campaign. In the years that followed, more and more activists, some politicians and religious leaders, and a few industrialists climbed on the climate change bandwagon and found it a powerful vehicle for advancing pet causes and concerns.

By the turn of the century, climate change as a man-made problem had become the default, politically correct, view of the matter and confirmed the ability of the human mind to rationalize and sweep confounding facts under the rug. The critical role of CO_2 in sustaining life, the enormous range of previous global temperatures and other climatic phenomena, the extent of Earth's natural destructive powers, and the role of adaptation in the evolution of living entities were all ignored in the rush to demonize carbon dioxide, fossil fuels, industrialization, capitalism, and modern civilization.

Al Gore insisted that anthropogenic global warming was an inconvenient truth and explained it on the basis of one distortion after another. By the time his film was distributed, otherwise sensible people, while perhaps not wholly comfortable with his message, nevertheless decided that it would be impolitic to say so. Worried that the little science they could remember from high school would not support them in a discussion with one of the movement's zealots, most people decided that discretion was the better part of valour and were unprepared to call the zealot's bluff and insist on some verifiable evidence. Thus, a significant share of the population found itself lulled to sleep as the feel-good wool of progressive politics and eco-fundamentalism was pulled over their eyes.

The advent of political interest in the UK

Alarmists in the UK were first out of the gate, and they succeeded in alerting Prime Minister Margaret Thatcher to their view of the potential dangers of global warming. Her concern followed a fateful 1984 discussion with diplomat-environmentalist Crispin Tickell.[3]

---

3. Tickell was an early convert to environmentalism. In 1984 he was an undersecretary in the British Foreign and Commonwealth Office. He subsequently became UK ambassador to the UN in New York. During a 1975-76 sabbatical at Harvard, he had researched the issues, become convinced that climate change was a major issue of global concern, and had written a book on the subject, *Climatic Change and Global Affairs* (Cambridge, MA: Harvard Center for International Affairs and the University Press of America, 1977). Tickell has devoted his retirement years to advancing anxiety about climate change and other environmental issues but in 1984 brought no special expertise to the issue; his degree at Oxford was in modern history.

He convinced her that climate change would provide her with an issue on which she could exercise global political leadership. He subsequently provided her with more of his thoughts and in 1988 suggested that she use an upcoming speech to the Royal Society to advocate that only urgent political action would stave off a crisis within a generation.[4] In response, Thatcher asked him to help draft the speech, and place the issue within the broader context of science and public policy. A quarter century later, her words to the Royal Society seem rather cautious and, when placed in the broader context of the speech, not quite a call to action: "the increase in the greenhouse gases ... has led some to fear that we are creating a global heat trap which could lead to climatic instability. We are told that a warming effect of $1°C$ per decade [sic] would greatly exceed the capacity of our natural habitat to cope. ... The Government espouses the concept of sustainable economic development. Stable prosperity can be achieved throughout the world provided the environment is nurtured and safeguarded. Protecting this balance of nature is therefore one of the great challenges of the late Twentieth Century and one in which I am sure your advice will be repeatedly sought." Often forgotten is the fact that she also told the assembled scientists: "Whatever international action we agree upon to deal with environmental problems, we must enable our economies to grow and develop, because without growth you cannot generate the wealth required to pay for the protection of the environment."[5]

It was hardly an endorsement of the alarms raised by climate activists. Her emphasis on the need to maintain strong economies was also at odds with the prevailing environmental wisdom that industrialization and economic growth were the root cause of environmental degradation and needed to be curbed. In Thatcher's view, only a full marshalling of modern technology and innovation with-

---

4. Tickell provides some background to the speech in a long, rambling interview with Malcolm McBain for the Cambridge University British Diplomatic Oral History Programme, January 28, 1999. More detail is provided in a Rupert Darwall interview with Tickell: *The Age of Global Warming: A History* (London: Quartet, 2013), 122-3.

5. Margaret Thatcher, Speech to the Royal Society, September 27, 1988. Her reference to a $1°C$ increase per decade indicates that the speech had not been vetted by a scientist. Not even the most alarmist scientist claimed a number this high. The worry was about a rise of about $0.2°C$ per decade. Thatcher had by this time also been influenced by John Houghton, the head of the UK's Meteorological Office, who had been appointed chair of the IPCC's Working Group I responsible for assessing the science of global warming.

in robust economies could successfully address environmental problems. Nevertheless, her words were more than any political leader had been prepared to utter to date, and a campaign was born, one to which every British prime minister since has been committed. Furthermore, Thatcher followed up with more speeches raising the climate issue and added it to issues to be discussed with foreign leaders. A year later she enlarged her scope and range at a special session of the UN General Assembly on the Environment. There she pronounced that "the problem of global climate change is one that affects us all and action will only be effective if it is taken at the international level. ... Before we act, we need the best possible scientific assessment: otherwise we risk making matters worse. ... The United Kingdom has agreed to take on the task of coordinating such an assessment within the Inter-governmental Panel on Climate Change, an assessment which will be available to everyone by the time of the Second World Climate Conference next year. ... I believe we should aim to have a convention on global climate change ready by the time the World Conference on Environment and Development meets in 1992." But, once again, she emphasized, as she had a year earlier, "But as well as the science, we need to get the economics right."[6]

In 1990 in one of her last speeches before being forced from office, she addressed the Second World Climate Conference in Geneva just after the IPCC had issued its first report. She praised the IPCC for its thorough report, written largely by British scientists and officials under the direction of Sir John Houghton, but also warned that "we don't know all the answers ... major uncertainties and doubts remain," pointing to the report itself to emphasize her point. She also noted that climate change was a natural phenomenon but that population pressure could accelerate and exacerbate problems. At the same time, she believed that uncertainty should not be used as an excuse to delay some precautionary steps at the international level, particularly "no-regrets" policies such as improving energy efficiency and developing alternative sources.[7]

6. Margaret Thatcher, Speech to the UN General Assembly, November 8, 1989.

7. In the early stages of climate change policy development, politicians found it easier to support so-called no-regrets policies than the more radical measures advocated by activists. Basically, these policies involved using taxes and regulations to reduce consumption of energy which, presumably, would lead to economic efficiency gains as well as a mitigation of greenhouse gas emissions. Regardless of how they are sold, such measures involve using the coercive power of the state to change consumer behaviour. A typical example is the gradual ban

Again, she emphasized that the answer lay in innovation and vigorous economies. "We have to recognize the importance of economic growth of a kind that benefits future as well as present generations everywhere. We need it not only to raise living standards but to generate the wealth required to pay for protection of the environment. It would be absurd to adopt polices which would bankrupt the industrial nations, or doom the poorer countries to increasing poverty."[8]

Thatcher's bachelor of science degree in chemistry helped to convince fellow leaders at, for example, G-7 economic summits that she understood the science, a fact that led to their adoption of calls for action in official communiqués. Her conviction led to her support for the establishment of the Hadley and Tyndall Centres at the University of East Anglia, which became critical centres for research on climate change and for generating the intellectual basis for the global warming alarmist movement. In many ways it allowed British scientists to gain sufficient funding to catch up with work in the United States. British scientists associated with these two centres became the core of the IPCC specialists responsible for Working Group I, the physical science of climate change. British officials, including Tickell at the UN, were active in 1988-89 in supporting the establishment of the UN's Intergovernmental Panel on Climate Change and in proceeding the following year with the negotiation of the UN Framework Convention on Climate Change.[9] By the time Thatcher left office in the fall of 1990, climate change had been thoroughly embedded in UK domestic and international politics.

What was *not* embedded in either UK or international discussions, however, was her emphasis on addressing climate change

---

of incandescent light bulbs in favour of fluorescent, LED, and other technologies, all of which cost more, create worse hazards, and in many applications provide inferior light. They are not the natural choice of most consumers. See the discussion in Thomas Gale Moore, *Climate of Fear: Why We Shouldn't Worry about Global Warming* (Washington: Cato Institute, 1998), 130-1.

8.   Margaret Thatcher, Speech to the Second World Climate Conference, Geneva, November 6, 1990.

9.   See Loren R. Cass, *The Failures of American and European Climate Policy* (Albany: State University of New York Press, 2006), 23-31. Cass suggests that "the Thatcher government utilized climate policy as a tool to improve its electoral position on environmental issues and as a justification for its broader policy goals, such as expansion of nuclear energy and its EC reform initiatives. ... The government supported American attempts to slow the international process to assure that the international response did not proceed more quickly than domestic constraints would allow." 31.

and other environmental problems in an economically responsible manner, relying on innovation, technology, entrepreneurship, and economic growth to provide solutions and a basis for assuming the costs of mitigation and adaptation policies. In her later years, in part because of the utopian, command-and-control approach adopted at the UN, she regretted her early support. In her last book she minced no words about what had happened in the 15 years since her 1988 speech. She devoted ten pages in *Statecraft* to a section called "Hot Air and Global Warming" and concluded with excerpts from statements by former US President Bill Clinton, former US Vice-President Al Gore, and former UK Labour Foreign Secretary Robin Cook to illustrate her point that climate change had become the latest anxiety promoted by global doomsters and had become central to the agendas of left-of-centre parties and governments.[10]

Thatcher's willingness to champion the global warming cause and support its pursuit at the UN stands out as a departure from an otherwise consistent political outlook. During her leadership she resisted all efforts to advance, for example, the pretensions of the UN's NIEO, using economic summits and other venues to stiffen the spines of other leaders in opposing UN ambitions.[11] She consistently pointed out that for developing countries their hope for the future lay in adopting and pursuing sound, market-based policies rather than in relying on the top-down, utopian plans favoured by the UN and its "norm" entrepreneurs. Similarly, she was scornful of doom mongers like the Club of Rome and was particularly dismissive of their predilection for world government and collectivism. It is surprising, therefore, that she found in Tickell, a fully paid-up member of this group of dreamers, a valuable adviser. As Thatcher wrote in her memoirs, "the desire to achieve grand utopian plans often poses a grave threat to freedom."[12] Though Rupert Darwall makes a convincing case that her concern was genuine and reflected a keen interest, it did not last.[13] Her later assessment of the issue was much more consistent with her political philosophy, making her brief flirtation with global salvationism an aberrant episode.

---

10. *Statecraft* (London: HarperCollins, 2003), 441-50.

11. See, for example, the discussion in Margaret Thatcher, *The Downing Street Years* (London: HarperCollins, 1993), 168-9.

12. Margaret Thatcher, *Statecraft*, 327.

13. Rupert Darwall, *The Age of Global Warming* (London: Quartet, 2013), 131.

While Thatcher may have been convinced in 1988, the British public had not yet climbed on board, nor had the quality press. Nigel Hawke, science correspondent for the *Times* of London, was withering in his dismissal of her conversion: "Computer models predicting temperature rises very much smaller than their proven margins of error are being used by a prime minister who claims to be a scientist as grounds for imposing economic sacrifices on the entire world. A couple of cold winters will take the froth off the debate, and allow us the time we need to discover whether or not the earth is really warming up."[14] Those cold winters came, but not before the movement had gained more momentum with even the so-called quality press climbing on board.

Thatcher's successor, John Major, who held office from November 1990 until the Labour victory in May 1997, did not share his predecessor's enthusiasm for solving the climate crisis. Major was a more pragmatic and less ideological politician than Thatcher. He inherited the climate file from her and dutifully carried it to a satisfactory conclusion with his signature of the UNFCCC on behalf of the UK at the Rio Summit. Thereafter, he was preoccupied with more pressing foreign policy issues, and climate change gradually receded into the background. He was prepared to let John Gummer, his environment secretary, carry the ball on the issue. Gummer turned out to be a climate change policy enthusiast of the first order and maintained UK involvement for the rest of Major's time in office but without the political leadership enjoyed under Thatcher. By that time, UK climate change policy was largely preoccupied with trying to limit the appetite of the EU Commission, which blunted the UK's ability to play a leadership role in post-Rio climate talks.

EU efforts to enlarge its competence in climate policy and related tax and energy policies troubled both prime ministers. Thatcher's answer had been in part to pursue the file through the UN, but after Rio the EU Commission became more aggressive and thus reduced Major's enthusiasm for climate issues.[15] British leadership on climate change gained new life, however, with the election of the Labour government of Tony Blair. Blair was an enthusiastic Europhile and accepted EU leadership on the climate file. In the de-

---

14. Nigel Hawke, "Is this really a scientist speaking?" *Times*, November 8, 1990, as quoted in Darwall, *The Age of Global Warming*, 135.

15. Loren Cass, "The Indispensable Partner: the United Kingdom in European Climate Policy," in Richard Harris, ed., *Europe and Global Climate Change: Politics, Foreign Policy and Regional Cooperation* (London: Edward Elgar, 2007), 69ff.

velopment of the EU's negotiating mandate for the Kyoto Protocol soon after Blair's election, the UK and Germany together were instrumental in the EU's aggressive stance, each benefitting from fortuitous circumstances. The UK's conversion of much of its electrical generation from coal to gas had significantly lowered its emissions profile and allowed it to insist that others catch up. Germany, as discussed below, benefitted from the unification of its two halves and the mothballing of much of East Germany's inefficient and emissions-rich industries and electrical generation. By the new century, climate change had become firmly embedded in the UK's domestic and foreign policy to an extent that Lady Thatcher deeply regretted.

Lighting the global warming torch in the United States

Climate change entered US public consciousness with the highly staged Senate testimony of James Hansen in the summer of 1988, aided and abetted by two alarmist Democratic senators, Timothy Wirth[16] and Al Gore. At the time, Hansen was director of NASA's Goddard Institute for Space Studies. He had started there as a specialist studying the atmosphere of Venus, but, as the global warming issue gained salience, he turned his attention to the earth's atmosphere and directed two major projects: modelling the earth's atmosphere and developing a global temperature series. Both projects convinced him of the dangerous role of greenhouse gases in driving Earth's temperature to unprecedented levels. He became director of GISS in 1981, succeeding Robert Jastrow,[17] and soon after

---

16. Wirth was the junior senator from Colorado and had previously served in the House, elected in a Denver district. During the first Clinton-Gore administration he served in the US State Department as the first Undersecretary for Global Affairs and led the US delegation during the negotiation of the Kyoto Protocol. In 1998 he became the first president of the UN Foundation and the Better World Fund, both funded by Ted Turner. In preparation for Hansen's testimony, Wirth and his aides ensured that the windows in the hearing room would be open, raising the temperatures and, as Hansen warned senators of the impending disaster of global warming, causing all in the room to break out in sweat in front of the cameras. In an interview for PBS's *Frontline*, Wirth proudly admitted the gamesmanship used to prepare the room for Hansen's testimony. *Frontline* interview with Senator Wirth, January 27, 2007.

17. Robert Jastrow, one of NASA's pioneers and the leader of the science dimension of NASA's lunar probes, was one of the founders of the Marshall Institute (together with Frederick Seitz and William Nierenberg), dedicated to a more conservative perspective on controversial issues such as ozone depletion, acid rain, and global warming, positions for which all three have been vilified by the alarmist movement.

testified before a Congressional committee about the dangers of global warming. This testimony, and repeat performances in 1986 and 1987, had little impact other than to annoy some of his superiors. [18] His 1988 testimony, however, captured media interest and served the purpose of those who had invited him. He asserted that "the earth is warmer in 1988 than at any time in the history of instrumental measurements. ... Global warming is now large enough that we can ascribe with a high degree of confidence a cause and effect relationship to the greenhouse effect. And ... our computer climate simulations indicate that the greenhouse effect is already large enough to begin to affect the probability of extreme events such as summer heat waves." [19] His testimony was widely reported in the major media. From the perspective of Senators Gore and Wirth, it served its purpose: dangerous global warming was now on the national radar screen.

Hansen may have "lit the bonfires of the greenhouse vanities" in 1988, as sceptical scientist Patrick Michaels has suggested, [20] but within the US scientific community and the scientific agencies of government, the torch had been lit since at least the early 1970s. Hansen was not a lone wolf. Others at NASA, NCAR, the Geophysical Fluid Dynamics Laboratory (GFDL) at Princeton University, and the Energy Department were engaged in similar research, even if remaining more cautious than Hansen in his testimony. [21] Officials in the Nixon Administration had established a US Climate Program to coordinate on-going research. A few years later, President Carter had signed the Climate Program Act, providing congressional funding and authorization for more research. Discussion among scientists was far from monolithic but was trending towards what would become official science in the United States. [22] In 1979 the US Na-

18. See Hansen interview by Spencer Weart for the Physics Oral History Project for the American Institute of Physics, November 27, 2000.

19. A transcript of his testimony can be found at image.guardian.co.uk/sys-files/ Environment/documents/2008/06/23/ClimateChangeHearing1988.pdf.

20. Patrick Michaels, "Inhaling a decade of hot air vapours," Washington *Times*, June 28, 1998. In addition to the daily media, news magazines such as *Time*, *Newsweek*, and even *Sports Illustrated* featured cover stories on global warming that summer.

21. Alan D. Hecht and Dennis Tirpak, "Framework agreement on climate change: a scientific and policy history," *Climatic Change*, 29 (1995), provide an overview of US government involvement in the evolution of US climate research.

22. Hecht and Tirpak observe that "there was intense debate on whether sufficient scientific evidence existed to justify policy actions to arrest climate change. A

tional Research Council had established a panel chaired by MIT's Jule Charney to look into the issue. Its report concluded on the basis of the crude models then available that a doubling of atmospheric CO_2 would lead to an increase "near 3°C with a probable error of ± 1.5°C," an estimate that has not changed much in the intervening years despite billions spent on further research and the development of much more sophisticated models.[23] While the Charney report attracted very little media or public attention, it exerted considerable authority among scientists. The more politically engaged among the scientists may have been disappointed in the panel's cautious approach to mitigation but were pleased it recognized their need for more funding.

Until his 1988 testimony, Hansen had been a relatively obscure scientist. He had not been involved in the series of meetings sponsored by the UN and other organizations and was not part of what would become the IPCC establishment. Following his testimony, however, he became one of the better known and more extreme advocates of policy to combat climate change, particularly in the United States, speaking out freely and frequently, influencing policy makers well beyond the United States. He remained less engaged with other climate scientists than many others in the movement, but his activism gave him significant media exposure and added measurably to public concern about global warming.

As electrifying as the media and alarmists found Hansen's testimony, it had little impact on US policy. Within government agencies the issues were well known and had long been debated. It did give a boost to a number of politicians in Congress pushing climate-related legislation, none of which made it past the committee stage. More importantly, it boosted media and public interest. Within the administration, the testimony proved useful to those in the interagency process championing US leadership at the UN (e.g., State and the EPA) but had little impact on the economic agencies (e.g., Treasury, Commerce, and Energy), and was dismissed at the

---

steady stream of reports and assessments were prepared by national and international expert groups." "Framework agreement on climate change," 379. See also William N. Butos and Thomas J. McQuade, "Causes and Consequences of the Climate Science Boom," forthcoming in *The Independent Review.*

23. The report is available at atmos.ucla.edu/~brianpm/download/charney _report.pdf. The high end of the estimate, an increase of 4.5°C, came from Hansen's model, which many scientists have since suggested projects too much warming.

Reagan White House. As Hecht and Tirpak recall, within the Reagan Administration "the mood of senior officials then in Washington was that the underlying scientific evidence for global warming was inconsistent, contradictory, and incomplete and did not justify policy actions that likely would be expensive."[24]

When George H.W. Bush assumed the presidency in 1989, he showed more interest in the cause and appointed William Reilly, at the time President of the World Wildlife Fund, to take the reins at the EPA. Reilly worked hard to convince Bush to follow his heart and did so with the support of Secretary of State James Baker and Energy Secretary James Watkins, but they could not offset the advice of Bush's closest and most trusted advisers: chief of staff John Sununu, Dick Darman, the Director of the Office of Management and Budget, Allan Bromley, the President's Science Advisor, and Michael Boskin, Chair of the Council of Economic Advisors. Sununu had earned a PhD in mechanical engineering, specializing in thermal applications. He was, therefore, thoroughly familiar with the arguments advanced by alarmist scientists and found them wanting. Bromley agreed that alarmist conclusions were premature, and Boskin and Darman weighed in with analysis of the economic consequences.[25] Despite his strong environmentalist background, Reilly was clearly outgunned. Such specialized knowledge is rare at the centre of public policy. Unlike Thatcher, Bush was counselled by advisers with competing perspectives and, thus, developed a policy tempered by discussion.

The debate within the Bush administration, echoing to some extent earlier discussion during the final years of the Reagan administration, framed the issue as a trade-off between environmental and economic considerations.[26] As discussed in chapter nine, only the US government had by this time examined the economic conse-

---

24.   Hecht and Tirpak, "Framework agreement on climate change," 380-1. US scepticism led to the suggestion of a mechanism to study the issue further, leading to the establishment of the IPCC, a proposal around which it was possible to forge interagency consensus. Further, they report, interest was building among a group of progressive Senators, as reflected in their requests for studies.

25.   Bromley was a Canadian nuclear physicist trained at Queen's University in Kingston and, like Sununu, fully capable of explaining the science of global warming to President Bush.

26.   Then Senate Majority Leader George Mitchell expresses his frustration with the Reagan and Bush Administrations' insistence that there was a trade-off between environmental and economic goals in *World on Fire: Saving an Endangered Earth* (New York: Scribner's 1991).

quences of mitigation measures and had found them troubling. Thatcher may have emphasized the need to balance economic and environmental considerations, but British officials took no steps to assess costs for nearly twenty years (the 2006 Stern Report). In Europe, Canada, and elsewhere economic considerations were given equally short shrift.

US administrations also needed to take account of congressional realities. No matter how convinced a president might be, any meaningful mitigation measures would require legislative action and thus the support of an electorally more sensitive Congress. As a result, the political debate was framed in much more cautious terms than elsewhere. Loren Cass characterizes the US position in international climate discussions as "intransigent" and out of step with emerging international norms.[27] A more accurate description would be cautious and realistic. Over the next twenty years, the rhetoric of administrations might change, but the innate caution evident from the beginning would remain and act as a restraint on the ambitions of European and UN "norm entrepreneurs."[28] Former German climate official Herman Ott reflects the European frustration with the US's perceived slowness in advancing the alarmist cause: "The United States perceives itself as the natural leader in world affairs, but is severely handicapped by a constitutional structure and lifestyle preferences that make leadership on this issue very unlikely for many years to come."[29]

The immediate issue for Bush was US participation in the negotiation of the UN framework agreement, the president's attendance at the 1992 Rio Summit, and his signatures on the two conventions on climate and biodiversity. Throughout his presidency, Bush agonized over the three issues, torn between Reilly's advice and that of Sununu. In the end, he compromised. Bush agreed that the US would push for a climate agreement but stipulated that it contain no commitments on emissions, a position that was sustained, and he agreed, at the last moment, to attend Rio and sign the agree-

---

27. International relations scholars like to refer to such concepts as "international norms" or the "views of the international community." Both terms, of course, refer to the views of progressive elites. Conservative views, by definition, would never reach such an exalted status within the academic establishment.

28. See Cass, *The Failures of American and European Climate Policy*, 33-42.

29. Herman E. Ott, "Climate Change: An Important Foreign Policy Issue," *International Affairs* 77:2 (April 2001), 295.

ments.[30] Hecht and Tirpak summarized the US position in the negotiations as follows: "Largely based on projected economic impacts, the US rejected any form of targets and timetables. Instead US policy focused on three main principles: actions taken by governments should be based on a 'no regrets' policy, namely involving policies and programs that are useful in their own right; actions should reflect a 'comprehensive approach,' namely including all greenhouse gases and all sources and sinks; and actions should be voluntary with non-binding targets and timetables."[31]

Bush's successor, Bill Clinton, proved a much more adroit and less conflicted manager of the climate file. He started with an important advantage. As a Democrat, the media and the environmental movement assumed that he would be sympathetic to environmental issues in general and global warming in particular. Criticism of his failure to pursue the issue with any vigour remained muted throughout his period in office. He was also prepared to make policy statements and proposals that he knew could not be implemented.[32] In Al Gore he had a vice-president who had begun his campaign to save the planet as far back as 1976 with congressional hearings and whose 1992 book, *Earth in the Balance*,[33] underlined that he shared the views of such doomsters as Ehrlich and Holdren. Clinton's appointment, at Gore's request, of Carol Browner to head the EPA solidified his claim as a climate "realist," committed to it and related issues. Browner had strong credentials as a hard-nosed environmentalist and committed progressive, had worked as a legisla-

---

30. See the discussion in Darwall, *The Age of Global Warming*, 117-25 and 136-54, and Sununu's brief discussion in "The Politics of Global Warming."

31. Hecht and Tirpak, "Framework agreement on climate change," 376.

32. See the discussion in Moore, *Climate of Fear*, 132-5.

33. *Earth in the Balance: Ecology and the Human Spirit* (Boston: Houghton Mifflin, 1992). The late social critic Alexander Cockburn captured Gore's persona to a T. He wrote: "As a denizen of Washington since his diaper years, Gore has always understood that threat inflation is the surest tool to plump budgets and rouse voters. By the mid-'90s he'd positioned himself at the head of a strategic alliance formed around 'the challenge of climate change,' which stepped forward to take Communism's place in the threatosphere essential to political life. The foot soldiers in this alliance have been the grant-guzzling climate modellers and their Internationale, the United Nations' Intergovernmental Panel on Climate Change, whose collective scientific expertise is reverently invoked by devotees of the fearmongers' catechism. The IPCC has the usual army of functionaries and grant farmers and the merest sprinkling of actual scientists with the prime qualification of being climatologists or atmospheric physicists." "Who Are the Merchants of Fear?" *The Nation*, May 28, 2007.

tive assistant to Gore, and would fare better in dealing with opposition from officials within economic agencies.

Over the course of his eight years in office, Clinton sent a number of climate and energy-related initiatives to Congress, including a BTU tax, but they received a frosty reception. Congressional hearings demonstrated that the issue was not as straightforward as proponents maintained. Kathryn Harrison points out: "The fraction of scientists testifying at Congressional hearings who sided with the views of the IPCC declined from 100 per cent at the time of the Earth Summit in Rio in 1992, to 50 per cent, evenly balanced with climate change sceptics, in 1997 in the lead up to Kyoto."[34] Clinton then turned to voluntary action with predictable results. There were many rhetorical flourishes and statements supporting binding emission controls but little that would meet the requirements of the activist lobby.[35] Executive action could only go so far, and Clinton preferred using his political coinage with Congress on other matters Congress debated climate issues throughout the 1990s, but the only consensus that emerged was the Byrd-Hagel Resolution, the 95-0 vote in the Senate which told the president not to enter into any agreement lacking burden-sharing among all countries or harming US economic interests, i.e., going beyond what had been agreed by Bush and incorporated into the UNFCCC. The House, which flipped to Republican control in the 1994 election for the first time in 40 years, was even less inclined to support legislation encumbering the US economy by restricting greenhouse emissions. The 1994 election proved an important lesson to the pragmatic Clinton: he was able to work well with the Republican-controlled House and Senate. His pragmatism was less compatible with the ideological zeal of those around him, including the vice-president, first lady, and EPA administrator. But in the end US support for efforts to "stop" global warming would be more rhetorical than real for the rest of his administration.

---

34. Kathryn Harrison, "The Road not Taken: Climate Change Policy in Canada and the United States," *Global Environmental Politics* 7:4 (November 2007), 100.

35. In his first Earth Day speech (April 21, 1993), Clinton announced his government's "commitment to reducing our emissions of greenhouse gases to their 1990 levels by the year 2000. I am instructing my administration to produce a cost effective plan ... that can continue the trend of reduced emissions." Nothing came of this commitment. Hecht and Dennis Tirpak, "Framework agreement on climate change," 396-7.

Internationally, Clinton was prepared to pursue an equal number of symbolic acts, including aggressive action under the UNFCCC. Once it came into force in 1994, the Secretariat immediately set to work organizing annual Conferences of the Parties (COPs). The first, in Berlin, tackled the knotty problem of common but differential treatment, i.e., that any future obligations would weigh more heavily on developed than developing countries. This had been a sticking point for the Bush administration and for the US Senate and had been finessed by making the Convention wholly voluntary.[36] Everyone knew, of course, that without mandatory measures extending to all countries, the Convention would never meet its objectives. At Berlin, the US bowed and accepted what became known as the Berlin mandate, i.e., mandatory measures would be limited to the so-called Annex I or industrialized countries.

At COP 3 in Kyoto in 1997, the US delegation initially pursued a relatively cautious approach, wary of the US Senate resolution, seeking acceptance of greater developing country participation only over time. This was categorically rejected by developing countries and dropped. The US also sought acceptance of various flexibility provisions to make any mandatory measures easier to implement, an approach also favoured by Canada, Australia, and Japan but strongly opposed by the EU. Finally, the US sought to move the first commitment period from 2000, as set out in the Convention, to 2010 and beyond, again with the support of Canada, Australia and Japan and opposed by the EU. All nuance in the US position went out the window, however, with the arrival of Vice-President Gore near the end of the Conference. He gave an inspirational speech that fully undercut his own delegation, highlighting full US support for the science of climate change and the need to take aggressive action.[37]

---

36. "The US delegation faced intense and bruising pressure from business and industry lobby groups not to agree to the exclusion of developing countries from this negotiating mandate. The decision of the US delegation to nevertheless fall in line with the consensus among its Western allies, and most of the rest of the world, represented a clear shift towards multilateral engagement on climate change. The positive engagement of the US was confirmed at COP 2 in July 1996, when the head of delegation announced support for legally binding emission targets, coupled with an emissions-trading system." Joanna Depledge, "Against the Grain: The United States and the Global Climate Change Regime," *Global Change, Peace & Security* 17:1 (February 2005), 15. Depledge was a UN official assigned to the UNFCCC Secretariat.

37. By the time of Gore's arrival in Kyoto, the conference had "degenerated into a mix of revival meeting and guerrilla warfare. ... The halls were swarming with young, earnest types ...who were preaching the gospel of an energy-free world.

The result was a US commitment to cut its emissions to 93 per cent of its 1990 level by the 2008-12 first commitment period, slightly higher than the 92 per cent EU commitment and slightly lower than those of Canada and Japan at 94 per cent.

The United States had succeeded on a number of its objectives. The commitment period was pushed forward by a decade, and the Protocol accepted a range of flexibility mechanisms, some of which would require further negotiation. But the US had failed in making the commitments extend to all emitters and in limiting its own to a level that would not harm the US economy. The result was predictable. Clinton signed the Protocol but did not even forward it to the Senate for ratification. The Senate, in turn, showed no inclination to take up ratification hearings. For Clinton, signature of the Protocol was little more than a symbolic act. The Clinton-Gore-Browner team left office in 2001 having talked the talk but not having walked very far.

European political engagement

No single event triggered European concern. Global warming alarm emerged concurrently with public awareness in the UK and the US but then developed more rapidly and beyond earlier discussions for both ideological and institutional reasons. On the continent, environmental sensitivities follow the typical European divide between northern and southern preferences. Northern members of the EU[38] – from Sweden and Denmark to Germany and the Netherlands – generally display a higher level of regard for environmental and other progressive issues, consistent with their higher levels of income and with the greater political success of left-of-centre parties, including green parties. The southern members – Spain, Italy, Greece, and Portugal – are more pre-occupied with economic de-

---

Abstinence or, in modern technology, conservation was the only road to salvation. ... Those who questioned the need for a treaty could be counted on one hand while those who thought no treaty would be strong enough to save the world were legion." Gore's sermon was crafted to appeal to this audience. Thomas Gale Moore, *Climate of Fear: Why We Shouldn't Worry about Global Warming* (Washington: Cato Institute, 1998), 139-40.

38. The term European Union (EU) should only be used for the entity that existed after entry into force of the Maastricht Treaty in 1993. Until 1967, the proper term is the European Economic Community (EEC). From 1967-1993, it was European Communities (EC) following the merger of the EEC, the European Coal and Steel Community, and the European Atomic Energy Community. Since its establishment in 1957 by the Treaty of Rome, the European integration movement has grown from the original six to the current 28 member states. Throughout, the Commission has been the EEC/EC/EU executive arm.

velopment and less concerned about environmental matters.[39] With the implosion of the Soviet Empire in 1989 and the eventual absorption of its former satellites into the EU, Europe-wide politics became even more complicated. In the lead-up to the conclusion of the UNFCCC and its signature at the 1992 Rio Summit, EC climate policy was dominated by the concerns of Germany, the UK, and France, with supporting roles played by the then remaining nine smaller members.

The environment had not been a central concern during the Community's formation but gradually gained in importance in the 1970s and led to the establishment of a small environment directorate. In the 1986 Single European Act, members agreed to extend the directorate's competence, providing it with a mandate "based on the principles that preventive action should be taken, that environmental damage should be rectified at source, and that the polluter should pay."[40] The 1992 Maastricht Treaty, which transformed the European Communities into the European Union, went a step farther, making the environment an explicit EU policy responsibility and giving the Commission greater powers to represent the EU in international negotiations, in discussions with third parties, and in implementing international agreements. The treaty's subsidiarity principle assured that many environmental decisions would remain at the local and national levels, but subsequent practice gradually strengthened the Commission's hand in dealing with member-state environmental policies, particularly those arising from the implementation of international agreements and EU-wide policy. Schreurs and Thiberghien argue that "as a whole, ... this structure allows for multiple leadership points. Far from creating deadlock, this decentralized multi-polar structure has allowed for competitive leadership and mutual reinforcement to take place on climate change."[41]

---

39. On the continent, proportional representation allows minor parties to gain parliamentary seats, often leading to coalition governments that may include members of green parties. See Ferdinand Müller-Rommel and Thomas Poguntke, *Green Parties in National Governments* (New York: Routledge, 2002).

40. Paragraph 2, article 130r of the 1986 *Single European Act*. Competence is the EU term for 'powers.' See John Vogler and Charlotte Bretherton, "The European Union as a Protagonist to the United States on Climate Change," *International Studies Perspectives* 7:1 (2006).

41. Miranda A. Schreurs and Yves Tiberghien, "Multi-Level Reinforcement: Explaining European Union Leadership in Climate Change Mitigation," *Global Environmental Politics* 7:4 (November 2007), 27.

During the lead up to the 1992 Rio Summit, Commission ambitions were well-supported by the media, the environmental community, and the public. Polling indicated a much higher level of support on the continent at this early stage than was apparent in the UK or in North America and Oceania. "According to a top official at the DG [Directorate General] Environment, climate change is an issue that has reached such a level of social and political acceptability across the EU that it enables (indeed, forces) the EU Commission and national leaders to produce all sorts of measures, including taxes."[42] European ENGOs are often supported by the Commission and are much less dependent on voluntary public contributions. They are another manifestation of the corporatist/statist/progressive political perspective allowing their elites to take positions much more supportive of both Commission and member-state policies than is the case in the Anglosphere. In the UK, US, Canada, Australia, and New Zealand, ENGOs typically oppose government policy as a matter of strategy in helping them raise funds.

Unlike the UK, which in the Thatcher-Major years was suspicious of EU Commission ambitions on the environment, and other files, most continental governments were more willing – at least until the full accession of the first wave of eastern European members in 2004 – to let the Commission take the lead on international environmental discussions, particularly on climate change. Among these countries, Germany had the greatest influence and, as the years progressed, boasted a sufficient number of alarmist climate and social scientists to become prominently engaged in the IPCC process and in other ways to influence the intellectual foundations and public discussion of international climate-change policy.

In Germany, climate change entered political consciousness about the same time as in the UK and United States.[43] Surprisingly, Germany did not have a federal department responsible for environmental affairs until 1986, before then parcelling out management of environmental issues among several ministries with the lead responsibility assigned to its transportation department. In the fall of 1987, however, the German parliament established an enquiry commission to look into "preventive measures to protect the earth's

---

42.   Schreurs and Tiberghien, "Explaining European Union Leadership," 30.

43.   Interestingly, Chancellor Helmut Schmidt had signalled potential interest as early as 1979, but little came of it. See Michael T. Hatch, "The Politics of Global Warming in Germany," *Environmental Politics* 4:3 (Autumn 1994), 415. Hatch provides a detailed account of the evolution of German policy.

atmosphere."[44] In addition to greenhouse gases, its mandate extended to acid rain and ozone, i.e., to concerns that had much more political salience in the 1980s than global warming. It issued its first report a year later without a dissenting voice, thereby enhancing its authority and credibility and providing German politicians with a strong "expert" basis for pursuing an aggressive climate change agenda. The report was alarmist in tone and content and warned that a rise of 1-2°C, due to earlier and future GHG emissions, was now inevitable and that steps needed to be taken urgently to prevent even more emissions and a consequent rise in temperature. Its absolutist tone became a template for subsequent German governmental and quasi-governmental reports on climate change.[45]

While economists participated in the commission's work, the report contained little analysis of costs and benefits. The commission issued two more reports before completing its work in 1990. It recommended that Germany's contribution to mitigating global climate change should involve a 30 per cent reduction in CO_2 emissions from a 1987 base by 2005, 50 per cent by 2020, and 80 per cent by 2050.[46] These were among the most aggressive targets announced anywhere at that point. The commission's findings and recommendations were wholly consistent with Germany's long-standing commitment to the precautionary principle,[47] which had developed as a canon of German thinking about the environment over the previous 20 years. Given the unanimity with which both politicians and experts comprising the commission had spoken, its

---

44. In Germany, enquiry commissions are established by the Bundestag or Parliament to look into long-term policy issues and involve parliamentarians from all parties as well as experts. See Jeannine Cavender-Bares and Jill Jäger with Renate Ell, "Developing a Precautionary Approach: Global Risk Management in Germany," in William C. Clark, et al., eds., *A Comparative History of Social Responses to Climate Change, Ozone Depletion, and Acid Rain* (Cambridge, MA: MIT, 2001), 81.

45. See for example, Umweltbundesamt (Federal Environmental Agency), *The Future in Our Hands: 21 Climate Policy Statements for the 21st Century* (Berlin, 2006), and German Advisory Council on Global Change (WGBU), *World in Transition: A Social Contract for Sustainability* (Berlin: WGBU, November 2013).

46. Rie Watanabe, *Climate Policy Changes in Germany and Japan: A Path to Paradigmatic Policy Change* (New York: Routledge, 2011), 72-4.

47. The precautionary principle's origins lay in widespread adoption in the 1980s of the German concept of *Vorsorgeprinzip*. See Sonja Boehmer-Christiansen, "The Precautionary Principle in Germany – enabling Government," in Timothy O'Riordan and James Cameron, eds., *Interpreting the Precautionary Principle* (London: Earthscan, 1994), 31-60.

findings carried tremendous weight and ensured that Germany would become Europe's leading voice in pressing for aggressive mitigation measures.

The influence of the enquiry commission on German policy was immediate. German scientists had not played a prominent role in the pre-1988 international discussions and had not gained a leadership role in the IPCC. The German science community and the government, however, soon caught up with the US and the UK and thus added their voices to international discussions. In order to provide an intellectual base for German participation in the discussions, the government helped in founding and funding new research centres, including the Wuppertal Institute for Climate, Environment and Energy in 1991, the Helmholtz Centre for Environmental Research in 1991, and the Potsdam Institute for Climate Impact Research (PIK) in 1992. By the end of the decade, these research centres would rival American and British research organizations in pushing the global warming threat and contributing directly to German and European policy development.

While the enquiry commission's work proceeded, re-unification was dominating German politics. The sorry state of East German industry and its environment brought home to politicians in the West the devastating impact that man could wreak on his natural environment – perhaps without always considering that central planning had been the handmaiden of this assault. East Germany's decrepit industry and its dilapidated energy sector also created an opportunity that would allow German politicians to provide leadership on a politically popular basis.[48] If Germany's two halves were to be united, the East would have to undergo a rapid and costly refit of its industrial sector, and, with that, a much cleaner energy sector and more efficient industry.[49] The modernization process would result, thus, in a rapid decline in Germany's aggregate CO_2 emissions and provide the government with a strong moral advantage in both intra-EU and international discussions.

Unlike follow-up efforts in both the UK and the US, interagency study and deliberations in Germany following the enquiry commission were both thorough and comprehensive, a fact which further

---

48. This interagency effort, however, paid little heed to the economic dimension. It was dominated by environment and foreign policy officials.

49. See Sonja Boehmer-Christiansen et al., "Ecological restructuring or environment friendly deindustrialization: The fate of the East German energy sector and society since 1990," *Energy Policy* 21:4 (April 1993), 355-73.

underpinned Germany's leading role in both intra-European and international discussions. An agreed mandate of a 25 per cent cut in German emissions provided German officials with the ability to insist that the rest of the EU agree to a strong mandate and that the EU pursue an international agreement with mandatory and verifiable emissions reductions and a framework within which this mandate could be achieved. By the early 1990s, therefore, Germany had developed a full-fledged climate policy subsystem flowing from its rapid accumulation of scientific knowledge, its participation in a series of international scientific as well as political conferences, and a growing climate awareness in its public.[50]

As a result of these fortuitous circumstances, Germany was able to establish itself as the leading norm entrepreneur. It rejected British and American caution and insisted on European leadership. It dismissed economic concerns as of decidedly secondary importance in the fight to save the planet. By the mid-1990s, physicist Hans-Joachim Schellnhuber, founder of the Potsdam Institute – with major infusions of cash from the federal government, the EU Commission, and German industry – became the leading voice of German climate alarmism, and with direct access to German politicians in his role as the first chair of the German Advisory Council on Global Change (WGBU) and as a member of the IPCC (Assessment Reports 2-5). His position immediately disposed him to rather grandiose pronouncements about climate issues going well beyond any scientific expertise he may have accumulated. In a 2003 piece for the *Guardian*, for example, he characterized "the consumption of cheap fossil fuels as a lifestyle of mass destruction. This very lifestyle, which confounds mobility with liberty, is unfortunately a mantra of modern civilization and it may need a hundred green Gorbachevs to bring about ecological perestroika."[51]

Concurrent with Germany's aggressive position on the necessity and feasibility of emission reductions, a number of other member states also announced national mitigation plans.[52] In 1989 the Neth-

---

50. Watanabe, *Climate Policy Changes in Germany and Japan*, 76.

51. "Action stations: As heatwaves hit Europe, John Schellnhuber warns that global warming is here to stay and we must start adapting now," *The Guardian*, August 6, 2003. For a profile of Schellnhuber, see Donna Laframboise, "Who Is Hans-Joachim Schellnhuber?" *Nofrakkingconsensus*, December 21, 2012.

52. Schreurs and Tiberghien note: "As medium to small-sized states within the EU, the political and economic influence of Austria, Belgium, Denmark, Finland, Luxembourg, the Netherlands, and Sweden is limited on an individual basis. Combined, their greenhouse gas emissions in 1990 were less than two-thirds

erlands issued its First National Environmental Policy Plan, calling for the stabilization of industrialized countries' CO_2 emissions at 1989/90 levels by 2000. The following year the Dutch government went a step farther by announcing its intentions to cut CO_2 emissions by 3 to 5 per cent by 2000 calculated on a 1989/90 baseline. Similarly, Denmark agreed on a reduction of CO_2 emissions by 20 per cent on 1988 levels by 2005, and Austria, to a 20 per cent reduction on the same terms.[53] France agreed to reductions but argued that its high reliance on nuclear energy – higher than anywhere else, supplying up to 80 per cent of its electricity generation – placed it in a different category because it had the lowest per capita emissions among OECD members. The southern tier adopted a pragmatic stance: aggressive EU-wide climate policies could only be implemented within the context of EU burden-sharing and regional-development policy.

Germany's strong leadership, with the support of the Netherlands, Denmark and Austria, proved of tremendous help to the EU Commission. Commission officials saw in the UNFCCC negotiations an opportunity to further centralize EU climate and energy policies and thus establish its dominant role in future climate negotiations. According to a subsequent president of the Commission (José Manuel Barroso, 2004 to 2014), during climate negotiations the EU had "worked hard to be worth listening to. We are maturing, speaking with a unified voice more often and on a broader range of issues."[54] Both Germany and the Commission wanted to use the negotiations to put pressure on the United States and Japan to take global climate change more seriously and accept specific, binding commitments. The UK straddled the US-EU divide, in part to frustrate Commission efforts to take the lead on the issue. In the end, the final UNFCCC negotiating session marginalized Germany and the EU in favour of a US-UK compromise, but the Convention itself

---

those of Germany. In the area of climate change, however, these states have often formed coalitions in support of aggressive action." "Explaining European Union Leadership," 38.

53. Austria – along with Sweden and Finland – did not become an EU member until 1995 but wanted to register its strong support for German leadership in the context of its accession negotiations. As a practical matter, all three countries were already coordinating their policies with the EU and were involved in discussions that would eventually result in the agreement on the European Economic Area and their eventual accession to the EU.

54. Quoted in Vogler and Bretherton, "The European Union as a Protagonist," 1.

set the stage for the future with a framework providing for an annual requirement to meet and move its agenda forward.

Germany and the Commission saw the climate issue as a planetary-scale threat which, with the right policies, could be averted. The United States saw it as an issue not yet ripe for far-reaching measures, which it believed would have repercussions well beyond the climate system. Cass notes: "Germany's emphasis on the precautionary principle, its electoral system that created openings for the Greens, and the ability of the Enquiry Commission to provide a scientific consensus for action created a fertile ground for supporters of efforts to address climate change. On the other hand, the fragmented American political system, the lack of scientific consensus, and the strong influence of economic interests created numerous obstacles to action in the American case."[55] In any event, with ratification and the entry into force of the UNFCCC in 1994, the Commission gained a strong position in any future negotiations. From then on, the EU participated in climate negotiations as a single player, coordinating its position in private and represented by its rotating six-month presidency. Much of the heavy lifting in this arrangement fell to the Commission. The result was an EU that could be the agenda setter and a norm entrepreneur in international climate negotiations.

Its first opportunity came in the negotiation of the Kyoto Protocol. At the first two COPs in Berlin and Geneva, the EU had set out an ambitious agenda for the first commitment period and had supported the developing countries in their effort to ensure that they would not be required to make any mandatory emission reductions. As Schreurs and Tiberghien observed at the time, "the EU has clearly been a leader in the climate change area along a number of fronts. The EU has functioned as a classic norm entrepreneur. It has been a powerful backer of the precautionary principle in relation to climate change... It has embraced the notion embodied in the [UNFCCC] that the industrialized states have the responsibility to act first given their historic contributions to anthropogenic greenhouse gas emissions. It has defined climate change action as a moral and ethical issue that must transcend narrow economic interests."[56]

At COP 3 in Kyoto, the EU continued with its aggressive position. It called for stringent measures by all industrialized countries.

---

55. Cass, *The Failures of American and European Climate Policy*, 51.

56. Schreurs and Tiberghien, "Explaining European Union Leadership," 23-4.

Internally, it promoted a reduction of as much as 85 per cent of 1990 levels by 2005. Internally, members had agreed that the burden of cutting emissions would be shared among members on the basis of their economic capacity to assume emissions cuts. The Commission claimed that existing EU programs would already lead to a 10 per cent reduction. In any event, the EU agreed to a rather modest 8 per cent cut by 2008-2012 to 92 per cent of 1990 emissions, a level that could easily be achieved on the basis of existing German, Danish, UK, and a few other national policies.[57] It reluctantly agreed to the demands of the US and others for flexibility provisions, expecting that future negotiations would lead to more stringent emissions reductions for all participants.

Schreurs and Tiberghien argue that "EU policy toward climate change often has been couched in terms of an ideational agenda, namely the representation of the EU as a different kind of polity, one more concerned with international law, institution-building, and a normative vision. Through their global policy-making actions the EU elites seek to increase public support for EU integration."[58] The EU approach provides a classic example of Europe's supposedly more sophisticated assessment of foreign policy and environmental issues, based less on national than on global interests whose pursuit would benefit all. EU "leadership" on climate is in part facilitated by the simple fact that EU policy-makers do not face direct elections. While the Strasbourg Parliament is elected, its mandate remains limited.

President Bush's science adviser, Allan Bromley, saw the other side of this sophistication during the Noordwijk conference in 1989.

---

57. EU members agreed to share the burden required to achieve its total of a cut of 8 per cent below 1990 levels for the EU as a whole as follows:

Austria	–13.0	Belgium	– 7.5	Denmark	–21.0
Finland	0.0	France	0.0	Germany	–21.0
Greece	+25.0	Italy	–6.5	Luxembourg	–28.0
Monaco	–8.0	Netherlands	–6.0	Portugal	+27.0
Spain	+15.0	Sweden	+4.0	UK	–12.5

Upon their accession to the EU in 2004, the ten Eastern European transition economies all accepted the EU level of –8 per cent of their 1990 levels with the exception of Hungary, which accepted a –6 per cent obligation. All of these commitments were higher than what had already been achieved as a result of the closing of Soviet-era industries, thus further easing any remaining burden on the other EU members.

58. Schreurs and Tiberghien, "Explaining European Union Leadership in Climate Change Mitigation," 26.

He recalls: "The lack of economic analysis was astonishing ... I asked the head of one of the major European delegations how exactly his country intended to achieve the projected emissions goals and was told, 'Who knows – after all it's only a piece of paper and they don't put you in jail if you don't actually do it.'"[59] Such a perspective reflects the reality that on the continent treaty obligations do not mean the same thing as they do in common law countries. In the United States, Canada, and elsewhere, an international agreement once ratified becomes binding on the state. On the continent, the state is above civil law and can thus ignore treaty obligations at will.[60] Hecht and Tirpak explain: "Why some EC countries supported targets and timetables knowing that these targets could not be met without some costs is complicated and subject to many different interpretations. At the time, most of the analyses in Europe were based on simple linear programming models that produced results showing substantial savings over the lifecycle of more energy efficient technologies. Few macroeconomic models had been run in Europe. Our assertion is that setting targets and timetables became for many European governments symbolic of showing political leadership by challenging the US."[61] For the rest of the world, EU global leadership on the climate file created problems that over time would overwhelm even the Europeans.

Russia: scepticism and opportunism

Over the past twenty-five years, through tumultuous internal political and economic upheavals, Russia has maintained a hard-nosed, national interest-based role in climate change discussions at the UN and elsewhere. It was prepared to go along only if it could extract sufficient ancillary benefits to offset any problems for its pursuit of other priorities, particularly economic development. It has demonstrated no interest in ideational foreign policy or in the progressive agenda. From the perspective of the more enthusiastic supporters of global climate change action, Russia has been a consistent disappointment.

---

59. Bromley, *The President's Scientist,* 144-5, as quoted in Darwall, *The Age of Global Warming,* 139.

60. See, for example, Richard Tol, "Economics versus Climate Change: A Comment," in Roger Guesnerie and Henry Tulkins, eds., *The Design of Climate Policy* (Cambridge, MA: MIT, 2008), 217.

61. Hecht and Tirpak, "Framework agreement on climate change," 388.

During the Soviet era, Russia had remained largely aloof from the UN's many ambitions, maintaining that the issues addressed at various UN-sponsored conferences were problems endemic to capitalism that had either been resolved or did not arise under communism. With the collapse of the Soviet Empire, this was a difficult perspective to maintain. Russia now had to engage, and with a more nuanced approach. During the critical decade from 1989-99 Russia had other priorities, most importantly the building of the institutions of post-Soviet governance and the transformation of its economy to a market basis. Global warming was far down on its list of priorities. Nevertheless, as a former great power, it felt that it deserved a prominent seat in multilateral discussions. Russia could also boast scientists with qualifications in atmospheric physics and related fields who could not be denied participation in the work of the IPCC. In the decade of the 1990s, Russia's ambivalence provided a dampening influence on the work of the IPCC and on UN climate negotiations, one that climate activists worked hard to isolate as the decade wore on.

Through the end of the Soviet era, the Yeltsin years, and into the Putin-Medvedev governments, the chief Russian spokesman was Yuri Izrael, director of the Institute of Global Climate and Ecology in the Russian Academy of Sciences. He was well known in the West and had served as a first vice-president of the WMO. When the UN established the IPCC, Izrael was elected one of its vice-chairmen – a position he held until 2008 – and was appointed co-chair of Working Group II, responsible for assessing the science of climate change impacts.[62] Unfortunately for Bert Bolin and his colleagues, Izrael was a sceptic, agreeing that the climate does change but that the human role had yet to be scientifically established. In his view, "much uncertainty remains in climate changes forecasts [sic]. Climate change is obvious, but science has not yet been able to identify the causes of it. ... Nature is complex and there are many chance factors that cannot be predicted."[63] If a few Russian scientists accepted the anthropogenic thesis, most agreed with Izrael.

---

62. While he remained a vice-chair, Izrael was later replaced as chair of WG 2 by US official Robert Watson.

63. As quoted by Tatyana Sinitsina, "The G8 Weather Report," *Ria Novosti*, July 5, 2005. See also Yuri Izrael, "Climate change: not a global threat," *Ria Novosti*, June 23, 2005 and "Climate: putting panic in perspective," *Ria Novosti*, April 18, 2007.

Three factors help to explain Russia's position. First, unlike most of the OECD countries, Russian policy emerges from within a highly centralized executive needing to pay only symbolic deference to the legislature, particularly following the end of the chaotic Yeltsin years. Second, the Russian people are preoccupied by many issues, and environmental worries are even farther down the list than they are in other countries. Lilianna Andova concurs: "Norm diffusion has had a ... limited impact on Russia's climate politics due to the weak position of environmental groups and low public concern with climate change. The most significant climate campaigns in Russia have been carried out by transnational NGOs. ... Russian NGOs remain largely of a 'decorative character', while the public remains undivided in its indifference to climate change as citizens are faced with more urgent economic problems."[64] Third, the Russian science establishment agreed largely with Izrael that climate change was a minor issue at best, a position that it confirmed in a number of subsequent reports, including a 2004 study by the Russian Academy of Sciences.[65]

The implosion of the Russian economy, however, provided Russia with the same advantage as Germany: a marked reduction in GHG emissions over the course of the 1990s as inefficient and uncompetitive Soviet-era industries were mothballed, many of them never to be replaced. Russia was thus in a position to demand aggressive reductions from others without requiring its own remaining industries to adapt in ways placing them at a competitive disadvantage. At the same time, Russia's large resources of fossil fuels and growing dependence on exports to earn foreign exchange to modernize its economy did not make aggressive reductions in emissions elsewhere attractive, a perspective Russia shared with the OPEC countries. Russia saw no benefits arising from a major transition to renewables, either at home or abroad.

Two results of the Russian position are worth noting. First, in the negotiation of the UNFCCC, Russia insisted that the treaty establish differences among countries' obligations. This was a position strongly supported by the developing countries as well as by Canada and Australia but one that undermined US insistence that

---

64. Liliana B. Andonova, "The climate change regime and domestic politics: the case of Russia," *Cambridge Review of International Affairs,* 21:4 (December 2008), 492-3.

65. Andonova, "The climate change regime and domestic politics: the case of Russia," 491-2.

all countries contribute to mitigation efforts. Second, Izrael's role at the IPCC had a dampening effect on its first assessment report, particularly the report of WG2. It was largely through his influence that WG2's first report provided a balanced assessment of climate change impacts. Nevertheless, to show good faith and demonstrate its commitment to the UNFCCC process, Russia was among the first to ratify the treaty. But over the next two decades, it maintained its sceptical attitude and manoeuvred to ensure that it could gain maximum economic advantage on the basis of a minimum of commitments.

At Kyoto, Russia possessed enough bargaining power to be a major factor. As a significant economy, its participation was essential for the agreement to succeed, but economic considerations ensured it would not accept any obligations beyond, perhaps, stabilization at 1990 levels. Russia's economy and GHG emissions had shrunk by 39 per cent and 34 per cent respectively between 1990 and 1997. It found itself more aligned with the US than with the EU, favouring, for example, unrestricted use of flexible economic mechanisms such as emissions trading and joint implementation. Under the right conditions, it knew it could expect to profit from the sale of "hot air," i.e., reductions that it did not need to meet its commitments as a result of its already large reductions in GHG emissions compared to the base year 1990. In the end Russia agreed to an emissions cap at the 1990 level which, based on 1997 levels, would allow it either to increase or sell emissions credits up to the equivalent of about 34 per cent of its 1990 base.[66]

Despite signing the Protocol, Russia was, nevertheless, reluctant to ratify it. Not convinced that the Protocol served any useful purpose, it was prepared to play its cards carefully in order to maximize credit for ratifying the agreement, cards that gained in strength when it became clear that the US would not ratify and that only Russian ratification could bring the Protocol into force. Russia's scepticism about the science was reinforced by its view of the treaty's economic implications. Because President Putin's principal economic adviser, Andrei Illarionov, frequently briefed western journalists on the Protocol's countless shortcomings, many believed that the Protocol would die.[67] In the end, however, Putin made a strate-

---

66. Andonova, "The climate change regime and domestic politics: the case of Russia," 489.

67. See Quinn Schiermeier and Bryon McWilliams, "Crunch Time for Kyoto," *Nature* 431 (September 2, 2004), 12-3.

gic decision. In return for EU assurances of support for Russia's accession to the WTO, Russia would ratify and thus allow the Protocol to come into force. The depth of that commitment, however, was rather shallow. Little needed to be done to comply, and sales of hot air turned out to be anaemic. In December 2011, at COP 17 in Durban, South Africa, Russia joined Japan and Canada in their determination not to sign up for a second commitment period.[68]

Japan takes a few tentative steps

Japan's interest in climate change developed late and was largely influenced by a foreign-policy-driven desire not to be isolated from international discussions, a concern not unique to environmental issues.[69] This desire was complicated by the collapse of Japan's rapid, post-war growth in 1990, coincident, in turn, with the emergence of climate change as a major preoccupation of international diplomacy. Climate change discussions were an unwelcome complication during this traumatic decade of change. The Japanese government found itself torn between two policy imperatives: a broadly perceived need to be an active and constructive participant in international discussions and a more immediate and pressing need to restart the economic engine.

A second hindrance to the development of Japanese policy emerged out of the differences between the EU and the United States. The US relationship remained the most important, but in the early post-Cold War years, relations with others gained in prominence, particularly with the EU, China, and Southeast Asia. Consistent with Japanese desires to be a constructive participant in all multilateral discussions, Japanese officials also saw the discussions through the lens of their impact on Japan's bilateral interests. An aggressive position on climate change would benefit relations with the EU but complicate relations with the United States, particularly

---

68. See "Russia supports Canada's withdrawal from Kyoto protocol," *The Guardian*, December 16, 2011.

69. Jasuko Kameyama writes: "The upshot is that climate change *as foreign policy* has been the most influential driving force in Japan's response to climate change. That is, being involved in and even trying to lead the climate change debate has in some respects become Japan's foreign policy, quite apart from global warming and climate change per se." "Climate change as foreign policy: from reactive to proactive," in Paul Harris, ed., *Global Warming and East Asia: The domestic and international politics of climate change* (London: Routledge, 2003), abstract.

if the result was a commitment to domestic measures that would undermine Japanese export interests.

Japanese industrial leaders had a perspective similar to that of their US counterparts: without full participation by all countries in any international reduction efforts, industry in advanced economies would be placed at a disadvantage that would come back to haunt them. Voluntary and no-regrets measures were one thing, mandatory emissions reductions quite another. This was the position advanced by the Keidanren, the most important voice of business in Japan. As elsewhere, there were minority views. Less energy intensive industries thought an international regime could be in Japan's long-term interest; others saw an advantage in aligning with Europe and gaining better access to EU markets; still others saw support of international coordination as boosting corporate Japan's image.[70] These minority voices, however, had little impact on Japanese policy development.

In Japanese political culture policy emerges out of multi-perspective, middle-management discussion. Senior management receptivity or not, nothing will happen if middle managers are opposed or indifferent, whether in business or in government. In the case of climate change, internal conflict found institutional expression in the different perspectives of officials at the Ministry of Foreign Affairs (MOFA), with support from those in the weak Environment Agency, and of officials in the powerful Ministry of International Trade and Industry (MITI), with support from officials in other economic agencies.[71]. As a result, van Asselt, Kanie, and Iguchi conclude, "due to the substantive differences in position between the three ministries involved and the lack of effective coordination mechanisms, the government's foreign policy on climate change has been both ambiguous and fragmented. ... Each ministry has its own networks ... and each ministry is protecting its own

---

70. See the discussion in Yves Tiberghien and Miranda A. Schreurs, "High Noon in Japan: Embedded Symbolism and Post-2001 Kyoto Protocol Politics," *Global Environmental Politics* 7:4 (November 2007), 70-91.

71. MITI's broad responsibility for all aspects of industry, including energy and pollution, had placed it in a much more powerful position than that of the environment agency, not elevated to the level of a ministry until 2001, when it assumed some of the responsibilities of MITI for pollution and other environmental issues. MITI was recast in 2001 as the Ministry of Economy, Trade and Industry (METI), with the merger of MITI and the Economic Planning Agency and a few other smaller agencies.

realm of competence."[72] MITI officials were extremely protective of the Japanese business community; they feared that domestic action on climate change would have an adverse effect on Japan's economy.

Japanese scientists had not been actively engaged in the WMO-sponsored scientific discussions in the 1980s. To the extent that they were aware of the emerging international scientific consensus, they had made no effort to engage the government and influence domestic policy development. As a result, during the 1988 Toronto Conference, Prime Minister Noburu Takeshita was surprised to learn that climate change was an emerging diplomatic issue.[73] Only upon his return did the Japanese government begin preparing to participate, making it the last major government to join the discussions. Initially, Japan opposed common emissions standards for good reason: it had an efficient industrial base and a relatively low level of per capita emissions vis-à-vis OECD countries, second only to France among the major countries.[74] Japan was not a major player in the UNFCCC negotiations but, consistent with its foreign policy interests, signed and ratified the agreement and only then initiated serious inter-agency discussions to develop a coherent policy position that would allow its participation in on-going discussions to develop an international emissions regime.

Kameyama argues that over the subsequent few years Japan's position evolved from a reactive to a proactive stance. That shift became most apparent when it hosted COP 3 in Kyoto. Until the start of negotiations resulting in the Kyoto Protocol, MITI had succeeded in framing climate change as largely an environmental or energy issue.[75] Consistent with this view, the Japanese government had formulated a voluntary Action Plan to Arrest Global Warming in Oc-

---

72. Harro van Asselt, Norichika Kanie, and Masahiko Iguchi, "Japan's position in international climate policy: navigating between Kyoto and the APP," *International Environmental Agreements: Politics, Law and Economics* (June 2009), 3.

73. Kameyama, "Climate change as foreign policy," 137.

74. Tiberghien and Schreurs argue that "in terms of its per capita CO_2 emissions, Japan (9.3 tonnes per capita in 2000) was doing considerably better than the OECD average (11.3) and somewhat better than Germany (10.0). It was performing far better than the US (20.0) and Canada (16.8) although its emissions were still substantially above those of France (6.0), where nuclear energy accounts for approximately three-quarters of the country's electricity." "High Noon in Japan," 72-3.

75. Noriko Sugiyama and Tsuneo Takeuchi, "Local Policies for Climate Change in Japan," *Journal of Environment and Development* 17:4 (December 2008), 423.

tober 1990 just before the Second World Climate Conference. The plan stipulated that it would seek to stabilize CO_2 emissions at 1990 levels by 2000. By the conclusion of the Kyoto negotiations, however, climate change had morphed into a foreign policy matter critical to broader Japanese interests in international affairs, particularly since the Protocol would bear the name of Japan's ancient capital city.[76] For the moment, anyway, foreign policy officials had the upper hand in developing Japan's position. MITI and its allies in the Keidanren, however, were not convinced and, in the absence of a strong domestic environmental lobby, the two opposing forces would continue to push Japanese climate change policy back and forth between these two poles.

By the turn of the century, the issue was less a matter of climate change than of Japanese honour. Mindful of the Protocol's name the Japanese public and the government could not imagine Japan's not ratifying and honouring the agreement. The economic arguments advanced by the Keidanren and its members, while powerful, would have to be addressed in the manner in which Japan agreed to implement its obligations, but every effort had to be made to reduce Japanese emissions to 94 per cent of the level reached in 1990 – a tall order. As elsewhere, Japan had only pursued voluntary measures to date – with little impact – and had developed no internal consensus on what would be required. Japan's most pressing domestic need was to boost economic growth, not to depress it with measures to curb emissions. The nuclear accident in Fukushima would make the task impossible.[77]

Canada and Australia: two contrasting perspectives

In Canada and Australia preoccupation with climate change policy started quite early but followed different paths. As important resource economies, both federal governments were keenly aware that the choice of mitigation as the preferred policy response would have a greater impact on them than on other countries. It also placed them among the highest per capita producers of anthropogenic GHGs – Australia and Canada rank second and third as sources of per capita GHG emissions – and made them vulnerable to activist campaigns. Carbon emissions on a GDP basis were also

---

76. See Kameyama, "Climate change as foreign policy," for a more complete discussion of this theme.

77. See Roger A. Pielke, Jr., "Mamizu climate policy: an evaluation of Japanese carbon emissions reduction targets," *Environmental Research Letters* 4 (2009), 1-6.

substantially higher than those in Europe, levels that would make the cost of any mitigation measures high as well. Both countries are large and sparsely populated, adding to the difficulty of reducing emissions. In 1998 and 2000 surveys, 53.7 per cent and 49.0 per cent of respondents in Australia saw climate change as 'very serious'. In Canada the results were 43.5 per cent and 44.6 per cent, levels that had remained consistent throughout the decade, even if, as Gallup warns, "public opinion with respect to the environment is typically characterized by a high level of support for action but low levels of attention."[78]

Both countries were initially at the forefront of scientific and political efforts to put climate change on the international agenda, but domestic institutional, economic, and political factors conspired to make them laggards internationally once discussions turned serious and required real-world commitments. Those factors included divisions among the federal and state and provincial governments and resistance from powerful actors in the energy sector and industry more broadly. As Heather Smith laments: "In spite of the creative leadership of Canadian bureaucrats and scientists in the agenda-setting phase of the climate change issue, political commitment and leadership were tentative at best."[79] And concerning Australia, Aynsley Kellow adds: "Norms and science are powerful and can undoubtedly help create pressure to negotiate and play a part in reconstituting interests, but while they are powerful cards which will continue to be played, interests appear to be trumps, and norms and science will be most effective when they coincide with interests."[80]

In both Canada and Australia climate politics is complicated by the limited constitutional authority exercised by the federal governments over environmental and resource matters that fall largely within the jurisdiction of the provinces, states, and territories. Federal governments have three tools at their disposal: the power of the purse, moral suasion, and international agreements. The creation of Environment Canada by Prime Minister Trudeau, for example, did

---

78. Loren R. Cass, "A climate of obstinacy: symbolic politics in Australian and Canadian policy," *Cambridge Review of International Affairs* 21: 4 (December 2008), 472, and Harrison, "The Road not Taken," 94.

79. Heather Smith, "Political Parties and Canadian Climate Change Policy," *International Journal* 64:1 (Winter 2008-09), 48.

80. Aynsley Kellow, "Australia in the Greenhouse: Science, Norms and Interests in the Kyoto Protocol," *Energy & Environment* 10:3 (May 1999), 288.

not transfer authority for many environmental issues to the federal government. As Harrison points out: "Although the federal government has indirect means to influence resource use, any such incursions are guaranteed to provoke opposition from the provinces, which jealously guard their control of natural resources and thus their local economies."[81] At its inception, therefore, the department and its ministers enjoyed only such relatively minor remits as the weather service and northern environmental issues and a role as hector of provincial policies and gadfly at international gatherings. The latter made for good press but little concrete action. Negotiating and ratifying the UNFCCC were fully within federal competence because the treaty relied on voluntary measures. The commitments made in Kyoto were a different matter and could only be met with the full cooperation of the provinces.

The Australian federal government faces a similar constitutional reality. There is no specific grant of authority over environmental issues, which, as a result, have become an area of joint jurisdiction. The states and territories, however, do have primary responsibility over resources. In response, Australia has developed some pragmatic intra-Australian arrangements to facilitate cooperation among the states, territories, and the federal government, reducing conflict over climate change issues. Over much of the 1990s, several Australian states were ahead of the federal government in their willingness to pursue climate change-related policies.

In Canada, Progressive Conservative Brian Mulroney was prime minister (1984-1993) during the emergence of climate change as a political issue. For Mulroney international climate politics presented an opportunity to define Canada's role as a responsible citizen in the international arena. To that end he fully supported the engagement of Canadian scientists in the early stages of discussion, opened the Toronto Conference, strongly endorsed the work of the Brundtland Commission and the Rio Summit, and quickly ratified the UNFCCC, a decision made easier because it required no action by the provinces. Despite his conservative politics, environmentalists had little to complain about during his leadership. Nevertheless, as Steve Bernstein observes, "Canada may often lead in identifying problems and supporting multilateral solutions, [but] it is frequent-

---

81. Kathryn Harrison, "The Road not Taken," 97.

ly unable to follow through with deep commitments owing to domestic constraints."[82]

Mulroney's Conservatives were succeeded by a Liberal government led first by Jean Chrétien (1993–2003) and then by Paul Martin (2003-06). Because of the Kyoto requirements, both were faced with the tougher challenge of translating international commitments into meaningful domestic policy. The first steps had been taken without much notice at home or much thought to economic and other implications. After Rio, both international and domestic forces led to the need for a more realistic assessment and greater engagement. Voluntary and no-regrets policies were not getting the job done. The governments of Alberta, British Columbia, Saskatchewan, Nova Scotia, and Newfoundland were all wary of any commitments, while Ontario worried about the impact of commitments on the competitiveness of its industries. Only Quebec was supportive of action, knowing that its reliance on hydro-electricity would limit possible economic consequences.

Much more than Australia, Canada also had to pay close attention to the evolution of climate policy in the United States. The need for climate policy emerged at a time when bilateral trade between the two countries had reached its highest historical levels, hovering well above 80 per cent of total trade as a result of the 1989 bilateral Canada-US Free Trade Agreement and the 1994 trilateral North American Free Trade Agreement. Canada had also become the number one source of US fossil-fuel imports. It could not pursue any policies far out of step with the United States. An aggressive US position would require Canada to follow suit, while a more pragmatic posture would require Canada not to get too far ahead and risk losing competitiveness. Australia, on the other hand, enjoyed more diversified export markets and relied largely on China, Japan, and other East Asian economies as markets for its coal. None of these was likely to take any precipitous action to reduce fossil-fuel use or to curb its imports from Australia as part of efforts to mitigate climate change.

Throughout this period, the instincts of Canadian environmental and foreign policy officials remained the same. Support for sustainable development in all its manifestations had become one of the touchstones of their advice, and as a responsible member of the

---

82. Steven Bernstein, "International institutions and the framing of domestic policies: The Kyoto Protocol and Canada's response to climate change," *Policy Sciences* 35 (2002), 213.

international community they insisted that Canada had to accept the need to address environmental and development problems. Chrétien relied on this advice and was content to pursue symbolic politics, particularly since the United States was similarly engaged. With the decision by the second Bush administration not to seek ratification of the Kyoto Protocol, there was little likelihood that the president would pursue aggressive domestic policies. The provinces as well as officials in economic departments had made it crystal clear that there was no way that Canada could deliver on its Kyoto commitments, particularly in the absence of a commitment by the United States. Chrétien's own Minister for Natural Resources, Herb Dhaliwall, quite candidly admitted that Canada had "no intention of meeting the conditions of the Kyoto Protocol on greenhouse gases even though the government hopes to ratify it this fall."[83] The government produced a number of action plans which, of necessity, had to be largely voluntary and symbolic in nature due to strong opposition from the resource-dependent provinces.[84] As Kathryn Harrison points out, "Canada's failure to follow through on [Chrétien's] commitment reveals the fragility of politicians' values in the absence of institutional and especially electoral support."[85]

The Chrétien government's cavalier attitude betrayed a fundamental departure from Canada's approach to the negotiation of international agreements. Canadian negotiators typically are well prepared, pay close attention to the detail of the negotiations, and have a first-rate record of convincing the government to faithfully implement the results. Consequently, the country has an excellent history of fulfilling its treaty obligations. The Kyoto negotiations betrayed that record on all fronts. The negotiators were ill-prepared, lacked a mandate from the provinces, and were eager to compro-

83. Bruce Cheadle, "Canada to sign Kyoto, but won't abide by it," Toronto *Star*, 5 September 2002.

84. For a detailed overview of the evolution of Canadian climate policy, see Jeffrey Simpson, Mark Jaccard, and Nic Rivers, *Hot Air: Meeting Canada's Climate Change Challenge* (Toronto: McClelland and Stewart, 2007). See also Robert C. Paehlke, *Some Like It Cold: The Politics of Climate Change in Canada* (Toronto: Between the Lines, 2008). Both are from an alarmist perspective, Paehlke much more so than Simpson et al.

85. Kathryn Harrison, "The Struggle of Ideas and Self-Interest in Canadian Climate Policy," in Harrison and Lisa McIntosh Sundstrom, *Global Commons, Domestic Decisions: The Comparative Politics of Climate Change* (Cambridge, MA: MIT, 2010), 195.

mise. The result was not surprising: an inability to implement and live up to the Protocol.

Stephen Harper led the Conservative Party to a minority government in 2006 and to a majority government after 2008. Officials' aspirations were frustrated by the Harper government, which showed no inclination to pursue the symbolic politics of the Liberals. To Harper, symbolic politics on climate change were more likely to harm than enhance Canada's international reputation. From that perspective, the fact that the Liberals had failed to come even close to delivering on their commitments had made a mockery of treaty obligations. Harper's solution lay in pursuing policies independently of the UNFCCC and in eventually emerging from under the Kyoto commitments.[86]

In Australia, Labor Prime Minister Bob Hawke (1983–1991) was leading the government when climate change became a diplomatic issue, and his Labor successor, Paul Keating, was prime minister (1991-1996) through the UNFCCC and the lead-up to the Kyoto negotiations. Both governments were supportive of international efforts but took few steps to address climate change in domestic legislation, relying instead on hortatory statements and voluntary measures. Hawke, for example, announced in 1990 that Australia would reduce its GHG emissions by 20 per cent below the 1988 level by 2005. Like Mulroney, Hawke embraced the Brundtland report and looked for ways to make it a touchstone of Australian environmental policy. Foreign Minister Gareth Evans was a strong upholder of progressive UN causes and good international citizenship and promoted the symbolic politics of left-wing causes. At the UNFCCC negotiations, Australia aligned itself with other progressive members of the international community.[87] Keating, keenly aware of strong business opposition to the costs of any emission measures, took a more cautious stance, retreating from Hawke's commitments and looking for a differentiated approach to international commitments, i.e., targets that would take into account the

---

86. See Michael Hart, "Stephen Harper's Foreign Policy: Prudence on the Climate Front," *Policy Options* 33:4 (April 2012), 48-53.

87. Loren R. Cass, "A climate of obstinacy: symbolic politics in Australian and Canadian policy," *Cambridge Review of International Affairs* 21: 4 (December 2008), 475-7. See also Simon Lightfoot, 'A good international citizen? Australia at the world summit on sustainable development', *Australian Journal of International Affairs* 60:3 (2006), 457–471. Lightfoot notes that "Keating did not share Hawke's perceived interest in environmental issues." 459.

different capabilities and the different economic burdens that those targets would have on states.[88]

Following the 1996 Liberal/National coalition victory, John Howard (1996–2007) led Australia through the Kyoto negotiations, the US rejection of Kyoto in 2001, and into the post-Kyoto discussions. Howard was not impressed by calls to pursue good international citizenship and other progressive nostrums, instead insisting on pursuing Australia's national interests.[89] His government negotiated as good a deal in the Kyoto Protocol as any Annex I government but then refused to ratify it, arguing that the issue was overblown and its implementation not in Australia's interest. Minister for the Environment and Heritage, David Kemp, explained that "the chorus of critics ... which joined the Kyoto conga line before it had the facts at its disposal, would do better to stop pretending that the Protocol will solve the global greenhouse problem. Kyoto will deliver at best around one per cent of abatement; fails to cover 75 per cent of global greenhouse emissions and does not involve developing countries, who will soon emit over half the world's greenhouse gases. ... This government's policies ... will not sacrifice Australian jobs and investment for the sake of looking green rather than delivering real results."[90] The title of the government's 1997 White Paper on Australia's Foreign and Trade Policy, *In the National Interest*, summed up its approach, including its position in international environmental negotiations.[91] As with Hawke and Keating, domestic responses remained largely rhetorical and voluntary, the 1998 National Greenhouse Strategy representing a good example.

To underline further its dissatisfaction with the EU-dominated UNFCCC process and its affinity with the US position, Australia

---

88. "When Hawke lost a leadership battle with his treasurer Paul Keating in 1991, the environment fell as an issue of government concern, with the dismantling of key institutions and processes despite public protest. ... Policy shifted under Keating, at the first Conference of Parties (COP1) to the UNFCCC in Berlin, in 1995, with Australia fighting targets, emphasizing costs, calling for differentiation, and no longer accepting developed nation responsibility." Harrison, "The Road not Taken," 127-8.

89. Cass, "A climate of obstinacy," 477-8.

90. 'Expert Debunks Government Critics on Kyoto', media release (16 September 2002), as quoted in Frank J. Jotzo, "Developing Countries and the Future of the Kyoto Protocol," *Global Change, Peace & Security* 17:1 (February 2005), 79.

91. "In its pursuit of international action which contributes to sustainable development, Australia must be active in protecting its fundamental national interests." *In the National Interest*, para 114.

joined with the US, Japan, Korea and China in 2005 to establish the Asia-Pacific Partnership for Clean Development and Climate in order to pursue solutions outside the UNFCCC framework, relying on technology sharing and voluntary measures rather than on mandatory targets. As Australian officials saw it, the EU was attempting to globalize its own cost structures by negotiating international environmental agreements. As stated in the government's 2003 White Paper, the European Union was "seeking increasingly to use international environmental negotiations to advance regulatory plans modelled on its own domestic arrangements. Such arrangements can, if applied globally, be excessively prescriptive and often unnecessarily costly to implement, and they may not always result in significant environmental outcomes."[92]

Despite generally positive polling favouring climate action, Australian politicians had not been swayed to the point of taking meaningful domestic action, maintaining that the economic costs were too great. Additionally, Australia's public discussion of climate issues featured an unusually knowledgeable and articulate group of scientists and policy thinkers, many with previous service as government officials. They opposed the global warming mantra and explained why emission control policies would be a fool's paradise for Australia, thus helping to offset the calls for action by greens and other pro-action groups. Over the years Australia became one of the few countries that benefitted from a serious and informed public debate.

The tenor of that debate changed substantially with the return of Labor, first under Kevin Rudd (2007-10), and then Julia Gillard (2010-13), and Rudd again (2013). Their placing of climate change policy at the centre of Australian politics led in 2013 to the return of the Liberal/National coalition under Tony Abbott. Rudd and Gillard embraced climate change mitigation and, in a highly symbolic act, Rudd made ratification of the Kyoto Protocol his first official act, promising that Australia would live up to its obligations. He established a climate commission, welcomed the Garnaut report, and otherwise ensured that Australia would be at the forefront of international climate discussions. Following a 2010 palace coup, Gillard implemented a controversial carbon tax and the promise of a carbon trading scheme, policies that were at the centre of the 2013 federal election and resulted in the subsequent rejection of Labor.

---

92. Department of Foreign Affairs and Trade, *Advancing the National Interests: Australia's Foreign and Trade Policy White Paper*, 2003, 67.

Abbott has started the process of dismantling the previous government's policies and has shown no interest in a successor protocol to the by-now expired Kyoto Protocol.

Developing countries: playing the UN north-south card

For developing countries climate change negotiations present a daunting internal contradiction. As envisioned by the IPCC, climate change will lead to an existential problem of global proportions; without radical action to curb emissions of greenhouse gases, particularly carbon dioxide, life on Earth will become unbearable for many species including humans. Even more to the point, alarmists claim that climate change's malign effects will be experienced more immediately and destructively in developing countries. Low-lying islands and coastal plains will be inundated, and agricultural production will become ever more fragile leading to widespread starvation.[93] At the same time, as discussed in chapters eight and nine, mitigating climate change by reducing fossil fuel use threatens more immediate harm by slowing or even reversing economic growth. Little wonder that developing countries have taken the view from the start that climate change is a problem created by industrialized countries and should be addressed by them. Whatever the long-term existential threat, the threat of economic decline is more immediate.

To complicate matters, climate change emerged as a global political issue coincident with an extraordinarily rapid period of economic growth in the developing countries. Triggered by a small group of countries in East Asia and spreading from there to the rest of Asia and Latin America, the two decades from 1980-2000 witnessed an unprecedented growth spurt. As Indian economist Surjit Bhalla observed, "no matter what statistic is used, ... we have just witnessed the 20 best years in world history – and doubly certainly the 20 best years in the history of poor people."[94] Because this peri-

---

93. Scientists and officials from developing countries had at best played a marginal role in the development of the science and its implications. Few countries had the resources to fund that kind of science. In the work of the IPCC, a few token, developing-country scientists played a role. When those from India, China, and a few other countries began to look at the science, they were as likely to poke holes in the theory as to find supporting evidence, thereby threatening the validity of IPCC modelling exercises. This was not difficult to do.

94. Bhalla, using a more generous base for calculating absolute poverty than that used by the World Bank, concludes that the share of the world's population living below that line at the turn of the millennium was down to less than 15 per cent. In absolute numbers, he calculates that the world's poor peaked at about two billion in 1980, and reached about 1.15 billion by 2000. Using more strin-

od of growth involved rapid industrialization based on the use of fossil fuels, primarily oil and coal, developing countries, by their growth, were becoming significant contributors to rising atmospheric GHG levels and climate change. If the prospect of catastrophic global warming was as imminent as the IPCC claimed, mitigation measures needed to be taken immediately and on a global basis. The United States and Europe were clearly the major sources of GHGs in 1990, the chosen base-line year, but projections indicated that China and India were rapidly catching up, with other developing countries from Brazil and Mexico to Indonesia and South Africa not far behind. By the end of the Kyoto Protocol's first commitment period, developing countries' total emissions might well surpass those of developed countries, more than wiping out any benefits from mitigation measures adopted by developed countries. Not surprisingly, therefore, from the perspective of officials in China, India, Brazil, and elsewhere, the politics of climate change appeared to have a strong tinge of eco-colonialism to it, threatening to throttle their development just as it was taking off.[95]

Nevertheless, developing country leaders were not averse to riding the climate change horse to see if there were opportunities that could be exploited. They were prepared to talk the talk if that would help convince developed countries to walk the walk. As Matthew Paterson and Michael Grubb saw it at the beginning of climate deliberations: "For many developing countries climate change is a Northern issue, both because it is the North that is primarily interested in it and because the North created the problem. ... many developing countries are interested in climate change only to the extent that they see the North as committed to substantial

---

gent definitions of absolute poverty yields even lower numbers. Surjit Bhalla, *Imagine There's No Country: Poverty, Inequality, and Growth in the Era of Globalization* (Washington: Institute for International Economics, 2002), 202 and Tables 9.1 and 9.4.

95. One sub-group within the developing country caucus, of course, saw the issue very differently. The OPEC countries were of the view that the IPCC case had little merit and that weaning the world off fossil fuels by whatever means would threaten their further economic development. Led by Saudi Arabia, they maintained a consistent rear-guard strategy to limit damage to their interests. Similarly, another sub-group led by Brazil and Indonesia maintained that their forestry practices were not up for negotiation. Whatever contribution logging and clear-cutting were making to climate change, decisions about these matters remained within their sovereign authority. As they remonstrated, resource sovereignty had long been a principle of international law and they were not about to abandon it.

transfers of finance and technology to help the South develop without increasing their emissions."[96] As discussions evolved, developing countries fell back on the familiar features of the earlier NIEO discussions, relying on North-South rhetoric rather than on substantive engagement.

While the UN system was adopting the concept of sustainable development as the organizing principle for much of its activity, developing countries made it crystal clear that, for them, the issue remained development. Making development more sustainable was a worthy goal, but not at the expense of economic development, a position they had maintained from the beginning. As Indian legal scholar Lavanya Rajamani notes, the impasse between developed and developing countries at Stockholm in 1972 was only settled "through a tenuous compromise that recognized that environmental protection is not necessarily incompatible with economic development."[97] The emergence of climate change as the principal vehicle for advancing the environmental agenda threatened to undo this compromise. Pakistani academic Adil Najam explains that "the principal and unchanged interest of the South has remained development and a better quality of life for its people; its principal fear, that the North is using environmental issues as an excuse to pull up the development ladder behind it."[98]

That threat became palpable when it was clear that developed countries were determined to pursue a mitigation rather than an adaptation strategy to address the climate issue. For many developing country governments, particularly those from island and coastal states, the need for massive adaptation measures was of more immediate concern than mitigation strategies. To that end, the devel-

---

96. Matthew Paterson and Michael Grubb, "The international politics of climate change," *International Affairs* 68:2 (1992), 297. See also Grubb, "Seeking fair weather: ethics and the international debate on climate change," *International Affairs* 71:3 (1995), 463-96, and the discussion in Bradley C. Parks and J. Timmons Roberts, "Inequality and the global climate regime: breaking the north-south impasse," *Cambridge Review of International Affairs* 21: 4 (December 2008), 621-48.

97. Lavanya Rajamani, "The Changing Fortunes of Differential Treatment in the Evolution of International Environmental Law," *International Affairs* 88: 3 (2012), 607.

98. Adil Najam, "An Environmental Negotiating Strategy for the South," *International Environmental Affairs* 7 (1995), as quoted in Frank J. Jotzo, "Developing Countries and the Future of the Kyoto Protocol," *Global Change, Peace & Security,* 17:1 (February 2005), 77.

oped countries, as the principal source of the GHGs threatening climate stability, had a moral obligation to provide the resources needed to address coastal erosion and flooding, to adopt more efficient and modern sources of energy, and to strengthen the resilience of developing country agricultural production. If the UN talks could deliver that kind of assistance, they would play along.

Pursuant to these considerations, developing countries maintained a coherent strategy that held together quite well throughout the negotiation of the UNFCCC and Kyoto. Sustainable development was a matter of equity and, if there was a burden to be shared, the developed countries needed to sustain their support for a long time before any commitments needed to be made by developing countries.[99] From this perspective, any burden-sharing should take account of two incontrovertible facts: historically, developing countries had contributed only a small fraction of current atmospheric GHG levels, and, on a per capita basis, their emissions would remain well below those of industrialized countries well into the foreseeable future. As Australian development scholar Frank Jotzo sums up, "developing countries are home to five out of the six billion people, but historically have contributed only around one-quarter of the greenhouse gases from energy consumption now accumulated in the atmosphere. Energy use and emissions per person in developing countries on average are still only around one-quarter of those in industrialized countries. Resources for economic restructuring are much more limited in developing countries, with average per capita incomes less than one-quarter of those in the industrialized world. Moreover, adverse impacts of climate change are likely to affect developing countries particularly badly."[100]

---

99. See Adil Najam, "Developing Countries and Global Environmental Governance: From Contestation to Participation to Engagement," *International Environmental Agreements* 5 (2005), 303–321, for a good discussion of the extent of developing country coherence and the evolution of developing country participation in global environmental negotiations from Stockholm through Kyoto.

100. Jotzo, "Developing Countries and the Future of the Kyoto Protocol," 77. Adil Najam, Saleemul Huq, and Youba Sokona point out that "the disparity between the average American who emits just under 20 t[onnes] of carbon dioxide per year while the average Indian emits less than 1 t and the average Chinese around 1.34 t, becomes unimportant. ... From a southern perspective, as the desire for efficiency overwhelms both equity and responsibility, the distinction between 'luxury' and 'survival' emissions is lost and any discussion of global or generational fairness becomes all but mute." "Climate negotiations beyond Kyoto: developing countries' concerns and interests," *Climate Policy* 3 (2003), 223.

During the UNFCCC negotiations, developing countries maintained that climate change needed to be approached within the framework set out by the Brundtland Commission, a position that sat well with the UN and developed countries given that the Convention was one of the two principal results of the Rio Earth Summit. To that end, they had insisted on an equity clause: "The Parties should protect the climate system for the benefit of present and future generations of humankind, on the basis of equity and in accordance with their common but differentiated responsibilities and respective capabilities. Accordingly, the developed country Parties should take the lead in combating climate change and the adverse effects thereof."[101] On that basis, developing countries rapidly ratified the convention and at COP 1 in Berlin in 1995 successfully proposed what became the Berlin Mandate: Parties to the Convention would "not introduce any new commitments for Parties not included in Annex I, but reaffirm existing commitments in Article 4.1 and continue to advance the implementation of these commitments in order to achieve sustainable development."[102] For the next decade and longer, developing countries rigorously maintained that under the UNFCCC, any measures adopted should be "in accordance with their common but differentiated responsibilities and respective capabilities" and that no commitments needed to be taken by non-Annex I countries.

Developing countries had thus successfully shielded themselves from the threat of any mandatory emissions commitments. Even raising the issue became illegitimate and smacked of negotiating in bad faith. For them, all that was left was to negotiate possible adaptation assistance and mechanisms to encourage technology transfers, financial assistance, and joint implementation provisions. In this endeavour they achieved limited success. Various mechanisms were negotiated, including the Clean Development Mechanism, but all were voluntary and, as developing countries learned, promises made at conferences rarely translated into the kind of action and assistance implied. To veterans of the NIEO and related UN negotiations, this was familiar territory: the politics of symbolism, a technique that brought conferences to an end and set matters up for the next iteration. As scholars at the Indian Centre for Science and Environment lamented, "Southern nations continue to be help-

---

101. UNFCCC, article 3:1.

102. COP 1, Decision 1, paragraph 2 (b).

less bystanders in these negotiations . . . Their negotiating strategy has had two simple components: to squeeze small commitments on technology transfer and additional aid from the North; and to use these invariably unfulfilled commitments to stall future negotiations."[103]

By insisting that they would not accept any commitments, developing countries cut off the prospect of selling their unused emissions quotas to those in developed countries with insufficient quotas. To work, a global emissions trading regime required universal participation, a global cap and trade regime, and a world price for carbon. If developing countries had been willing to accept generous emissions caps well above their current levels, this might have been possible, and they could then have sold the excess to countries with tight caps.[104] To ENGOs, this would have been an elegant way to address sustainable development and thus provide for significant resource transfers to developing countries as global emissions were gradually ratcheted down to a more sustainable level. Developing countries did not see it that way and ensured that they would not be caught up in a regime that might gradually throttle their economic development. By 2000, the UN and its members could boast of a treaty and a protocol with nearly universal membership of whom eighty per cent were largely bystanders with a limited stake in their operation. Would the twenty per cent carry on?

#####

---

103. Anil Agarwal et al., eds., 'Introduction', in *Global Environmental Negotiations 2: Poles Apart* (New Delhi: Centre for Science and Environment, 2001), as quoted in Jotzo, "Developing Countries and the Future of the Kyoto Protocol," 79.

104. See the discussion in Jean-Charles Hourcade, P.R. Shukla, and Sandrine Mathy, "Untying the Climate-Development Knot: Economic Options in a Politically Constrained World," in Roger Guesnerie and Henry Tulkins, eds., *The Design of Climate Policy* (Cambridge, MA: MIT, 2008), 75-99.

13 | Rhetoric vs. Reality

I find it far scarier than any "climate change" that leaders of advanced western nations now go around sounding like the kind of apocalyptic loons who used to wander the streets wearing sandwich boards and passing out homemade leaflets.[1]

Mark Steyn, columnist to the world

By the turn of the century the climate movement could look back with some satisfaction at the progress it had made. The United Nations had fully embraced its cause and had made it the central organizing principle for its broader progressive agenda of global equity, social justice, and sustainable development. Virtually all of the world's governments had signed on to the movement and had committed to a Framework Convention on Climate Change under the UN's auspices (UNFCCC). They had also adopted the greenhouse gas theory as official climate science, had declared that the science was settled and that CO_2 was a "pollutant," and had agreed that it was appropriate for governments to pursue policies to mitigate the alleged malign effects of climate change. The media were generally on board and regularly published stories to keep the public alarmed. Progressive elites reinforced the media's message and sought to buttress the view that only a few benighted souls refused to accept "the science." Church leaders, university professors, schools, and even business leaders were supporting the campaign.

---

1.   Mark Steyn, "The Pause and the Cause," *National Review*, February 21, 2014.

Following the 1997 Kyoto Conference, all but a few governments had signed on to a protocol to the UNFCCC that committed them to a global regime to reduce GHG emissions. Compared with previous global salvationist campaigns, the extent of progress in little more than two decades was a singular achievement. The major problem was that only 20 per cent of governments had signed on to reduce emissions, and, even if achieved, their commitments added up to at best a token of what was required to make a difference in total GHG emissions and in their feared climatic impact. Much more was required.

Beginning of the end or end of the beginning?

With the fall of the Soviet Empire and the collapse of the post-war international order, concern about climate change had in many ways filled an ideological vacuum. The looming climate catastrophe had added urgency and a rationale for the transformative steps required to achieve the UN's sustainable development goals. The emerging, human-caused threat to the planet's survival had thus been framed as an existential crisis only to be avoided by a radical recasting of civilization. Similar to the UN's answer to wars, conflicts, and failed states – the responsibility of all members to come to the aid of vulnerable people and protect them wherever conflict broke out – sustainable development transcended national interests. Indeed, for the progressive movement, national interest had no place in a properly organized world order. In this brave new world, enlightened leaders would be guided by norms developed by experts and enshrined in international law. The so-called "international community" would finally come into its own.

Nevertheless, the movement was far from satisfied. A careful examination of its achievements suggested that they were more a matter of rhetoric than of serious policy steps. The UNFCCC was a weak agreement dependent almost entirely on voluntary commitments. Climate science had been declared settled, and most of the world's media treated it as such; yet a growing number of well-credentialed scientists were pointing to the problems with the science. Public support had peaked; the more educated public was the least convinced.[2] More generally, governments were learning that public acceptance was more evident as an abstraction than as a mat-

---

2. See, for example, the work done by Yale's Dan Kahan, "Climate Science Communication and the Measurement Problem," *Advances in Political Psychology* 36:S1 (February 2015), 1-43.

ter of support for the carbon taxes and other measures on the movement's agenda. The Kyoto Protocol had been signed, but it was clear that the US would not ratify it. All the world's governments had ratified the UNFCCC, but the vast majority of the signatories were developing countries that had made their participation conditional on the Berlin Mandate, i.e., acceptance by the industrialized countries that developing countries need not make any commitments to reduce GHG emissions. The Protocol had been widely hailed as a first step towards stabilizing emissions, but most analysts concluded that it would have little impact on the climatic future.[3] Many supportive analysts warned that only a much more serious and universal commitment to reducing emissions would have the desired impact. Some analysts concluded that the emphasis on mitigation was the wrong strategy and that only a reorientation towards a more balanced approach with more emphasis on adaptation – now also called resilience – stood any hope of ever being fully implemented.[4] Such analysis, of course, assumed that climate stability was the ultimate goal. But for many activists climate was little more than a vehicle for an ambitious agenda aimed at establishing global governance on progressive principles.

Ominously for them, national interest had once again come to the fore. In reality, it had always been there, but it had been treated as a retrograde force, one that needed to be tamed and eventually banished from a properly organized global order. Academic analysts, in particular, regarded it as a regressive sentiment.[5] As it

---

3. NCAR scientist Tom Wigley, one of the leading IPCC authors, calculated in 1998 that if every nation met its obligations under the Kyoto Protocol, the earth's temperature in 2050 would be 0.07°C lower as a result. The Cato Institute's Pat Michaels agreed: "A useless appendage to an irrelevant treaty," July 29, 1998, in testimony before the US House Committee on Small Business.

4. See, for example, Roger Pielke, Jr., "Rethinking the role of adaptation in climate policy," *Global Environmental Change*, 8:2 (1998), 159-70, and many subsequent articles; Aynsley Kellow and Sonja Boehmer-Christiansen, *International Environmental Policy: Interests and the Failure of the Kyoto Process* (Cheltenham, UK: Edward Elgar, 2002); and Gwyn Prins and Steve Rayner, "Time to ditch Kyoto," *Nature* 449 (October 25, 2007), 973-7. The idea was fully articulated in the 2010 Hartwell paper: Gwyn Prins et al., *The Hartwell Paper: A new direction for climate policy after the crash of 2009* (Oxford: Institute for Science and Innovation, 2010).

5. As Irving Kristol astutely observed regarding an earlier generation of academics: "Though they continue to speak the language of 'progressive reform,' in actuality they are acting on a hidden agenda: to propel the nation from that modified version of capitalism we call 'the welfare state' toward an economic system so stringently regulated in detail as to fulfil many of the anti-capitalist aspira-

turned out, the most successful post-national movement proved to be Islamism and its handmaiden international terrorism. And when it came to taming climate change, the developing countries' embrace was only for exploitive purposes. Though Russia had blatantly toyed with it, it did so only to pursue its own interests. Governments in Japan, Canada, and Australia had been briefly tempted, but domestic political and economic realities had overwhelmed the desire to sacrifice national interests on the altar of global citizenship. The United States, in gradually coming to terms with the emerging post-Cold War order, had also been making its accommodations. It soon became apparent, however, that the kind of regime that US political forces could accept was not in the cards; a regime that might pass through the Senate could never gain developing country support. Two Democratic administrations were prepared to buy into the new order, but the Senate, whether controlled by Democrats or Republicans, could not be convinced. That left the European Union. Even within the EU, however, regressive national interests had not been wholly vanquished, and, as the pain of implementing even the modest requirements of Kyoto became clear, doubts began to grow. So did resistance to the EU Commission's goals and ambitions. By the second decade of the new century, political and economic reality was gradually overwhelming the prospects for a new norm-based global order.

These and other tensions were gradually undermining support for the alarmist movement and disposing governments to go slowly on full implementation of the legislative and programmatic steps. Time, however, was not in the movement's favour. Alarmists were having to face up to the very inconvenient fact that the global climate was failing to live up to model projections; the earlier, 20-year-observed rise in temperature had ended in 1997-8 and had yet to resume. Efforts to explain the hiatus were not convincing. Alarmist climate scientists had been hoisted by their own petard; they had chosen the global average temperature anomaly as the key metric pointing to anthropogenic warming and insisted that their confidence in that metric was warranted. That metric, however, had refused to do more than fluctuate within a narrow band averaging out to a trend line of zero (see figure 13-1). Meanwhile, the supposed principal driver of global warming, CO_2, continued to increase, breaking the 400 ppm barrier by the end of 2013.

---

tions of the Left." "On Corporate Capitalism in America," in Kristol and Nathan Glazer, eds., *The American Commonwealth* (New York: Basic Books, 1976), 134.

Figure 13-1: Remote Sensing Systems (RSS) global temperature anomaly, 1997-2015, calibrated to a tenth of a degree

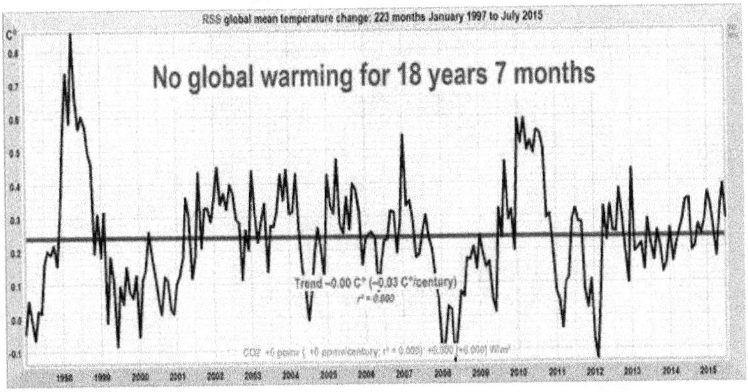

Source: Christopher Monckton, based on RSS monthly satellite global mean surface temperature, *WattsUpWithThat*, August 6, 2015.

At the same time, many of the economies of the more advanced developing countries were growing rapidly, largely due to fossil fuels, particularly coal, the most "polluting" form of energy. The steps taken in the 1990s by many industrialized countries to curb their emissions were falling short of the committed targets, largely because global energy consumption continued to grow at a steady pace, reaching 12,730.4 million tonnes of oil equivalent (Mtoe) by 2013. Renewables accounted for 8.9 per cent of the total, with hydro accounting for 75 per cent of their consumption. (See figure 13-2) Efficiency and other gains had maintained OECD consumption between 5,454.5 (1987) and 5,723 (2006) Mtoe between 1987 and 2012, but non-OECD consumption was rising steadily from 4,143.4 (1987) to 7,197.3 (2013) Mtoe. As a result, global carbon emissions continued to grow (Figure 13-3). Rather than approaching the target of 20 per cent of primary global energy coming from wind, solar, and other modern renewables by 2020, three per cent had proven to be the best that could be achieved, a woefully inadequate number. Investment in renewables in industrialized countries was also disappointingly small. Global GHG emissions were climbing more rapidly than modelled in the 1990s, particularly in newly industrialized countries, but appeared to be having no impact on global temperatures, thus bringing the whole theory into question.

Figure 13-2: 25 Years of World Primary Energy Consumption, 1987-2013

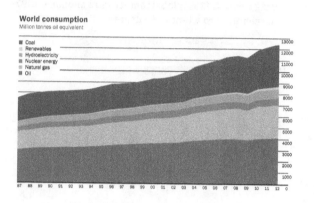

World primary energy consumption continued to grow steadily even as governments agreed on the need to reduce consumption. Oil remains the world's leading fuel, followed closely by rapidly growing coal consumption. Hydroelectric output and power generation from other renewables also continued to grow but from a very low base, reaching record shares of 6.7% and 2.2%, respectively, of global primary energy consumption. Source: *BP Statistical Review of World Energy 2013*, 42.

In the absence of increasing temperatures, movement scientists now began to insist that the climate was becoming *unstable*, that, due to human interference, the globe was seeing more *extreme* weather. Supportive media joined in, obligingly insisting that every hurricane, tropical storm, tornado, heavy snowfall, prolonged drought, severe downpour, or heat wave was the direct result of increased atmospheric greenhouses gases. More sophisticated movement scientists argued that the occurrence of weather extremes was part of nature but that the human contribution added to its extent and frequency, thus leading to more intense storms, rain, snow, droughts, heat waves, and other weather phenomena. The fact that there were no known ways to verify or falsify this new wrinkle in the alarmist theory made it all the more useful as a feature to keep the public concerned and to maintain the pressure on governments for more research and public policy countermeasures. Conveniently ignored was the fact that reliable data to substantiate the claims of more intense and more frequent extremes were limited to only some regions of the globe and did not support this contention. Therefore, movement leaders, in order to convince the public and political leaders that the crisis remained real, pressed for governments to take a much more aggressive approach. The growing penchant for public policy symbolism – rather than effective action – was a wor-

Figure 13-3: Emissions of CO₂ from fossil fuel use and cement production, 1990-2013 (million tons CO^2)

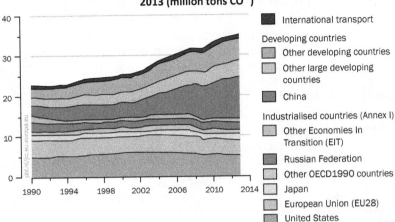

Emissions of CO_2 from fossil fuel consumption have risen steadily since the IEA began to keep statistics. They declined slightly only during the 1981-82 and 2008-09 global recessions. Any reductions in emissions from industrialized countries and post-Soviet transitional economies have been more than replaced by emissions from developing countries, particularly since 1995, when industrialization gained momentum in countries such as China and India.

Source: Netherlands Environmental Assessment Agency, December 16 2014, at www.pbl.nl/en/publications/trends-in-global-co2-emissions-2014-report.

rying development and, if not reversed, would condemn the movement to the same fate as earlier utopian initiatives championed by the UN. Nigel Lawson's conclusion is on the mark:

> The gap between rhetoric and reality when it comes to global warming, between the apocalyptic nature of the alleged threat and the relative modesty of the measures so far implemented (not to mention the sublime disregard of international obligations solemnly undertaken), is far greater than I can recall with any other issue in a lifetime of either observing or practicing politics. ... While the consequence in terms of political posturing may be distasteful, at least it has so far mitigated (to coin a phrase) the damage that would have been done had the more strident governments' deeds matched their extravagant words.[6]

Kyoto's sad reality

In 2005 the movement had achieved a critical milestone, implementation of the Kyoto Protocol. By successfully twisting Russian presi-

---

6.   Lawson, *An Appeal to Reason: A Cool Look at Global Warming* (London: Overlook Press, 2008), 103. If anything, symbolism has become even more prevalent.

dent Putin's arm and promising to clear the way for Russian accession to the WTO, the EU had ensured the number of ratifications required to bring the protocol into force, which it did on February 15, 2005. With that hurdle out of the way, the UN, the UNFCCC Secretariat, the IPCC, ENGOs, and the rest of the movement could concentrate their considerable resources on convincing the signatories to the Framework Convention to negotiate either a further, more ambitious protocol or a new agreement with sufficient teeth to make a difference. To that end every annual conference of the parties to the UNFCCC (COP) from 2005 on was dedicated to advancing that agenda. Signatories to the Protocol were urged to take whatever steps were necessary to implement the required measures and to create an improved base on which to take the next steps.

As this effort unfolded, the climate movement was confronted with the reality of the rule of the lowest common denominator: multilateral negotiations are limited in their ability to craft an agreement attracting universal support by the *least* enthusiastic player. Jon Hovi and Detlef Sprinz conclude that environmental agreements are particularly prone to this limitation.[7] Any climate mitigation agreement, by definition, addresses a problem of the global commons and cannot be resolved satisfactorily on any but a universal basis, as had become painfully clear by the time Kyoto entered into force. With only industrialized countries committed to reducing emissions, the program was bound to fail. Ironically, at the same time that the Protocol came into force, the combined emissions of developing countries began growing much more rapidly and surpassed those of Annex I countries within a few years. Kyoto was further undermined by the US failure to ratify it. Given the absence of Senate approval and continued opposition, a wholly new treaty was necessary. Such a treaty might be able to avoid the stumbling block of the Berlin mandate. Because the US was not a signatory to Kyoto, the UN needed to embark on a double track: a track engaging all UNFCC members in an effort to negotiate a treaty capable of attracting universal support, and a second track limited to Kyoto signatories negotiating either a successor to or an extension of the Protocol. These were not mutually exclusive processes and added further confusion to what was already a convoluted approach.

A treaty that is to be binding on democratic nations is always a bold project. By insisting on a treaty that would impose burdensome

---

7.   Jon Hovi and Detlef F. Sprinz, "The Limits of the Law of the Least Ambitious Program," *Global Environmental Politics* 6:3 (August 2006), 28-42.

mitigation requirements, the UN, the IPCC, and the climate movement more broadly had set ambitious objectives. Adding to those objectives the UN's desire to use the treaty as an instrument to advance its broader, utopian agenda took what was already a daunting task and doomed it to failure.[8] As John Fund observes, many climate activists "despair that the kind of radical change they advocate probably won't result from the normal democratic process. It's no surprise then that the rhetoric of climate-change activists has become increasingly hysterical."[9] To put it more kindly, the architects of the climate agenda were pursuing a path that was more likely to result in symbolic statements than in effective, enforceable commitments, a harsh reality they could not afford to admit.

Many academics have devoted considerable energy to analysing, describing, and proposing global regimes to address climate change. As two Harvard academics, Joseph Aldy and Robert Stavins, put it, their goal is "to help identify the key design elements of a scientifically sound, economically rational, and politically pragmatic post-2012 international policy architecture for addressing the threat of climate change."[10] Their efforts are focused on mitigation-based agreements. The problem is that not one such scheme is negotiable for the simple reason outlined by the Research Council of Norway: measures that are politically feasible are ineffective and measures that would be effective are politically infeasible. Put into economic terms, as we saw in chapters eight and nine, the reality is that measures that are affordable are ineffective and measures that are effective are not affordable.[11] Various other criteria have been proposed by academic analysts to assess possible recommendations,

---

8. At COP 18 in Doha, Qatar, UNFCCC Executive Secretary Christiana Figueres reminded the press that "what is occurring here, not just in Doha, but in the whole climate change process is the *complete transformation of the economic structure of the world.*" [emphasis added] Press Conference, Doha, Qatar, November 26, 2012.

9. John Fund, "The Crumbling Climate-Change Consensus," *National Review Online*, September 21, 2014.

10. Joseph E. Aldy and Robert N. Stavins, eds., *Post-Kyoto International Climate Policy: Implementing Architectures for Agreement* (Cambridge: Cambridge University Press, 2010), 2.

11. Research Council of Norway, August 13, 2014. See a review of a range of proposals in Joseph E. Aldy, Scott Barrett, and Robert N. Stavins, "Thirteen Plus One: A Comparison of Global Climate Policy Architectures," *Climate Policy*, 3:4 (2003), 373-397. See also Roger Guesnerie and Henry Tulkens, *The Design of Climate Policy* (Cambridge, MA: MIT Press, 2008).

and almost all invariably turn out to be applicable only to the make-
believe world of models. All these analyses are preoccupied with
such issues as:

- *The right environmental outcome,* i.e., a lower, steady-state level of
  atmospheric concentrations of GHGs, a stable climate, and a
  stable biosphere – essentially unknowable and unnatural results
  given the dynamic and chaotic characteristics of the carbon cycle, climate patterns, and other earth systems; outcomes can be
  modelled on the basis of a wide range of assumptions none of
  which approximate real-world circumstances; modelling tends
  to inflate confidence in the ability of governments to successfully negotiate agreements.

- *Full participation and compliance* – given the extent to which the
  problem has been determined to be global in scope, only near-
  universal participation will ensure an effective result; failure to
  attract widespread and meaningful participation will weaken
  compliance and effectiveness; as experience to date has made
  clear, the fact that developing countries have clearly stated their
  unwillingness to commit to mitigation measures of any kind
  makes the pursuit of a universal agreement virtually impossible.

- *Equity* – Climate policy raises considerations of both intergenerational, international, and intra-national equity; a whole discipline has developed to parse the various nuances of climate equity; to many environmentalists, the most equitable solution is
  to saddle the current population of advanced economies with
  significant costs that may benefit future generations in both developed and developing countries, most of which will be better
  off than today's generation and have access to more advanced
  technologies. Such a proposed outcome, which may score high
  among academics, environmentalists, and climate alarmists, is
  virtually non-negotiable.

- *Dynamic efficiency* – Any public policy that entails intervention
  in the economy will lead to both gains and losses. In the case of
  environmental agreements, a dynamically efficient outcome ensures that the planet and its human population will, in aggregate, gain net benefits from the measures implemented to limit
  climate change; again, measuring such an outcome depends on
  unknowable variables. As we saw in earlier chapters, there are
  too many variables, leading to high levels of uncertainty, to

make it possible to calculate meaningful costs and benefits over the required time period.

- *Cost-effectiveness,* i.e., ensuring that any agreement focuses on the least costly among policy alternatives, e.g., a carbon tax rather than a cap-and-trade measure. Conventional economic analysis has long identified such measures but faces continuing criticism from environmental and other interests that value certainty over cost-effectiveness. In a sober analysis of the political economy of the Kyoto Protocol soon after it was negotiated, Scott Barrett presciently pointed out that "the Kyoto Protocol must produce for its parties a favourable benefit-cost ratio or else it will either never enter into law or it will collapse."[12] As time would show, the naïve belief that economics should not influence environmental policy doomed the Protocol from the start and will continue to bedevil environmental policy making.

- *Policy flexibility in the face of new information* – Given the large number of imponderables that feature in the science and economics of climate change, policy analysts insist that any agreement needs to be crafted in such a way as to ensure that the regime can accommodate new developments in science, economics and, most of all, technology, that point to the need for significant changes. Experience demonstrates that inertia features as a major characteristic of public policy, whether implemented internationally or domestically; governments will only reluctantly admit that they are wrong and need to shift strategy; interests committed to the status quo will fight tenaciously to maintain it.

The Kyoto Protocol fails on all of these basic criteria and was still the best that could be negotiated in 1997. Political will has declined since then. Kyoto's broad support among alarmists and environmentalists more generally was largely due to its symbolic value and to the hope that it might lead to something more meaningful. Since then, prospects of that happening have become increasingly slim with every passing year as governments have faced the reality that the modest climate change experienced to date has not constituted a crisis and that there is unlikely to be one in the foreseeable future. Resorting to the precautionary principle has been an effort to evade serious risk-cost-benefit analysis. Whatever problems emerge in the coming years, local and national adaptation is increasingly

---

12. Scott Barrett, "Political Economy of the Kyoto Protocol," *Oxford Review of Economic Policy* 14:4 (1998), 37.

recognized to be more than sufficient. The need for a top-down, highly intrusive, transformative order commands support only among a dwindling herd of hard-core alarmists and their utopian progressive cheerleaders.

The fundamental problem with much of the academic analysis is that it *assumes* that there is a serious climate problem – if not in the immediate future, then within the politically foreseeable medium term – one that needs to be addressed with a far-reaching treaty to curb emissions by changing energy and other consumption patterns over a foreseeable time frame. None of the academic analysts are prepared to accept the fact that the planet faces *at worst* a marginal, gentle increase in warming superimposed on a norm of natural cyclical patterns. There is no real-world evidence that the modest increase in atmospheric greenhouse gases since the middle of the 20th century has had any malign effects on the climate system. On the contrary, there is only evidence of the benign impact of the slight post-1950 warming. Indications of extensive malign impacts come from flawed climate models fed improbable assumptions which, in turn, are then fed into highly imaginative and creative impact studies, again based on computer models.

The road to Copenhagen

In order to succeed, therefore, the movement realized that it would need to turn up the heat. There was obviously not enough grass-roots support for the previously proposed agreement. The most immediate obstacle was the US administration of George W. Bush. Over his remaining three years, he was subjected to a relentless campaign from the media, from some of his G-8 colleagues, from the progressive wing of the Congress, ENGOs, the alarmist science community, and business leaders, all hoping to benefit from opportunities that could emerge with the right policies.

While this campaign was widely supported, Bush was not without allies. Internationally, many of his reservations were shared by three of his G-8 colleagues, Canada, Japan, and Russia. In Canada, the election of Conservative Stephen Harper in 2006 had put an end to the ineffective policies of his predecessors. Early in his mandate, Harper made it clear that the Liberals, by accepting targets at Kyoto but doing nothing about them, had put Canada in an impossible position. Therefore, the new federal government would make good-faith efforts to pursue responsible energy policies but would run out the Kyoto clock and not sign up to any successor. Japan, which had ratified the Protocol in order to preserve national self-esteem, was in no position to make good on its commitments and, like Canada, would eventually withdraw from Kyoto. Russia, which had ratified for reasons unrelated to the Protocol's objectives, made clear during this period that its goal was to grow its economy by using its rich storehouse of fossil fuels, both at home and for export to earn foreign exchange; it had no interest in any further commitments, either for itself or for others. Thus the much-vaunted G-8 solidarity amounted to little more than happy talk from the European members and the Commission. Pressure on Bush was similarly limited to the European members, principally Tony Blair and later Gordon Brown, and the EU Commission president. In Washington, any urging from major developing countries such as India, China, and Brazil had little impact, given their determination not to accept any disciplines on their own energy consumption and choices.

Various hearings and legislative initiatives were pursued in both the US House and Senate, but none gained sufficient support to pass. Legislation sponsored by Democratic Congressmen Henry Waxman and Ed Markey, the most ambitious of the various bills considered, passed the House by a narrow margin in 2010 after much log-rolling by Speaker Nancy Pelosi but was never considered by the Senate. A similar bill sponsored by Senators John Kerry and Barbara Boxer was never considered by the Senate as a whole because the Democratic leader of the Senate, Harry Reid, knew he did not have the votes. With the House under Republican leadership after the 2010 mid-term election, no climate legislation has since reached the floor of either the House or Senate. The Obama administration has talked bravely about the need to do so but so far has made little effort other than to take increasingly aggressive executive actions, particularly through the EPA, actions that stretch the president's authority and widen the rift between the president and the Congress. [13] They can all be rescinded with a stroke of the pen by the next president.

The 2008 election had promised to bring the stalemate to an end in the United States. Both presidential candidates had expressed strong concern about the perceived fast-approaching crisis and vowed to take steps to bring the US onside. Their ignorance, how-

---

13. See Alan Carlin, *Environmentalism Gone Mad: How a Sierra Club Activist and Senior EPA Analyst Discovered a Radical Green Energy Fantasy* (Mt. Vernon, WA: Stairway Press, 2015) for an account at the frontline of Obama green policy making.

ever, was as breath-taking as that of Al Gore, albeit neither could achieve Gore's levels of rhetorical hysteria and hypocrisy; and neither was addicted to his lavish, self-indulgent lifestyle. With the election of Barack Obama, movement leaders believed it might be possible to get supporting legislation through Congress. Obama's appointments of a quartet of movement sympathisers – Carol Browner (White House environmental adviser), John Holdren (White House science adviser), Steven Chu (Energy Secretary), and Lisa Jackson (Director of the Environmental Protection Agency) – were perceived by many as breathing new life into the file. In any event, these appointments turned out, again, to be more symbolic than real with little impact on long-term public attitudes. Without congressional support, Obama soon learned what was possible: he could use his office to signal his support for the alarmist movement but could not effect the fundamental changes it sought. Efforts to use the regulatory power of the EPA and presidential administrative power have resulted in significant pushback from industry and the public and a growing wariness in Congress and industry in response to the mounting impact of a smothering regulatory state.

To show his support for the movement, Obama injected billions of dollars of stimulus funding into the development of wind, solar, and other alternative energy projects, only to discover that many of them were far from viable. Starting in the second half of 2011, an increasing number of participating companies filed for bankruptcy, sending both the stimulus funds and the hopes for lower emissions down the drain. In the discussions leading up to Copenhagen, US negotiators benefitted from somewhat more flexible instructions than had been the case up until 2008, but at the UNFCCC meetings it became clear that Obama was prepared to accept only what could be achieved politically.

In Europe, the EU had the bit fully in its teeth, with the Commission setting ever more unrealistic targets and sending out directives to reduce consumption and switch to renewables. Denmark and Germany took it all to heart, building windmills and solar arrays and closing down coal thermal plants. The same was true in the UK. Following their European counterparts, first Tony Blair, then Gordon Brown, and more surprisingly, even David Cameron, became committed to tackling global warming. The extent of British political enthusiasm was on full display in the debate on the 2008 Climate Bill; there was very little opposition except from a small band of well-informed sceptics with backgrounds in science. Rather than being toned down during debate, the bill was made even more unrealistic, with MPs enthusiastically endorsing, 465 to 5, a legal obligation to cut British emissions to 80 per cent below 1990 levels by 2050 which, given normal economic growth, would require a 90 per cent or more reduction in current levels. Over the next few years first Brown and then Cameron announced various plans to meet these obligations at a cost of billions to UK taxpayers and consumers for years to come.[14] Not surprisingly, British public support had peaked by this point and began its steady descent following passage of the Climate Act.

ENGOs and climate scientists also turned up the heat, releasing a barrage of model studies, press releases, and similar propaganda. In 2005-6, Al Gore embarked on a lecture tour to convince the public of the approaching Armageddon. In order to reach a wider audience, the tour became the basis of a documentary and a book: *An Inconvenient Truth: The Planetary Emergency of Global Warming and What We Can Do About It*. Both were widely distributed, particularly in schools, and, despite their exaggerations, had the desired impact of raising public concern. The New York *Times* called it "lucid, harrowing and bluntly effective" and believed it "could goad the public into reading more scholarly books on the subject, and it might even push awareness of global warming to a real tipping point – and beyond."[15] Adding to its value as propaganda, Hollywood decided that the movie deserved an Oscar, while the Swedish Nobel Committee awarded Gore, along with the IPCC, the Nobel Peace Prize.[16] The theme of the alarmist movement had a new resonance. It was now "worse than we thought," a phrase that became a regular media feature. The fact that virtually all of these prognostications originated in computer-based model studies was rarely mentioned. In the lead-up to the 2009 Copenhagen COP, media-fed climate change derangement syndrome reached a fever pitch.

---

14. See Christopher Booker, *The Real Global Warming Disaster* (London: Continuum, 2009), 11-12, and Rupert Darwall, *The Age of Global Warming* (London: Quartet, 2013), 267-9. Both authors provide detailed accounts of developments in the years leading up to Copenhagen, particularly in the UK.

15. Michiko Kakutani, "Al Gore Revisits Global Warming, With Passionate Warnings and Pictures," New York *Times*, May 23, 2006.

16. Gore has also trained and certified thousands of volunteers to present his *Inconvenient Truth* lecture, including UNFCCC Executive Secretary Christiana Figueres. All are trained not to respond to questions from sceptics, as I learned at a 2009 church basement session.

The release of the Stern Review in 2006 followed by the IPCC's fourth assessment report in 2007 added further to the movement's credibility with the media and some governments, both emphasizing that the science was settled and that the time for action was now. Nevertheless, criticism of the deceptions, fabrications, and exaggerations featured in both reports gradually became widely known and led to a hint of balance in some media accounts and, as discussed in chapter five, prompted various official enquiries into the movement's scientific and economic credibility.

Ironically, both reports lent increasing credibility to the sceptical perspective. While ignored by the mainstream media, conservative media such as Fox News and the *Wall Street Journal* in the United States, the UK's *Daily Telegraph* and *Spectator,* Canada's *Financial Post,* and Australia's *Herald Sun* and *Daily Telegraph* reported on the growing number of sceptical scientists. Coverage began, for example, of their meetings at the annual Conference on Climate Change sponsored by the Heartland Institute, their contributions to the reports of the Non-governmental International Panel on Climate Change (NIPCC), their participation in weblogs, and their publication of books and articles. The divide within the business community was also becoming more apparent, with one group seeing opportunity and profit in climate change measures and others finding the movement bizarre and capable of destroying whole industrial sectors, even undermining the global trade and payments system. The publication of Nigel Lord Lawson's 2008 book, *An Appeal to Reason: A Cool Look at Global Warming,* provided both sceptical business leaders and politicians with a carefully researched and reasoned assessment of the claims of anthropogenic climate change proponents in language that they could understand. Lawson went on to found the Global Warming Policy Foundation, bringing together prominent political and business leaders, scientists, and economists dedicated to providing the media, business, and political elites with a reasoned, evidence-based alternative to the propaganda provided by governments, ENGOs, and the alarmist movement.

The campaign for an ambitious treaty took on a surreal tone when, on September 15, 2008, Lehman Brothers failed, ushering in a more urgent crisis. For the next few years, global leaders and legislators needed to pay attention to something much more serious. Meetings of the G-8 and G-20 now focused on addressing a potentially global, financial disaster. Whatever their anxieties about warming, the financial crisis now tended to overshadow, if not obscure it. In the United States, Gallup's regular poll on climate change began its steady descent. It has stayed on that course. As US pundit George Will averred: "On graphs tracking public opinion, two lines are moving in tandem and inversely: The sharply rising line charts public concern about the economy, the plunging line follows concern about the environment. A recent Pew Research Center poll asked which of 20 issues should be the government's top priorities. Climate change ranked 20th."[17] In recent polling Americans continued to rank climate change at low levels, with economy-related issues near the top. Results in Canada, Australia, the UK and other countries, were similar.

In the lead up to Copenhagen, chances of reaching a full-fledged treaty looked more and more unrealistic, largely because of the continuing North-South divide, one apparent from the beginning at COP 1 in Berlin and repeated at COP 2 in Montreal. That divide had been reiterated at the beginning of the second phase of negotiations at COP 11 in Montreal (2005) and papered over with various procedural resolutions at subsequent meetings. At COP 13 in Bali (2007), governments had agreed to the Bali roadmap, which had set up the two-track approach to the negotiations of future commitments. Developing countries were mostly interested in track one while developed countries set their sights on track two because, from their perspective, only a comprehensive agreement could solve the global problem. Developed countries could agree to transitional differentiated commitments (as called for by the UNFCCC) but insisted that the next phase needed to include global targets that extended to all participants. Extending the Kyoto Protocol sidestepped that requirement by staying within the Berlin mandate.

At Poznan in December 2008 (COP 14), as well as at a series of preparatory meetings both before and after Poznan, officials had failed to make any meaningful breakthroughs, the many positive press releases from the UN and ENGOs notwithstanding.[18] Brazil, South Africa, India, and China, the so-called BASIC group, continued to insist that Annex I parties agree to a second commitment period under Kyoto and opposed any agreement requiring reductions of their own emissions. Little progress had been made in narrowing those differences, and the draft texts going into Copenhagen were

---

17. George Will, "Apocalypse Now? Highly Unlikely," Washington *Post*, February 15, 2009.

18. See Daniel Bodansky, "The Copenhagen Climate Change Conference: A Post-Mortem," *American Journal of International Law*, 104 (2010).

heavily bracketed, making an agreement there virtually impossible. The thrust of the draft treaty suggested that UN ambitions continued to be focused on a highly interventionist, transformative agreement rather than on a more focused agreement limited to immediate efforts to reduce global emissions.[19]

Up to Copenhagen, movement scientists argued that the warming was taking place faster than had been thought and increasing at an even greater speed. Their evidence could be found in the models, once again manipulated to achieve the required results. Their claims could be easily dismissed. There had been no measurable global warming for ten years, and cold weather records were being established around the world. There was resort now to the use of a new word in the alarmist vocabulary, *abnormal* climate change. Its introduction raised even greater suspicions among all but the fully convinced.

On the eve of COP 15, the issue of climate science received another blow. Five months earlier, scientists at the University of East Anglia, home of the Climatic Research Unit (CRU), had baldy admitted that the raw data from which the CRU's climate data series had been built – and which provided the basis for much of the work of the IPCC – had been destroyed and could no longer be retrieved in order to verify the credibility of the homogenized data. At the end of November, a hacker, possibly an insider reacting to the chicanery that had characterized CRU responses to Freedom of Information requests, released a voluminous file of e-mails and documents retrieved from a CRU server. As discussed in chapter five, the e-mails revealed that respondents were behaving abominably in order to further their cause while, in addition, assiduously discrediting non-conforming analysis and blocking its publication. Even the more supportive branches of the media had to admit that such behaviour had been egregious and had undermined the credibility of the science.

At Copenhagen, despite the attendance of more than 100 heads of government and more than 45,000 delegates, advisers, media, and ENGOs, leaders could do no more than cobble together a last-minute political agreement, the Copenhagen Accord. As the *Wall Street Journal* editorialized: "the idea that global warming represents the gravest threat to humanity has become totemic in much of the

---

19. The draft text prepared for ministerial consideration at Copenhagen, UN document FCCC/AWGLCA/ 2009/INF.2, extended to 172 pages and included extensive alternative proposals.

world, a belief invested with religious fervour and barely susceptible to rational discussion, let alone debate. Yet it remains telling
how quickly a sense of reality has reasserted its cold grip in light of
the choices Copenhagen now brings starkly into view."[20]

Key elements of the Accord included the following, all of which
would be virtually impossible to translate into meaningful and enforceable treaty language, reflecting similarly unrealistic but politically appealing pledges set at the national level in many developed
countries:

- A commitment to limit increases in global temperature to no
  more than 1.5° or 2° Celsius above pre-industrial levels, a symbolic goal that had become integral to public discourse;
- A set of three overlapping goals expressing the desire to reduce
  atmospheric emissions and concentrations of greenhouse gases
  by: 1) limiting atmospheric concentrations of greenhouse gases
  to 350 or 450 ppm – a rather imprecise but symbolic target, particularly since concentrations passed 400 ppm a few years later;
  2) a long-term goal to reduce global emissions of carbon dioxide
  by 50 per cent below 1990 levels by 2050; and 3) a hope to
  achieve a target date for the peaking of global emissions. All
  three could only be achieved if both developed and developing
  countries took drastic steps to reduce emissions sooner rather
  than later; and
- A "collective commitment" for developed countries to provide
  "new and additional resources ... approaching $30 billion" in
  "fast-start" money for the 2010-2012 period, balanced between
  adaptation and mitigation, and a longer-term collective "goal"
  of mobilizing $100 billion per year by 2020 from all sources
  (public and private, bilateral and multilateral). The pledges
  were linked to "meaningful mitigation actions and transparency
  on implementation" by developing countries.[21]

Instead of adopting the Accord, the Conference *took note* of it,
leaving its future status unresolved. Many members did subsequently make national pledges, including leading developing countries such as China and India, which both adopted carbon *intensity*
targets: pledges to reduce emissions per unit of GDP by 40-45 per

---

20. "The Copenhagen Concoction: The UN's climate confab runs into the reality of
    costs and science," *Wall Street Journal,* December 8, 2009.

21. The text of the Accord, UN document FCCC/CP/2009/L.7, 18 December 2009,
    can be found at unfccc.int/resource/docs/2009/cop15/eng/l07.pdf.

cent (China) and by 20-25 per cent (India) from 2005 levels by 2020, targets that would be met in any event as both economies modernized. These pledges, however, were political and had a rather weak legal status. Similar pledges had become an important part of international climate theatre: symbolically important and functionally meaningless. While the Accord provided a basis for concluding the meeting, no substantive issues were resolved. It introduced none of the outstanding issues that would have required a future agreement. It rather proposed what amounted to a third track defined by voluntary arrangements. Even if officials were to continue to concentrate on track 1, extending the Protocol, and track 2, negotiating a new agreement, only the third track might, politically, have realistic prospects. The Accord also allowed President Obama to hurry home before an early winter snowstorm had closed down Andrews Air Force Base outside Washington, the beginning of another harsh winter.

The outcome was a huge blow to EU leaders, whose collective egos had thought that a more ambitious agreement could be reached. They had little input into the Accord negotiated by Obama with the leaders of Brazil, South Africa, India, and China. The UN made the best of its failure to conclude an agreement by acting as a clearing house for the voluntary pledges called for by the Accord. After 21 years of high-energy negotiations, it finally dawned on European leaders what had been clear to others for some time, that the UNFCCC process was not likely to achieve its goals. If anything were going to be accomplished, it would have to be on the basis of a fresh start, that is, building on whatever unilateral measures the leading countries were prepared to undertake. Talks would continue, as they always do at the UN, but the UFCCC/Kyoto process was looking increasingly threadbare. If the media continued to flog governments for their lack of ambition, the politicians, in their turn and with straight faces, could only insist that they were making progress. The game would go on, but future COPs would feature non-political officials. Leaders vowed to stay home, and only the occasional minister ventured to make an appearance at later meetings. As the years went by, each COP shrank a little more in the public eye. Like UNCTAD a generation earlier, the UNFCCC may well become another moribund part of the UN.

Not all in Europe were disappointed with the outcome. Dominic Lawson at *The Sunday Times* urged others to join him in toasting the negotiators: "By failing so spectacularly, they have presented us with a wonderful Christmas present. All we have to do is open it."[22] Over at *The Telegraph,* its leading political pundit was even more dismissive: Copenhagen "has marked the beginning of the landslide that is collapsing the whole AGW imposture. The pseudo-science of global warming is a global laughing stock and Copenhagen is a farce. In the warmist camp the Main Man [Pachauri] is a railway engineer with huge investments in the carbon industry. ... Al Gore, occupant of the only private house that can be seen from space, so huge is its energy consumption, wanted to charge punters $1,200 to be photographed with him at Copenhagen. There is a man who is really worried about the planet's future."[23] On the other side of the Atlantic, Walter Russell Mead was even more dismissive: "The leading green political strategy – to stop global warming by a treaty that gains unanimous consent among 190 plus countries and is then ratified by 67 votes in a Senate that rejected Kyoto 95-0 – is and always has been so cluelessly unrealistic as to be clinically insane."[24] Margaret Wente at Canada's *Globe and Mail* was equally blunt: "After two weeks of chaos, the talks collapsed in a smouldering heap of wreckage. ... Copenhagen was not a political breakdown. It was an intellectual breakdown so astonishing that future generations will marvel at our blind credulity. Copenhagen was a classic case of the emperor with no clothes."[25]

Since Copenhagen, negotiations have continued on all three tracks. The Kyoto track, conducted in the Ad Hoc Working Group on Further Commitments for Annex I Parties under the Kyoto Protocol (AWG-KP), was initiated in 2005 and does not include the United States. The second track is being pursued in the Ad Hoc Working Group on Long-Term Cooperative Action under the Convention (AWG-LCA), with the mandate to develop a comprehensive outcome. This includes: a shared long-term vision, mitigation commitments, or actions by developed countries; nationally appropriate mitigation actions by developing countries; financial arrangements, measures to address adaptation and technology trans-

---

22. "The Copenhagen farce is glad tidings for all," *The Sunday Times,* December 20, 2009.

23. Gerald Warner, "Copenhagen climate summit: 'most important paper in the world' is a glorified UN press release," *The Telegraph,* December 19, 2009.

24. Walter Russell Mead, "The Greening of Godzilla," *The American Interest,* August 28, 2010.

25. Margaret Wente, "Can environmentalism be saved from itself?" *Globe and Mail,* November 27, 2010.

fers, and a system for measurement, reporting and verification. The UN continues to collect and publicize Copenhagen pledges. Little progress has been made, however, although rhetoric and angst continue.

After Copenhagen

Concluding the UNFCCC and Kyoto had been facilitated by the unique, geopolitical moment of the collapse of the Soviet Empire and the end of the Cold War. That moment did not last much beyond the turn of the century. The Europeans, by overreaching, had lost the full participation and leadership of the United States and earned its suspicion. The developing countries had similarly lost the goodwill of the United States, goodwill that had been essential to many of their post-war gains. Trying to bully the United States had been a recipe for failure and ensured that the climate movement would not last without a thorough rethinking. Burden sharing can work, but only when the principal bearer of the burden enjoys a hegemonic position and is willing to take it on. The US carried the burden of much of the research for the first 30 years of the climate movement's campaign, but the ensuing 20 years of criticism and overreach had led to the systematic alienation of many US political leaders.

At Copenhagen, the rift between developed and developing countries was also on full display. Obama was correct in looking to the leaders of the BASIC group to determine whether something could be agreed. European leaders had become thoroughly compromised by their adherence to their "save-the-planet" mantra. Progressives in the developed countries wanted to save the planet by emphasizing the UN's program of sustainability. Developing countries demurred. The rift had become increasingly clear in the post-1992 period, starting at COP 1 when the two groups reached an uneasy truce in Berlin. The Western progressive elites and developing country leaders were also at odds over population control, an article of faith among the first but regarded with great suspicion by the latter.

Despite the debacle at Copenhagen, the UN kept trying, not only at successive COPs, but also at other events. Twenty years after the Rio Earth Summit, for example, the UN convened a follow-up Earth Summit, dubbed Rio+20, again in Rio de Janeiro. Some 45,000 flocked there for the ten-day mega-conference. The international environmental community built up a considerable head of steam, insisting that Rio 2012 opened up an opportunity for a new resolve to the commitments made at Rio 1992. All the major international EN-

GOs and many national ones devoted considerable resources to hyping the summit, insisting that the time had come for rhetorical commitment to be translated into action.

Forewarned, the media largely gave the summit a miss, having learned that interest had peaked at Copenhagen and that subsequent gatherings were now largely serving to encourage the faithful.[26] While a considerable number of heads of government did come for the final two days, the only G-8 leader in attendance was France's newly elected socialist president, François Hollande. Leaders from Canada, Germany, Italy, Japan, the UK, and the US were conspicuous by their absence. As is typical for such events, UN officials and national delegates had prepared a long and ambitious agenda and a far-reaching document built on the 1992 Conference's Agenda 21: "The Future We Want." The draft version, tabled at the beginning of the conference, a 49-page resolution, and a 283-paragraph wish list, reads like a dream come true for every progressive cause in the world, except that most of those groups lambasted it as insufficient and lacking in detail and specificity.[27] In any event, political leaders neither adopted nor rejected the document, an ambiguous result allowing them to go home and NGOs to interpret the results however they wanted.[28]

The weather seemed also to be on the side of the sceptics. Europe and North America experienced one harsh winter after another with record snowfalls and low temperatures, all of which made a

---

26. *The Economist*, for example, concluded: "Despite the presence of scores of heads of state, Rio+20, as the summit is known, was expected to produce the weakest imaginable commitment to greening the global economy. On many counts – including a wishy-washy commitment to look beyond GDP as a measure of progress – it represented little or no progress on the summit's more illustrious forebear, the 1992 Earth Summit, at which important agreements on combating climate change and biodiversity loss were reached." "Green business," June 22, 2012.

27. See, for example, Jonathan Watts and Liz Ford, "Rio+20 Earth Summit: campaigners decry final document," *The Guardian*, June 23, 2012.

28. Even if adopted, implementing the UN's "Future" strategy would have required both global governance and the development of a new man, one prepared to believe in the full alarmist creed and to forego all individual needs and desires. Al Gore preaches the warming gospel, but his lifestyle shouts his failure to live up to it. Agenda 21 represented the culmination of a series of lower profile UN conferences. As noted earlier, the UN has learned to advance its agenda on the basis of frequent conferences, each of which moves the agenda forward ever so slowly. National governments have not done much to implement the Agenda, but many local governments use it to justify local interventionist programs. Most people admit they know little about it.

mockery of long-term forecasts provided by the IPCC and other official sources. At the same time, weather extremes failed to materialize. Global hurricane intensity declined, and the last major Atlantic
storm had been Katrina (2005) nearly a decade before. Sandy (2012)
had been demoted to a tropical storm when it made landfall. Chip
Knappenberger suggests in summarizing that "there are a lot more
cases of non-extreme weather than there are of extreme weather,
and as many or more cases to be made for weather catastrophes *averted* by conditions 'consistent with global warming' than
caused by it. So if you want to play the all-weather-is-influenced-
by-global-warming game, you are going to lose. Best bet would be
to stick with the science, which for most types of extreme weather
events and for most places indicates that a definitive link between
event characteristics and human-caused climate change has not
been established."[29]

Five more COPs have met since 2009: Cancun (16), Durban (17),
Doha (18), Warsaw (19), and Lima (20). Each conference involved
extensive preparatory meetings of all the parties as well as meetings
of smaller technical working groups. The 2015 Paris Conference
(COP 21) is now being hailed as the next Copenhagen, the conference that will finally adopt an agreement putting the world on a
permanent path to sustainability. Don't bet on it. Cancun basically
confirmed the dual-track process and set up a Green Climate Fund,
mainly on the basis of unfulfilled pledges. At Durban, parties to the
Kyoto Protocol – by the end of the conference without the United
States, Canada, Russia, and Japan – agreed to extend its life beyond
2012 but did not make any new commitments. The Conference also
adopted the Durban Platform for Enhanced Action, which calls for
parties to develop a protocol, another legal instrument, or an agreed
outcome with legal force under the convention applicable to all parties. This new treaty is scheduled to be adopted at COP 21 and implemented by 2020. At Doha, Qatar, delegates devoted most of their
energies to housekeeping measures, completing some of the preparatory work needed for COP 21, finalizing rules for Kyoto's second
commitment period, and otherwise keeping the process alive. Kyoto
was extended to 2020 when, based on the UN agenda, the new
agreement will come into force. In Warsaw in 2013 (COP 19), delegates agreed to the establishment of a loss-and-damage mechanism

---

29. Chip Knappenberger, "Should We Credit Global Warming When Disasters
     Don't Happen?" Center for the Study of Science, Cato Institute, September 19,
     2014.

obliging developed countries to help pick up the tab for damage which had resulted from severe climate events, an important step towards achieving a climate agreement in Paris but little more than recognition of current practice. Lima was devoted to preparing for the "breakthrough" anticipated the following year in Paris, which remains on a "then-a-miracle-occurs" schedule. In anticipation, COP 19 watered down expectations by agreeing that only those countries that are "ready" will be required to make commitments in 2015.[30]

In preparation for Paris, all participants in the negotiations are expected to table submissions with the UNFCCC secretariat setting out their visions of how the 2015 agreement should look. Among the major players, the UK and the US have tabled their visions. The UK released its submission in September 2014 in a glitzy brochure setting out some ambitious goals, not only for itself but also for other participants.[31] It wants to be confident that all countries will commit to making emissions cuts "consistent with their national situation, the opportunities available to them, and both their relative past and future contributions to climate change" based on their different levels of economic development. The report suggests that the EU, US, Japan, Canada, and Australia cut their emissions by 2030 from 34 to 74 per cent of their 1990 levels. Similar reductions are suggested for China (33 per cent from 2010 to 2030 levels) and India (the peaking of emissions by 2030). Each country is to set its own targets and choose a set amount each year. The document insists that the deal must be legally binding.

The US vision is a more prosaic typescript calling for every country to outline:

- its timeframe for emissions cuts,
- the year that cuts will be measured against, known as the base year,
- the sectors and greenhouse gases included in the pledge,
- the level of cuts as a percentage,
- details on how it will measure the reductions, and
- what policies it already has in place to help cut emissions.

---

30. Each of these conferences maintains its own website replete with its own logo, hopeful statements, and useful information for delegates.

31. *Paris 2015: Securing our prosperity through a global climate change agreement*, Written Statement to Parliament, September 9, 2014.

The US wants each country to review its emissions reduction pledges on a regular basis and believes that the deal should be "designed to promote ambition, gain widespread participation, and be durable in the long term."[32] The US submission is vague on the arrangement's legal nature, but President Obama has already urged an accord that would not require Senate ratification.[33] This novel approach would allow him to neatly sidestep Senate approval, serve a symbolic purpose, and provide cover for continued aggressive executive action. Rupert Darwall concludes that the US is now pursuing an agreement containing "no legally binding commitments but voluntary pledges, notified under the auspices of the 1992 convention and underpinned by a regime of 'naming and shaming' those who don't live up to them. There is a big problem with this. It has already been tried, and it failed."[34] It is difficult to imagine why other countries would accept such a proposal, particularly countries with legal systems that make international agreements entered into by the executive binding on the government.

Other countries have yet to make their own submissions, but press reports are not promising. Canada, Japan, Russia, and Australia, while continuing to participate in the UNFCCC process, have not shown any signs of warming up to a binding agreement with ambitious obligations. All four leaders declined Ban Ki-Moon's invitation to attend the September 2014 Climate Summit and sent lower ranked officials in their place. All have averred that the only arrangement they will consider is one that includes fair obligations for all signatories, not at some point in the future but at the time it comes into force. Canada has been more specific than other countries. [35] Its terms for committing to a negotiated settlement require an agreement that: 1) includes meaningful and transparent commitments from *all* major emitters leading to constructive and ambitious global action; 2) balances environmental protection with economic prosperity; and 3) maintains a long-term focus.

An indication of Japan's priorities is its pledge in 2010 under the Copenhagen Accord to *cut* its emissions by 25 per cent by 2020. Three

32. US submission to UNFCC, September 2014, at Unfcc.int.

33. Coral Davenport, "Obama Pursuing Climate Accord in Lieu of Treaty," New York *Times,* August 26, 2014.

34. Rupert Darwall, "Goodbye, Treaty," *National Review Online,* August 29, 2014.

35. Canada's Priorities at COP 19 at climatechange.gc.ca/default.asp?lang=En&n=FAC4B67D-1.

years later, it revised the pledge to a 3.8 per cent *increase* by 2020. Russia continues to play cat and mouse, remaining engaged in the process but coy about any future commitments. Australia's prime minister, Tony Abbott, has been busy undoing previous climate policy commitments and continues to lack enthusiasm for the UN process. He told reporters during a June 2014 visit to Ottawa that "Stephen Harper and I are like-minded on this. The argument is not about climate change – the argument is about the best means to respond to climate change and I believe that carbon taxes and emissions trading schemes are the wrong way to go."[36]

The BASIC group of countries continues to have no interest in any immediate emissions reduction commitments on their part. All four have initiated policies aimed at making the energy sector more efficient and have included climate change considerations in their national policies. None, however, are inclined to have their economic development hobbled by international agreements. All have been courted by the US and the EU, but none of these *démarches* have provided much comfort for the movement. Narendra Modi, the newly elected prime minister of India, ran a campaign based on a vision of Indian prosperity: more power, electrified cities, and more prosperous citizens. He is unlikely, therefore, to soften India's negative stance on emissions reduction commitments. Modi has been enigmatic in his press comments on climate change, stating that "at best, you have the right to milk nature. You can milk a cow, but you can't kill the cow. Climate change? Is this terminology correct? The reality is this that in our family, some people are old. ... They say this time the weather is colder. And, people's ability to bear cold becomes less. We should also ask is this climate change or have we changed."[37] The other BASIC countries have been equally cautious. It is unlikely that there will be sufficient change by 2015 in Paris to break the 30+ years of deadlock.

Smaller developing countries continue to hold out the forlorn hope that the negotiations will provide them with more financial aid and some mitigation measures. They have the force of numbers on their side – there are as many as a hundred such countries – but the politics of international negotiations does not provide much confidence that whatever is achieved will be as helpful as they would like.

---

36. "Tony Abbott finds climate comfort in Canada," *The Australian,* June 9, 2014.

37. "Modi waxes eloquent on women and goddesses," *The Hindu,* September 2, 2014.

EU strategists continue to hold out for a binding treaty embracing all major emitters. They see the annual cycle of UNFCCC negotiating conferences as a means to gradually fragment the coalition of developing nations with promises of billions of dollars of climate aid. They hope that India and China, with substantive commitments extended to them, will eventually buckle under international pressure and sign on to a comprehensive treaty. More sober analysts see this as a strategy to keep the Commission engaged on a project rapidly losing public support.[38] Over the past year or two, one European member government after another has softened its stance on climate change and has begun the politically painful task of beginning to dismantle the misguided policies mandated by the Commission and implemented at the national level. Walter Russell Mead observes that "the EU, where disingenuous politicians are forced to demagogue green issues because addlepated proportional representation rules empower the lunatic eco-fringe in key countries, ratified Kyoto, and Americans were then treated to years of vainglorious Euro-puffery about the nobility, the wisdom and the self-sacrificial idealism of the cutting edge eco-warriors of the Green Continent."[39]

That puffery is rapidly waning as European leaders learn that the expensive programs initiated have had little impact on CO_2 emissions and none on the climate but are rapidly leading to the de-industrialization of Europe. Growing awareness of the crisis hung like a pall over the 2014 Davos meeting. Benny Peiser, director of the UK-based Global Warming Policy Foundation, reports that [former] EU climate commissioner Connie Hedegaard is increasingly isolated. "It is becoming obvious that Europe's unilateral climate policy, the whole green agenda of the last 20 years, has turned into an unmitigated fiasco. Even EU leaders are beginning to accept reality."[40] Spain, a pioneer in subsidizing renewable energy, can no longer afford it and is reneging on its commitments to investors.[41] Germany's Economics Minister and Vice Chancellor to Angela Merkel, Sigmar Gabriel, announced in an April 2014 speech that the country's once highly ballyhooed transformation to renewable en-

---

38. "Worse than useless; European climate policy," *The Economist*, January 25, 2014.

39. Walter Russell Mead, "Kyoto Fraud Revealed," *The American Interest*, October 14, 2010.

40. Michael Bastasch, "EU retreats on global warming while US pushes ahead," *Daily Caller*, January 1, 2014.

41. "Spanish Lesson for Obama: Green Energy Transition Unaffordable, May Soon Crash," Global Warming Policy Foundation, May 31, 2014.

ergy, the so-called *Energiewende* model adopted by a number of countries worldwide, is "on the verge of failure."[42] The *Sunday Times'* Dominic Lawson was told by a former EU commissioner that Germany's energy policy "is the stupidest policy ever proposed by any post-war German government – unless, that is, the purpose is to destroy Germany's hard-won competitiveness." Lawson adds that UK prime minster David Cameron, in an unguarded moment, told a group of Conservatives in November 2013 that he would be focusing on getting rid of all the "green crap."[43] *Investors' Business Daily* concludes that "the Europeans made nearly a $100 billion wrong bet on renewable energy, and their economies and citizens have taken a big hit. Now they've awakened to their mistakes. The shame is Washington is still slumbering."[44]

Robert Bryce, now a senior fellow at the Manhattan Institute, details the extent to which Europe's energy policy – similar to that of Ontario in Canada – has hobbled the economy of each member state. He points out that "in 2012, the average household price of electricity among the 27 members of the European Union was $0.26. In Denmark – a country that many wind-energy proponents admire – a kilowatt-hour of electricity for residential customers cost $0.41. In Germany – by far, Europe's biggest economy, largest electricity consumer, and most important manufacturer – the cost was $0.35. In Spain, another country that has provided huge subsidies to the renewable-energy sector, it was $0.29. Meanwhile, in the US, the average residential cost of electricity in 2012 was about $0.12."[45] Europe's share of global CO_2 emissions is steadily declining, less because of its climate policies and more because of its steady deindustrialization.

Changes at the Commission may portend a less aggressive stance by the end of 2015. The new Commission (2014-19) is headed by Luxembourg's Jean-Paul Juncker, long-serving prime minister of Luxembourg and the first Commission president elected initially as

42. P. Gosselin, "Angela Merkel's Vice Chancellor Stuns, Declares Germany's 'Energiewende' to be on the 'The Verge of Failure'!" *No Tricks Zone*, April 27, 2014.

43. Dominic Lawson, "Admit it, greenies: the game's up for renewable energy," *The Sunday Times*, January 27, 2014.

44. Editorial, "Europe Starts To Run Not Walk Away From Green Economics," *Investor's Business Daily*, February 5, 2014.

45. Robert Bryce, *Maintaining the Advantage: Why the US Should Not Follow the EU's Energy Policy*, Center for Energy Policy and the Environment, Manhattan Institute, February 2014, 1.

a member of the European Parliament under the banner of the centre-right European People's Party and then selected by the Parliament as its preferred candidate.[46] His Commission includes an experienced conservative Spanish politician, Miguel Arias Cañete, as commissioner for both climate and energy. It may be too early to tell, but the appointment suggests a less dogmatic stance on climate change. Cañete appears to be more pragmatic than his predecessor, Hedegaard, a Danish centre-right politician who was prone to giving Al Gore-like statements but proved ineffective as a political leader.

In these circumstances, it is not surprising that each conference continues to be dominated more by rhetoric than by action, and the public knows it. This is not to suggest that billions of dollars have not been spent on a host of programs and policies in each of the OECD countries and in a number of developing countries, many of them of questionable utility. Governments, however, have focused on incremental policies and programs and have avoided taking the kinds of steps that the alarmist community believes are necessary. The Kyoto Protocol, which was hailed as a major first step, has barely survived and has not produced an ambitious successor to take its place. Cap and trade and carbon taxes that have been implemented have failed to achieve more than the symbolic sacrifices that they largely are. Philip Stott points out that "the real killer crunch with the public will be the ever-widening credibility gap between the rhetoric of the politicians and their appalling hypocrisy and abject failure to be able to reduce CO_2 emissions in any meaningful manner. As coal-fired power stations are re-established around the world – without carbon-capture-and-storage (CCS), of course; as natural gas flourishes; as tar-oil sands are developed; as car ownership continues to grow; and as politicians, and pie-in-the-sky academics, fly to ever more conferences, the public will call time on the Great 'Global Warming' Charade. After all, people now have real problems to face, like losing their jobs and being unable to pay the rent or the mortgage."[47]

---

46. The EPP is centre-right by European political standards, similar to the contemporary Conservative leadership in the UK. In North American terms, Juncker would be a Centrist Liberal or Democrat. José Manuel Barrosso, Juncker's predecessor, was also affiliated with the EPP, despite his background as a Maoist student leader.

47. Philip Stott, "Crunch Time for 'Global Warming'," March 29, 2009, at his now defunct weblog, *Clamour of the Times*.

Table 13-1: UN MyWorld global poll

UN MyWorld Poll, based on six priorities chosen from among 16 choices

	Total Votes	Per Cent
A good education	5,120,912	67.6
Better healthcare	4,320,099	56.2
Better job opportunities	4,194,514	54.6
An honest and responsive government	3,521,132	45.8
Affordable and nutritious food	3,129,071	40.7
Protection against crime and violence	3,003,180	39.1
Access to clean water and sanitation	2,850,407	37.1
Support for people who can't work	2,587,761	33.7
Better transport and roads	2,453,008	31.9
Equality between men and women	2,420,680	31.5
Reliable energy at home	2,277,806	29.7
Freedom from discrimination and persecution	2,159,182	28.1
Political freedoms	2,104,775	27.4
Protecting forests, rivers and oceans	2,000,005	26.0
Phone and internet access	1,891,937	24.6
Action taken on climate change	1,585,879	20.6
Total Votes Cast at Poll	7,680,657	100.0

Source: MyWorld Analytics at data.myworld2015.org, July 18, 2015.

Waning public interest

UN officials are well aware that the climate issue has rapidly faded as a global priority. Following Rio+20, the UN sponsored a poll on people's priorities from around the world, largely in developing countries, asking them to list six issues from a list of sixteen which are most important to them and their families. As of July 18, 2015, 7,680,657 votes had been cast. A good education, better health care, and better job opportunities ranked one, two, and three. Action on climate change was running dead last, included by only one in five people among their six priorities.[48] (Table 13-1) Based on the analytics available at the website, people from 196 countries have voted, with a similar pattern from all countries.

Recent public opinion polls in North America and Europe similarly continue to place global warming dead last among public concerns. Popular culture's awareness of climate change/global warming peaked around 2007 and has declined steadily since then. (See Figure 13-4) Following his 2012 re-election, President Obama changed

---

48. "MYWorld is a United Nations global survey for citizens. Working with partners, we aim to capture people's voices, priorities and views, so world leaders can be informed as they begin the process of defining the next set of global goals to end poverty." data.myworld2015.org.

Figure 13-4: Climate Change in Popular Culture

Mentions of 'Global
Warming' and 'Cli-
mate Change' in
87,000 Movies and
TV Shows, 1970-
2013.
Source: Francie Di-
ep, "Pop Culture
Mentions of Global
Warming Have
Plummeted Since
2007," *Popular Sci-
ence,* September
18, 2014.

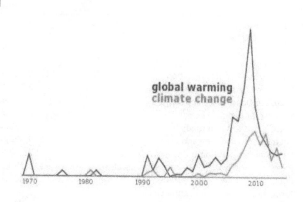

his science, energy, and environment team, retaining only Holdren. Undeterred by lack of progress during his first term and frustrated with his failure to get Congress to act, he announced a major new Climate Action Plan in June, 2013, that would bypass the Congress and rely on administrative action. In his promotional speech, his rhetoric also suggested that he had caught a new variant of climate derangement syndrome. After delivering a litany of exaggerations and falsehoods to a student audience at Georgetown University, Obama urged them: "to educate your classmates, your colleagues, your parents, your friends. Tell them what's at stake. Speak up at town halls, church groups, PTA meetings. Push back on misinformation. Speak up for the facts. Broaden the circle of those who are willing to stand up for our future."[49] In response, the *Wall Street Journal* editorialized, "President Obama's climate speech on Tuesday was grandiose even for him, but its surreal nature was its particular hallmark. Some 12 million Americans still can't find work, real wages have fallen for five years, three-fourths of Americans now live paycheque to cheque, and the economy continues to plod along four years into a quasi-recovery. But there was the President in tony Georgetown, threatening more energy taxes and mandates that will ensure fewer jobs, still lower incomes and slower growth."[50]

49. Speech at Georgetown University, June 25, 2013, announcing the President's Climate Action Plan.

50. "The Carbonated President," *Wall Street Journal,* June 26, 2013.

A year later, it was US Secretary of State John Kerry's turn to reveal his full commitment to the climate change and sustainability mantras.[51] His audience was a group of Indonesian students. Given his dismal record in dealing with real problems within his own portfolio, from the Russian invasion of Ukraine and the threat of Iran's becoming a nuclear nation to the rise in Syria, Iraq, Yemen, and Libya of a more violent and threatening form of Islamist terrorism, Kerry's re-enlistment in the war on climate change suggested a man looking for a post-political career.[52] Without any knowledge of the science but with a deep commitment to the ideology of sustainability, Kerry told the students that "the science of climate change is leaping out at us like a scene from a 3D movie. It's warning us; it's compelling us to act. And let there be no doubt in anybody's mind that the science is absolutely certain. It's something that we understand with absolute assurance of the veracity of that science." He then went on to exhibit the depth of his ignorance: "Try and picture a very thin layer of gases – a quarter-inch, half an inch, somewhere in that vicinity – that's how thick it is. It's in our atmosphere. ... In modern times, as human beings have emitted gases into the air that come from all the things we do, that blanket has grown thicker and it traps more and more heat beneath it, raising the temperature of the planet." He then went on to repeat a litany of exaggerations, assuring students that solutions were at hand if only political leaders would show the necessary resolve, and then adding a new rhetorical excess: "climate change can now be considered the world's largest weapon of mass destruction, perhaps even, the world's most fearsome weapon of mass destruction."[53] Kerry appears to be wholly out of step with political sentiments in the rest of the world, which are rather tamping down expectations and stepping back

---

51. Kerry, together with his wife Teresa Heinz, founded *Second Nature* in 1993, an organization devoted to making sustainability a campus issue. Its mission "is to proactively build a sustainable and positive global future through initiating bold commitments, scaling successful actions, and accelerating innovative solutions among leadership networks in higher education." His understanding of sustainability is that of the UN, a movement dedicated to the economic and social transformation of the world along progressive lines. It is the sponsor of the American College and University Presidents' Climate Commitment, signed by 685 presidents by the end of July 2015. Its publications track the wildest claims of the climate movement. See www.secondnature.org.

52. The extent of Kerry's failure on the diplomatic front is well detailed in Seth Mandel, "Foggy Mind at Foggy Bottom," *Commentary*, September 1, 2014.

53. Secretary of State John Kerry, "Remarks on Climate Change," Jakarta, Indonesia, February 16, 2014.

from many of the ruinous policies already adopted. Indonesia's president, having been forewarned, declined to meet with him.[54]

Ironically, by the end of 2012, US emissions of greenhouse gases had fallen to approximately their 1990 levels, a fact further undercutting the sense of urgency advanced by the President and his environmentalist constituency.[55] The reason lay not in effective conversion to alternative sources of energy but in the continuing impact of an anaemic economy and the widespread availability of natural gas, available now through the increasing use of fracking technology. Its development has occurred on private land. Vastly more could be found on state lands that remain closed to exploration. As a result, US utilities are rapidly converting from coal to gas, which is more efficient than coal and releases less carbon dioxide. The market response to both phenomena has been predictable: the price of natural gas has steadily declined, depressing the price of other fossil fuels and reducing market interest in alternative sources of energy.

The President's decision to go all out on a new climate initiative was all the more curious in the face of faltering certainty about climate change among even supportive media. A week before his speech, the *New Republic* had carried an article by Nate Cohn pointing out that "the so-called scientific consensus on global warming doesn't look much like consensus when scientists are struggling to explain the intricacies of the earth's climate system, or uttering the word 'uncertainty' with striking regularity. ... In the current political climate, debates about things like climate change are carried out in broad-brush assertions. The challenge for scientists is that the more they understand the climate system, the more complex it gets, and the harder it gets to model with precision – not to mention making the kinds of sweeping statements the news cycle requires."[56] In commenting on Cohn's article, the *Economist*, which has long been a staunch propagator of alarmism, warned that "as a rule, climate scientists were previously very confident that the planet would be warmer than it is by now, and no one knows for sure why it isn't. This isn't a crisis for climate science. This is just the way sci-

---

54. Michael Gordon and Coral Davenport, "Kerry Implores Indonesia on Climate Change Peril," New York *Times,* February 16, 2014.

55. US Energy Information Administration (USEIA), "Energy-related carbon dioxide emissions declined in 2012," *Today in Energy,* April 5, 2013.

56. Nate Cohn, "Explaining the Global Warming Hiatus," *New Republic,* June 18, 2013.

ence goes. But it is a crisis for climate-policy advocates who based their arguments on the authority of scientific consensus. ... There's no way around the fact that this reprieve for the planet is bad news for proponents of policies, such as carbon taxes and emissions treaties, meant to slow warming by moderating the release of greenhouse gases. ... They will become harder, if not impossible, to sell to the public, which will feel, not unreasonably, that the scientific and media establishment has cried wolf."[57]

Even one of Obama's former climate advisers, Steven Koonin, Undersecretary for Science in the Energy Department during Obama's first term, has muted his former enthusiasm, pointing out that the science that is settled is not germane to policy making while the science that remains unsettled is critical to determining an appropriate policy path. Koonin, a physicist, wrote: "the idea that 'Climate science is settled' runs through today's popular and policy discussions. Unfortunately, that claim is misguided. It has not only distorted our public and policy debates on issues related to energy, greenhouse gas emissions and the environment. But it also has inhibited the scientific and policy discussions that we need to have about our climate future."[58] Koonin adds that alarmist models differ on such a large scale that they are worse than useless and mislead both scientists and policy makers. Ross McKitrick agrees, observing that "the claim that the 'science is settled' never made sense for climate, and is now completely untenable in light of the wild profusion of new theories to explain the mismatch between models and data."[59] Unfortunately, as Richard Lindzen explains, "climate alarm belongs to a class of issues characterized by a claim for which there is no evidence, that nonetheless appeals strongly to one or more interests or prejudices. Once the issue is adopted, evidence becomes irrelevant. Instead, the believer sees what he believes. Anything can serve as a supporting omen."[60] One would hope, probably in vain, that these and other informed opinions would make their way into the climate folders prepared for Obama and Kerry.

---

57. "A Cooling Consensus," *The Economist,* June 20, 2013.

58. Steven Koonin, "Climate Science is not Settled," *Wall Street Journal,* September 19, 2014.

59. Ross McKitrick, "Climate change and the false case for haste," *Orange County Register,* September 17, 2014.

60. Richard Lindzen, "Reflections on Rapid Response to Unjustified Climate Alarm," Center for the Study of Science, Cato Institute, September 18, 2014.

In these circumstances, the wise course would be to accept faltering political momentum and take a breather in order to give scientists time to sort out the many contradictions and contraindications that have become all too evident over the past decade. As McKitrick insists, "The arguments for hasty action on greenhouse gases do not hold up. This is a case where there is a positive value to waiting for the policy-relevant scientific information we know will be emerging in the next few years, before committing to a long-term course of action."[61] Unfortunately, however, that has been the problem from the beginning. The movement insisted that it knew enough in 1988 to set the policy train in motion and by now has too much invested to admit that it may be on the wrong track. For the UN and the progressive movement, climate change was never about the science; rather, it was always about blazing a trail to utopia.

Georgia Tech's Judith Curry, addressing a Washington audience, cautioned that the movement needs to address two fundamental problems of its own creation: From the beginning, movement scientists oversimplified the complexity of the climate system in order to promote politically attractive, simple solutions that could be pursued to resolve global warming; and, working with supporters at the UN and elsewhere, movement scientists put the policy cart in front of the scientific horse. They decided on a story line and then started gathering facts to support it. Now that much more is known about the science, the solutions look increasingly inappropriate and even unnecessary. In her words, "the 1992 UNFCCC treaty was signed before the balance of evidence suggested even a discernible human influence on global climate, which was assessed by the IPCC Second Assessment Report in 1995. The 1997 Kyoto Protocol was implemented before we had any confidence that most of the warming was caused by humans." Most scientists still do not have that confidence.[62]

Neither the UN nor the alarmist movement, however, is prepared to acknowledge that it has painted itself into a corner. Both must carry on and see if they can convince enough governments to take the next step. As Rupert Darwall explains: "The game now is to keep the process going indefinitely. [Western governments] have committed their countries to immensely costly de-carbonization policies. Without the prospect of coordinated global action, any objective justification for them vanishes. ... From the phase of trying and

---

61. Ross McKitrick, "Climate change and the false case for haste."
62. Judith Curry, "JC at the National Press Club," *Climate Etc.*, September 16, 2014.

failing, the climate-change talks have evolved to creating a phony impression of a horizon lit with the prospect of global agreement to justify costly de-carbonization programs at home. Developing nations are happy to go along with this as long as it doesn't hurt their economies. In other words, the talks have become an exercise in deception."[63]

In the lead-up to COP 21 in Paris, the UN has begun another all-out push to conclude a comprehensive, top-down climate treaty, replicating the Kyoto Protocol but extending it to more countries and with tougher emissions reduction targets. For the annual 2014 September opening of the UN General Assembly, Secretary-General Ban Ki-moon invited leaders to join him in the exercise of leadership on the climate file. As Darwall saw it, "at a perilous juncture in world affairs and with the international system visibly breaking down – the first forcible annexation of European territory since Hitler's war; a bunch of fanatics and psychopaths, perpetrators of a double genocide, seizing control of a vast swath of the Levant, and American leadership exhausted – the UN secretary general, Ban Ki-moon, is convening a summit of world leaders to discuss, of all things, global warming."[64] Obama accepted the invitation, as did the UK's David Cameron, but leaders from Russia, China, India, Canada, Germany, and Australia declined. All have begun to sense that the UN's handling of the file is irresponsible. At the UN, climate change may be the defining issue for our time, but elsewhere more pressing problems are demanding serious political attention.

On a positive note, Ban Ki-moon announced that he had recruited another Hollywood celebrity as the newest UN Messenger for Peace. On the eve of the Summit, Leonardo DiCaprio joined Al Gore and hundreds of other celebrities to lead a People's Climate Parade, choking the streets of New York and snarling traffic in order to impress leaders of the urgency of the moment.[65] Similar parades took

---

63. Rupert Darwall, "The UN's Climate-Summit Charade," *National Review Online,* September 18, 2014.

64. Darwall, "The UN's Climate-Summit Charade."

65. Like Gore, DiCaprio symbolizes the hypocrisy of the movement. Passionately arguing for regulations to force people to live simpler lives, DiCaprio forgets to mention that he "took at least 20 trips across the nation and around the world [in the first eight months of 2014] alone - including numerous flights from New York to Los Angeles and back, a ski vacation to the French Alps, another vacation to the French Riviera, flights to London and Tokyo to promote his film *Wolf of Wall Street,* two trips to Miami and a trip to Brazil to watch the World Cup. And those were just the trips where he was spotted in public. Additionally, Di-

place in many cities around the world. As the science has become more troubled, hysteria has risen, not only among the useful idiots who populate marches and rallies but even among some world leaders and elites. At the New York gathering, according to Christopher Booker, an "increasingly soporific audience was treated to an endless queue of world leaders, ... [making] leaden little appeals for humanity to take urgent action to halt global warming. The purpose of this special meeting, summoned by that dim little nonentity Ban Ki-moon, was to issue a desperate last-minute call for a legally binding treaty in Paris next year, whereby they would all agree to save the planet through an 80 per cent cut in those CO_2 emissions, which are inseparable from almost all the activities of modern civilization."[66]

In keeping with the growing unreality of global climate talks, G-7 leaders boldly pledged, once again, at their 2015 meeting, to "decarbonize" the world by the end of the century, with a few equally symbolic intermediate goals for 2030 and 2050. None are likely to be in office in 2030, let alone later. Peter Foster saw their pledge as "a clear signal – if any more signals are needed that December's COP meeting is doomed to failure: that is, failure to agree to crippling and likely pointless policies. Such failure was also adumbrated by yet another pre-COP draft text-writing bash in Bonn this week, where the G77 and the Alliance of Small Island States bemoaned the lack of progress in shipping them more cash."[67] Judith Curry agreed, "fiddling with distant targets is a great way to generate headlines, but doesn't do much to affect policy and emissions themselves; at best it's marginally irrelevant, at worst it lets people feel good without doing anything."[68]

Caprio owns at least four homes: two apartments in New York and mansions in Hollywood and Palm Springs. He also recently sold an estate in Malibu for $17 million. And this summer, he spent his World Cup vacation on the fifth largest yacht in the world, a 482-foot behemoth owned by Mansour bin Zayed Al Nahyan - a billionaire oil tycoon from the UAE." Michael Zennie, "'You can either make history or be vilified by it:' Leo DiCaprio lectures UN on climate change," *Mail Online*, September 24, 2014. Other luminaries in attendance were Bill McKibben, Naomi Klein, and Vermont Senator Bernie Sanders. Klein has found another cause to help push her nihilist, anti-capitalism message.

66. Christopher Booker, "Dreary climate summit was surely their saddest fiasco yet," *Daily Telegraph*, September 27, 2014.

67. Peter Foster, "G7 kicks the climate can off the planet," *Financial Post*, June 10, 2015.

68. Judith Curry, "What matters (and doesn't) in the G7 Climate Declaration," *Climate Etc.*, June 11, 2015.

The EU and the UN continue to hope for a miracle; LDC leaders pray that it will lead to a pot of gold at the end of the rainbow. Four of the five major emitters, however, are willing to continue to talk the talk but at the same time ensure that the outcome is no more than a variation on what happened in Copenhagen in 2009. The UN never appears to learn, and the EU leaders, to hope, that their vision will be shared by the rest of the world. By the end of 2015, at the Paris conference, Obama will be near the end of his term, Putin will still have more immediate, and disturbing, ambitions, and neither India nor China will want restrictions on their economic development efforts anymore than they did six years ago. In sum, it will be harder for everyone to sound convinced that the climate crisis is still with us.

If there is a case for medium-term climate action, it is one limited to adaptation by businesses and local governments to ensure that infrastructure and business plans vulnerable to any significant changes are reviewed and necessary steps taken. In many parts of the world, this will involve very little. In other parts, particularly the poorer areas, more may need to be done, including perhaps some international assistance. Even here, however, there needs to be much more evidence than has to date been presented by the reports from the UN or the climate alarmist community. Both are convinced that there have already been momentous impacts from less than 20 years of minor warming from 1977-1997 and none since. Whether from the perspective of science, economics, or politics, the case for action remains weak.

Nevertheless, the fall of 2015 will see renewed efforts to gain public and political support ahead of the Paris conference. As with the lead up to the Copenhagen conference, studies aimed at raising alarm are being released for consumption by a gullible media. The keepers of the land-based temperature series have been busy adjusting their algorithms, lowering earlier temperatures in order to demonstrate that the planet has continued to warm over the past fifteen years. The keepers of the satellite-based series have stubbornly resisted. By the end of July 2015, both series showed that the pause was now well into its 18th year (see figures 4-7 and 13-1).

In September, UN Secretary-General Ban Ki-Moon hosts another climate summit in New York to raise the political temperature. This time he has succeeded in enticing Pope Francis to join world leaders in order to help him in emphasizing the moral imperative of concluding a climate treaty by the end of the year. The Pope has also declared September 1ˢᵗ to be a World Day of prayer for the Care of Creation, an occasion he urges other religious leaders to support and make into an annual event. [69] Unlike many of his liberal colleagues, who have welcomed the pope's surprisingly progressive outlook on popular new age issues, Washington *Post* columnist George Will sees him, more perceptively, as a man "who stands against modernity, rationality, science and, ultimately, the spontaneous creativity of open societies in which people and their desires are not problems but precious resources." In his view, "Pope Francis embodies sanctity but comes trailing clouds of sanctimony. With a convert's indiscriminate zeal, he embraces ideas impeccably fashionable, demonstrably false and deeply reactionary. They would devastate the poor on whose behalf he purports to speak — if his policy prescriptions were not as implausible as his social diagnoses are shrill." [70]

Meanwhile, negotiators are busy working out the details of a "binding" treaty with "voluntary" commitments, thus setting a new bar for UN-based political double-speak. They are coalescing around a deal that would commit every country to restricting their GHG emissions but would bind none to any specific targets. As Walter Russell Mead sees it: "It is, in its way, a perfect solution. It is a legally binding agreement to disagree about carbon. Each country is legally bound to do exactly what it wants. It is an elegant escape from the climate treaty dilemma; the diplomats can report success, and the world can turn away from a misbegotten negotiation that has already caused humiliation ... and can only yield impasse as long as it is pursued. To produce a failure but to call it success is one of the oldest political tricks in the book; this is probably now the best case scenario for the global treaty movement." [71]

✦✦✦✦✦

---

69.   Letter of His Holiness Pope Francis for the establishment of the "World Day of Prayer for the Care of Creation," *Libreria Editrice Vaticana,* August 6, 2015.

70.   George Will, "Pope Francis fact-free flamboyance," Washington Post, September 18, 2015.

71.   Walter Russell Mead, "Greens Prep for Talks as Dips Plan a Fudge," The American Interest, July 18, 2015. See also Matt Ridley, "The Paris Climate Summit," *The Times,* July 30, 2015.

14 | Immorality Pretending to Virtue

Climate alarm belongs to a class of issues characterized by a claim for which there is no evidence ... [and which is] characterized by profound immorality pretending to virtue.[1]

Climate Scientist Richard Lindzen

Of all tyrannies, a tyranny sincerely exercised for the good of its victims may be the most oppressive. It may be better to live under robber barons than under omnipotent moral busybodies. The robber baron's cruelty may sometimes sleep, his cupidity may at some point be satiated, but those who torment us for our own good will torment us without end for they do so with the approval of their own conscience.[2]

C.S. Lewis, *God in the Dock* (1949)

In announcing the latest iteration of his Clean Power Plan, President Obama told the media that he is "convinced that no challenge poses a greater threat to our future and future generations than a changing climate."[3] Fully implemented, his plan might lead to a 0.03°C reduction in global temperature by the end of the century. It would also destroy thousands of jobs and increase costs throughout the US economy. Such is the two-edged sword of the climate change file: political theater, costly programs, and little if any perceptible impact on a climate that is largely determined by natural forces. Climate change's greatest danger, therefore, will come from the solutions being imposed rather than from the impact of whatever

---

1. Richard Lindzen, "Reflections on Rapid Response to Unjustified Climate Alarm," Center for the Study of Science, Cato Institute, September 18, 2014.

2. C.S. Lewis, "The Humanitarian Theory of Punishment," in *God in the Dock* (Grand Rapids, MI: Eerdmans, 1970), 292.

3. "Remarks by the President in Announcing the Clean Power Plan," at *Whitehouse.gov*, August 3, 2015.

changes may be in store as the result of forces far beyond human control. The most serious threat to the future lies in the extent to which the alarmist movement has undermined public confidence in reputable science and scientists and in the extent to which gullible governments led by fearful politicians have been prepared to let the climate change story be used to undermine national economies, public welfare, and democratic governance.

After 25 years of endless talk, it should by now be clear to all but the morally blind that alarmist scientists, mendacious environmentalists, utopian progressives, and overreaching politicians are trying to steer the world towards a dark future, offering economy-destroying solutions for at worst a marginal problem. Their quest is to reverse the development of human freedoms and well-being in order to "save" a planet that has withstood more than 4.5 billion years of assaults much more momentous than the gentle rise in atmospheric levels of a benign gas critical to all of life. Their willingness to condemn two-thirds or more of the planet's people to perpetual poverty belies a stunning moral blindness. Their fixation on an abstract and largely imaginary problem and their insistence on radical solutions without reference to wider ethical issues or to political and practical feasibility have confused a large segment of the population in the very countries that have most benefited from the application of cheap energy to satisfy basic human needs and desires. Their moral obtuseness extends as far as counselling that we give up on democratic politics and human ingenuity and settle for Malthusian moralizing. It is time to end this charade and channel global energies and resources into more pressing and rewarding scientific research. It is time for public officials to turn their attention to issues that matter and undo the damage already done.

The alarmism that drives much of the public discussion of climate change is based on dubious assumptions, divided science, and disputed evidence claiming that something must be done – and done quickly – to halt and even reverse changes in the global climate and avoid catastrophic harm to the biosphere. Ethicist Thomas Derr suggests that "talk of global warming has become pervasive – and pervasively one-sided. Churches of all varieties have signed on to the issue as a moral cause. Corporations, including former doubters, have adopted anti-warming language, either from new conviction or for convenient public image. The denizens of the annual Davos pilgrimage organized by the World Economic Forum, with a wary eye to the zeitgeist, added climate change to their list of major concerns in 2007. Politicians, with few exceptions, dare not openly deny that there is a problem, though their responses may vary."[4] The pessimism driving alarmists' apocalyptic claims, however, is more than matched by the optimism that underpins their assessments of proffered solutions. Some alarmists are even prepared to argue that their solutions can be win-win, i.e., good for the planet and good for the economy, an optimism that is wholly without foundation.

The rise of *official* science and climate change

The climate change saga is embedded in what has become one of the most troubling forces in modern society, *official science*, a corrupting form of science in which political and other goals rather than scientific curiosity have become the principal objectives of scientific enquiry. Over the last forty some years public policy has become more and more dependent on the findings of science to address societal concerns with risk, mostly involving safety, health, and the environment. Government willingness to use the regulatory power of the state to respond to perceived as well as real risks has awakened whole armies of activists determined to save us from one fear after another. They, in turn, have relied on scientists willing to shape their research to reinforce activist goals. The pressure on scientists – whether in industry, academia, think tanks, ENGOs, or elsewhere – to tailor their research programs to support activist claims is fed by the need to secure ever more difficult-to-obtain funding. To succeed they must publish and announce newsworthy results. They have learned that alarm is a potent media and activist draw. Much of science continues to serve society well, but a growing percentage has succumbed to the allure of fame and fortune, particularly in softer, observation-based science such as public health and the environment. The result is a mounting scandal of scientific misconduct and retractions. A rising share of published science – by some estimates more than half – is proven wrong or misleading within five years, often sooner. To governments, this kind of uncertainty threatens to undermine political support of regulatory systems. Their response is to encourage official science, i.e., government-sanctioned science that supports political agendas. Climate science may be the most obvious and abused sector, but it is far

---

4.   Thomas Sieger Derr, "The Politics of Global Warming," *First Things*, August/ September 2007.

from unique. Other examples abound in pharmaceuticals, the regulation of chemicals, and many environmental claims.

In the case of climate change, scientists by the end of the 1980s
had succeeded in constructing a Kuhnian paradigm, i.e., defining
normal climate science to be the work of those scientists who shared
the assessment that the climate system could be largely understood
as a matter of radiative balance, based on forcings and feedbacks,
and principally controlled by the greenhouse effect, other factors being of secondary importance. They focused on finding a human or
anthropogenic cause to explain recent climate change while underestimating the extent to which climate is always changing on all
spatial and temporal scales. Over the next quarter century, many
scientists not part of the dominant group – i.e., those not committed
to official science – made significant strides in understanding the
role of *other, natural* factors, concluding that they were underspeci-
fied in the official understanding of the climate system. Given the
growing political importance that the official view had captured,
governments acquiesced in efforts to demonize and ostracize those
who failed to adhere to the politically driven consensus. To that end,
the proponents of official science used all the tools at their disposal,
from funding and publication decisions to public discourse that
demonized non-conforming scientists, all at a great cost to the integrity of the scientific process.

Both groups of scientists agree that global climate change is real,
part of the chaotic and unpredictable interaction of various natural
cycles, including cycles in the earth's rotation on its axis and around
the sun, cycles in the sun's energy output, and cycles in ocean surface temperatures and currents. The minority view, however, points
to the significant body of scientific research indicating that the extent of recent change – both warming and cooling – is both modest
and fully within previous human experience; over human time
there have been many larger changes in climate. The current phase
of benign climate is of relatively recent origin (less than 12,000
years), and within that time frame there have been at least eight cycles of both warming and cooling, largely unaided by human activity. The recent relative warming is part of the reversal of what paleoclimatologists refer to as the Little Ice Age (ca. 1350-1800), which in
turn succeeded the Medieval Climate Optimum (ca. 800-1200).

The minority group emphasizes that the climate system is extremely complex and only beginning to be understood. Both the
immediate and long-term drivers of climate change remain matters

of scientific enquiry and debate and are not sufficiently well understood to justify a wholesale re-orientation of modern economies away from reliance on fossil-fuel-based energy. The role of carbon dioxide (CO_2), the focus of concern in official science, is arguably marginal, and higher levels of atmospheric CO_2 may well have more beneficial effects than problematic ones. The minority points out that CO_2 is *not* a pollutant despite misguided calls to treat it as one. The diversity of opinion in the community of so-called climate sceptics is a desirable indication of vigorous on-going research and discussion. It is a healthy contrast with claims of a "consensus" by the alarmist community that is suggestive of the latter's lack of willingness to engage in debate and pursue further research to strengthen theories and accumulate observation-based evidence.

Despite their lack of funding and inadequate access to the leading publications, independent scientists are making significant progress in understanding the climate system, progress that is barely reflected in the work of the UN's Intergovernmental Panel on Climate Change (IPCC) and other official bodies. The satellite era has allowed scientists to begin gathering data on a uniform and global basis; it is proving immeasurably more reliable and useful than earlier data. That information continues to undermine confidence in the models favoured by the IPCC, which rely on hypotheses and assumptions derived from earlier, less reliable data. The gap between model results and observations has continued to widen, pointing to a fundamental need to reassess many of the assumptions that are driving official science. The fact that virtually all models overestimate the extent of the warming points to the commitment of their creators to political rather than scientific goals. Observation-based estimates of the sensitivity of the climate system to a doubling of atmospheric CO_2 have steadily declined, and yet IPCC scientists have stubbornly clung to their earlier and now more and more questionable view. Similarly, understanding has grown concerning the complicated interaction of coupled atmospheric-oceanic cycles, of cyclical patterns in solar output, and of the impact of cosmic rays on cloud formation and surface warming.

Other areas of potential importance, moreover, will require further investigation, including the complicated interactions of Earth's atmosphere, biosphere, lithosphere, cryosphere, and hydrosphere and their impact on the carbon cycle and climate patterns. Scientific understanding of the planet's radiative balance is improving but remains to be fully factored into the work of the scientists committed to the alarmist perspective. As this work proceeds, many scientists are becoming more aware of what climate scientist Judith Curry calls the uncertainty monster that bedevils resolving this "wicked" problem. The universe of known unknowns is increasing while that of unknown unknowns is, hopefully, shrinking.

Humans do contribute to climate change, more at the local and regional than at the global level, through land use, agricultural and forestry practices, urbanization, and industrialization. The extent of this anthropogenic influence is difficult to measure, though it pales in comparison with the impact of natural influences. Nevertheless, to the extent that some of these influences are malign, governments should continue to take such steps as may prove prophylactic and/or ameliorative, for example, pollution abatement or research and development of alternative – and economically realistic – forms of energy. Evidence from the past thirty years of research indicates that, at the most, there is a basis for judicious spending on research and on adaptation to emerging negative effects, such as sea-coast erosion or vulnerability to the weather extremes that are an integral part of earth systems.

The bogus science of impacts

Even more problematic than the science of climate change itself is the growing field of impact studies pursued by biologists, environmentalists, geographers, ecologists, and sociologists. Much more than the work of natural scientists, the work of these researchers is marred by extensive *a priori* assumptions. Impact studies occupy two-thirds of the energy and resources of the IPCC and have spawned an industry devoted to producing a cornucopia of alarmist books, articles, studies, and papers, many of them of highly questionable quality and provenance. Similar to the work of alarmist climate scientists, these studies rely to a large extent on computer models as well as on high doses of creativity and imagination. And with governments failing to question the credibility of this material, they have marched in lockstep towards the make-believe world of decarbonated economies ostensibly leading to global sustainability.

Effective public policy discussion of the projected impacts of global warming has been compromised by the tendency to systematically overestimate negative impacts, to discount natural adaptation and technological developments, and to attribute issues arising due to population or other pressures to global warming. The burden of proof should not lie with those who believe that adaptation will be sufficient to address gradual warming – or cooling – but with those convinced that the impacts will be catastrophic and unmanageable and require immediate and radical solutions with unpredictable results. Further, political scientists have supported some of these questionable studies, arguing that climate change can be managed through dubious political and institutional reforms that sacrifice democracy and individual freedoms to the utopian ideals of a world ruled by technocrats.

Both natural and social science investigators have exhibited many of the pathologies of modern science, i.e., failing to report negative findings, ignoring counter evidence, relying on correlation rather than on more rigorous tests for causation, failing to pursue replication studies that would confirm – or refute – earlier results, misusing data, attributing higher levels of accuracy to their data than warranted, and torturing statistics in order to produce more useful outcomes. Much of this abuse is driven by the competition for funding and prestige that characterizes modern academic research, but some of it is driven by ideological preferences. American essayist Marilynne Robinson aptly observes in her *Terry Lectures:* "There are inevitable problems with para-scientific arguments. At best, arguments based on science, no matter what their source, are vulnerable over the medium term, at least, on account of the very commendable tendency of science to change and advance. At this point, the para-scientific genre feels like a rear-guard action, a nostalgia for the lost certitudes of positivism."[5] Rupert Darwall similarly notes that "genuine scientific inquiry is degraded when science becomes politicized. The standards that have prevailed since the Scientific Revolution conflict with the advocacy needs of politics, and AGW would be finished as the basis of a political program if confidence in its scientific consensus were undermined. Its advocates' evasion of rigorous falsifiability tests points to AGW's current weakness as a science."[6] Evidence of such politicization of climate science and of its descent into post-normal or para-science has become overwhelming and should be raising serious concerns among governments who have become the primary clients of this science.

✦✦✦✦✦

---

5.   Marilynne Robinson, *Absence of Mind: The Dispelling of Inwardness from the Modern Myth of the Self* (New Haven, CT: Yale University Press, 2010), 38.

6.   Rupert Darwall, "An Unsettled Climate," *City Journal,* Summer 2014.

The troubling public policy response

The world's governments – Canada's included – have for more than thirty years been officially committed to the *idea* that the planet is warming, that much of this warming is anthropogenic, that the impact of warming will be catastrophic, and that it is necessary – and possible – to pursue mitigation strategies. All but the first of these assertions is without supporting evidence. The establishment of the IPCC in 1988, the negotiation of the UN Framework Convention on Climate Change in 1992, and the conclusion of the Kyoto Protocol in 1997 were all predicated on these ideas. Scientific assessment of the issue was collectively outsourced to the IPCC with a mandate to confirm the *human* fingerprint, to ignore the natural dimension, and to recommend ways to address the anthropogenic factor. To date, no *public* effort has been dedicated to determining the extent of natural forces in climate change, to analysing whether AGW constitutes a serious threat, or to determining whether mitigation strategies are feasible from a scientific, technical, or political-economic perspective. Governments have decided not to seek a second opinion on an issue bringing into question the viability of the planet and of human civilization. Given the lack of evidence, it is not surprising that progress in implementing mitigation strategies and in moving towards a carbon-free economy has been anaemic.

Governments remain confident that the technology exists to replace fossil-based fuels with renewables within a relatively short time and at reasonable cost. There is no basis for this confidence. Governments and entrepreneurs have now logged thirty years of experience with renewables, and the only conclusion that can be drawn is that current renewables will never play more than a niche role in some special applications. They have been proven to be costly, unreliable, and counterproductive as a means of generating less atmospheric CO_2. Nuclear and hydro are at this point the only reliable, non-fossil-fuel-based baseline generators of electricity but still require the complement of fossil-fuel-based generators. In many applications, oil and gas-based power remains the only viable and reliable form of energy. Even if there are revolutionary breakthroughs in the ability, for example, to harness solar energy more efficiently, effectively, and economically, it will take two or three generations to replace current energy infrastructure.

Official analyses of the *economics* of climate change policy are based on heroic and unreliable assumptions; they reveal a need to spend billions with questionable results, billions that could be devoted to much more productive and reliable solutions to other, real global problems. Academic analyses of the economics range far and wide, but most demonstrate the high cost of the transition to a non fossil-fuelled energy future and require the assumption that renewable and other "clean" energy sources will become much more efficient and reliable than they currently are. Analyses relying on guessing circumstances far into the future, from technology to population, approach the realm of science fiction.

Much of the policy analysis, discussion, and negotiating activity pursued by the UN, the IPCC, and governments has placed the policy cart before the scientific, economic, and technological horses. Until there is a firmer basis in science for understanding the evolution of the climate system, these discussions amount to the wish fulfilment of utopian impulses and can spawn disastrous consequences for current and future generations. The idea that human policy measures can alter fundamental natural climate patterns is sheer hubris. Humans can adapt, as they have in the past, but they do not know enough or command the resources to change the impact of natural forces far beyond their control, especially since the impact of foreseeable warming – whether anthropogenic or natural – is, on balance, likely benign. There may be need for some adaptation, but, based on past experience, human ingenuity and resilience are fully up to the challenge. Past experience also points to the much more lethal impact of cooling on a global scale, as was evident after both the Roman and Medieval periods of global warming.

Much global warming alarmism has taken on all the characteristics of a religious cult. As historian Paul Johnson observes, "global warming, like Marxism, is a political theory of actions, demanding compliance with its rules. ... Marxism, Freudianism, global warming. These are proof – of which history offers so many examples – that people can be suckers on a grand scale. To their fanatical followers they are a substitute for religion. Global warming, in particular, is a creed, a faith, a dogma that has little to do with science. If people are in need of religion, why don't they just turn to the genuine article?"[7] People are free to hold any belief that appeals to them, but they should not expect others to join the stampede in order to indulge their preferences. The fact that international agencies are engaged in promoting this cult is immaterial. It is not the first, and will not be the last, cause that has found internationalism to be a convenient vehicle

---

7. Paul Johnson, "The Nonsense of Global Warming," *Forbes*, October 6, 2008.

for gaining attention and gaming the system. It is irresponsible, however, for politicians and the media to fall naively into the traps set for them. For some climate and social scientists, the motivation appears to be more sinister: a means to extract research funds from gullible government agencies and foundations or to subsidize economic activities that cannot find a place in the market on their own merits.

The often repeated call for governments to do *something* just in case the alarmists are right – a general application of the pernicious precautionary principle – betrays a lack of seriousness. "Doing something" is not without cost. Reconstituting modern industrial society on a basis other than that of fossil fuels will require a disruption of lives and create major societal and individual upheaval. The prospect of a substantive decline in living standards in advanced countries is real, but reversing economic development in poorer countries would be tragic, unprincipled, and immoral. No government should entertain policy choices with such momentous negative consequences without a much firmer basis in both science and economics as well as a thorough cost-benefit analysis and open public debate. The fact that alarmists insist that the "crisis" is too great and solutions too urgently needed to allow time for sober analysis and informed debate speaks volumes about the merits of their case. Their claim to virtue corrupts the language.

While alarmists and environmentalists have cloaked themselves in the mantle of morality and virtue, the truth is that their project reeks of the same immorality as eugenics and other earlier population control movements. It took the evils of Hitler's quest for Aryan purity to open people's eyes to the pernicious assumptions of eugenics. What will it take to open people's eyes to the immorality of climate alarmism? Indo-British economist Deepak Lal reminds us in his latest book, *Poverty and Progress,* that "the greatest threat to the alleviation of the structural poverty of the Third World is the continuing campaign by western governments, egged on by some climate scientists and green activists, to curb greenhouse emissions, primarily the CO_2 from burning fossil fuels.... [I]t is mankind's use of the mineral energy stored in nature's gift of fossil fuels ... accompanying the slowly rolling Industrial Revolution, [that] allowed the ascent from structural poverty which had scarred humankind for millennia."[8] Nigel Lord

---

8.   Deepak Lal, *Poverty and Progress: Realities and Myths About Global Poverty* (Washington: Cato Institute, Kindle edition, 2013), ch. 10, para 1. Lal's thesis is explored more fully in David Legates and Cornelis van Kooten, *A Call to Truth,*

Lawson, one of the most perceptive critics of the climate alarm mantra puts it more bluntly: "Global warming orthodoxy is not merely irrational. It is wicked."[9]

Combating climate change or ushering in Utopia?

But an even larger question arises: is climate change the real issue? It may have been the immediate issue to some, but as the years have gone by and the climate system has not responded as predicted, it becomes ever more apparent that climate alarmism is a stalking horse for a more ambitious agenda on the part of the UN and its progressive supporters. As Father Raymond de Souza observes, "the settled science ... fits together all too neatly with the agenda of those arguing for ever greater control of economic life."[10] In that sense, climate alarmists have become no more than useful idiots in a much bigger game. As the ideological fabric of climate change becomes more and more tattered, those committed to the transformational politics of the left will start to look for a new vehicle and the game will continue. Deep ecology and climate change replaced the Cold War as the consuming concern of politicians in the developed world. Hysteria about climate change followed earlier alarms about run-away population growth, impending food shortages, declining natural resources, and precipitous declines in species diversity, none of which were borne out but all of which emerged from the same paradigm that saw man as the enemy of nature and a parasite on the planet. There is a need to provide new intellectual leadership based on a more positive – and realistic, evidence-based – assessment of man's place in nature. Historian Patrick Allitt provides a compelling assessment of the role of the successive waves of overhyped alarms that have plagued western society over the past sixty plus years. He documents the extent to which these alarms were exaggerated and the extent to which they proved manageable with conventional policy responses.[11]

Robert Zubrin concludes that as a result of the growth in anxiety and the rise in radical responses, modern society faces a choice be-

Prudence, and Protection of the Poor 2014: The Case against Harmful Climate Policies Gets Stronger (Burke, VA: Cornwall Alliance, September, 2014).

9. Nigel Lawson, "Cool it: an essay on climate change," in Alan Moran, ed., *Climate Change: The Facts* (Melbourne, Australia: Institute of Public Affairs, 2015), kindle edition.

10. Raymond de Souza, "The Religion of Planet Earth," *National Post,* March 4, 2015.

11. Patrick Allitt, *A Climate of Crisis: America in the Age of Environmentalism* (New York: Penguin, 2014).

tween a humanism based on freedom of choice, ingenuity, and prosperity and an anti-humanism that demands ever tighter controls upon human aspirations. As he says: "If the idea is accepted that the
world's resources are fixed with only so much to go around, then
each new life is unwelcome, each unregulated act or thought is a
menace, every person is fundamentally the enemy of every other person, and each race or nation is the enemy of every other race or nation.
The ultimate outcome of such a view can only be enforced stagnation,
tyranny, war, and genocide."[12] The only antidote is to pursue a world
based on continuing faith in the virtues of human capacity for creativity, invention, and entrepreneurship and a deep respect for the digni-
ty and freedom of each individual.

The story of the past quarter century is that more and more people, particularly those in authority, have chosen the path of superstition and self-destruction, justifying their choices on the basis of misguided morals, false religious beliefs, and pseudo-scientific analysis.
In the fall of 2015, the evidence for that choice will look increasingly
tattered. The big issue now is whether or not, after so much has been
invested in the science and politics of global warming, it is possible to
reverse course. Reflecting on the excesses of the climate change
bandwagon, American pundit Walter Russell Mead trenchantly asks:
"How long will it be before serious people who seriously care about
the environment realize that the clowns, poseurs and hotheads currently shaping the movement's public agenda constitute a grave and
urgent threat to the health of the only planet we've got? How high a
price must the world pay for green folly? How many years will be
lost, how much credibility forfeited, how much money wasted before
we have an environmental movement that has the intellectual rigor,
political wisdom and mature, sober judgment needed to address the
great issues we face?"[13]

Beyond Europe, governments have largely limited their policy
responses to rhetoric and symbolic gestures, many of them annoying
but having little or no impact on climate or on the composition of the
atmosphere. They have, nevertheless, been considered politically
necessary in order to satisfy the pressure brought to bear by the environmental movement, both nationally and internationally. Light
bulbs that provide inferior light, toilets that have to be flushed twice,
garbage that needs to be sorted into ever-more specific piles, and un-

---

12.   Robert Zubrin, *Merchants of Despair* (New York: New Atlantis Books, 2012), 252.

13.   "Kyoto Fraud Revealed," *Via Meadia*, October 14, 2010.

sightly windmills that decimate local and migrating bird populations and plague local residents: they all form part of the symbolism of modern green politics. Ordinary people grumble but have managed to live with the annoyances. Europeans have been forced to endure a higher and more costly level of indoctrination but are tiring of the game as dithering politicians, from Angela Merkel to David Cameron, consider how to reverse what they have started.

The conversion of Pope Francis to the climate mantra in the lead-up to the December 2015 Paris Conference at which governments will try once again to conclude a climate treaty with teeth indicates the extent to which this madness has permeated modern sensibilities. The Pope's June encyclical reads like a primer of the global salvationist catechism and has transformed him into a modern media star, finding favour with all the "right" people, including all those who had long condemned Rome for its stubborn failure to commit to the modern secular agenda. Wittingly or not, Francis appears to have committed the Church to much of the new, post-Christian eco-morality that integrates fear of markets, condemnation of human progress and ingenuity, and belief in the power of technocrats to solve all the world's problems, including that of so-called over-population.

As rosily as the UN and its fellow travellers may paint their utopia of global governance and collectivist central direction by technocrats, both the desire and its fruits are grounded in the impulse to control and organize people's lives for their own good; the result will inevitably be the destruction of liberal democracy and the triumph of totalitarian democracy, as understood by the Israeli historian Jacob Talmon.[14] Looking back on half a century of barbarism, Ludwig von Mises observed: "The intellectual leaders of the peoples have produced and propagated the fallacies which are on the point of destroying liberty and Western civilization. The intellectuals alone are responsible for the mass slaughters which are the characteristic mark of our century. They alone can reverse the trend and pave the way for a resurrection of freedom. ... What is needed to stop the trend towards socialism and despotism is common sense and moral courage."[15]

Little has changed except the fables with which western intellectuals have tried to lull us into the false comfort of a world of om-

---

14. Jacob Talmon, *Origins of Totalitarian Democracy* (London: Secker and Warburg, 1955).

15. Ludwig von Mises, "The Alleged Inevitability of Socialism," in *Planned Chaos*, volume three of *Socialism: an Economic and Sociological Analysis* (New Haven, CT: Yale University Press, 1951), 592.

nipotent governance. As von Mises, Talmon, Popper, Hayek, and other survivors of the barbarism of the mid-twentieth century counselled then and would counsel today, only the ancient virtues of common sense and moral courage will reverse the tide. Their contemporary heir, Thomas Sowell, observes in *Intellectuals and Society* that it is striking "how difficult it is to think of benefits [intellectuals] have conferred on anyone but their own circles – and how painfully apparent it is how much they have in fact cost the rest of society at large, not only economically but in many other ways. ... [and yet] despite formidable weapons wielded by the intelligentsia in their crusades for cultural, moral, and ideological hegemony, they are not always able to neutralize the countervailing force of facts, experience and common sense."[16]

It will be cold comfort to future generations when their leaders finally realize how badly they have been fooled and how deeply they have embedded global warming hysteria into their cultural and governing norms, from tax policy to education programs. They will wonder, along with Richard Lindzen, why "the early twenty-first century's developed world went into hysterical panic over a globally averaged temperature increase of a few tenths of a degree, and, on the basis of gross exaggerations of highly uncertain computer projections combined into implausible chains of inference proceeded to contemplate a rollback of the industrial age."[17] Now is the time for governments to begin the painful task of dismantling a movement that they have helped to create and that now threatens much more than the integrity of science.

*****

16. Thomas Sowell, *Intellectuals and Society* (New York: Basic Books, 2009), 301 and 314.

17. Richard Lindzen, as quoted by Vaclav Klaus, "Human Freedom, Not Climate, is at Risk," *Environment & Climate News*, September 2007.

Notes

Index

Acronyms and Abbreviations

°C	degrees centigrade
°F	degrees Fahreneit
°K	degrees Kelvin
AGW	anthropogenic global warming
AMO	Atlantic Meridional Oscillation
APS	American Physical Society
AR 1-5	Assessment Reports of the UNIPCC, 1(1990), 2(1995), 3(2001), 4(2007), 5(2013-14)
AWG-KP	Ad Hoc Working Group on Further Commitments (UNFCCC negotiating process)
AWG-LCA	Ad Hoc Working Group on Long-Term Cooperative Action under the Convention (UNFCCC negotiating process)
BASIC	Brazil, South Africa, India, China
BEST	Berkeley Earth Surface Temperature project
BOPs	Balance of Payments
BTU	British Thermal Unit
CBDR-RC	Common but differentiated responsibilities and respective capabilities
CCS	carbon capture and sequestration
CDM	Clean Development Mechanism (UN)
CERN	European Organization for Nuclear Research
CFCs	chlorofluorocarbons
CIHR	Canadian Institutes of Health Research
CITES	Convention on International Trade in Endangered Species
CO_2	carbon dioxide
CO_{2e}	Carbon dioxide equivalent, i.e., the impact of all GHGs in terms of CO_2
COP	Conference of the Parties to the UNFCCC
CRO	contract research organization
CRU	Climatic Research Unit at the University of East Anglia
CRU	Climatic Research Unit at UEA
CSIRO	Commonwealth Scientific and Industrial Research Organisation (Australia)
CSWS	Commission for a Sustainable World Society sponsored by Socialist International
DDT	dichlorodiphenyltrichloroethane
EIA	(US) Energy Information Administration
ENGO	environmental non-government organization
ENSO	El Niño Southern Oscillation
EPA	Environmental Protection Agency (US)

ERF	effective radiative forcing
ETS	European Emissions Trading System
EU	European Union
FAO	Food and Agriculture Organization (UN)
FDA	Food and Drug Administration (US)
GARP	Global Atmospheric Research Program of the WHO
GATT	General Agreement on Tariffs and Trade
GCM	Global climate model or general circulation model
GDP	gross domestic product
GFDL	Geophysical Fluid Dynamics Laboratory at Princeton University
GHGs	greenhouse gases
GISS	Goddard Institute for Space Studies
GLOBE	Global Legislators Organization for a Balanced Environment
GM	genetically modified
GMST	Global Mean Surface Temperature
GWPF	Global Warming Policy Foundation (UK)
HadCRUT	Hadley Centre and Climate Research Unit at the University of East Anglia
IAC	InterAcademy Council
IAM	Integrated Assessment Model
ICSU	International Council for Science (formerly International Council of Scientific Unions)
IEA	International Energy Agency
IGY	International Geophysical Year
IMF	International Monetary Fund
IPC	Integrated Program for Commodities (UNCTAD)
IPCC	See UNIPCC
LIA	Little ice age
MCO	Medieval climate optimum
MEA	multilateral environmental agreement
MERs	Market exchange rates
METI	Ministry of Economy, Trade and Industry (Japan), formerly MITI, Ministry of International Trade and Industry
MOFA	Ministry of Foreign Affairs (Japan)
MSU	microwave sounding units
Mtoe	Metric tonne equivalent
MW	megawatt
MWh	megawatt hour
NAO	North Atlantic Oscillation
NASA	National Aeronautics and Space Administration
NCAR	National Center for Atmospheric Research
NGO	non-governmental organization
NIEO	New International Economic Order
NIH	National Institutes of Health

NIPCC	Non-governmental International Panel on Climate Change
NOAA	National Oceanic and Atmospheric Administration
NRC	National Research Council (Canada)
NSERC	Natural Sciences and Engineering Research Council of Canada
NSF	National Science Foundation
OECD	Organization for Economic Cooperation and Development
PCB	polychlorinated biphenyl
PDO	Pacific Decadal Oscillation
Ph level	measure of the acidity or alkalinity of ocean water
PP	precautionary principle
ppmv	parts per million by volume
PPPRs	Purchasing power parity (exchange) rates
RSS	Remote Sensing System
SCEP	Study of Critical Environmental Problems
SEJ	Society of Environmental Journalists
SEPP	Science and Environment Policy Project
SI	Socialist International
SPM	Summary for Policymakers (IPCC AR)
SPPI	Science and Public Policy Institute
SR	Synthesis Report (IPCC AR)
TS	Technical Summary (IPCC AR)
TSI	total solar insolation
TWh	terawatt hour
UAH	University of Alabama at Huntsville
UCAR	University Corporation for Atmospheric Research
UEA	University of East Anglia (UK)
UHI	urban heat island effect
UNCTAD	United Nations Conference on Trade and Development
UNEP	United Nations Environment Program
UNFCCC	United Nations Framework Convention on Climate Change
UNIPCC	United Nations Intergovernmental Panel on Climate Change
USA	United States of America
WCC	World Climate Conference
WCP	World Climate Program
WG1	Working Group I of the UNIPCC dealing with the Physical Science Basis of Climate Change
WG2	Working Group II of the UNIPCC dealing with Climate Change Impacts, Adaptation and Vulnerability
WG3	Working Group III of the UNIPCC dealing with Mitigation of Climate Change
WGBU	German Advisory Council on Global Change

WHO	World Health Organization
WMO	World Meteorological Organization
WTO	World Trade Organization
WUWT	What's up with that, a popular science blog
WWF	World Wide Fund for Nature, formerly World Wildlife Fund

❦❦❦❦❦

The Author

Michael Hart is emeritus professor of international affairs at the Norman Paterson School of International Affairs at Carleton University, where he previously held the Simon Reisman Chair in Trade Policy. He taught courses on the laws and institutions of international trade, Canadian foreign policy, and the politics of climate change. He held the Fulbright-Woodrow Wilson Center Visiting Research Chair in Canada-US Relations in 2004-05. Concurrently, he was also a Scholar-in-Residence in the School of International Service and a Senior Fellow in the Center for North American Studies at American University in Washington.

He is a former official in Canada's Department of Foreign Affairs and International Trade, where he specialized in trade policy and trade negotiations. He was involved in the Canada-US Free Trade Negotiations, the North American Free Trade Negotiations and various GATT, textile, and commodity negotiations.

He was founding director of Carleton's Centre for Trade Policy and Law and stepped down in September 1996 after a second term as director. As a distinguished fellow of the Centre, he led training and advisory missions to Russia, Vietnam, the Caribbean, and Central America, and taught training courses for the Organization of American States, the Inter-American Development Bank, the Monterey Institute for International Studies, and various universities in Canada and the United States.

He holds an MA in medieval history from the University of Toronto and is the author, editor, or co-editor of more than a dozen books and numerous articles and chapters in books on international trade and public policy issues, including *A Trading Nation,* short-listed for the Donner Prize (Public Policy), the J.W. Dafoe Prize (History), the Donald V. Smiley Prize (Political Science), and Purvis Prize (Economics) in 2003, and *Decision at Midnight,* short-listed for the Gelber Prize and the Canadian Business Book Award in 1995.

♦♦♦♦♦

Other books by Michael Hart

From Pride to Influence: Towards a New Canadian Foreign Policy (Vancouver: UBC Press, 2008).

A Trading Nation: Canadian Trade Policy from Colonialism to Globalization (Vancouver: UBC Press, 2002).

Fifty Years of Canadian Tradecraft: Canada at the GATT 1947-1997 (Ottawa: Centre for Trade Policy and Law, 1998).

Multilateral Negotiations: Lessons from Arms Control, Trade and the Environment with Fen Hampson (Baltimore: Johns Hopkins University Press, 1995).

Also Present at the Creation: Dana Wilgress and the United Nations Conference on Trade and Employment at Havana (Ottawa: Centre for Trade Policy and Law, 1995).

Decision at Midnight: Inside the Canada-US Free Trade Negotiations with Bill Dymond and Colin Robertson (Vancouver: University of British Columbia Press, 1994).

What's Next: Canada, the Global Economy and the New Trade Policy (Ottawa: Centre for Trade Policy and Law, 1994).

Trade – Why Bother? (Ottawa: Centre for Trade Policy and Law, 1992).

A North American Free Trade Agreement: The Strategic Implications for Canada (Centre for Trade Policy and Law and Institute for Research on Public Policy, 1990).

Canadian Economic Development and the International Trading System: Constraints and Opportunities (University of Toronto Press, 1985).

Some Thoughts on Canada-United States Sectoral Free Trade (Halifax: Institute for Research on Public Policy, 1985).

Edited Books

With Fen Hampson and Martin Rudner, *A Big League Player? Canada Among Nations 1999* (Toronto: Oxford, 1999).

Finding Middle Ground: Reforming the Antidumping Laws in North America (Ottawa: Centre for Trade Policy and Law, 1997).

With Debra Steger, *In Whose Interest? Due Process and Transparency in International Trade* (Ottawa: Centre for Trade Policy and Law, 1992).

With Richard Dearden and Debra Steger, *Living With Free Trade: Canada, the Free Trade Agreement and the GATT* (Ottawa and Halifax: Centre for Trade Policy and Law and Institute for Research on Public Policy, 1990).